Mathematical Methods for Physics

This detailed yet accessible text provides an essential introduction to the advanced mathematical methods at the core of theoretical physics. The book steadily develops the key concepts required for an understanding of symmetry principles and topological structures, such as group theory, differentiable manifolds, Riemannian geometry, and Lie algebras. Based on a course for senior undergraduate students of physics, it is written in a clear, pedagogical style and would also be valuable to students in other areas of science and engineering. The material has been subject to more than 20 years of feedback from students, ensuring that explanations and examples are lucid and considered, and numerous worked examples and exercises reinforce key concepts and further strengthen readers' understanding. This text unites a wide variety of important topics that are often scattered across different books, and provides a solid platform for more specialized study or research.

Esko Keski-Vakkuri received his Ph.D. in physics from Massachusetts Institute of Technology in 1995, and is currently a senior faculty member at the University of Helsinki. He has previously held positions at the California Institute of Technology and Uppsala University. His research is focused on string theory, black holes, holographic duality, and quantum information.

Claus K. Montonen received his Ph.D. from the University of Cambridge in 1974 and later held positions at the Université de Paris XI (CNRS), the Research Institute for Theoretical Physics, Helsinki, and CERN. He held various senior faculty positions in the Department of Physics at the University of Helsinki from 1978 until his retirement in 2011, where he was responsible for curriculum design in theoretical physics. His research interests are in S-matrix theory, string theory, and quantum field theory, having made major contributions to early string theory and duality in field and string theory.

Marco Panero received his Ph.D. in physics from the University of Turin in 2003, after which he held postdoctoral positions at the Dublin Institute for Advanced Studies, the University of Regensburg, ETH Zurich, the University of Helsinki, and the Autonomous University of Madrid. Since 2014 he has been an associate professor in physics at the University of Turin. His main research interests are in lattice field theory and in theoretical high-energy physics.

Mathematical Methods for Physics

An Introduction to Group Theory, Topology, and Geometry

ESKO KESKI-VAKKURI
University of Helsinki

CLAUS K. MONTONEN
University of Helsinki

MARCO PANERO
University of Turin

CAMBRIDGE
UNIVERSITY PRESS

CAMBRIDGE
UNIVERSITY PRESS

Shaftesbury Road, Cambridge CB2 8EA, United Kingdom

One Liberty Plaza, 20th Floor, New York, NY 10006, USA

477 Williamstown Road, Port Melbourne, VIC 3207, Australia

314–321, 3rd Floor, Plot 3, Splendor Forum, Jasola District Centre, New Delhi – 110025, India

103 Penang Road, #05–06/07, Visioncrest Commercial, Singapore 238467

Cambridge University Press is part of Cambridge University Press & Assessment, a department of the University of Cambridge.

We share the University's mission to contribute to society through the pursuit of education, learning and research at the highest international levels of excellence.

www.cambridge.org
Information on this title: www.cambridge.org/9781107191136

DOI: 10.1017/9781108120531

First published 2022

A catalogue record for this publication is available from the British Library

Library of Congress Cataloging-in-Publication data
Names: Keski-Vakkuri, Esko, author. | Montonen, C. (Claus), author. |
Panero, Marco (Professor of physics), author.
Title: Mathematical methods for physics : an introduction to group theory,
topology and geometry / Esko Keski-Vakkuri, Claus Montonen, Marco Panero.
Description: Cambridge ; New York, NY : Cambridge University Press, 2022. |
Includes bibliographical references and index.
Identifiers: LCCN 2022016948 (print) | LCCN 2022016949 (ebook) |
ISBN 9781107191136 (hardback) | ISBN 9781108120531 (epub)
Subjects: LCSH: Mathematical physics. | BISAC: SCIENCE / Physics /
Mathematical & Computational
Classification: LCC QC20 .K437 2022 (print) | LCC QC20 (ebook) |
DDC 530.15–dc23/eng20220628
LC record available at https://lccn.loc.gov/2022016948
LC ebook record available at https://lccn.loc.gov/2022016949

ISBN 978-1-107-19113-6   Hardback

To Marika, Anne, and Juha

—Esko

To Leone

—Claus

To Saija, and to Mariangela and Giovanni

—Marco

Contents

1   Introduction

This textbook presents an introduction to a set of mathematical tools that are extensively used in modern physics, and is mainly aimed at advanced undergraduate, graduate, and doctoral students in physics, engineering, and mathematics. The reader is ideally accompanied on a journey through a number of different, albeit related, topics.

Chapter 2 introduces group theory and related notions, including, in particular, group homomorphisms and isomorphisms, before discussing in detail the group of permutations and some other particularly interesting finite groups. Then, the formalism of Young diagrams is introduced, and an alternative definition of groups in terms of their presentation is given. The rest of the chapter is devoted to continuous groups and to groups acting on a set.

The next chapter, Chapter 3, discusses the different representations that groups can have: After a brief reminder of linear-algebra concepts, the definition of group representations is formulated. Then, the discussion focuses on the concept of reducibility of group representations, which leads to a classification of irreducible representations. Group characters are introduced and their use in the classification of inequivalent irreducible representations is explained. The chapter also discusses the properties of the regular representation, which is induced by the action of the group on itself through a translation. The final part of the chapter introduces dual vectors and tensors, and an example of application of these notions for a spin-chain system of relevance in quantum physics and condensed-matter theory.

In Chapter 4 we first introduce the concepts that allow one to endow a generic set with a topology, then we define manifolds and, finally, differential manifolds. The chapter discusses in detail calculus on manifolds, differential forms and their integration, and finally presents a formulation of classical mechanics in terms of differential forms.

The following chapter, Chapter 5, is devoted to Riemannian geometry: Topics such as metric tensors, the induced metric, affine connections, connection coefficients, and their transformation properties under coordinate changes are discussed in detail. The chapter presents a thorough exposition of the concepts relevant for the general relativity theory of gravitation and for gauge theories, including parallel transport and holonomy, covariant derivatives, geodesics, curvature, and torsion. The final sections of the chapter are devoted to the discussion of isometries and Killing vector fields.

Chapter 6 presents a discussion of semisimple Lie algebras (highlighting their relevance for different physical applications, from quantum mechanics to the theory of elementary particles in and beyond the Standard Model) and their unitary representations. After defining the Lie algebras of the generators of Lie groups, we introduce the concepts of roots, weights, and Cartan generators, and present the systematic classification of the algebras associated with the classical and exceptional simple Lie groups with the corresponding Dynkin diagrams. The chapter also discusses in detail the explicit construction of irreducible representations for Lie algebras of special unitary groups, using tensor methods and Young diagrams. The last part of the chapter describes the representations of products of unitary groups (such as those that describe the gauge interactions between fundamental particles), and the Lorentz and Lorentz–Poincaré groups relevant for the theory of special relativity and for quantum field theory.

Finally, Appendix A presents detailed solutions for a subset of the problems included at the end of each of the previous chapters. Other solutions are made available to the course instructors through the website of Cambridge University Press.

The book is ideally suited for a university course on mathematical methods for physics; the main emphasis is on geometrical and topological concepts, which are essential for the understanding of the symmetry principles and topological structures in modern physics. The book is largely self-contained, but some important mathematical prerequisites are assumed: In particular, it is assumed that the reader is already familiar with the basics of real and complex analysis and linear algebra.

In writing this book, we put a very strong emphasis on the *pedagogical aspects*. The book is primarily targeted at physics and engineering students and, following M. David Merrill's *application principle* in instructional design (which states that learning is promoted when the learner applies the new knowledge), its goal is to enable them not only to learn a collection of fundamental notions in different branches of mathematical physics, but also to directly *apply* these tools to concrete problems. To this purpose, the final section of each chapter includes a large collection of original problems and exercises. As in actual scientific research, some of these problems stimulate the readers to combine tools which are relevant for the different subjects presented in the various chapters, and to keep a broad perspective – rather than adopting a narrow, hyperspecialized approach.

There are numerous sources that we have used in the preparation of this textbook. Our main inspiration and influence comes from this short list of classic works that we highly recommend for further reading on the subjects covered herein. We list them here, mentioning the chapters for which they are most relevant and highly recommended for further reading on the subjects:

- Important references for Chapters 2 and 3 are the books by Hugh F. Jones [6], by Michael Tinkham [13], and by Morton Hamermesh [4].
- For further reading about the topics of Chapters 4 and 5, we recommend the books by Mikio Nakahara [10], by Charles Nash and Siddhartha Sen [11], by John M. Lee [8], and by Jeffrey M. Lee [7], as well as the books on the theory of general relativity by Charles W. Misner, Kip S. Thorne, and John A. Wheeler [9], by Robert M. Wald [14], and by Sean M. Carroll [1].
- Useful references for further reading on the topics discussed in Chapter 6 are the books by Howard Georgi [2] and by Francesco Iachello [5]. The applications for elementary particle theory can be found in many excellent textbooks; our discussion of the symmetry representations of the Standard Model is closest to that in the book by Mark Srednicki [12].

We conclude this Introduction by thanking all of the students who have taken the course, on which this textbook is based, for helping us to present the topics of this book as clearly as we could. We also thank our former teaching assistants and our colleagues Antti Kupiainen and Jouko Mickelsson, who also taught these subjects, for additional inspiration, as well as Klaus Larjo and Niko Jokela for assistance in converting our original handwritten notes into electronic format. We also thank Jarkko Järvelä and Saga Säppi, who generously shared their solutions to some of the problems. In addition, we are indebted to many colleagues with whom we have discussed about and from whom we have learned the topics presented herein. Finally, we acknowledge with gratitude the careful work and patience of the editorial staff at Cambridge University Press, especially Sarah Armstrong, Elliot Beck, Henry Cockburn, Sapphire Duveau, Kuttappan Suresh Kumar, Sarah Lambert, Beverley Lawrence, Róisín Munnelly, Aparna Nair, Mathew Rohit, and in particular Nicholas Gibbons, who followed this book project from its beginning to its completion.

Group Theory

The first part of this chapter introduces the basic notions of group theory. Then we present a detailed discussion of some interesting finite groups. Next, we introduce Young diagrams and an alternative definition of a generic group in terms of its presentation. Finally, we discuss continuous groups and groups acting on a set.

2.1 Groups

2.1.1 Definitions: Groups, Abelian Groups, and Related Concepts

Consider an arbitrary set $G = \{a, b, \ldots\}$, and a composition law that, for every $a \in G$ and for every $b \in G$, assigns to the ordered pair (a, b) an element $a \cdot b$, which is also an element of G. Then, we define $(G, \cdot)$ to be a *group* if the following conditions simultaneously hold:

G1 (associativity): for all $a, b, c \in G$, $a \cdot (b \cdot c) = (a \cdot b) \cdot c$;

G2 (existence of the unit element): there is an element $e \in G$ such that for all $a \in G$: $a \cdot e = e \cdot a = a$;

G3 (existence of the inverse): for all $a \in G$ there is an element $a^{-1} \in G$ such that $a \cdot a^{-1} = a^{-1} \cdot a = e$.

In that case, the composition law is usually called the group *law of multiplication* (or *product*). We could have added an item **G0**: the composition law must be well defined. An attempt to define a product may not be consistent or well defined for all elements of the set. We will see examples of this later.

Moreover, we define an *Abelian group* as a group for which an additional condition holds:

AG4 (commutativity): for all $a, b \in G$: $a \cdot b = b \cdot a$.

Abelian groups are named after the Norwegian mathematician Niels Henrik Abel.

It is interesting to note that some different mathematical structures that share some properties with but are more general than groups can also be defined: For example, when the composition law is such that $\forall a \in G$ and $\forall b \in G$, one has $a \cdot b \in G$, but the properties **G1**–**G3** are not necessarily satisfied, then $(G, \cdot)$ is called a *magma*. A magma for which the associativity condition **G1** holds is called a *semigroup*. In addition, a semigroup for which also the condition **G2** is satisfied (i.e., a semigroup in which a unit element exists) is called a *monoid*. It turns out that groups have a much richer mathematical structure than magmas, semigroups, and monoids. In addition, groups have many more physics applications than magmas, semigroups, and monoids; hence, we will not discuss these more general structures in detail in this book, and we will primarily focus on groups.

Example   Consider the set $G = \{e, a\}$ and the multiplication $\cdot$ defined as follows: $e \cdot e = e, a \cdot e = e \cdot a = a$, and $a \cdot a = e$. This is compatible with the group structure, e is the unit element, and a is the inverse of itself. The $(G, \cdot)$ group is called the *cyclic group of order* 2. It is an Abelian group, and is usually denoted as $\mathbb{Z}_2$. We will present a more general definition of cyclic groups in Section 2.4.

In the following, we will also use the simpler notation G rather than $(G, \cdot)$ to denote the group.

The number of elements in a set X is denoted by $|X|$. The number of elements $|G|$ in a group G is called the *order of the group*. If $|G|$ is finite, then G is said to be a *finite group*. We adopt some notations. We often drop the product symbol and write gh instead of $g \cdot h$ if there is no confusion. We also use the notation $g^n = g \cdots g$ for the product where g appears n times. For example, $g^2 = gg = g \cdot g, g^3 = ggg = g \cdot g^2$, etc. We then define the *order of the element* g to be the smallest positive number n such that $g^n = e$. For example, in $\mathbb{Z}_2$ the order of the element e is 1 and the order of a is 2.

The smallest finite group is called the *trivial group*: it contains only the unit element e, and thus has order 1. We denote the trivial group by $\mathbb{Z}_1$. The only multiplication that can be done among elements of this group is $e \cdot e = e$. There cannot be a group of order 0, because, in order to satisfy the property **G2**, a group must necessarily contain at least the unit element. A possible way to characterize a generic finite group is by means of its *Cayley table* (or *multiplication table*), named after the British mathematician Arthur Cayley. It is a square table, of size equal to the order of the group, listing all the products of the group elements. In particular, the entries of the Cayley table of a finite group are the products $p \cdot q$, where p is one of the group elements listed in the column on the left of the table, and q is one of the group elements listed in the row at the top of the table. Table 2.1 gives the Cayley table of $\mathbb{Z}_2$ as an example.

If $|G|$ is not finite, but G is a discrete set (i.e., a set whose elements can be put in one-to-one correspondence with the natural numbers), then $(G, \cdot)$ is called a *discrete group*. Conversely, when G is a continuous set, $(G, \cdot)$ is a called a *continuous group*.

Comments

1. The unit element of any group is unique. If both e and e' are unit elements, then $ee' = e'$ (because e is an unit element) and at the same time $ee' = e$ (because e' is an unit element). Therefore $e' = ee' = e$, so $e = e'$.

Table 2.1. The Cayley table of $\mathbb{Z}_2$		
	e	a
e	e	a
a	a	e

2. For a given group, the inverse element of any element is unique. If both b and b' are inverse elements of a, then $ba = e$ and $ab' = e$. But then $bab' = eb'$, hence $b = b'$.

3. Note that, by definition, the unit element commutes with all elements of the group; however, the unit element is not necessarily the only element with this property. The subset of group elements that commute with all elements of the group is called the *center of the group*. The center of a group is always an Abelian group; if it contains only the unit element, then the group is said to have a *trivial center*.

4. The definition of a group, essentially, is the definition of the group multiplication among all possible ordered pairs of group elements, while the nature of the elements in the set G does not necessarily have to be specified.

The latter point means that the definition of a group is abstract. Specifying the nature of the set of group elements corresponds to defining a particular *realization* of the group. More precisely, a realization of an abstract group G is a map from G to a particular set, on which an internal binary operation exists, that satisfies the properties defining the group.

Example   Consider the cyclic group of order 2 introduced above, $\mathbb{Z}_2$, with e as the unit element and a as the other element. Examples of realizations of $\mathbb{Z}_2$ include the following:

- Take $e = 0$, $a = 1$, and addition modulo 2 to be the group multiplication. It is trivial to show that $(\{0, 1\}, + \bmod 2)$ is an explicit realization of the abstract $\mathbb{Z}_2$ group structure defined above; it has the multiplication table as shown in Table 2.1.
- Take $e = 1$, $a = -1$, and the ordinary multiplication as the group multiplication. Also in this case, one can immediately show that $(\{1, -1\}, \cdot)$ is a realization of the $\mathbb{Z}_2$ group.
- Consider the set of truth variables of Boolean algebra, {FALSE, TRUE}, and the binary operator XOR ("exclusive or") as the group multiplication. Recalling that p XOR q is TRUE when p is TRUE and q is FALSE, or vice versa, while it is FALSE when p and q are both TRUE or both FALSE, one can explicitly check that the $\mathbb{Z}_2$ group structure is realized by identifying $e = $ FALSE and $a = $ TRUE.

A particularly important class of realizations of a group are those in which the group elements are associated with linear transformations among the elements of a vector space: Such realizations are called *representations*, and will be discussed thoroughly in Chapter 3.

It is also interesting to consider groups that are constructed by taking Cartesian pairs of elements from other groups. Given two groups G_1 and G_2, their *direct product* $G_1 \times G_2$ is defined as the set of all ordered pairs (g_1, g_2), with $g_1 \in G_1$ and $g_2 \in G_2$, with the multiplication $(g_1, g_2) \cdot (g_1', g_2') = (g_1 g_1', g_2 g_2')$, where $g_1 g_1'$ is computed using the group multiplication of G_1, while $g_2 g_2'$ is computed using the group multiplication of G_2.

It is straightforward to prove that $G_1 \times G_2$ is a group itself; in particular, its unit element is (e_1, e_2), where e_1 is the unit element of G_1 and e_2 is the unit element of G_2, and the inverse element of a generic element (g_1, g_2) is (g_1^{-1}, g_2^{-1}). It is also trivial to see that, if both G_1 and G_2 are finite, so is $G_1 \times G_2$, and its order is $|G_1 \times G_2| = |G_1||G_2|$.

Similarly, one can also define the direct product of three or more groups.

Partly for historical reasons, several groups are defined from (and, sometimes, named after) their explicit realization in sets of numbers or of transformations of a vector space, by a process of *abstraction*, i.e., by focusing on the way the group multiplication acts on ordered pairs of elements, leaving the nature of the group elements (or, in fact, of the group multiplication itself) unspecified.

2.1.2 Examples of Groups

The definitions introduced above can be easily elucidated by some examples of groups:

- $\mathbb{Z}$, the set of integers, with + (addition) as the multiplication law, is a discrete Abelian group. The "unit" element is 0, and, for any element $a \in \mathbb{Z}$, the inverse is $-a$.

- Similarly, $\mathbb{R}$, the set of real numbers, with the addition as the multiplication law, is a continuous Abelian group. Again, the "unit" element e is 0.

- $\mathbb{R}_0 = \mathbb{R} \setminus \{0\}$, the set of real, nonzero numbers, with $\cdot$ (multiplication) as the multiplication law, is a continuous Abelian group. The unit element is $e = 1$, and the inverse of a generic element g is $1/g$. In this case, the set of group elements was defined by removing 0, in order to ensure that all elements have an inverse.

- For any positive integers n and m, the set of $n \times m$ matrices with real entries, with the matrix addition as the group multiplication, forms a continuous Abelian group. The unit element of this group is the $n \times m$ matrix whose entries are all equal to 0, while, for a generic matrix M of elements M_{ij}, the inverse element has entries $-M_{ij}$. Note that, in practice, this group consists of nm independent copies of $(\mathbb{R}, +)$.

- For any positive integer n, the set of real square matrices of size n and nonvanishing determinant, with the matrix product as the group multiplication, forms a continuous group denoted by $\text{GL}(n, \mathbb{R})$. The unit element is the identity matrix $\mathbb{1}$ (of elements $\mathbb{1}_{b,c} = \delta_{b,c}$), and the fact that the set includes only matrices M with $\det M \neq 0$ implies that the inverse matrix M^{-1} exists for any M in the set. In contrast to the previous example, this group is not Abelian (except for $n = 1$).

- Given a regular polygon of n sides, the set of geometric transformations that leave the polygon invariant forms the dihedral group D_n. This group includes n rotations and n reflections. For n odd, the reflections leaving the polygon invariant are about axes going through each of the polygon vertices, the center, and the midpoint of the opposite side, whereas for n even, there are $\frac{n}{2}$ reflections about axes going through opposite vertices and $\frac{n}{2}$ reflections about axes going through the midpoints of opposite sides. The dihedral groups D_n are finite groups of order $|D_n| = 2n$, and they are non-Abelian for all $n > 2$.

- Consider linear transformations of an orthonormal reference frame in the two-dimensional real vector space $\mathbb{R}^2$. They can be represented by real matrices of size 2×2, having (the components of) the vectors of the transformed reference frame as columns. If the transformations are required to preserve the orthonormality of the reference frame, then the columns of the matrix have to be orthogonal to each other, and normalized to 1. The most general form of the matrix is then

$$\begin{pmatrix} \cos \alpha & \sin \alpha \\ \mp \sin \alpha & \pm \cos \alpha \end{pmatrix}, \qquad \alpha \in \mathbb{R}. \tag{2.1}$$

In the set of these transformations one can define an internal composition law, as the operation of applying two such transformations one after the other; it corresponds to the matrix product and satisfies the requirements of a group multiplication. The unit element of the group is the 2×2 identity matrix, and the inverse of a generic element of the group is obtained by taking the transpose of the original matrix. This is the orthogonal group of dimension 2, denoted as $O(2, \mathbb{R})$ (or, more concisely, as $O(2)$); it is the group of transformations of the two-dimensional Euclidean space, that preserve a fixed point and the length of vectors.

2.1.3 Examples of Sets That Are Not Groups

It is also instructive to list some examples of structures $(G, \cdot)$ that are *not* groups:

1. $\mathbb{N}$, the set of natural numbers, with addition as the "multiplication" is not a group, because no element (except for 0) admits an inverse in the group. $(\mathbb{N}, +)$ is, however, a monoid, because the addition is internal in $\mathbb{N}$ (the sum of any two natural numbers is a natural number) and is an associative operation, which admits the number 0 as the "unit element."

2. Consider the three-dimensional real vector space $\mathbb{R}^3$, and the multiplication law defined as the cross-product of vectors, namely $a \cdot b = a \times b$, that is, $(a \cdot b)_i = \sum_{j,k=1}^{3} \epsilon_{ijk} a_j b_k$, where $\epsilon_{ijk} = 1$ for $i = 1, j = 2$, and $k = 3$, and it is totally antisymmetric under the interchange of any pair of the indices (which, in particular, implies that $\epsilon_{ijk} = 0$ when at least two indices are equal). Since this multiplication law is internal in $\mathbb{R}^3$, it endows $\mathbb{R}^3$ with the structure of a magma. Given that this multiplication law is not associative, this magma is not a semigroup.

3. $\mathbb{R}^3$ with the multiplication law defined by the scalar product of vectors, namely $(a \cdot b) = \sum_{i=1}^{3} a_i b_i$, is not even a magma (because the multiplication law is not an internal operation in $\mathbb{R}^3$; the result of $a \cdot b$ is a real number, not a three-dimensional real vector).

4. For any positive integer n, the set of real square matrices of size n and nonvanishing determinant, with matrix addition as the group multiplication, is not a magma. For a generic element M (with entries $M_{b,c}$) in this set, the matrix N of entries $N_{b,c} = -M_{b,c}$ has determinant $(-1)^n \cdot \det M$, which is nonvanishing because $\det M \neq 0$, hence N also belongs to the set. But $(M + N)$ is the zero matrix, which is not in the set, because its determinant vanishes.

2.2 Subgroups

The concept of subgroup is a particularly important one in the theory of groups (and in its mathematical and physical applications). In short, a subgroup is a subset H of a group G, that is itself a group, with the same composition law as G.

More formally, a subset H of the group G is called a *subgroup* of G if it is closed under the group multiplication, i.e., $\forall h_1, h_2 \in H$ one has $h_1 \cdot h_2 \in H$, and if the inverse of each of its elements is also in H, i.e., $\forall h \in H$ also $h^{-1} \in H$.

Note that every subgroup of G must contain at least the unit element e of G.

A subgroup H of a group G is said to be a *trivial subgroup* if it contains only the unit element ($H = \{e\}$) or if it coincides with G itself ($H = G$). Every group admits at least these two trivial subgroups as subgroups. (For the trivial group $\mathbb{Z}_1$ the two coincide.)

Conversely, a subgroup H of a group G is said to be a *proper subgroup* if it is not trivial, i.e., if $H \neq \{e\}$ and $H \neq G$. If H is a proper subgroup of a finite-order group G, then $1 < |H| < |G|$.

A subgroup H of a group G is said to be a *normal subgroup* when, for all $h \in H$ and for all $g \in G$, the product $g \cdot h \cdot g^{-1}$ is also an element of H.

A group G that does not have proper normal Abelian subgroups is said to be a *semisimple group*; if it does not have proper normal subgroups, then it is said to be a *simple group*. Clearly, every simple group is also semisimple (but the converse is not true, as there exist semisimple groups which are not simple).

2.3 Group Homomorphisms and Isomorphisms

Two finite groups are "the same" (up to a relabeling of the elements of the groups) if they have the same Cayley table. We will introduce another technical notion to decide when two groups can be identified. For the comparison, we define maps from one group to another that preserve the group structure, mapping products to products of image elements. These maps are called group homomorphisms and group isomorphisms. Before defining them and discussing their properties, we introduce some further notions.

Given an arbitrary non-empty set X, a binary relation (denoted by $\diamond$) among elements of X is said to be an *equivalence relation* when it is simultaneously reflexive ($\forall x \in X: x \diamond x$), symmetric ($\forall x, y \in X: x \diamond y$ implies $y \diamond x$), and transitive ($\forall x, y, z \in X$: if $x \diamond y$ and $y \diamond z$, then $x \diamond z$).

Given a set X and an equivalence relation $\diamond$ among its elements, the *equivalence class* of a generic element $a \in X$ is the subset of X containing all elements x such that $x \diamond a$ holds, and is denoted as $[a]$. Then, any element belonging to $[a]$ is called a *representative* of that equivalence class. Note that, given any element $a \in X$, the equivalence class $[a]$ is non-empty, because the reflexivity of the equivalence relation $\diamond$ implies that $[a]$ contains at least a itself. Moreover, the following fact holds:

Theorem 2.1 *Any equivalence relation $\diamond$ defined in a set X partitions it into mutually disjoint equivalence classes.*

Proof Consider two distinct equivalence classes $[a]$ and $[b]$. Then, unless $[a]$ is a strict subset of $[b]$, there exists at least one element, say c, which belongs to $[a]$ (so that $c \diamond a$) but not to $[b]$. If $[a]$ and $[b]$ are non-disjoint, then $[a] \cap [b] \neq \emptyset$. Then, let d be an element of $[a] \cap [b]$: Since $d \in [a]$, it follows that $d \diamond a$. Then, the symmetry and transitivity of $\diamond$ imply that $c \diamond d$. On the other hand, since $d \in [b]$, it also follows that $d \diamond b$. Then, the transitivity of $\diamond$ implies that $c \diamond b$, i.e., $c \in [b]$, in contradiction with the assumption that c is not in b. Finally, if $[a]$ is a strict subset of $[b]$, the same argument can be applied by interchanging $[a]$ and $[b]$. $\qquad\qquad$ □

Given a set X and an equivalence relation $\diamond$ among the elements of X, the *quotient space* induced by $\diamond$ is defined as the set of all equivalence classes into which $\diamond$ partitions X, and is denoted as $X/\diamond$.

Given two arbitrary, non-empty sets X and Y, let $\mathrm{Map}(X, Y)$ denote the set of functions (or "mappings") from X to Y:

$$\mathrm{Map}(X, Y) = \{f : X \to Y \,|\, \forall x \in X : \exists! f(x) \in Y\}. \qquad (2.2)$$

Within $\mathrm{Map}(X, Y)$, there exist special types of functions.

- A function $f : X \to Y$ is called an *injection* (or a *one-to-one function*) if $f(x) \neq f(x')$ $\forall x \neq x'$.
- A function $f : X \to Y$ is called a *surjection* (or an *onto function*) if $\forall y \in Y$ there exists at least one $x \in X$, such that $f(x) = y$.
- A function $f : X \to Y$ is called a *bijection* if it is both an injection and a surjection.

If a function f is a bijection, then it is *invertible*, namely one can construct the *inverse function* $f^{-1} : Y \to X$, defined by the property that, for any $x \in X$, one has $f^{-1}(f(x)) = x$. In particular, the fact that f is a surjection implies that, for any $y \in Y, f^{-1}(y)$ can be defined, while the fact that f is an injection implies that $f^{-1}(y)$ is *uniquely* defined. Furthermore, f^{-1} is a bijection, too, and the inverse function of f^{-1} is f.

Example   Consider a set of apples, $A = \{a_1, a_2, a_3\}$, and a set of oranges, $O = \{o_1, o_2, o_3\}$. A mapping $f(a_1) = f(a_2) = o_1$, $f(a_3) = o_2$ is not an injection, a mapping $g(a_1) = o_2$, $g(a_2) = o_3$, $g(a_3) = o_1$ is both an injection and a surjection, hence a bijection. The inverse map is $g^{-1}(o_1) = a_3$, $g^{-1}(o_2) = a_1$, $g^{-1}(o_3) = a_2$.

In general, a mapping $f : X \to Y$ is generally called a *homomorphism* when it preserves some structure. A homomorphism that is a bijection is called an *isomorphism*. In the following, we will be particularly interested in homomorphisms and isomorphisms between groups, which can be defined as follows.

Given two groups $(G, \cdot)$ and $(H, \bullet)$, a homomorphism $f : G \to H$ is called a *group homomorphism* if it preserves the group multiplication, i.e., if $\forall g_1, g_2 \in G$ one has $f(g_1 \cdot g_2) = f(g_1) \bullet f(g_2)$. When a group homomorphism f is bijective, it is called a *group isomorphism*. Two groups G and H are said to be *isomorphic* $(G \cong H)$ if there exists at least a group isomorphism between them.

The relation of isomorphism among groups is an equivalence relation in the set of groups, because the relation of isomorphism among groups is reflexive (every group is isomorphic to itself; the identity mapping is the isomorphism that proves this), symmetric (if $f : G \to H$ is a isomorphism, then $\exists f^{-1} : H \to G$, which is also an isomorphism), and transitive (given two isomorphism $f : G \to H$ and $l : H \to K$, the composite map $l \circ f$ is an isomorphism from G to K).

Isomorphic groups have the same structure, so they can be identified. More precisely, each abstract group can be identified with an equivalence class defined by the equivalence relation of group isomorphism.

Example   Take the two groups $G = (\mathbb{R}_+, \cdot)$ and $H = (\mathbb{R}, +)$. Define the mapping

$$f : G \to H, \quad f : x \to f(x) = \ln x. \tag{2.3}$$

Note that f is a group homomorphism, because $f(xy) = \ln(xy) = \ln x + \ln y = f(x) + f(y)$. In fact, f is also a group isomorphism, because it is a bijection, the inverse mapping being $f^{-1}(x) = e^x$.

2.4 The Smallest Finite Groups

Finite groups have several applications in physics. A classic example is in solid state physics, where they are used to classify general crystal structures (the so-called crystallographic point groups). In addition, they also have applications in classical mechanics, where they can be used to reduce the number of relevant degrees of freedom in systems with certain symmetries, as well as in many different areas of modern physics.

For certain (sufficiently small) finite sets, the requirements that a binary operation on the set elements has to satisfy, in order to be a group multiplication, are so constraining that they uniquely define the group. In this section, we present the list of all groups of finite order $N \leq 6$. Note that, since every group must contain at least the unit element, the order of the group is at least 1.

- **Order $N = 1$:** This is the trivial group $G = \{e\}$, containing only the unit element (which, by definition, is the inverse of itself).
- **Order $N = 2$:** In this case $G = \{e, a\}$, with $a \neq e$. The definition of the unit element implies that $e^2 = e$, $ea = ae = a$. The only remaining multiplication is a^2: To ensure that the multiplication is a closed operation in the group, the result must be either e or a. However, if $a^2 = a$, then $a = ae = a(aa^{-1}) = a^2a^{-1} = aa^{-1} = e$, in contradiction with the assumption that a and e are distinct. So the only possibility is $a^2 = e$. Accordingly, the Cayley table of the group of order 2 is uniquely fixed to be

$$
\begin{array}{c|cc}
  & e & a \\
\hline
e & e & a \\
a & a & e
\end{array} \ . \tag{2.4}
$$

This is the $\mathbb{Z}_2$ group that we already introduced. One of its realizations, in addition to those already mentioned, is in terms of one the symmetric groups (that will be defined and discussed in detail in Section 2.5 and that contain permutations interchanging the group elements), the symmetric group of degree 2, $S_2 = \mathrm{Perm}(\{1, 2\})$. Clearly, S_2 contains only two permutations: The identity permutation E, which leaves the order of the elements unchanged, and the permutation that interchanges them, which can be denoted as A, and which, in cycle notation, can be written as $(1, 2)$. It is easy to prove that S_2 is isomorphic to $\mathbb{Z}_2$: For example, considering the realization of $\mathbb{Z}_2$ in terms of $\{1, -1\}$ (with the ordinary product as the group multiplication), one can define the mapping $f : \mathbb{Z}_2 \to S_2$, such that $f(1) = E, f(-1) = A$. It is easy to see that f is a group isomorphism, so $\mathbb{Z}_2 \cong S_2$.

- **Order** $N = 3$: Consider the set $G = \{e, a, b\}$, assuming that both a and b are distinct from the unit element e, and that $a \neq b$. It turns out that, again, there is only one possible abstract group of order 3. This abstract group can be determined by working out its Cayley table:

$$
\begin{array}{c|ccc}
 & e & a & b \\
\hline
e & e & a & b \\
a & a & ? & ? \\
b & b & ? & ?
\end{array}
\tag{2.5}
$$

Consider ab, and suppose it is equal to b. But then $a = a(bb^{-1}) = (ab)b^{-1} = bb^{-1} = e$, which contradicts the assumption that e, a, and b are all different. Similarly, if $ab = a$, then one would have $b = (a^{-1}a)b = a^{-1}(ab) = a^{-1}a = e$, again in contradiction with the assumption that e and b are different. So it must be $ab = e$. Analogously, one can prove that $ba = e$. These two equalities imply that a is the inverse of b (and vice versa). Then, consider a^2: If it were equal to a, then one would have $a = a(aa^{-1}) = a^2a^{-1} = aa^{-1} = e$, in contradiction to the assumption that $a \neq e$. Similarly, if $a^2 = e$, then one would have $b = eb = a^2b = a(ab) = ae = a$, contradicting the assumption that $a \neq b$. Thus, it must necessarily be $a^2 = b$. Similarly, one can show that $b^2 = a$. The complete Cayley table for the group of three elements reads as follows:

$$
\begin{array}{c|ccc}
 & e & a & b \\
\hline
e & e & a & b \\
a & a & b & e \\
b & b & e & a
\end{array}
\tag{2.6}
$$

This group is called $\mathbb{Z}_3$. Since $b = a^2$, one has $\mathbb{Z}_3 = \{e, a, a^2\}$. $\mathbb{Z}_3$ and $\mathbb{Z}_2$ (and, in fact, also the trivial group containing only the unit element) are examples of *cyclic groups*.

More in general, the *cyclic group* of order N, denoted as $\mathbb{Z}_N$, is defined as a finite group in which each element can be written as a power of a single *generating element a*.

Thus, the generic cyclic group of order N is

$$
\mathbb{Z}_N = \{e, a, a^2, \ldots, a^{N-1}\},
\tag{2.7}
$$

where the unit element is $e = a^0$, while the inverse of a generic element a^p is a^{N-p}. Note that a cyclic group can be defined for any order $N \geq 1$. Since $a^p \cdot a^q = a^{p+q(\mathrm{mod}\, N)} = a^q \cdot a^p$ for any p and q, all cyclic groups are Abelian. One realization of cyclic groups is given by the complex Nth roots of 1: $\mathbb{Z}_N = \{\exp(2\pi i k/N),\ \text{for } k = 0, 1, \ldots, N-1\}$, with the usual product as group multiplication. Thinking about the representation of the roots of 1 in the complex plane, this realization also reveals a geometric interpretation of $\mathbb{Z}_N$: It is the symmetry group of rotations of a regular directed polygon with n sides. It is trivial to prove that the set $\{0, 1, \ldots, N - 1\}$, with the addition modulo N as the group multiplication, provides another realization of $\mathbb{Z}_N$.

- **Order $N = 4$:** Since cyclic groups $\mathbb{Z}_N$ are defined for any positive integer N, we can immediately construct a finite group of order 4, namely $\mathbb{Z}_4 = \{e, a, a^2, a^3\}$. Its Cayley table is

$$
\begin{array}{c|cccc}
 & e & a & a^2 & a^3 \\
\hline
e & e & a & a^2 & a^3 \\
a & a & a^2 & a^3 & e \\
a^2 & a^2 & a^3 & e & a \\
a^3 & a^3 & e & a & a^2
\end{array}
. \tag{2.8}
$$

However, there exists also a different group of order 4: the direct product $\mathbb{Z}_2 \times \mathbb{Z}_2$. Denoting $\mathbb{Z}_2 \times \mathbb{Z}_2 = \{f, b\} \times \{g, c\}$, with f and g the unit elements of the two groups, $b^2 = f$ and $c^2 = g$, the set of elements of this group can be written as $\{(f, g), (f, c), (b, g), (b, c)\}$. Note that $\mathbb{Z}_2 \times \mathbb{Z}_2$ is an Abelian group, but it is different from $\mathbb{Z}_4$, because it is not a cyclic group. In particular, the elements (f, c) and (b, g) are not powers of the same element. Neither of the two is a power of the other, and they are not powers of the remaining two elements of the group either (by definition, any power of the unit element (f, g) is equal to itself, while all even powers of (b, c) are equal to the unit element, and all odd powers are equal to (b, c) itself). The Cayley table of $\mathbb{Z}_2 \times \mathbb{Z}_2$ can easily be worked out to be

$$
\begin{array}{c|cccc}
 & (f, g) & (f, c) & (b, g) & (b, c) \\
\hline
(f, g) & (f, g) & (f, c) & (b, g) & (b, c) \\
(f, c) & (f, c) & (f, g) & (b, c) & (b, g) \\
(b, g) & (b, g) & (b, c) & (f, g) & (f, c) \\
(b, c) & (b, c) & (b, g) & (f, c) & (f, g)
\end{array}
\tag{2.9}
$$

and it is different from that of $\mathbb{Z}_4$. The $\mathbb{Z}_2 \times \mathbb{Z}_2$ group is also called *Vierergruppe* (and denoted by V_4) or *Klein four-group*, after the German mathematician Christian Felix Klein. Considering that the trivial group can be identified with $\mathbb{Z}_1$, the Klein four-group is the smallest noncyclic finite group. It is possible to show that $\mathbb{Z}_2 \times \mathbb{Z}_2$ and $\mathbb{Z}_4$ are the only two groups of order 4.

- **Order $N = 5$:** The only finite group of order $N = 5$ is the cyclic group $\mathbb{Z}_5 = \{e, a, a^2, a^3, a^4\}$.

- **Order $N = 6$:** At order $N = 6$, there exist two non-isomorphic finite groups: One of them is the cyclic group $\mathbb{Z}_6 = \{e, a, a^2, a^3, a^4, a^5\}$, whose Cayley table is

$$
\begin{array}{c|cccccc}
 & e & a & a^2 & a^3 & a^4 & a^5 \\
\hline
e & e & a & a^2 & a^3 & a^4 & a^5 \\
a & a & a^2 & a^3 & a^4 & a^5 & e \\
a^2 & a^2 & a^3 & a^4 & a^5 & e & a \\
a^3 & a^3 & a^4 & a^5 & e & a & a^2 \\
a^4 & a^4 & a^5 & e & a & a^2 & a^3 \\
a^5 & a^5 & e & a & a^2 & a^3 & a^4
\end{array}
. \tag{2.10}
$$

Interestingly, the $\mathbb{Z}_6$ group turns out to be isomorphic to the direct product $\mathbb{Z}_2 \times \mathbb{Z}_3$. Denoting $\mathbb{Z}_2 = \{f, b\}$ (with f the unit element and $b^2 = f$) and $\mathbb{Z}_3 = \{g, c, c^2\}$ (with g the unit element and $c^3 = g$), the Cayley table of $\mathbb{Z}_2 \times \mathbb{Z}_3$ reads

$$
\begin{array}{c|cccccc}
 & (f,g) & (f,c) & (f,c^2) & (b,g) & (b,c) & (b,c^2) \\
\hline
(f,g) & (f,g) & (f,c) & (f,c^2) & (b,g) & (b,c) & (b,c^2) \\
(f,c) & (f,c) & (f,c^2) & (f,g) & (b,c) & (b,c^2) & (b,g) \\
(f,c^2) & (f,c^2) & (f,g) & (f,c) & (b,c^2) & (b,g) & (b,c) \\
(b,g) & (b,g) & (b,c) & (b,c^2) & (f,g) & (f,c) & (f,c^2) \\
(b,c) & (b,c) & (b,c^2) & (b,g) & (f,c) & (f,c^2) & (f,g) \\
(b,c^2) & (b,c^2) & (b,g) & (b,c) & (f,c^2) & (f,g) & (f,c)
\end{array}
\qquad (2.11)
$$

An isomorphism relating this group to $\mathbb{Z}_6$ is the one mapping the generic element $(b^p, c^q) \in \mathbb{Z}_2 \times \mathbb{Z}_3$ to the element $a^{(3p+4q) \bmod 6} \in \mathbb{Z}_6$: So, for example, $(f, c) = (b^0, c^1)$ corresponds to a^4, while $(b, c^2) = (b^1, c^2)$ is mapped to a^5. The product $(f, c) \cdot (b, c^2) = (b, g) = (b^1, c^0)$ is mapped to a^3, and this is consistent with the fact that the isomorphism preserves the group product, as $a^4 \cdot a^5 = a^3$ in $\mathbb{Z}_6$.

In addition to $\mathbb{Z}_6 \cong \mathbb{Z}_2 \times \mathbb{Z}_3$, there exists another, non-isomorphic, finite group of order 6: the symmetric group S_3, which is the smallest non-Abelian group (and which coincides with D_3, the group of symmetries of an equilateral triangle). Its elements can be written as: $\{e, a, b, aba, ab, ba\}$ and they satisfy $a^2 = b^2 = (ab)^3 = (ba)^3 = e$, with e the unit element. Note that these properties imply $aba = bab$. The Cayley table of S_3 is

$$
\begin{array}{c|cccccc}
 & e & a & b & aba & ab & ba \\
\hline
e & e & a & b & aba & ab & ba \\
a & a & e & ab & ba & b & aba \\
b & b & ba & e & ab & aba & a \\
aba & aba & ab & ba & e & a & b \\
ab & ab & aba & a & b & ba & e \\
ba & ba & b & aba & a & e & ab
\end{array}
\qquad (2.12)
$$

Note that the non-Abelian nature of the group is reflected in the fact that the Cayley table is not symmetric under reflection about the diagonal.

- **Order** $N = 7$: The only finite group of order $N = 7$ is the cyclic group $\mathbb{Z}_7 = \{e, a, a^2, a^3, a^4, a^5, a^6\}$.

- **Order** $N = 8$: There exist five non-isomorphic finite groups of order 8; three of them are Abelian: $\mathbb{Z}_8$, $\mathbb{Z}_4 \times \mathbb{Z}_2$ and $\mathbb{Z}_2 \times \mathbb{Z}_2 \times \mathbb{Z}_2$. The remaining two groups of order 8 are non-Abelian. One of them is the dihedral group D_4, which can be interpreted as the symmetry group of a square: It consists of four rotations by angles which are integer multiples of $\pi/2$, two reflections about axes going through the midpoints of pairs of opposite sides, and two reflections about the diagonals of the square. The group elements can be written as $\{e, a, a^2, a^3, b, ab, a^2b, a^3b\}$, where e is the unit element and $a^4 = b^2 = (ab)^2 = e$. In terms of transformations that leave a square invariant, a can be interpreted as a rotation by $\pi/2$ and b as a reflection about the axis going through the midpoints of two opposite sides. The Cayley table of D_4 is

	e	a	a^2	a^3	b	ab	a^2b	a^3b
e	e	a	a^2	a^3	b	ab	a^2b	a^3b
a	a	a^2	a^3	e	ab	a^2b	a^3b	b
a^2	a^2	a^3	e	a	a^2b	a^3b	b	ab
a^3	a^3	e	a	a^2	a^3b	b	ab	a^2b
b	b	a^3b	a^2b	ab	e	a^3	a^2	a
ab	ab	b	a^3b	a^2b	a	e	a^3	a^2
a^2b	a^2b	ab	b	a^3b	a^2	a	e	a^3
a^3b	a^3b	a^2b	ab	b	a^3	a^2	a	e

$$(2.13)$$

The last finite group of order 8 is the non-Abelian *quaternion group* $Q_8 = \{\pm e, \pm i, \pm j, \pm k\}$, where e is the unit element, and $i^2 = j^2 = k^2 = ijk = -e$. The Cayley table of Q_8 is

	e	$-e$	i	$-i$	j	$-j$	k	$-k$
e	e	$-e$	i	$-i$	j	$-j$	k	$-k$
$-e$	$-e$	e	$-i$	i	$-j$	j	$-k$	k
i	i	$-i$	$-e$	e	k	$-k$	$-j$	j
$-i$	$-i$	i	e	$-e$	$-k$	k	j	$-j$
j	j	$-j$	$-k$	k	$-e$	e	i	$-i$
$-j$	$-j$	j	k	$-k$	e	$-e$	$-i$	i
k	k	$-k$	j	$-j$	$-i$	i	$-e$	e
$-k$	$-k$	k	$-j$	j	i	$-i$	e	$-e$

$$(2.14)$$

If one identifies i, j, and k with the unit vectors defining a right-handed reference frame in the three-dimensional real vector space $\mathbb{R}^3$, then the group multiplication between these elements is consistent with the cross-product when they are distinct (and do not differ simply by a sign): $ij = k$, and cyclic permutations thereof. Note, however, that in $\mathbb{R}^3$ the cross-product of a vector with itself (or with its opposite) vanishes, whereas this is not the case for the group multiplication in Q_8. Q_8 has four proper subgroups

$$\{e, -e\} \cong \mathbb{Z}_2, \tag{2.15}$$

$$\{e, i, -e, -i\} \cong \mathbb{Z}_4, \tag{2.16}$$

$$\{e, j, -e, -j\} \cong \mathbb{Z}_4, \tag{2.17}$$

$$\{e, k, -e, -k\} \cong \mathbb{Z}_4, \tag{2.18}$$

all of which are normal subgroups.

2.5 Permutations, the Symmetric Group, and Cayley's Theorem

Consider again $\text{Map}(X, Y)$, the set of mappings from a generic set X to another generic set Y. If the two sets coincide, then $\text{Map}(X, X)$ can be endowed with a semigroup structure, by taking the composition of maps as the composition law:

$$fg = f \circ g, \qquad (f \circ g)(x) = f(g(x)) \quad \text{for } \forall x \in X. \qquad (2.19)$$

Note that the composition is well defined only if $X = Y$, because g is a map from X to Y, but the domain of f is X.

Given a non-empty set X, a bijective function $f : X \to X$ is called a *permutation* of X.

Example Let X be the set of 52 cards of a deck. Shuffling the deck executes a permutation of X.

The set of permutations of X is denoted as Perm(X). In this set one can introduce the composition of permutations (i.e., the operation consisting in applying one permutation after the other) as the group multiplication. Composition of maps is associative. Then, the identity map $E : X \to X$, $E : x \to E(x) = x$ for all $x \in X$ is the unit element of Perm(X). Finally, since permutations are bijections, every $f \in$ Perm(X) has an inverse, so Perm(X) is a group.

Note that the group multiplication in the set of permutations is defined according to the convention that the order in which they are applied is "from right to left," meaning that first one performs the rightmost permutation in a given expression, then continues with the next one to its left, and so on. This convention is inherited from that of composite mappings, for which, for example, $(fg)(x)$ means $f(g(x))$. In general, the result obtained by applying first one permutation, then another, is not the same that one obtains by multiplying the same permutations in the opposite order. Therefore, in general, Perm(X) is not an Abelian group.

Example Let X be a set of three elements, say $X = \{a, b, c\}$. Let f be the permutation that interchanges the second and the third element of X, leaving the first unchanged, i.e., $f(a) = a$, $f(b) = c$, $f(c) = b$, and let g be the permutation that interchanges the first and the second element, leaving the third unchanged, i.e., $g(a) = b$, $g(b) = a$, $g(c) = c$. Then, for instance, $(f \cdot g)(a) = f(g(a)) = c$, while $(g \cdot f)(a) = g(f(a)) = b$. Hence $f \cdot g \neq g \cdot f$, so Perm(X) is not an Abelian group.

When X has a finite number N of elements, its group of permutations is called the *symmetric group* (or *permutation group*) *of degree N* and is denoted by S_N. We leave it as an exercise (see Problem 2.6) to prove that S_N contains $N!$ elements.

The smallest symmetric groups are isomorphic to groups that we already mentioned. In particular, $S_1 = \{e\}$ is the trivial group $\mathbb{Z}_1$, containing only the unit element, while S_2 is isomorphic to $\mathbb{Z}_2$. Both of them are Abelian groups. On the contrary, S_3 is a non-Abelian group. It is isomorphic to D_3, the dihedral group of order 6, which describes the symmetries of an equilateral triangle.

A possible way to denote the elements of $S_N = $ Perm($\{1, 2, \ldots, N\}$) is in terms of matrices of size $2 \times N$, in which each row contains all the numbers from 1 to N, and the permutation acts by mapping each number in the first row to the one in the second row and in the same column. So, for example, the permutations in S_2 are

$$E = \begin{pmatrix} 1 & 2 \\ 1 & 2 \end{pmatrix}, \qquad A = \begin{pmatrix} 1 & 2 \\ 2 & 1 \end{pmatrix}, \qquad (2.20)$$

so that E leaves the order of the two elements unvaried, while A interchanges them.

A more compact and more convenient way to denote permutations, however, is the one in terms of cycles. Given a permutation $P \in S_N$, a *cycle* is defined as an ordered sequence of labels, which are subsequently mapped one to another by P, with the last element of the cycle being mapped to the first.

A cycle can be considered as a permutation acting only on its labels (leaving the others unchanged). They are often written by listing the sequence of labels in parentheses, e.g., (132). In a given permutation P, any label that is mapped to itself (i.e., left invariant) by P can be considered as a cycle of unit length, e.g., (4). This means that N cycles of unit length, which act on only one label, leaving it unchanged, correspond to the unit element of S_N.

Clearly, the order of the labels within a cycle of length larger than one is relevant: Cycles containing the same labels but in a different order describe different permutations. So, for instance, (132) and (123) are different. However, cycles which differ only by a cyclic permutation of their labels are equivalent: Thus, (132) and (321) describe the same permutation. By virtue of this latter property, one can for example define the convention that, in each cycle, the smallest label appears in the leftmost position.

A generic permutation (different from the unit element of S_N) can be written as the product of its disjoint cycles of length larger than 1. Note that, when a permutation is written as the product of cycles acting on disjoint sets of labels, the order in which such cycles are multiplied is irrelevant. Thus, for example,

$$(132)(45) \qquad \text{and} \qquad (45)(132) \tag{2.21}$$

describe the same permutation in S_5. (Thus, one can uniquely fix the expression of a permutation as a product of cycles, for example by imposing the convention that the cycles are ordered so that the smallest label in each cycle is always increasing, when going from the leftmost to the rightmost cycle factor.)

Given two permutations in cycle notation, the cycle notation of their product can be expressed as follows.

1. Write the first label (say, x) appearing in the first cycle of the permutation that is applied first.
2. Starting from x, read the label that follows it (say, y) in the permutation that is applied first.
3. Read the label (say, z) that follows y in the permutation that is applied as second, and write it in the cycle notation of the permutation product.
4. Iterate the procedure, starting from z in the permutation that is applied first, until a cycle is completed.
5. Repeat the procedure for the other labels in the cycles of the permutation that is applied first.

If a generic permutation P is written as the product of nonoverlapping cycles, the inverse permutation P^{-1} can be expressed by reversing the order of the labels within each cycle of P (up to cyclic permutations of the labels within each cycle).

The cycle notation for permutations is best illustrated by examples. For instance, the permutation of six labels

$$P = \begin{pmatrix} 1 & 2 & 3 & 4 & 5 & 6 \\ 2 & 5 & 6 & 4 & 1 & 3 \end{pmatrix} \tag{2.22}$$

decomposes into the product of three disjoint cycles $1 \to 2 \to 5 \to 1, 3 \to 6 \to 3$ and $4 \to 4$. Thus it can be written as $(125)(36)(4)$, or, omitting cycles of unit length, as

$$P = (125)(36). \tag{2.23}$$

As an example of product of permutations, the algorithm described leads to

$$(146)(253) \cdot (13)(26)(45) = (12)(34)(56). \tag{2.24}$$

Note that, in the product appearing on the left-hand side of this equation, the permutation that is applied first is the rightmost one, i.e., $(13)(26)(45)$.

Finally,

$$(142)(35) \quad \text{and} \quad (124)(35) \tag{2.25}$$

are an example of permutations that are the inverse of each other. (Note that the labels in the cycles of the permutation on the right-hand side have been written in the opposite order with respect to those of the permutation on the left-hand side, then – for notational convenience – they have been reordered by a cyclic label permutation within each cycle, in order to have each cycle start with the smallest label it contains.)

The cycle notation provides a convenient way to list the elements of S_N, according to their cycle structure, i.e., by the number and length of cycles they contain (and denoting the trivial permutation, i.e., the unit element, of S_N as E). For illustration, we list the first permutation groups S_N, for $1 \le N \le 4$:

$$S_2 = \{E, (12)\} \tag{2.26}$$
$$S_3 = \{E, (12), (13), (23), (123), (132)\} \tag{2.27}$$
$$S_4 = \{E, (12), (13), (14), (23), (24), (34), (12)(34), (13)(24), (14)(23),$$
$$(123), (132), (124), (142), (134), (143), (234), (243),$$
$$(1234), (1243), (1324), (1342), (1423), (1432)\}. \tag{2.28}$$

The cycle notation makes it easy to write down all the permutations in a concise and systematic way (one which, if necessary, can be readily automated in numerical implementations).

The simplest nontrivial permutations are the 2-cycles, which interchange two labels. In fact, it is possible to show that *any* permutation can be built from products of overlapping 2-cycles; a generic cycle of length r can be written as the product of $r - 1$ overlapping 2-cycles:

$$(n_1 n_2 \ldots n_r) = (n_1 n_2)(n_2 n_3) \ldots (n_{r-1} n_r). \tag{2.29}$$

Then, since any permutation is a product of cycles, it can always be written as a product of 2-cycles. According to the number of 2-cycles they can be decomposed into, a generic permutation can be classified as "even" or "odd": More precisely, a permutation is said to be an *even* permutation if it can be factored into a product of an even number of 2-cycles. Conversely, an *odd* permutation is one that factorizes into the product of an odd number of 2-cycles. The *signature* of a permutation P, denoted as $\text{sgn}(P)$, is defined as

$$\text{sgn}(P) = \begin{cases} 1 & \text{if } P \text{ is an even permutation} \\ -1 & \text{if } P \text{ is an odd permutation} \end{cases}. \tag{2.30}$$

Note that a generic r-cycle is even if r is odd (and vice versa), because it can be decomposed into the product of $r - 1$ overlapping 2-cycles.

Also, note that the unit element of S_N, i.e., the identity permutation E, is even (it contains no 2-cycles).

The *alternating group* A_N is defined as the subgroup of even permutations of S_N. Its order is $|A_N| = |S_N|/2 = (N!)/2$, thus for any $N > 2$ the alternating group is a proper subgroup of S_N.

Note that the set of *odd* permutations is not a subgroup of S_N: In fact, the multiplication of two odd permutations is an even permutation, so the multiplication is not a closed operation in the set of odd permutations. This implies that the set of odd permutations is not even a magma.

The reason why groups of permutations have a special status among finite groups is because of the following theorem, named after Arthur Cayley and first proven by the French mathematician Marie Ennemond Camille Jordan.

Theorem 2.2 (Cayley's theorem)   *Every finite group of order N is isomorphic to a subgroup of S_N.*

Proof   Let $(G, \cdot)$ be a generic finite group, with $|G| = N$. For any element $g \in G$, consider the function $f_g \in \mathrm{Map}(G, G)$ defined as "multiplication by g on the left":

$$f_g : x \to f_g(x) = g \cdot x. \tag{2.31}$$

Note that f_g is an injection, because $f_g(x_1) = f_g(x_2)$ means $g \cdot x_1 = g \cdot x_2$; multiplying both sides of this equation by g^{-1} on the left, this implies $x_1 = x_2$. Furthermore, f_g is also a surjection, because, for any $y \in G$, there exists an element $x \in G$ such that $f_g(x) = y$: such x is simply $g^{-1} \cdot y$. Thus, $f_g : G \to G$ is a bijection, so it is a permutation of G. Now, consider the set of functions

$$K = \left\{ f_g : g \in G \right\}. \tag{2.32}$$

Defining the composition of functions as the group multiplication, K is endowed with group structure. Indeed, K is closed under the composition of functions (for any g_1 and $g_2 \in G$, the action of the composite map $f_{g_2} f_{g_1}$ on a generic $x \in G$ is defined as $f_{g_2}(f_{g_1}(x)) = f_{g_2}(g_1 \cdot x) = g_2 \cdot g_1 \cdot x = f_{g_2 \cdot g_1}(x)$, hence $f_{g_2} f_{g_1} = f_{g_2 \cdot g_1} \in K$, because $g_2 \cdot g_1 \in G$), which is an associative operation, f_e (with e the unit element of G) is the unit element of K, and the inverse of a generic $f_g \in K$ is $f_{g^{-1}}$. Note that this definition of group multiplication in K is the same as the definition of group multiplication in $\mathrm{Perm}(G)$, since the operation of multiplying two permutations has been defined as applying one after the other (in particular, applying the one on the right first – in agreement with the rule of composition of functions). Thus, K is a set containing permutations of G, i.e., a subset of $\mathrm{Perm}(G)$, and a group with the same group multiplication as $\mathrm{Perm}(G)$. This means that K is a subgroup of $\mathrm{Perm}(G)$. Also, note that, since G is completely generic, apart from being a finite group of order N, $\mathrm{Perm}(G)$ is simply the group of permutations of a set with N elements, i.e., S_N. Next, it is trivial to show that there exists a group homomorphism $\mathcal{H}$ between G and K: Such homomorphism is just the function

$$\mathcal{H} : G \to K, \qquad g \to \mathcal{H}(g) = f_g, \tag{2.33}$$

on which the definition of K is based. $\mathcal{H}$ is a group homomorphism because, as we showed previously, the product $\mathcal{H}(g_2)\mathcal{H}(g_1) = f_{g_2}f_{g_1}$ in K is equal to $f_{g_2 \cdot g_1} = \mathcal{H}(g_2 \cdot g_1)$, hence $\mathcal{H}$ preserves the group product. Furthermore, $\mathcal{H}$ is an injection: $\mathcal{H}(g_1) = \mathcal{H}(g_2)$ means that the two functions f_{g_1} and f_{g_2} are equal, i.e., that, for any $x \in G$, one has $f_{g_1}(x) = f_{g_2}(x)$, that is $g_1 \cdot x = g_2 \cdot x$. Multiplying both sides of this equality by x^{-1} on the right, one gets $g_1 = g_2$. Finally, $\mathcal{H}$ is also a surjection, because K is defined as the set of functions f_g, that is, $\mathcal{H}(g)$, for all $g \in G$. This implies that the group homomorphism $\mathcal{H}$ is a bijection, i.e., it is a group isomorphism. We conclude that any generic finite group G of order N is isomorphic to a subgroup K of S_N. □

2.6 Partitions, Young Diagrams, and Multisets

2.6.1 Partitions

We noticed that the elements of S_N fall into subsets where the permutations have a similar cycle structure (or they are of similar *cycle type*).[1] For example, in S_4 we had the following types:

(1234), etc.	one 4-cycle	4
(123)(4), etc.	one 3-cycle, one 1-cycle	3 + 1
(12)(34), etc.	two 2-cycles	2 + 2
(12)(3)(4), etc.	one 2-cycle, two 1-cycles	2 + 1 + 1
(1)(2)(3)(4), etc.	four 1-cycles	1 + 1 + 1 + 1.

The right-hand column above lists the lengths of all cycles, including 1-cycles, for permutations of four elements. We notice that adding up the lengths always gives 4. More in general, in permuting N elements, the sum of lengths of all cycles must be N for the permutation to map all of the N elements. In the above, the different sums are different *partitions* of 4.

A *partition* of N is defined as a sum

$$N = \sum_i n_i, \tag{2.34}$$

where all $n_i \in \mathbb{Z}_+$. The number of different partitions of N (different ways of breaking N into a sum of type (2.34)), denoted $p(N)$, is called the *partition function*.

For example, $p(4) = 5$. One way to compute $p(N)$ is to use a generating function. One can show that the following identity holds:

$$\sum_{N=0}^{\infty} p(N)x^N = \prod_{k=1}^{\infty} \left(\frac{1}{1 - x^k}\right). \tag{2.35}$$

Now, expanding all factors on the right-hand side as Taylor series:

$$(1 - x)^{-1} = 1 + x + x^2 + x^3 + \cdots$$
$$(1 - x^2)^{-1} = 1 + x^2 + x^4 + x^6 + \cdots$$

---

[1] We will learn later that the subsets are the so-called *conjugacy classes* of S_N.

$$(1 - x^3)^{-1} = 1 + x^3 + x^6 + x^9 + \cdots,$$

etc., $\qquad\qquad$ (2.36)

then

$$\prod_{k=1}^{\infty} \left(\frac{1}{1 - x^k} \right) = (1 + x + x^2 + \cdots)(1 + x^2 + x^4 + \cdots)(1 + x^3 + \cdots)\cdots$$

$$= 1 + x + 2x^2 + 3x^3 + \cdots. \qquad\qquad (2.37)$$

Matching the coefficients of x^N on both sides of (2.35) gives the values of $p(N)$: $p(0) = 1$, $p(1) = 1$, $p(2) = 2$, $p(3) = 3$, ... For large values of N, Godfrey Harold Hardy and Srinivasa Ramanujan derived the asymptotic formula

$$p(N) \sim \frac{1}{4N\sqrt{3}} \exp\left(\pi \sqrt{\frac{2N}{3}} \right), \quad N \to \infty. \qquad\qquad (2.38)$$

2.6.2 Young Diagrams

The different partitions can be represented graphically with *Young diagrams* (sometimes also called *Young tableaux*), which are named after the British mathematician Alfred Young. Recall the partition sum (2.34). Assume that the summands have been indexed in descending order: $n_1 \geq n_2 \geq n_3 \geq \cdots$. Then draw a figure with n_1 adjacent boxes on the first row, n_2 boxes in the second row, and so on.

1	2	$\cdots$	$n_1 - 2$	$n_1 - 1$	n_1
1	2	$\cdots$	$n_2 - 1$	n_2	
1	2	$\cdots$	n_3		
$\vdots$	$\vdots$	$\ddots$			

The resulting figure is the *Young diagram* corresponding to the partition (2.34). For example, for $N = 4$ we have the following partitions and Young diagrams.

4

3 + 1

2 + 2

2 + 1 + 1

1 + 1 + 1 + 1

The Young diagrams then also represent graphically the different types of permutations (different conjugacy classes) of S_N.

2.6.3 Multisets

Permutations reorder the elements of a set X with *distinct* elements. If we represent the elements by alphabetic letters and each ordering as a *word*, then each permutation generates an *anagram*. For instance, for a set of four elements T, E, A, and M we get $4! = 24$ different anagrams:

TEAM

MEAT

MATE

ATEM

...

What about words where the same letter appears more than once, such as

ABRACADABRA

How many different anagrams would we generate now? Let us first define a *multiset* as a set where an element x_i can appear multiple times, specified by its *multiplicity* m_i. For example, in

$$X = \{x_1, x_2, x_2, x_2, x_3, x_3\} \tag{2.39}$$

the element x_1 has multiplicity $m_1 = 1$, x_2 has $m_2 = 3$, x_3 has $m_3 = 2$. The total number of elements of X is $N = \sum_i m_i$ ($N = 6$ in the above example), when we do not require that all the elements are distinct. The letters of the word ABRACADABRA form the multiset

$$X = \{A, A, A, A, A, B, B, R, R, C, D\} \tag{2.40}$$

with $m_A = 5$, $m_B = 2$, $m_R = 2$, $m_C = m_D = 1$, and $N = 11$. Different words formed by the letters are the different permutations (reorderings) of the multiset. A priori, N elements can be reordered $N!$ times. But there are $m_1!$ ways to reorder the elements x_1 with no effect, $m_2!$ ways to reorder the elements x_2, and so on. Thus the total number of *distinct* reorderings of the elements of X is

$$\binom{N}{m_1, m_2, \ldots, m_k} \equiv \frac{N!}{m_1! \, m_2! \cdots m_k!}, \tag{2.41}$$

where k is the number of distinct elements of X, and $\sum_{i=1}^{k} m_i = N$ is a partition of N. Equation (2.41) defines a *multinomial coefficient*, which is a generalization of the binomial coefficient. The name comes from the generalization of the binomial theorem to k variables, the *multinomial theorem*

$$(x_1 + \cdots + x_k)^N = \sum_{\{m_i\}} \binom{N}{m_1, m_2, \ldots, m_k} x_1^{m_1} x_2^{m_2} \cdots x_k^{m_k}, \tag{2.42}$$

where the sum is over all partitions $\{m_i\}$ of N.

Now we can compute how many different anagrams of ABRACADABRA there are. The answer is

$$\frac{11!}{5! \, 2! \, 2! \, 1! \, 1!} = 83160. \tag{2.43}$$

2.7 Free Groups, Presentations of Groups, and Braid Groups

This section introduces a new way to construct groups.

2.7.1 Free Groups and Presentations

We begin by defining *free groups*.

Let G be a group and $X = \{g_1, g_2, \ldots, g_n\}$ a subset of elements of G. If *every* element $g \in G \setminus \{e\}$ (excluding the unit element e) can be *uniquely* written as a product

$$g = g_{j_1}^{i_1} g_{j_2}^{i_2} \cdots g_{j_m}^{i_m} \tag{2.44}$$

of elements g_{j_k} taken only from the set X with exponents $i_k \in \mathbb{Z} \setminus \{0\}$ such that no two adjacent elements are equal (i.e., $g_{j_i} \neq g_{j_{i+1}}$), we say that G is a *free group* and X is a *free set of generators* (of G).

The elements of X are called *letters*. An *arbitrary* product of letters

$$g = g_{j_1}^{i_1} g_{j_2}^{i_2} \cdots g_{j_m}^{i_m}, \tag{2.45}$$

where the exponents $i_k \in \mathbb{Z}$ (note: $i_k = 0$ is allowed) is called a *word*. If it satisfies the additional conditions of the previous definition, $i_k \neq 0$ and $g_{j_i} \neq g_{j_{i+1}}$, the word is called a *reduced word*.

Note that this is otherwise like the familiar construction of words with letters (with $a^3 b^2 = aaabb$, etc.), except that group elements also have inverses. When all exponents are zero, the product is assumed to yield the unit element e. If a word is not a reduced word, one can perform a *reduction* to rewrite it in a reduced form (combining adjacent elements and removing unit elements). Note also that the product is in general not commutative: $ab \neq ba$.

Example Let $X = \{a, b, c\}$ be a collection of elements of a group G (excluding the unit element). For example, $g = a^3 b^{-1} c^2 b^4 c$ is a reduced word, while $h = c^{-1} b^3 b^{-2} a^0$ is a word, but not a reduced one. The reduction of h produces the reduced word $h' = c^{-1} b$.

Words can be joined together by forming a *product*. For example,

$$gh = a^3 b^{-1} c^2 b^4 c c^{-1} b^3 b^{-2} a^0. \tag{2.46}$$

The reduction of this gives the reduced word $(gh)' = a^3 b^{-1} c^2 b^5$. If we replace in the product the word h by its reduced form h' and then perform a reduction, we obtain the same reduced form: $(bh')' = (bh)'$. We can now define a free group G in an alternative way.

A *free group generated by* X is the set of of all reduced words formed from the letters of X and the empty word e, with products of words (joining of words) followed by reduction as the multiplication rule. To emphasize the generators X, we denote the free group generated by X by $F(X)$.

Example Let $X = \{a\}$. It generates the free group $F(X) = \{a^n | n \in \mathbb{Z}\}$, which is isomorphic to $(\mathbb{Z}, +)$. The isomorphism is $a^n \leftrightarrow n$, with $a^n a^m \leftrightarrow n + m$.

Example   Let $X = \{a, b\}$. In this case the free group generated by X is

$$F(X) = \{e, a^n, b^n, a^n b^m, b^n a^m, a^n b^m a^k, b^n a^m b^k, \ldots\}, \qquad (2.47)$$

where $n, m, k, \ldots$ are integer numbers.

We can define a constraint by setting a reduced word to be equal to the unit element by an equation

$$r \equiv g_{j_1}^{i_1} g_{j_2}^{i_2} \cdots g_{j_m}^{i_m} = e. \qquad (2.48)$$

Such a constraint is called a *relation*. There can be several independent relations r_1, $r_2, \ldots, r_n$.

Example   Let $X = \{a\}$, set $r \equiv a^n = e$. With this relation, the set X generates the cyclic group $\{e, a, a^2, \ldots, a^n\} \cong \mathbb{Z}_n$.

A definition of a group now consists of the set of generators $X = \{g_1, g_2, \ldots, g_n\}$ and the complete list of independent relations $r_1, r_2, \ldots, r_m$. We use the notation $\langle g_1, g_2, \ldots, g_n | r_1, r_2, \ldots, r_m \rangle$ to denote this group, called the *presentation* of the group. For the previous example, the presentation of the group is

$$\langle a | a^n \rangle \cong \mathbb{Z}_n. \qquad (2.49)$$

Additional Examples

1. Let $X = \{a, b\}$, $r = aba^{-1}b^{-1} = e$, and define the group $\langle a, b | aba^{-1}b^{-1} \rangle$. Note that the relation is equivalent to the equation $ab = ba$, meaning that the generators commute. Thus

$$\langle a, b, | aba^{-1}b^{-1} \rangle = \{a^n b^m | ab = ba; \ n, m \in \mathbb{Z}\} \cong (\mathbb{Z} \times \mathbb{Z}, +). \qquad (2.50)$$

2. The *dihedral group* D_4 is the group of symmetries of a square. Consider the operations $r =$ rotate the square by $\pi/2$ and $f =$ reflect the square about the symmetry axis passing through the midpoints of opposite sides. The following relations are easy to see: $r^4 = e$ (rotation by 2π) and $f^2 = e$ (reflecting twice). A bit less obvious is $rfrf = e$, which is illustrated in Fig. 2.1. One can check that there are only these three independent relations. The dihedral group has then the presentation $D_4 = \langle r, f | r^4, f^2, rfrf \rangle$. More in general, dihedral groups D_n describe the symmetries of regular polygons with n sides, and can be defined via the presentation $D_n = \langle r, f | r^n, f^2, rfrf \rangle$.

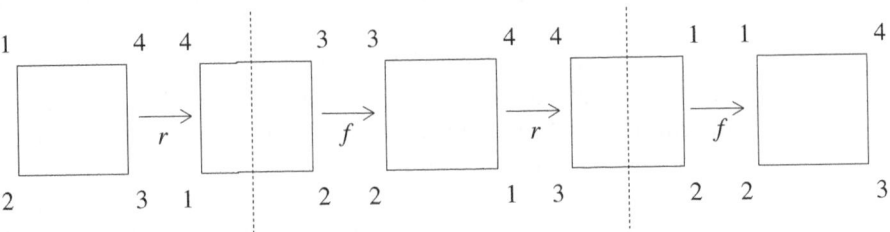

Fig. 2.1   Illustration of the relation $rfrf = e$, characterizing the dihedral group D_4.

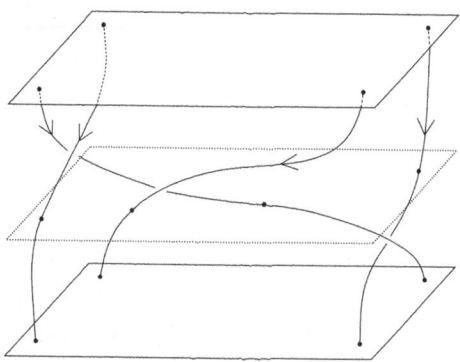

Fig. 2.2 Worldlines of four particles moving on a two-dimensional plane. Time runs from the top to the bottom, and the locations of the particles at the initial, at the final, as well as at a generic intermediate time (represented by the dotted plane) are shown by the black bullets.

3. The *Pauli group* G_1 is a finite group of order 16, consisting of the three Pauli matrices σ_1, σ_2, σ_3, and the identity matrix $\mathbb{1}_2$ multiplied by the factors $\pm 1, \pm i$:

$$G_1 = \{\pm \mathbb{1}_2, \pm i \mathbb{1}_2, \pm \sigma_1, \pm i\sigma_1, \pm \sigma_2, \pm i\sigma_2, \pm \sigma_3, \pm i\sigma_3\}. \tag{2.51}$$

Recalling the relations

$$\sigma_a^2 = \mathbb{1}_2 , \; \sigma_a\sigma_b = i\varepsilon_{abc}\sigma_c, \tag{2.52}$$

the Pauli group can be defined through the presentation

$$G_1 = \langle \sigma_1, \sigma_2, \sigma_3 | \sigma_a^2 = \mathbb{1}_2, \; \sigma_a\sigma_b = i\varepsilon_{abc}\sigma_c, \forall a \neq b, \; a, b \in \{1, 2, 3\}\rangle. \tag{2.53}$$

2.7.2 Braids

Next, we turn to consider something familiar from knitting. A *braid* consists of *strands* which run forward and can pass under or over each other. In a physics context, an important related situation is met when one considers worldlines of point particles moving in two space dimensions, as in Fig. 2.2. The particle worldlines then form strands that become entangled, just like knitting strands. This phenomenon gives rise to exotic quantum statistics for quantum particles in two dimensions,[2] in addition to the usual bosonic and fermionic statistics. Here we adopt a convention where the braid is drawn upright (another alternative would be to draw it sideways), with braiding of strands beginning from the top and proceeding downwards, as shown in Fig. 2.2. Braids are usually represented in a "flattened" form, as shown in Fig. 2.3.

One can imagine that the strands are like pieces of string or cord, and thus they can be moved and deformed continuously as shown in Fig. 2.4, or as in the example in Fig. 2.5.

In a braid of n strands, the strands can be labeled by an index i running from 1 to n, from left to right. A possible way to move strands in a braid, shown in Fig. 2.5, is known as the *Reidemeister move of the second type*.[3] To form braids, another

---

[2] The reason why this is not true in three or more space dimensions is that there the strands can then be disentangled. Likewise, all one-dimensional knots in higher dimensions become trivial.

[3] There exists also a move called the *Reidemeister move of the first type*, which is simply defined as untwisting a strand passing over itself.

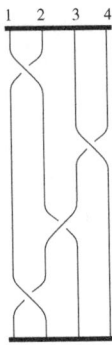

Fig. 2.3   A braid with four strands.

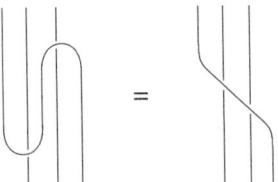

Fig. 2.4   The braid shown on the left is equivalent to the one on the right.

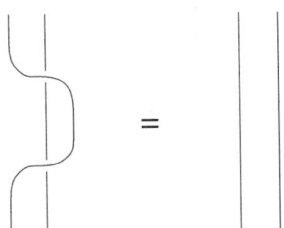

Fig. 2.5   The braid depicted on the left is equivalent to the one that is shown on the right.

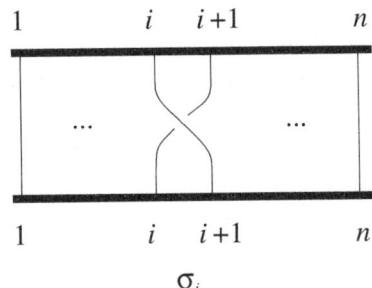

Fig. 2.6   The basic operation σ_i, which moves the ith strand over the $(i + 1)$th strand.

important basic operation, denoted as σ_i, consists in moving the ith strand over the $(i + 1)$th strand, as shown in Fig. 2.6.

The inverse operation σ_i^{-1} moves the $(i+1)$th strand over the ith strand, as depicted in Fig. 2.7.

A crucial rule to note is that after every operation, the strands are indexed again from left to right from 1 to n, and the next operation is performed on the strands labeled according to the new indexing. This is depicted in Figs. 2.6 and 2.7.

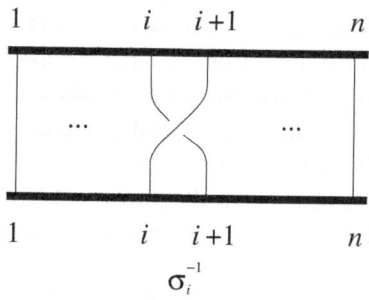

Fig. 2.7   The operation σ_i^{-1}, defined as the inverse of the one shown in Fig. 2.6.

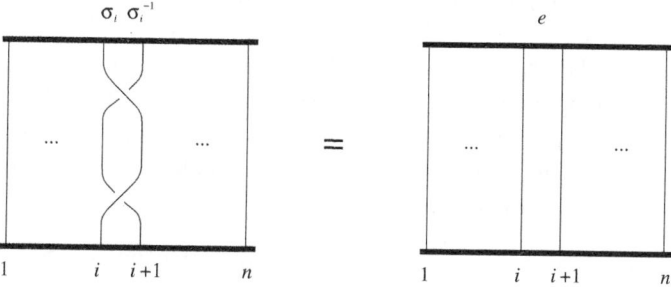

Fig. 2.8   The braid $\sigma_i\sigma_i^{-1}$ reduces to the trivial one (denoted as e).

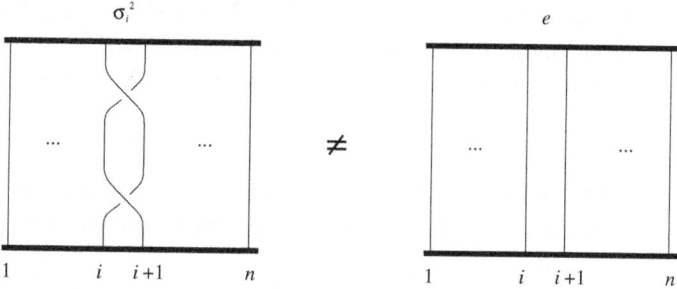

Fig. 2.9   The braid σ_i^2 is nontrivial.

To illustrate the indexing, consider the sequences of operations $\sigma_i\sigma_i^{-1}$ and $\sigma_i^2 = \sigma_i\sigma_i$. For $\sigma_i\sigma_i^{-1}$, after the first operation σ_i, the ith strand becomes the $(i+1)$th strand, and vice versa. With the new indexing, the next operation σ_i^{-1} then acts as in Fig. 2.7, and the final result for the $\sigma_i\sigma_i^{-1}$ operation is the braid on the left-hand side of Fig. 2.8. In turn, by a Reidemeister move of the second type shown in Fig. 2.5, $\sigma_i\sigma_i^{-1}$ reduces to the trivial braid e, the one that does not move any strand, as shown on the right-hand side of Fig. 2.8. Similarly, one can verify that $\sigma_i^{-1}\sigma_i = e$. By the same rules one can see that, in contrast, after applying $\sigma_i\sigma_i$, the strands become entangled so the final result is nontrivial, i.e., different from e; this is shown in Fig. 2.9.

After these elementary examples, we move to more general constructions. For a braid of n strands, the basic operations are $\sigma_1, \sigma_2, \ldots, \sigma_{n-1}$, that can be thought of as the letters of a set X_{n-1} (an *alphabet*). Thus, each *word*

$$\sigma = \sigma_{j_1}^{i_1}\sigma_{j_2}^{i_2}\cdots\sigma_{j_m}^{i_m} \tag{2.54}$$

with exponents $i_i \in \mathbb{Z}$ represents a braid. As before, when drawing the braid, one starts from the top, reads the letters of the word from left to right and performs the corresponding operation on the strands, continuing the drawing downwards. For two generic braids σ and σ', the product braid $\sigma\sigma'$ is defined as the braid with instructions from σ followed by the instructions of σ'. The inverse of the generic braid defined in Eq. (2.54) is obtained as the product of the inverse operations, in the opposite order (in agreement with the usual rule for the inverse of products):

$$\sigma^{-1} = \sigma_{j_m}^{-i_m} \cdots \sigma_{j_2}^{-i_2} \sigma_{j_1}^{-i_1}. \tag{2.55}$$

Example   The braid of four strands shown in Fig. 2.3 is $\sigma_1\sigma_3\sigma_2^{-1}\sigma_1$.

The group generated by X_{n-1} for $n > 2$ is not a free group. There are relations among the generators, because one can move and deform the strands continuously. One relation is due to the fact that the operations of braiding strands that are sufficiently apart from each other are independent operations, so the corresponding generators commute:

$$\sigma_i\sigma_j = \sigma_j\sigma_i \quad \text{when } |i - j| \geq 2. \tag{2.56}$$

For example, in the braid of Fig. 2.3 the first two operations σ_1 and σ_3 commute, and could be done simultaneously. Thus, both of them could be shown at the top of the figure (although, for clarity, in Fig. 2.3 we showed explicitly that in the $\sigma_1\sigma_3\sigma_2^{-1}\sigma_1$ product, the first operation to be carried out is σ_1).

A less obvious relation is the *Reidemeister move of the third type*,

$$\sigma_i\sigma_{i+1}\sigma_i = \sigma_{i+1}\sigma_i\sigma_{i+1}, \tag{2.57}$$

which is depicted in Fig. 2.10 and, in a simplified version, in Fig. 2.11 (which can be verified empirically by moving three straws from the configuration on the left to the one on the right).

It turns out that the diagram in Fig. 2.11 also appears in the context of exactly solvable problems in statistical mechanics and quantum field theories, where it is related to a *Yang–Baxter relation*. This is not an accident, since the conditions that determine the integrability of those models have a connection with the properties of braids that we are discussing.

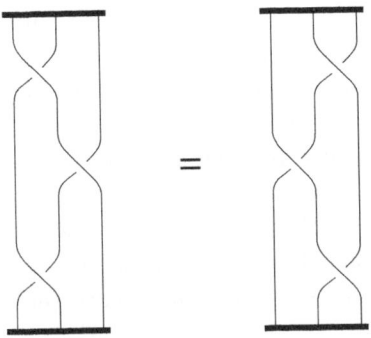

Fig. 2.10   A figure illustrating the relation $\sigma_i\sigma_{i+1}\sigma_i = \sigma_{i+1}\sigma_i\sigma_{i+1}$.

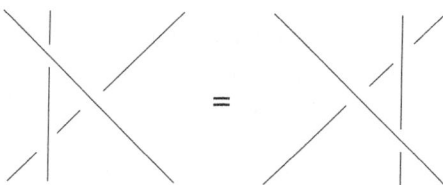

Fig. 2.11 An equivalent illustration of the relation $\sigma_i\sigma_{i+1}\sigma_i = \sigma_{i+1}\sigma_i\sigma_{i+1}$.

Fig. 2.12 A trefoil knot K in $\mathbb{R}^3$.

One can show that relations (2.56) and (2.57) are the only relations between the generators σ_i. This leads us to define the *braid group of n strands B_n* by its presentation

$$B_n = \langle \sigma_1, \sigma_2, \ldots, \sigma_{n-1} \mid \sigma_i\sigma_j = \sigma_j\sigma_i, \ \sigma_i\sigma_{i+1}\sigma_i = \sigma_{i+1}\sigma_i\sigma_{i+1} \rangle, \qquad (2.58)$$

where $|i - j| \geq 2$ in the first relation, and $1 \leq i \leq n - 2$ in the second relation.

We conclude by listing some properties of braid groups.

1. B_1 is the trivial group $\mathbb{Z}_1 = \{e\}$.
2. $B_2 \cong (\mathbb{Z}, +)$, so it is an Abelian group.
3. B_3 is non-Abelian. After introducing homotopy groups, one way to characterize B_3 is that it is isomorphic to the fundamental group $\pi_1(\mathbb{R}^3 \setminus K)$, where K is the *trefoil knot* shown in Fig. 2.12.
4. The braid groups form a nested sequence of subgroups: $B_n \subset B_{n+1}$ for every positive natural number n.

2.8 Continuous Groups and Lie Groups

Continuous groups are groups having an infinite and uncountable number of elements. Since the order of a continuous group is not a meaningful concept, the "size" of a continuous group is instead characterized in terms of its *dimension*.

The *dimension* of a continuous group G, denoted as $\dim G$, is the minimum number of real parameters needed to uniquely identify its elements.

The real parameters identifying an element of a continuous group are sometimes called its *coordinates*. Note that each coordinate may take values within the whole $\mathbb{R}$, or within a real interval.

The definition of group dimension introduced above is distinct from (and more general than) the definition of dimension of a linear vector space. Indeed, continuous groups may, but do not necessarily have to, be realized in terms of linear vector spaces.

As will be discussed more in detail in Chapter 4, given the product of two elements of a continuous group, $g'' = g'g$, the coordinates of g'' must be continuous functions of the coordinates of g and g'. This means that the set of coordinates parameterizing a group form a *manifold*, called the *group manifold*.

The most interesting continuous groups are the *Lie groups*, named after the Norwegian mathematician Marius Sophus Lie: A *Lie group* is a continuous group whose group manifold is *differentiable*, i.e., it is possible to take smooth partial derivatives of the group elements with respect to the coordinates.

Examples

- The set of real numbers $\mathbb{R}$, with the addition $+$ as the group product, is a realization of a continuous group of dimension $\dim \mathbb{R} = 1$.
- A straightforward generalization of the group of real numbers can be constructed, by taking the multiple direct product of n copies of $\mathbb{R}$:

$$\mathbb{R}^n = \{(r_1,\ldots,r_n) : r_i \in \mathbb{R}, i = 1,\ldots,n\} = \overbrace{\mathbb{R} \times \cdots \times \mathbb{R}}^{n \text{ times}}, \qquad (2.59)$$

with the group product

$$(r_1,\ldots,r_n) \cdot (r'_1,\ldots,r'_n) = (r_1 + r'_1,\ldots,r_n + r'_n). \qquad (2.60)$$

Obviously, $\dim \mathbb{R}^n = n$, since every element of $\mathbb{R}^n$ is written as an n-uplet of independent real numbers.

- The set of complex numbers $\mathbb{C}$ with addition as the composition law forms a realization of a continuous group of dimension $\dim \mathbb{C} = 2$: The two real parameters to describe a generic element $z \in \mathbb{C}$ can be taken to be its real and imaginary parts $z = x + iy$ (with $x \in \mathbb{R}$ and $y \in R$) or its modulus and phase $z = \rho e^{i\theta}$ (with ρ a nonnegative real number, and $\theta \in \mathbb{R}$ with, for example, $0 \le \theta < 2\pi$).
- The set of complex numbers of modulus one, $\{z \in \mathbb{C} : |z| = 1\}$, with multiplication as the composition law, is (a realization of) the continuous group called the *unitary group of degree* 1 and denoted as $U(1)$; $\dim U(1) = 1$, since any element can be identified using only one real parameter: $z = e^{i\theta}$, with $\theta \in [0, 2\pi)$. Note an important difference between $U(1)$ and $\mathbb{R}$: both groups have $\dim = 1$, but the group manifold of the former is the circle S^1, while the group manifold of the latter can be taken to be the infinite straight line. As will be discussed, the global *topological* properties of the group manifolds of $U(1)$ and $\mathbb{R}$ are different; in particular, in S^1 it is possible to define closed loops which cannot be continuously deformed into a point, while this is not the case in the group manifold of $\mathbb{R}$.
- A generalization of the unitary group of degree 1 can be constructed as the direct product of n of its copies:

$$U(1)^n = \{(z_1,\ldots,z_n) : z_i \in U(1), i = 1,\ldots,n\} = \overbrace{U(1) \times \cdots \times U(1)}^{n \text{ times}}, \qquad (2.61)$$

with the group product

$$(z_1, \ldots, z_n) \cdot (z'_1, \ldots, z'_n) = (z_1 z'_1, \ldots, z_n z'_n), \tag{2.62}$$

with $\dim U(1)^n = n$. The group manifold of $U(1)^n$ is an n-torus $\overbrace{S^1 \times \cdots \times S^1}^{n \text{ times}}$, Again, the topological properties of an n-torus are different from those of $\mathbb{R}^n$.

- The set of square matrices of size $n \times n$ with real entries, $\mathrm{Mat}(n, \mathbb{R})$, with matrix addition as the composition law, is a realization of a continuous group of dimension $\dim \mathrm{Mat}(n, \mathbb{R}) = n^2$. In particular, it is easy to prove that this group is isomorphic to $\mathbb{R}^{n^2}$.

2.8.1 Matrix Lie Groups

Other particularly interesting examples of Lie groups are those defined (by abstraction) from certain sets of matrices, with the matrix "row-by-column" product as the composition law.

- The group of general linear transformations with real coefficients $GL(n, \mathbb{R})$ is defined as the subset of matrices in $\mathrm{Mat}(n, \mathbb{R})$ having a nonvanishing determinant, with the matrix product as the group multiplication:

$$GL(n, \mathbb{R}) = \{A \in \mathrm{Mat}(n, \mathbb{R}) : \det A \neq 0\}. \tag{2.63}$$

The identity matrix of size $n \times n$ is the unit element of the group, and the existence of the inverse of any element of $GL(n, \mathbb{R})$ is guaranteed by the condition that the determinant is different from zero. Note that the determinant of a matrix is a continuous function of its entries. The dimension of $GL(n, \mathbb{R})$ is $\dim GL(n, \mathbb{R}) = n^2$, like that of $\mathrm{Mat}(n, \mathbb{R})$: Note that the constraint that the determinant of the elements of $GL(n, \mathbb{R})$ is *different* from zero does not reduce the number of independent real parameters of $GL(n, \mathbb{R})$. While $GL(n, \mathbb{R})$ and $\mathrm{Mat}(n, \mathbb{R})$ have the same dimension, their group manifolds have a different global structure. In particular, in contrast to $\mathrm{Mat}(n, \mathbb{R})$, for which the matrix entries are a set of suitable coordinates, in $GL(n, \mathbb{R})$ the condition $\det A \neq 0$ for a generic group element A implies that the coordinates parameterizing the group elements cannot take arbitrary values in $\mathbb{R}^{n^2}$, but rather in $\mathbb{R}^{n^2}$ with the exclusion of the hypersurface on which the matrix determinant vanishes (this hypersurface is a set of measure zero in $\mathbb{R}^{n^2}$). Thus the group manifold of $GL(n, \mathbb{R})$ is divided into two disconnected components (in which the determinant of a generic element is either positive or negative).

- The group of general linear transformations can be generalized, by allowing the matrix elements to take values in the complex field; this leads to the group

$$GL(n, \mathbb{C}) = \{A \in \mathrm{Mat}(n, \mathbb{C}) : \det A \neq 0\}, \tag{2.64}$$

again with matrix product as the group multiplication. As for $GL(n, \mathbb{R})$, the unit element is the identity matrix of size $n \times n$, and the existence of the inverse of any element of the group is guaranteed by the fact that the determinant is nonvanishing. Since each of the n^2 entries of a generic matrix $A \in GL(n, \mathbb{C})$ is a complex number, which can be parameterized by two real coordinates, and since the $\det A \neq 0$

constraint does not reduce the number of independent parameters, the dimension of $GL(n, \mathbb{C})$ is $2n^2$. Finally, note that $GL(n, \mathbb{R})$ is a proper subgroup of $GL(n, \mathbb{C})$. The group $GL(n, \mathbb{C})$ and its subgroups are called *matrix Lie groups*.

- The set of special linear transformations of degree n with real coefficients is denoted as $SL(n, \mathbb{R})$ and is defined as

$$SL(n, \mathbb{R}) = \{A \in \text{Mat}(n, \mathbb{R}) : \det A = 1\}. \qquad (2.65)$$

Note that, since $\det A = 1$ is a special case of $\det A \neq 0$, $SL(n, \mathbb{R})$ is a subset of $GL(n, \mathbb{R})$. Taking the matrix product as the composition law, and observing that the product of two matrices with determinant 1 also has determinant 1, that the inverse of any matrix of determinant 1 also has determinant 1, and that the determinant of the identity matrix is 1, too, it follows that $SL(n, \mathbb{R})$ is a proper subgroup of $GL(n, \mathbb{R})$. The dimension of $SL(n, \mathbb{R})$ is $n^2 - 1$: The condition that the determinant of each element *equals* 1 is expressed as an equation, that reduces the number of independent coordinates from n^2 down to $n^2 - 1$.

- The orthogonal group of degree n is defined as the group of real matrices (*orthogonal* matrices) whose inverse coincides with their transpose. For a given matrix A, the latter is defined as the matrix obtained from A by interchanging its rows with its columns, $(A^T)_{ab} = A_{ba}$:

$$O(n, \mathbb{R}) = \{A \in GL(n, \mathbb{R}) : A^T A = AA^T = \mathbb{1}_n\}, \qquad (2.66)$$

where $\mathbb{1}_n$ is the identity matrix of size $n \times n$. It is easy to prove that $O(n, \mathbb{R})$ (with the matrix product as the group multiplication) is a proper subgroup of $GL(n, \mathbb{R})$, since:

(a) the product of two elements of $O(n, \mathbb{R})$, A and B, is also an orthogonal matrix:

$$(AB)^T(AB) = B^T A^T AB = B^T B = \mathbb{1}_n; \qquad (2.67)$$

(b) the unit element for the matrix product is the identity matrix, which is an orthogonal matrix ($\mathbb{1}_n^T \mathbb{1}_n = \mathbb{1}_n$);

(c) the inverse of a generic orthogonal matrix A is also orthogonal:

$$\left(A^{-1}\right)^T A^{-1} = \left(AA^T\right)^{-1} = \mathbb{1}_n^{-1} = \mathbb{1}_n. \qquad (2.68)$$

Orthogonal matrices can be seen as linear transformations acting on the n-dimensional real vector space $\mathbb{R}^n$ and preserve the modulus of vectors in $\mathbb{R}^n$. Indeed, the modulus of a vector $v \in \mathbb{R}^n$ is given by $\sqrt{\sum_{a=1}^n v_a^2} = \sqrt{\vec{v}^T \vec{v}}$. Upon a transformation $A \in O(n, \mathbb{R})$, the vector $\vec{v}$ gets mapped to $A\vec{v}$, whose modulus is $\sqrt{(A\vec{v})^T(A\vec{v})} = \sqrt{\vec{v}^T A^T A \vec{v}} = \sqrt{\vec{v}^T \vec{v}}$, i.e., it is left invariant. As a consequence, the orthogonal group can be interpreted as a group of rotations and reflections in $\mathbb{R}^n$. To determine the dimension of $O(n, \mathbb{R})$, note that each element is expressed by a matrix with n^2 real entries, but they are not all independent, because the orthogonality condition restricts the number of independent coordinates. In particular, the requirement that $A^T A = \mathbb{1}_n$ implies that each diagonal entry of $A^T A$ must be equal to 1; this imposes n equations. In addition, all of the entries above the diagonal must vanish, and this gives further $\frac{n(n-1)}{2}$ independent equations. If the entries above the diagonal vanish, then so do those below the diagonal, because $A^T A$ is a symmetric matrix. The constraints imposed by the $AA^T = \mathbb{1}_n$ condition are equivalent. Hence, the orthogonality constraint reduces the number of independent real parameters down to $n^2 - n - \frac{n(n-1)}{2} = \frac{n(n-1)}{2}$. Finally, note that the orthogonality condition (and the fact that the determinant of a matrix and that of its transpose are equal) also implies that the determinant of any element of $O(n, \mathbb{R})$ is either 1 or -1, because $\det\left(A^T A\right) = \det \mathbb{1}_n$ implies $(\det A)^2 = 1$, hence $\det A = \pm 1$. The subset of $O(n, \mathbb{R})$ including the matrices of determinant 1 is the *special orthogonal group of degree n*:

$$SO(n, \mathbb{R}) = \{A \in GL(n, \mathbb{R}) : A^T A = AA^T = \mathbb{1}_n, \ \det A = 1\}. \tag{2.69}$$

The special orthogonal group of degree n is a proper subgroup of $O(n, \mathbb{R})$ and its dimension is $\dim SO(n, \mathbb{R}) = \frac{n(n-1)}{2}$, like $O(n, \mathbb{R})$ (because the determinant of elements of $O(n, \mathbb{R})$ can only take values ± 1, hence the condition that the determinant equals 1 for $SO(n, \mathbb{R})$ elements *does not* reduce the number of independent real coordinates). $SO(n, \mathbb{R})$ can be interpreted as the subgroup of $O(n, \mathbb{R})$ which preserves the orientation of a basis of vectors in $\mathbb{R}^n$: So, for example, acting with a matrix of $SO(3, \mathbb{R})$ on the vectors of a right-handed orthonormal basis of $\mathbb{R}^3$ results into a rotated basis of vectors, which is still right-handed. Note that, by contrast, the subset of $O(n, \mathbb{R})$ including the matrices with determinant -1 is not a subgroup (in fact, it is not even a magma, since $\det A = \det B = -1$ implies $\det AB = 1$). Often the orthogonal groups and their special subgroups are denoted as $O(n)$ and $SO(n)$.

- The generalization of the orthogonal group of degree n to the complex field is the *unitary group of degree n*, defined as

$$U(n) = \{A \in GL(n, \mathbb{C}) : A^\dagger A = AA^\dagger = \mathbb{1}_n\}, \tag{2.70}$$

where $A^\dagger = (A^\star)^T = (A^T)^\star$ is the *conjugate transpose* of A: $(A^\dagger)_{ij} = (A_{ji})^\star$, with $\star$ denoting complex conjugation. Note that $(AB)^\dagger = B^\dagger A^\dagger$. These matrices preserve the modulus of complex vectors $\vec{z} \in \mathbb{C}^n$, defined as $\sqrt{\sum_{a=1}^{n} z_a^\star z_a} = \sqrt{\vec{z}^\dagger \vec{z}}$. A generic element $A \in U(n)$ maps a vector $\vec{z}$ to $A\vec{z}$, whose modulus is $\sqrt{(A\vec{z})^\dagger A\vec{z}} = \sqrt{\vec{z}^\dagger A^\dagger A\vec{z}} = \sqrt{\vec{z}^\dagger \vec{z}}$, i.e., is left invariant by the action of A. Hence unitary matrices can be interpreted as "rotations" in $\mathbb{C}^n$. (More generally, a linear map $A : V \rightarrow W$ between two vector spaces V and W that preserves the lengths of vectors is called an *isometry*.) $U(n)$ is a proper subgroup of $GL(n, \mathbb{C})$. One can show that $\dim U(n) = n^2$: Starting from the dimension of $GL(n, \mathbb{C})$, which is $2n^2$ (where the factor 2 accounts for the fact that each matrix element has a real and an imaginary part), the dimension of $U(n)$ can be obtained by subtracting from it the number of independent constraints that the $U(n)$ matrix entries must obey. From the requirement that $A^\dagger A = \mathbb{1}_n$ follows, first of all, that each diagonal element of $A^\dagger A$ must be equal to 1:

$$\left(A^\dagger A\right)_{ii} = \sum_{k=1}^{n} A_{ik}^\dagger A_{ki} = (\mathbb{1}_n)_{ii} = \delta_{ii} = 1, \quad \text{for each } 1 \le i \le n. \tag{2.71}$$

Using the fact that $A_{ik}^\dagger = A_{ki}^\star$, Eq. (2.71) can be rewritten as

$$\sum_{k=1}^{n} A_{ki}^\star A_{ki} = \sum_{k=1}^{n} |A_{ki}|^2 = 1, \quad \text{for each } 1 \le i \le n. \tag{2.72}$$

Equation (2.72) corresponds to n constraints (not $2n$, because a sum of squared moduli of complex numbers is necessarily a real, nonnegative number, i.e., the imaginary part of each sum is already necessarily zero, without imposing any constraint on it). Conversely, for the $\frac{n(n-1)}{2}$ elements above the diagonal of the $A^\dagger A$ matrix, the conditions

$$\left(A^\dagger A\right)_{ij} = \sum_{k=1}^{n} A^\dagger_{ik} A_{kj} = 0, \quad \text{for } 1 \le i < j \le n, \tag{2.73}$$

enforce the vanishing of the real and of the imaginary part, and thus correspond to two independent constraints for each element. Finally, the constraints on the remaining $\frac{n(n-1)}{2}$ elements below the diagonal of the $A^\dagger A$ turn out to be equivalent to those expressed by Eq. (2.73). Also, one can check that the same constraints arise if one requires $AA^\dagger = \mathbb{1}_n$. We conclude that the dimension of the unitary group of of degree n is given by $\dim \mathrm{U}(n) = 2n^2 - n - 2\frac{n(n-1)}{2} = n^2$.

Note that, according to the definition (2.70), $\mathrm{U}(1) = \{a \in \mathbb{C} : a^\star a = 1\}$, so the group manifold of $\mathrm{U}(1)$ can be identified with the circle of unit radius on the complex plane.

- The subset of $\mathrm{U}(n)$ containing matrices of determinant equal to 1 constitutes the *special unitary group of degree n*:

$$\mathrm{SU}(n) = \{A \in \mathrm{U}(n), \ \det A = 1\}. \tag{2.74}$$

This is a proper subgroup of $\mathrm{U}(n)$, and its dimension can be computed by noting that, in addition to the constraints that every $\mathrm{U}(n)$ matrix must satisfy, the requirement $\det A = 1$ imposes one additional condition (not two, because from $A^\dagger A = \mathbb{1}_n$, which holds for every unitary matrix A, and from the property that $\det A^\dagger = (\det A)^\star$, already follows that $\det(A^\dagger A) = \det(A^\dagger)\det A = (\det A)^\star \det A = |\det A|^2 = \det \mathbb{1}_n = 1$, i.e., the real and imaginary parts of $\det A$ cannot be arbitrary numbers, since they already satisfy the constraint that $|\det A| = 1$). Thus $\dim \mathrm{SU}(n) = (\dim \mathrm{U}(n)) - 1 = n^2 - 1$.

- The subset of $\mathrm{Mat}(2n, \mathbb{R})$ containing the matrices M, called "symplectic matrices," that satisfy the property

$$M^T \Omega M = \Omega, \quad \text{with: } \Omega = \begin{pmatrix} 0_n & \mathbb{1}_n \\ -\mathbb{1}_n & 0_n \end{pmatrix}, \tag{2.75}$$

is a subgroup called the *symplectic group of degree 2n*, and denoted as $\mathrm{Sp}(2n, \mathbb{R})$, of dimension $\dim \mathrm{Sp}(2n, \mathbb{R}) = n(2n + 1)$. Note that the skew-symmetric matrix Ω has $\det \Omega = 1$ and its inverse is $\Omega^{-1} = \Omega^T = -\Omega$.

Let us define the *Pfaffian* of a generic skew-symmetric matrix A, of size $2n \times 2n$, as

$$\mathrm{Pf}\, A = \frac{1}{2^n n!} \sum_{P \in S_{2n}} \mathrm{sgn}(P) \prod_{k=1}^{n} A_{P(2k-1), P(2k)}, \tag{2.76}$$

where the sum runs over the permutations of $2n$ elements, and $\mathrm{sgn}(P)$ is the signature of the permutation P.

Consider now the Pfaffian of the matrix on the left-hand side of Eq. (2.75): Using the fact that it equals $\mathrm{Pf}\left(M^T \Omega M\right) = (\det M)(\mathrm{Pf}\,\Omega)$, and that $M^T \Omega M = \Omega$, it is evident that all symplectic matrices have $\det M = 1$. Finally, note that for $n = 1$ the condition (2.75) is equivalent to $\det M = 1$; therefore $\mathrm{Sp}(2, \mathbb{R}) = \mathrm{SL}(2, \mathbb{R})$.

More on the SU(2) Group

Being related to linear transformations in complex vector spaces, unitary groups are particularly important in modern physics. In particular, $\mathrm{U}(1)$ is the group of phase transformations in quantum mechanics. As we mentioned, it is an Abelian group of dimension 1, and its group manifold can be identified with the S^1 circle.

On the other hand, the non-Abelian group $\mathrm{SU}(2)$ plays a crucial rôle in the context of spin. To characterize this group better, it is interesting to look at its defining realization, in terms of complex unitary matrices of size 2×2 with determinant 1. A generic complex 2×2 matrix can be written as

$$A = \begin{pmatrix} \alpha & \beta \\ \gamma & \delta \end{pmatrix}, \tag{2.77}$$

with α, β, γ, and $\delta \in \mathbb{C}$ (for a total of eight real parameters). The requirement that A be a unitary matrix means $A^\dagger = A^{-1}$, where the inverse of A is simply the transpose of the matrix of cofactors, when $\det A = 1$. Hence

$$A^\dagger = \begin{pmatrix} \alpha^\star & \gamma^\star \\ \beta^\star & \delta^\star \end{pmatrix} = \begin{pmatrix} \delta & -\beta \\ -\gamma & \alpha \end{pmatrix} = A^{-1}, \tag{2.78}$$

which implies $\delta = \alpha^\star$ and $\gamma = -\beta^\star$. This corresponds to four independent conditions, which completely fix (for example) the real and imaginary parts of γ and δ. Finally, the requirement that $\det A = 1$ implies $\alpha\delta - \beta\gamma = 1$, i.e., $|\alpha|^2 + |\beta|^2 = 1$. Since the moduli of α and β are necessarily nonnegative real numbers, this corresponds to one further constraint, reducing the minimal number of independent real variables to parameterize $\mathrm{SU}(2)$ down to three. Thus, a generic element of $\mathrm{SU}(2)$ can be written as

$$A = \begin{pmatrix} \alpha & \beta \\ -\beta^\star & \alpha^\star \end{pmatrix}, \qquad \text{with } |\alpha|^2 + |\beta|^2 = 1. \tag{2.79}$$

Note that $|\alpha|^2 + |\beta|^2 = 1$ implies, in particular, that both the real and the imaginary part of each entry of A must be in the interval $[-1, 1]$. Writing explicitly the real and imaginary parts of α and β as $\alpha = x_4 + ix_3$, $\beta = x_2 + ix_1$ shows that $\mathrm{SU}(2)$ can be parameterized in terms of the four real variables x_1, x_2, x_3, and x_4, subject to the constraint $x_1^2 + x_2^2 + x_3^2 + x_4^2 = 1$. This equation defines the three-dimensional sphere of unit radius S^3 in the four-dimensional Euclidean space $\mathbb{R}^4$.

It is also convenient to introduce the following matrices, named after Wolfgang Pauli:

$$\sigma^1 = \begin{pmatrix} 0 & 1 \\ 1 & 0 \end{pmatrix}, \quad \sigma^2 = \begin{pmatrix} 0 & -i \\ i & 0 \end{pmatrix}, \quad \sigma^3 = \begin{pmatrix} 1 & 0 \\ 0 & -1 \end{pmatrix}. \tag{2.80}$$

The Pauli matrices satisfy

$$\sigma^a \sigma^b = \delta_{ab} \mathbb{1}_2 + i\epsilon_{abc} \sigma^c \tag{2.81}$$

(where δ_{ab} is the Kronecker delta and ϵ_{abc} is the totally antisymmetric Levi-Civita symbol, with $\epsilon_{123} = 1$) as well as the *completeness relation*

$$\sum_{i=1}^{3} \sigma_{ab}^{i}\sigma_{cd}^{i} = 2\delta_{ad}\delta_{bc} - \delta_{ab}\delta_{cd}. \tag{2.82}$$

An equivalent parameterization of a generic element $A \in \mathrm{SU}(2)$ can then be written in terms of the identity matrix and the Pauli matrices

$$A = e^{i\phi\hat{v}\cdot\vec{\sigma}} = (\cos\phi)\mathbb{1}_2 + i(\sin\phi)\hat{v}\cdot\vec{\sigma}, \tag{2.83}$$

with $\phi \in [0, 2\pi]$, and $\hat{v} \in \mathbb{R}^3$, with $|\hat{v}| = 1$. The explicit connection with the previous parameterization is

$$\alpha = \cos\phi + iv_3\sin\phi, \quad \beta = (v_2 + iv_1)\sin\phi. \tag{2.84}$$

We now define a *Hermitian matrix A* as a square matrix that equals its own conjugate transpose

$$A^{\dagger} = A. \tag{2.85}$$

Note that Hermitian matrices of a given size form a group under matrix addition (but not under matrix multiplication, since, in general, $(AB)^{\dagger} = B^{\dagger}A^{\dagger} = BA \neq AB$).

Conversely, a square matrix B satisfying

$$B^{\dagger} = -B \tag{2.86}$$

is said to be *anti-Hermitian*.

We note that the unit matrix and the Pauli matrices form a basis for Hermitian 2×2 matrices: Every Hermitian 2×2 matrix H can be expanded as

$$H = a\mathbb{1}_2 + b\sigma^1 + c\sigma^2 + d\sigma^3, \tag{2.87}$$

where $a, b, c, d \in \mathbb{R}$.

Other Lie Groups

Finally, we mention some more exotic Lie groups.

- One among several equivalent ways to define the algebra of *octonions* ($\mathbb{O}$), first discovered by John Thomas Graves (and, independently and nearly at the same time, by Arthur Cayley), is as the set of linear combinations of the elements of the magma $\mathfrak{C} = \{e_0, e_1, e_2, e_3, e_4, e_5, e_6, e_7\}$ with the internal, nonassociative composition law given by

$$\begin{cases} e_0 e_\mu = e_\mu e_0 = e_\mu & \forall \mu \in \{0, 1, \dots, 7\}, \\ e_i e_j = -\delta_{ij} e_0 + f_{ijk} e_k & \forall i, j, k \in \{1, \dots, 7\}, \end{cases}$$

where f_{ijk} is totally antisymmetric under exchange of any pair of its indices, and $f_{123} = f_{145} = f_{176} = f_{246} = f_{257} = f_{347} = f_{365} = 1$ (while all of its components, that are not related to these by exchanges of indices, vanish). The set $\mathfrak{C}$ with this composition law is called a *Cayley algebra*. Then a generic octonion x can be written as

$$x = x_0 e_0 + \sum_{i=1}^{7} x_i e_i. \tag{2.88}$$

For later convenience, it is useful to denote x_0, i.e., the coefficient of the e_0 component of x, as $R(x)$. Then, one can also define the $\mathbb{G}_2$ group as the subgroup of SO(7) containing the matrices M that satisfy

$$f_{ijk} = \sum_{a,b,c=1}^{7} f_{abc} M_{ai} M_{bj} M_{ck}, \tag{2.89}$$

i.e., as the group of transformations leaving the Cayley algebra invariant. $\mathbb{G}_2$ is a non-Abelian Lie group of dimension $\dim \mathbb{G}_2 = 14$, and is the first (and smallest) of five "exceptional groups."

- The *exceptional Jordan algebra* $\mathfrak{J}$ (named after Ernst Pascual Jordan) is defined as the set of Hermitian 3×3 matrices with entries in the Cayley algebra, endowed with the composition law $\circ$ defined by $X \circ Y = (XY + YX)/2$, for any matrices X and Y in $\mathfrak{J}$. A generic element $X \in \mathfrak{J}$ can be written in the form

$$X = \begin{pmatrix} \xi_1 & x_{12} & x_{13} \\ x_{12}^\star & \xi_2 & x_{23} \\ x_{13}^\star & x_{23}^\star & \xi_3 \end{pmatrix}, \quad \text{with } \xi_1, \xi_2, \xi_3 \in \mathbb{R}, \ x_{12}, x_{13}, x_{23} \in \mathfrak{C}, \tag{2.90}$$

so the dimension of $\mathfrak{J}$ is $3 + 3 \cdot 8 = 27$. In $\mathfrak{J}$, the determinant of a generic matrix X is defined as

$$\det X = \xi_1 \xi_2 \xi_3 + 2R(x_{23} x_{13}^\star x_{12}) - \xi_1 x_{23} x_{23}^\star - \xi_2 x_{13}^\star x_{13} - \xi_3 x_{12} x_{12}^\star. \tag{2.91}$$

Then, the $\mathbb{F}_4$ group is defined as the group of transformations leaving the Jordan algebra $\mathfrak{J}$ invariant. It is a non-Abelian exceptional Lie group, and its dimension is $\dim \mathbb{F}_4 = 52$.

- The determinant-preserving linear transformations of the exceptional Jordan algebra form the $\mathbb{E}_6$ exceptional Lie group, of dimension $\dim \mathbb{E}_6 = 78$.
- The $\mathbb{E}_7$ exceptional Lie group, of dimension $\dim \mathbb{E}_7 = 133$, is the group of symmetries of a certain space, called "quater-octonionic projective plane" and defined starting from quaternions and octonions. We omit the precise definition, which is quite technical.
- Similarly, the $\mathbb{E}_8$ exceptional Lie group, of dimension $\dim \mathbb{E}_8 = 248$, is the group of symmetries of a space, called "octo-octonionic projective plane" and defined in terms of octonions. Also in this case, the precise definition is quite technical and involved, hence we omit it.

2.9 Groups Acting on a Set

When discussing the orthogonal groups $O(n, \mathbb{R})$ in Section 2.8, we mentioned that their elements (in the defining representation as orthogonal matrices) can be considered as rotations and reflections in the Euclidean vector spaces $\mathbb{R}^n$. In order to make this notion more precise, it is convenient to introduce some new concepts.

Given a group G and a set X, a *left action* of G on X is a homomorphism

$$L : G \to \mathrm{Perm}(X), \; g \to L_g. \tag{2.92}$$

Since a left action is a group homomorphism, it preserves the group multiplication:

$$\forall g_1 \in G, \; \forall g_2 \in G, \; \forall x \in X : \; L_{g_2}\left(L_{g_1}(x)\right) = L_{g_2 \cdot g_1}(x). \tag{2.93}$$

Given a group G, a set X is called a *left group-space* of G (or a *left G-space*) if there exists a left action of G on X.

Similarly, given a group G and a set X, a *right action* of G on X is a function $R : G \to \mathrm{Perm}(X)$, $g \to R_g$ such that, for all g_1 and g_2 in G, and for all $x \in X$, one has $\left(R_{g_1} \cdot R_{g_2}\right)(x) = R_{g_2 \cdot g_1}(x)$. Then, a set X is said to be a *right group-space* of G (or a *right G-space*) if there exists a right action of G on X.

Given a group G, two left G-spaces X and X' can be identified, if there is a bijection $i : X \to X'$ such that, for any $g \in G$ and for any $x \in X$, one has $i\left(L_g(x)\right) = L'_g\left(i(x)\right)$, where L and L' are the left actions of G on X and on X', respectively. While here and in the following we will mostly discuss left actions and left G-spaces, the reader should bear in mind that analogous definitions and properties apply to right G-spaces.

Given a left G-space X of a group G, the *orbit* of a point $x \in X$ under the left action of G on X (denoted as L) is the set $O_x = \{L_g(x) | \, g \in G\}$.

In other words, the orbit of a point $x \in X$ is the subset of X containing all points that can be reached from x by acting on it with the left action of some element of G.

Note that, given a group G and a left G-space X, it is possible to introduce a relation (to be denoted as $\divideontimes_G$, where the subscript indicates that the definition of the relation depends on G) between pairs of elements of X, which is satisfied when they belong to the same orbit. The relation $\divideontimes_G$ is an equivalence relation.

Proof The fact that G contains the unit element (which is mapped to the identity permutation by L, because L is a group homomorphism) implies that $\divideontimes_G$ is a reflexive relation. The fact that any element $g \in G$, which is mapped to L_g by the group homomorphism L, has an inverse g^{-1}, which is mapped to $(L_g)^{-1}$ by L, implies $\divideontimes_G$ is a symmetric relation. Finally, the facts that the group multiplication is an internal operation in G and that L is a group homomorphism imply that $\divideontimes_G$ is a transitive relation. Being simultaneously reflexive, symmetric and transitive, $\divideontimes_G$ is an equivalence relation.   □

By their definition, the orbits are the equivalence classes into which $\divideontimes_G$ partitions X, i.e., the set of orbits is the quotient space of X defined by the $\divideontimes_G$ relation: $X/\divideontimes_G = \{O_x\}$. In the following, we will denote the set of orbits defined by a left action of G on X as X/G.

Given a group G, a left action of G on a set X is said to be *transitive* when all elements of X belong to the same orbit, i.e., X/G contains only one element.

In practice, a left action L of a group G on a set X is transitive when for any $x \in X$ and for any $y \in X$, there exists an element $g \in G$, such that $y = L_g(x)$, i.e., a left action is transitive if it connects every pair of elements in X.

A set X is said to be a *homogeneous space* under the action of G, when the action of G on X is transitive.

Example Let $G = U(1) = \{g = \exp(i\theta),\ \theta \in [0, 2\pi)\}$, $X = S^1 = \{x = (x_1, x_2) \in \mathbb{R}^2 : |x| = 1\}$, and the left action L defined as $L_g(x) = (x_1 \cos\theta - x_2 \sin\theta, x_2 \cos\theta + x_1 \sin\theta)$. Let $x = (x_1, x_2)$ and $y = (y_1, y_2)$ be two arbitrary elements of S^1; then, since their modulus is 1, they are necessarily of the form $x = (\cos\alpha_x, \sin\alpha_x)$ and $y = (\cos\alpha_y, \sin\alpha_y)$, with α_x and α_y both real. Defining $\theta = (\alpha_y - \alpha_x) \mod 2\pi$, and $g = \exp(i\theta)$, it is straightforward to show that $y = L_g(x)$; hence any two arbitrary elements of X belong to the same orbit. Therefore X/G contains only one orbit, i.e., the action is transitive.

Example Let $G = \mathbb{Z}_2 = \{e, a\}$ (with e the unit element and $a^2 = e$) and $X = \mathbb{R}$; consider the left action L such that $L_e(x) = x$, $L_a(x) = -x$. The orbit of a generic point $x \in X$ contains either two distinct elements $O_x = \{x, -x\}$ (if $x \neq 0$) or just x itself (if $x = 0$). Hence two generic elements of X, x and y, belong to the same orbit only if $x = \pm y$. The action is not transitive, because X/G contains infinitely many orbits (one for each nonnegative real number).

Example Consider $G = SO(2, \mathbb{R})$ and $X = \mathbb{R}^2$, parameterizing a generic element $g \in G$ as

$$g = \begin{pmatrix} \cos\theta & -\sin\theta \\ \sin\theta & \cos\theta \end{pmatrix}$$

and a generic element x of X as a column-vector

$$x = \begin{pmatrix} x_1 \\ x_2 \end{pmatrix}.$$

Define the left action L in terms of the matrix-times-vector product, i.e.,

$$L_g(x) = \begin{pmatrix} \cos\theta & -\sin\theta \\ \sin\theta & \cos\theta \end{pmatrix} \begin{pmatrix} x_1 \\ x_2 \end{pmatrix} = \begin{pmatrix} x_1 \cos\theta - x_2 \sin\theta \\ x_1 \sin\theta + x_2 \cos\theta \end{pmatrix}$$

for any $g \in G$ and for any $x \in X$. Note that this left action L preserves the modulus of vectors in X, i.e., $|L_g(x)| = |x|$. This implies that two elements x and y in X necessarily belong to distinct orbits if $|x| \neq |y|$. As a consequence, X/G does not contain only one orbit, thus the action is not transitive. In particular, a generic orbit O_x is a circle of radius $r = |x|$ centered on the origin of $\mathbb{R}^2$ (with the orbit of the origin of $\mathbb{R}^2$ containing only that point) and, since the modulus of a generic element $x \in \mathbb{R}^2$ can be any nonnegative real number, there exists an uncountable infinite number of orbits.

Given a group G, a left G-space X, and an element $x \in X$, the *isotropy group* of x is the subgroup of G containing all elements that leave x invariant: $G_x = \{g \in G | L_g(x) = x\}$.

The isotropy group is also called the *little group*, or the *stability group*, or the *stabilizer*.

For some points x it may turn out that $G_x = G$: $L_g(x) = x\ \forall g \in G$. In this case we call x a *fixed point* under the action of G.

Example Let $G = SO(2, \mathbb{R})$ act as anticlockwise rotations about the origin on the Euclidean plane $\mathbb{R}^2$. Then the origin $0 \in \mathbb{R}^2$ is a fixed point.

We may also be interested in the converse problem, that of finding all the points that are left invariant by the action of a given element g of a group G. Let g be an element of a group G acting on a set X. We denote by X^g the set of points left invariant by the (left) action of g:

$$X^g = \{x \in X | L_g(x) = x\}. \tag{2.94}$$

Note that for the unit element e, $X^e = X$ since e is always represented by the identity map.

A particularly interesting case of a group-space is the group itself, $X = G$. The simplest left and right actions are simply the *translations*, i.e., the actions defined by multiplication by a group element on the left or on the right (obviously, the two coincide for an Abelian group). The *left translation* of a group G is the group homomorphism $L : G \to \mathrm{Perm}(G)$, $g \to L_g$, defined as $L_g(g') = gg'$ for any $g' \in G$. Similarly, the *right translation* is the group homomorphism $R : G \to \mathrm{Perm}(G)$, $g \to R_g$, defined as $R_g(g') = g'g$ for any $g' \in G$.

Note that the left (or right) translation is always a transitive action: Every group element belongs to the orbit of the unit element e, since $L_g(e) = ge = g$ (or $R_g(e) = eg = g$), i.e., $O_e = G$. A more interesting way to define the group action on itself is by *conjugation*: The *conjugation* of a group G is the group homomorphism $L : G \to \mathrm{Perm}(G)$, $g \to L_g$, defined as $L_g(g') = gg'g^{-1}$ for any $g' \in G$. Then, a *conjugacy class* of an element of G is defined as its orbit under conjugation.

In other words, the conjugacy class of a generic element $g \in G$ is the subset of G containing all elements g' that can be written in the form $g' = xgx^{-1}$ for some $x \in G$. Then g and g' are said to be *conjugate*, and x is called the *conjugating element*.

In general, conjugation is not a transitive action.

It is also interesting to consider the action of a subgroup $H \subset G$ on G. In particular, the right action of H on G defined by translation, $R_h(g) = gh$, is not necessarily transitive if H is a proper subgroup.

Given a group G and a subgroup $H \subset G$, the *left coset* associated with an element $g \in G$ is the orbit of g under right translation by elements of H, i.e., the set $gH = \{g' \in G | \exists h \in H : g' = gh\}$.

The set of left cosets forms the quotient space $G/H = \{gH | g \in G\}$.

Similarly, one can define the left translation by elements of H as $L_h(g) = hg$ for any $g \in G$, and construct the right cosets Hg. The corresponding quotient space is denoted as $H \backslash G$. Obviously, when G is an Abelian group, the left and right cosets coincide.

Note that, if two elements g and $g' \in G$ can be obtained from each other via right translation by an element of the subgroup, i.e., $g' = gh$ with $h \in H$, then their left cosets coincide: $gH = g'H$.

The inverse is also true: If the left cosets of two elements g and $g' \in G$ coincide, then $g^{-1}g' \in H$, i.e., $\exists ! h \in H$, such that $g' = gh$.

Finally, one can prove that the elements of each coset can be put in a one-to-one correspondence with the elements of H itself.

Proof   For a given $g \in G$, the mapping

$$f_g : H \to gH, \quad h \to gh \tag{2.95}$$

is surjective, by definition of gH. It is also injective, because multiplying $gh_1 = gh_2$ by g^{-1} on the left it follows that $h_1 = h_2$. Being both injective and surjective, f_g is bijective, i.e., it is a one-to-one correspondence between H and the generic coset gH. □

In particular, if H is a finite subgroup, then the order of all cosets must equal $|H|$. This leads to the following theorem, due to Joseph-Louis Lagrange.

Theorem 2.3 (Lagrange's theorem) *The order $|H|$ of any subgroup H of a finite group G must be a divisor of $|G|$.*

Proof Under the right action of H, the group G is partitioned into mutually disjoint orbits gH, each having the same order as H. Hence $|G| = n|H|$ for some $n \in \mathbb{N}_0$. □

Corollary *If p is prime, then $\mathbb{Z}_p$ is the only abstract finite group of order p.*

Proof Consider a generic finite group G of order p, and let g be an element of G different from the unit element e. Let m denote the order of g; since $g \neq e$, it follows that $m > 1$. Then $H = \{e, g, \ldots, g^{m-1}\} \cong \mathbb{Z}_m$ is a subgroup of G, and Lagrange's theorem implies $p = nm$, for some $n \in \mathbb{N}_0$. Since p is prime and $m > 1$, this implies $n = 1$ and $p = m$, i.e., $|G| = |H|$. Then G coincides with H, which is isomorphic to $\mathbb{Z}_p$. Thus any finite group of order p is isomorphic to $\mathbb{Z}_p$, which is then the only abstract group of order p. □

Note that the quotient space G/H is a G-space, if we define the left action $L : G \to \mathrm{Perm}(G/H)$, $g \to L_g$ such that

$$L_g(g'H) = gg'H. \tag{2.96}$$

Note that any two arbitrary elements of G/H (say, g_1H and g_2H) always belong to the same orbit defined by L, because $y = g_2 g_1^{-1} \in G$, and $L_y g_1 H = g_2 g_1^{-1} g_1 H = g_2 H$, thus the action is transitive. The inverse statement is also true:

Theorem 2.4 *Every group G acting transitively on a set X admits a subgroup H, such that the quotient space G/H is in one-to-one correspondence with X.*

Proof For a given $x \in X$, consider its isotropy group G_x as the subgroup H. Consider the map $i : G/H \to X$ defined as $i(gH) = L_g(x)$. Note that i is well defined, i.e., $i(gH)$ does not depend on the representative (g) chosen in gH, because the definition of gH implies that for any other representative $g' \in gH$ there exists an $h \in H$, such that $g = g'h$. Thus $L_g(x) = L_{g'h}(x)$, and, since L is a group homomorphism, this implies $L_g(x) = L_{g'}(L_h(x))$. But $L_h(x) = x$, because $h \in H$ and H is the isotropy group of x. As a consequence, $L_g(x) = L_{g'}(x)$. Moreover, i is injective, because $i(gH) = i(g'H)$ means $L_g(x) = L_{g'}(x)$, which implies $x = L_{g^{-1}}\left(L_{g'}(x)\right)$. Since L is a homomorphism, the latter expression implies $x = L_{g^{-1}g'}(x)$, which means that $L_{g^{-1}g'}$ leaves x invariant, i.e., that $g^{-1}g' \in H$. This means that $g' = gh$, thus the cosets gH and $g'H$ are equal. Finally, i is also surjective; since G acts transitively, for any $x' \in X$ there exists a $g \in G$, such that $x' = L_g(x)$. But the right-hand side of the latter equation equals $i(gH)$, hence for any $x' \in X$ there exists a $g \in G$, such that $x' = i(gH)$. Thus there exists a bijective (i.e., one-to-one) correspondence i between X and G/H. □

Corollary *The orbit of a point $x \in X$ can be identified with G/G_x.*

Proof Since G acts transitively on its orbits, the statement of the corollary is a consequence of the previous theorem. □

Thus the orbits are determined by the subgroups of G; in other words, the action of G on X is determined by its subgroup structure.

Example Consider the left action of $G = \mathrm{SO}(3, \mathbb{R})$ on $\mathbb{R}^3$. The orbits are the spheres $|x|^2 = x_1^2 + x_2^2 + x_3^2 = r^2$, which can be identified with the S^2 sphere (up to a rescaling of coordinates) for any $r > 0$. Given the point $x = (0, 0, r)$ on each orbit $r > 0$, its isotropy group is

$$G_x = \left\{ \begin{pmatrix} A_{2\times 2} & 0 \\ 0 & 1 \end{pmatrix}, \text{ with } A_{2\times 2} \in \mathrm{SO}(2, \mathbb{R}) \right\} \cong \mathrm{SO}(2, \mathbb{R}). \tag{2.97}$$

By Theorem 2.4 and its corollary, it follows that $\mathrm{SO}(3, \mathbb{R}) / \mathrm{SO}(2, \mathbb{R}) = S^2$.

If G is a finite group acting on a finite set X, by a similar argument as the proof of Lagrange's theorem it follows that

$$|O_x| = \frac{|G|}{|G_x|}, \tag{2.98}$$

where $|O_x|$ denotes the number of elements in the orbit of x. This is known as the *orbit-stabilizer theorem*.

The following lemma, due to William Burnside, is sometimes useful in combinatorial problems, such as establishing how many different cubes can be obtained with n possible colour choices for its faces, or how many necklaces or bracelets can be built with a given choice of colored beads.

Lemma 2.5 (Burnside's lemma) *Let G be a finite group acting on a finite set X. The number of orbits, $|X/G|$, is*

$$|X/G| = \frac{1}{|G|} \sum_{g \in G} |X^g|. \tag{2.99}$$

Proof On the right-hand side of the equation there is a sum which can be rewritten, by reversing the order of counting as follows:

$$\sum_{g \in G} |X^g| = |\{(g, x) \in G \times X | L_g(x) = x\}| = \sum_{x \in X} |G_x|, \tag{2.100}$$

i.e., instead of counting first the elements x left invariant by an element g and then repeating the count for all g, we first count the elements g which leave a given x invariant and then repeat the count for all x. On the other hand, according to the orbit-stabilizer theorem

$$|G_x| = \frac{|G|}{|O_x|}. \tag{2.101}$$

In fact, all orbits A in X have the same number of elements: $|A| = |O_x|$ for all $A \in X/G$. Furthermore, we can break up the sum over all elements $x \in X$ into two separate sums: First sum over all elements x belonging to a given orbit A and then repeat this sum for all orbits A. So we can rewrite:

$$\sum_{x \in X} |G_x| = |G| \sum_{x \in X} \frac{1}{|O_x|} = |G| \sum_{A \in X/G} \sum_{x \in A} \frac{1}{|A|} = |G| \sum_{A \in X/G} 1 = |G||X/G|, \quad (2.102)$$

which concludes the proof.   □

Example   A bracelet is made by sliding colored beads to a string and tying its ends. How many different bracelets can you make with three red and three blue beads? Note that you can put the bracelet into your wrist in two possible ways and you can rotate it. To find the solution, one can think that the beads form a multiset $Y = \{R, R, R, B, B, B\}$ with R for red, B for blue. Sliding the beads to the string we can think of forming words from this multiset, and as we learned in Section 2.6 there are $\frac{6!}{3!3!} = 20$ words, i.e., arrangements of the beads. Now we tie the ends of the string together to form a bracelet. Let X be the set of 20 arrangements of the beads. What counts as different bracelets are arrangements that are different up to the cyclic rotations $e = r^0, r^1, \ldots, r^5$ and up to the reflection f (flipping the bracelet around). As we have seen in Section 2.1, these generate the dihedral group $D_6 = \{e, r, r^2, \ldots, r^5, f, rf, r^2f, \ldots, r^5f\}$ with $|D_6| = 12$. Thus, D_6 acts on the set of placements of the beads X, and each orbit counts as the same bracelet. The number of different bracelets is the number of orbits, $|X/D_6|$. So one can use Burnside's lemma. Clearly the neutral element keeps all the configurations fixed so that

$$|X^e| = 20. \quad (2.103)$$

For the rotations r^2 and r^4 there are two arrangements $RBRBRB$ and $BRBRBR$ that remain invariant, these are also invariant under three of the reflections. For all six reflections (about the six symmetry axes) there is also one invariant arrangement for each (with the blue beads next to each other and red beads next to each other, the reflection axis passing through the center red bead and center blue bead). The invariant arrangements are shown in Fig. 2.13. So one has $|X^g| = 2$ for $g = r^2$ and r^4, $|X^g| = 2$ for three reflections and $|X^g| = 1$ for three reflections, and $|X^g| = 0$ for $g = r, r^3$, and r^5. Hence by Burnside's lemma the total number of bracelets is

$$|X/D_6| = \frac{1}{12}(20 + 2 \cdot 2 + 3 \cdot 2 + 6 \cdot 1) = 3. \quad (2.104)$$

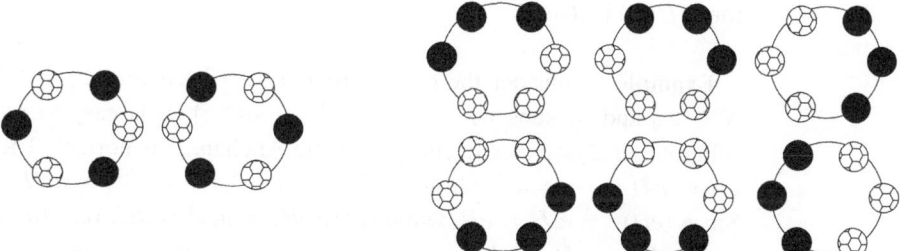

Fig. 2.13   The eight arrangements of beads that are symmetric under some elements of D_6 excluding e. The two arrangements on the left are both symmetric under two rotations and under three reflections, whereas each of the six arrangements on the right is symmetric under one reflection. The remaining 12 arrangements are not invariant under any of the symmetry transformations of D_6: they are obtained from a single chiral bracelet and its partner (related by reflection) by applying five rotations on each.

There is a generalized version of Burnside's lemma that applies to a wider variety of combinatorial problems, the *Pólya enumeration theorem*, named after the Hungarian mathematician George Pólya, but we leave it to the interested reader to find more about it elsewhere. We will instead return from this combinatorial sidetrack back to group theory.

Since the quotient space G/H is constructed out of a group and its subgroup, it is natural to ask if it can also be a group. The first guess for a multiplication law would be $(g_1 H)(g_2 H) = g_1 g_2 H$. This definition would be meaningful if the right-hand side were independent of the labeling of the cosets, but this is not always the case; the first factor on the left-hand side is such that $g_1 H = g_1 h H$ for any $h \in H$, so one could replace g_1 with $g_1 h$; however, on the right-hand side it is not necessarily true that $g_1 g_2 H$ and $g_1 h g_2 H$ coincide.

To endow G/H with a meaningful group multiplication, it is convenient to require H to be a normal subgroup (i.e., a subgroup $H \subseteq G$ such that, for all $g \in G$ and for all $h \in H$, one has $ghg^{-1} \in H$, as defined in Section 2.2).

If H is a normal subgroup, then the quotient space G/H can be endowed with group structure by defining the group multiplication as

$$(g_1 H)(g_2 H) = g_1 g_2 H, \qquad \forall g_1 H \in G/H, \ \forall g_2 H \in G/H. \qquad (2.105)$$

The fact that H is a normal subgroup makes the right-hand side uniquely defined, i.e., makes the definition of the group multiplication meaningful: On the left-hand side, g_1 and g_2 are defined only up to right translations by elements of H, so that the right-hand side of the equation above could be replaced by $g_1 h_1 g_2 h_2 H$. This is obviously equal to $g_1 h_1 g_2 H$, so the ambiguity in the definition of g_2 is not a problem. If H is a normal subgroup, then there exists an $h \in H$ such that $h_1 g_2 = g_2 h$; then, $g_1 h_1 g_2 H = g_1 g_2 h H$, which, in turn is equal to $g_1 g_2 H$, so the right-hand side of Eq. (2.105) is unaffected by the ambiguity in the definition of g_1 and g_2. It is trivial to show that this definition of group multiplication in G/H satisfies associativity (a direct consequence of the associativity of the product in G), that it admits a unit element (the coset $eH \cong H$) and that every element $gH \in G/H$ has $g^{-1} H$ as its inverse. With the multiplication defined by eq. (2.105), the quotient space G/H becomes a group, called the *quotient group*.

Note that, by Lagrange's theorem 2.3, if G is a finite group, then the order of G/H (or of $H \backslash G$) is $|G|/|H|$.

Example Consider the cyclic group $\mathbb{Z}_{2N} = \{e, a, a^2, \ldots, a^{2N-1}\}$, for a generic $N \in \mathbb{N}_0$ and its subset $H = \{e, a^2, a^4, \ldots, a^{2N-2}\}$. It is easy to prove that H is a subgroup of $\mathbb{Z}_{2N}$. Since cyclic groups are Abelian, H is normal. The two cosets are $eH = a^2 H = \cdots = a^{2N-2} H$ and $aH = \{a, a^3, a^5, \ldots, a^{2N-1}\} = a^3 H = \cdots = a^{2N-1} H$. Since $(eH)^2 = e^2 H = eH$, while $(aH)(eH) = aeH = aH$, $(eH)(aH) = eaH = aH$, and $(aH)^2 = a^2 H = eH$, the quotient group $\mathbb{Z}_{2N}/H$ is isomorphic to $\mathbb{Z}_2$.

Example Let $G = \mathrm{SU}(2)$ and $H = \{\mathbb{1}_2, -\mathbb{1}_2\} \cong \mathbb{Z}_2$. Obviously $A \in \mathrm{SU}(2)$ satisfies the property $A(\pm \mathbb{1}_2) = \pm \mathbb{1}_2 A$, hence H is a normal subgroup. Thus the quotient space $\mathrm{SU}(2)/\mathbb{Z}_2$ is a group; in particular, it turns out to be isomorphic to $\mathrm{SO}(3, \mathbb{R})$.

The $\mathbb{Z}_2$ subgroup in the latter example is the *center* of the SU(2) group, i.e., the subset containing the elements which commute with all elements of the group.

It is easy to prove that the center of a group is a subgroup, and obviously it is Abelian. Moreover, it also normal; hence, the quotient space $G/Z(G)$ is always a group. Note that, if the center is trivial, i.e., if $Z(G) = \{e\}$, then $G/Z(G) \cong G$.

Obviously, if G is an Abelian group, then $Z(G) = G$. On the other hand, the centers of some non-Abelian groups are as follows.

- $G = S_3$ has a trivial center: $Z(G) = \{e\}$;
- $G = D_4$ has center $Z(G) = \{e, a^2\} \cong \mathbb{Z}_2$; the quotient group $G/Z(G)$ is isomorphic to the Klein four-group $\mathbb{Z}_2 \times \mathbb{Z}_2$;
- $G = Q_8$ has center $Z(G) = \{e, -e\} \cong \mathbb{Z}_2$; the quotient group $G/Z(G)$ is isomorphic to the Klein four-group $\mathbb{Z}_2 \times \mathbb{Z}_2$;
- $G = SO(2k + 1, \mathbb{R})$, with $k \geq 1$, has a trivial center: $Z(G) = \{e\}$;
- $G = SO(2k, \mathbb{R})$, with $k \geq 2$, has center $Z(G) = \{e, -e\} \cong \mathbb{Z}_2$;
- $G = U(N)$ has center $Z(G) = \{\exp(i\theta), \ \theta \in [0, 2\pi)\} \cong U(1)$;
- $G = SU(N)$ has center $Z(G) = \{\exp(2k\pi i/N)\mathbb{1}_N, \text{ for } k \in \{0, 1, \ldots N - 1\}\}$, which is isomorphic to $\mathbb{Z}_N$;
- $G = Sp(2N, \mathbb{R})$ has center $Z(G) = \mathbb{Z}_2$;
- $G = \mathbb{G}_2$ has a trivial center: $Z(G) = \{e\}$;
- $G = \mathbb{F}_4$ has a trivial center: $Z(G) = \{e\}$;
- $G = \mathbb{E}_6$ has center $Z(G) = \mathbb{Z}_3$;
- $G = \mathbb{E}_7$ has center $Z(G) = \mathbb{Z}_2$;
- $G = \mathbb{E}_8$ has a trivial center: $Z(G) = \{e\}$.

We conclude this chapter by discussing another way of finding normal subgroups and quotient groups, in terms of the *kernel* and *image* of group homomorphisms.

Given the two groups G_1 and G_2 and the group homomorphism $\mu : G_1 \to G_2$, the *kernel* of μ, denoted as Ker μ, is the subgroup of G_1 containing the elements mapped to the unit element of G_2 by μ. On the other hand, the *image* of μ, denoted as Im μ, is the subgroup of G_2 containing the elements that can be obtained by applying μ to some element of G_1.

In other words, the kernel of μ is defined as the set

$$\text{Ker } \mu = \{g_1 \in G_1 : \ \mu(g_1) = e_2\}, \tag{2.106}$$

where e_2 denotes the unit element of G_2, while the image of μ is defined as the set

$$\text{Im } \mu = \{g_2 \in G_2 : \ \exists g_1 \in G_1 \text{ such that: } g_2 = \mu(g_1)\}. \tag{2.107}$$

It is easy to prove that Ker μ is a subgroup of G_1, and that Im μ is a subgroup of G_2.

Moreover, Ker μ is a normal subgroup: for any $k \in$ Ker μ and for any $g \in G_1$, one has $\mu\left(gkg^{-1}\right) = \mu(g)\mu(k)\mu\left(g^{-1}\right) = \mu(g)e_2\mu\left(g^{-1}\right) = \mu(g)\mu\left(g^{-1}\right) = e_2$, hence $gkg^{-1} \in$ Ker μ. Since Ker μ is a normal subgroup of G_1, it is possible to define the quotient group $G_1/$Ker μ. This quotient group turns out to be isomorphic to Im μ.

Theorem 2.6 *Given two groups G_1 and G_2 and the group homomorphism $\mu : G_1 \to G_2$, the quotient group $G_1/$Ker μ is isomorphic to Im μ.*

Proof To simplify the notation, let K denote Ker μ. Consider the homomorphism

$$i : G_1/K \to \text{Im}\ \mu, \quad gK \to i(gK) = \mu(g). \tag{2.108}$$

The definition is meaningful, because if $gK = g'K$ then there is a $k \in K$ such that $g' = gk$; then $i(g'K) = \mu(g') = \mu(gk)$. Since μ is a group homomorphism, this equals $\mu(g)\mu(k)$, and since $k \in K$ one has $\mu(k) = e_2$, so that $i(g'K) = \mu(g) = i(gK)$, so i is well defined. One can prove that i is injective: $i(gK) = i(g'K)$ means $\mu(g) = \mu(g')$; multiplying both sides of this equation by $[\mu(g)]^{-1}$ on the left, one obtains $e_2 = [\mu(g)]^{-1}\mu(g')$. Since μ is a group homomorphism, this implies $e_2 = \mu\left(g^{-1}\right)\mu(g') = \mu\left(g^{-1}g'\right)$. This means that $g^{-1}g' \in K$, namely that there exists an element $k \in K$ such that $g' = gk$. This implies that $g'K = gkK = gK$. In addition, i is also surjective. Indeed, Im μ contains all elements of G_2 which can be written as $\mu(g)$ for some $g \in G_1$. By definition of i, each of these elements is also the image of $gK \in G_1/K$. Being a group homomorphism which is simultaneously injective and surjective, i is a group isomorphism. This implies that $G_1/\text{Ker}\ \mu$ is isomorphic to Im μ.   □

Example Considering the previous example, one can show that $\text{SU}(2)/\mathbb{Z}_2 \cong \text{SO}(3, \mathbb{R})$ by constructing a surjective group homomorphism $\mu : \text{SU}(2) \to \text{SO}(3, \mathbb{R})$ with the property that Ker $\mu = \{\mathbb{1}_2, -\mathbb{1}_2\} \cong \mathbb{Z}_2$. One possible such homomorphism is the one mapping a generic $\text{SU}(2)$ matrix A to the $\text{SO}(3, \mathbb{R})$ matrix of components

$$\left[\mu(A)\right]_{ab} = \frac{1}{2} \text{tr}\left(\sigma^a A \sigma^b A^{-1}\right), \qquad \text{for } a, b \in \{1, 2, 3\}, \tag{2.109}$$

where σ^a and σ^b are two of the Pauli matrices defined in Eq. (2.80). It is straightforward to show that μ is a group homomorphism; given two $\text{SU}(2)$ elements A and B and their product $C = AB$, one has

$$\begin{aligned}
\left[\mu(A)\right]_{ab} &= \frac{1}{2} \text{tr}\left(\sigma^a A \sigma^b A^{-1}\right) \\
&= \frac{1}{2} \sum_{i,j,k,l=1}^{2} \sigma^a_{ij} A_{jk} \sigma^b_{kl} A^\star_{il},
\end{aligned}$$

where we used the fact that, since A is a unitary matrix, $(A^{-1})_{li} = A^\star_{il}$ and, similarly,

$$\begin{aligned}
\left[\mu(B)\right]_{bc} &= \frac{1}{2} \text{tr}\left(\sigma^b B \sigma^c B^{-1}\right) \\
&= \frac{1}{2} \sum_{m,n,p,q=1}^{2} \sigma^b_{mn} B_{np} \sigma^c_{pq} B^\star_{mq}.
\end{aligned}$$

Thus, the generic entry of the $\mu(A)\mu(B)$ product is

$$\begin{aligned}
\left[\mu(A)\mu(B)\right]_{ac} &= \sum_{b=1}^{3} \left[\mu(A)\right]_{ab} \left[\mu(B)\right]_{bc} \\
&= \frac{1}{4} \sum_{i,j,k,l,m,n,p,q=1}^{2} \sigma^a_{ij} A_{jk} A^\star_{il} B_{np} \sigma^c_{pq} B^\star_{mq} \sum_{b=1}^{3} \sigma^b_{kl} \sigma^b_{mn}.
\end{aligned}$$

Using the completeness relation, Eq. (2.82), one obtains

$$\left[\mu(A)\mu(B)\right]_{ac} = \frac{1}{2}\sum_{i,j,k,l,p,q=1}^{2}\sigma_{ij}^{a}A_{jk}A_{il}^{\star}B_{kp}\sigma_{pq}^{c}B_{lq}^{\star}$$

$$-\frac{1}{4}\sum_{i,j,k,m,p,q=1}^{2}\sigma_{ij}^{a}A_{jk}A_{ik}^{\star}B_{mp}\sigma_{pq}^{c}B_{mq}^{\star}$$

$$= \frac{1}{2}\sum_{i,j,p,q=1}^{2}\sigma_{ij}^{a}\sum_{k=1}^{2}A_{jk}B_{kp}\sigma_{pq}^{c}\sum_{l=1}^{2}A_{il}^{\star}B_{lq}^{\star}$$

$$-\frac{1}{4}\sum_{i,j,p,q=1}^{2}\sigma_{ij}^{a}\sum_{k=1}^{2}A_{jk}A_{ik}^{\star}\sum_{m=1}^{2}B_{mp}B_{mq}^{\star}\sigma_{pq}^{c}$$

$$= \frac{1}{2}\sum_{i,j,p,q=1}^{2}\sigma_{ij}^{a}\,(AB)_{jp}\,\sigma_{pq}^{c}\,(AB)_{iq}^{\star}$$

$$-\frac{1}{4}\sum_{i,j,p,q=1}^{2}\sigma_{ij}^{a}\sum_{k=1}^{2}A_{jk}A_{ki}^{\dagger}\sum_{m=1}^{2}B_{qm}^{\dagger}B_{mp}\sigma_{pq}^{c}$$

$$= \frac{1}{2}\sum_{i,j,p,q=1}^{2}\sigma_{ij}^{a}\,(AB)_{jp}\,\sigma_{pq}^{c}\,(AB)_{qi}^{\dagger} - \frac{1}{4}\sum_{i,j,p,q=1}^{2}\sigma_{ij}^{a}\delta_{j,i}\delta_{q,p}\sigma_{pq}^{c}$$

$$= \frac{1}{2}\sum_{i,j,p,q=1}^{2}\sigma_{ij}^{a}\,(AB)_{jp}\,\sigma_{pq}^{c}\,(AB)_{qi}^{\dagger} - \frac{1}{4}\sum_{i=1}^{2}\sigma_{ii}^{a}\sum_{p=1}^{2}\sigma_{pp}^{c}$$

$$= \frac{1}{2}\sum_{i,j,p,q=1}^{2}\sigma_{ij}^{a}\,(AB)_{jp}\,\sigma_{pq}^{c}\,(AB)_{qi}^{\dagger},$$

having used the fact that $AA^{\dagger} = B^{\dagger}B = \mathbb{1}_2$ (since A and B are unitary matrices) and the tracelessness of the Pauli matrices. Thus

$$\left[\mu(A)\mu(B)\right]_{ac} = \frac{1}{2}\sum_{i,j,p,q=1}^{2}\sigma_{ij}^{a}\,(AB)_{jp}\,\sigma_{pq}^{c}\,(AB)_{qi}^{\dagger}$$

$$= \frac{1}{2}\,\mathrm{tr}\left(\sigma^{a}\,(AB)\,\sigma^{c}\,(AB)^{-1}\right)$$

$$= \left[\mu(AB)\right]_{ac}.$$

Since this holds for any a and $c \in \{1,2,3\}$, it follows that $\mu(A)\mu(B) = \mu(AB)$, i.e., μ is a group homomorphism. It is possible to prove that μ is surjective. However, μ is not injective, because $\mu(A) = \mu(-A)$ for every $A \in \mathrm{SU}(2)$. In particular, $\mu(\mathbb{1}_2) = \mu(-\mathbb{1}_2) = \mathbb{1}_3$, so $\mathrm{Ker}\,\mu = \{\mathbb{1}_2, -\mathbb{1}_2\} \cong \mathbb{Z}_2$. Using Theorem 2.6, we conclude that $\mathrm{SU}(2)/\mathbb{Z}_2 = \mathrm{SO}(3,\mathbb{R})$.

This connection is central to the quantum mechanical treatment of rotationally invariant systems and spin representations.

Problems

2.1   Consider the group $G = \{e, x_1, x_2, x_3, x_4, x_5\}$, where

$$e = \begin{pmatrix} 1 & 0 & 0 \\ 0 & 1 & 0 \\ 0 & 0 & 1 \end{pmatrix}, \quad x_1 = \begin{pmatrix} 0 & 1 & 0 \\ 1 & 0 & 0 \\ 0 & 0 & 1 \end{pmatrix}, \quad x_2 = \begin{pmatrix} 0 & 0 & 1 \\ 0 & 1 & 0 \\ 1 & 0 & 0 \end{pmatrix},$$

$$x_3 = \begin{pmatrix} 1 & 0 & 0 \\ 0 & 0 & 1 \\ 0 & 1 & 0 \end{pmatrix}, \quad x_4 = \begin{pmatrix} 0 & 1 & 0 \\ 0 & 0 & 1 \\ 1 & 0 & 0 \end{pmatrix}, \quad x_5 = \begin{pmatrix} 0 & 0 & 1 \\ 1 & 0 & 0 \\ 0 & 1 & 0 \end{pmatrix},$$

and the composition law is the matrix multiplication. Show that G is isomorphic to a known group, and give an explicit construction of the isomorphism.

2.2   Let G be the group of discrete transformations that leave a rectangle invariant (with the composition law given by subsequent application of two transformations as the group product), including the trivial transformation which does not interchange any of its vertices (e), rotations by integer multiples of π (a), reflections about the vertical symmetry axis (b), and reflections about the horizontal symmetry axis c. Construct the Cayley table of G and identify which group it is.

2.3   Let S^2 denote a sphere of unit radius, centered at the origin of $\mathbb{R}^3$. What is the group of transformations that leave it invariant? Next, let $S^2_{(1)}$ be a sphere of unit radius, centered at the origin of three-dimensional space, with a point on its surface. What is the group of transformations that leave $S^2_{(1)}$ invariant?

2.4   Let $S^2_{(2)}$ denote a sphere of unit radius, centered at the origin of three-dimensional space, with two noncoincident points on its surface. What is the group of transformations that leave $S^2_{(2)}$ invariant? Does the group depend on the relative position of the two points? Does it depend on whether the two points are distinguishable or indistinguishable?

2.5   Is O(2) an Abelian or a non-Abelian group?

2.6   Prove that the order of the symmetric group of a set of N elements, S_N, is $N!$.

2.7   Prove that the only finite groups of order 4 are $\mathbb{Z}_2 \times \mathbb{Z}_2$ and $\mathbb{Z}_4$.

2.8   Prove that the cyclic group of order 3 does not have proper subgroups.

2.9   Prove that the groups $(\mathrm{Mat}(n, \mathbb{R}), +)$ and $\left(\mathbb{R}^{n^2}, +\right)$ are isomorphic.

2.10   Let N be a positive integer. Consider the relation $\circledast$ among pairs of integers $r, s \in \mathbb{Z}$ defined as $r \circledast s$ when $r - s$ is an integer multiple of N. Prove that $\circledast$ is an equivalence relation and identify the elements of the quotient space $Q = \mathbb{Z}/\circledast$. Prove that the composition law among equivalence classes $\odot$ defined as $[a] \odot [b] = [a + b]$ endows Q with group structure, and identify to which group $(Q, \odot)$ is isomorphic.

2.11   Consider the sets and the composition laws that are defined as follows:

- The set of 2×3 matrices with real coefficients, and the composition law defined as the entry-by-entry matrix sum.
- The set of 2×3 matrices with real coefficients, and the composition law defined as the row-by-column matrix multiplication.
- The set of 2×2 matrices with real coefficients, and the composition law defined as the row-by-column matrix multiplication.
- The set of real numbers with the composition law $\odot$ defined by

$$x \odot y = \frac{x}{y}.$$

- The set of real, strictly positive numbers with the composition law $\circ$ defined by

$$x \circ y = \frac{x}{y}.$$

Check whether each of them is a magma, a semigroup, a monoid, a group, an Abelian group.

2.12 Define three different functions $f_i : \mathbb{R} \to \mathbb{R}$, $i \in \{1, 2, 3\}$ in such a way that

- f_1 is an injection but not a surjection,
- f_2 is a surjection but not an injection,
- f_3 is a bijection.

2.13 Given three generic functions $f_i : \mathbb{R} \to \mathbb{R}$, $i \in \{1, 2, 3\}$, such that f_1 is an injection but not a surjection, f_2 is a surjection but not an injection, and f_3 is a bijection, consider the function

$$f : \mathbb{R} \to \mathbb{R}^3, \quad x \to \begin{pmatrix} f_1(x) \\ f_2(x) \\ f_3(x) \end{pmatrix}$$

and answer the following questions, providing arguments for the answers.

- Is f an injection?
- Is f a surjection?
- Is f a bijection?

2.14 Consider the group of permutations of the set $X = \{1, 2, 3, 4, 5, 6\}$. Compute the following products, where permutations are denoted using the shortened cycle notation, i.e., omitting one-element cycles, and with the trivial permutation denoted by E, the symbol "·" represents the group product, and the permutation on the right is the one which acts first:

- $(235)(46) \cdot (14)(265)$,
- $(1635)(24) \cdot (1536)(24)$,
- $(26)(35) \cdot (24536)$,
- $(165)(23) \cdot (13)(26)$,
- $(1326) \cdot (154)$,

2.15 Prove the isomorphism in Eq. (2.50), i.e., show that the presentation that can be written as $\langle a, b, |aba^{-1}b^{-1}\rangle$ is isomorphic to $\mathbb{Z} \times \mathbb{Z}$, with addition as the group multiplication.

2.16   Let X be the n-dimensional unit sphere $S^n = \{x \in \mathbb{R}^{n+1} : |x| = 1\}$ with antipodal points identified, and let A be the $(n + 1)$-dimensional real vector space without the point at the origin: $A = \mathbb{R}^{n+1}\backslash\{0\}$. Define an equivalence relation $\sim$ among elements of A, such that X can be identified with the quotient space of A with respect to $\sim$, i.e., $X = A/\sim$.

2.17   Consider the two groups $G_1 = SO(3)$ and $G_2 = U(1)$ and a surjective mapping $\mu : G_1 \rightarrow G_2$. Discuss whether, in general, μ is a homomorphism, or not.

2.18   Let x and y be two elements of the five-dimensional complex vector space $\mathbb{C}^5$, and let A be a set, such that each of its elements can be written as $(\alpha x + \beta y)$, with α and β two complex numbers. Discuss the existence and properties of a possible subgroup of $U(5)$ leaving A invariant.

2.19   The default privacy settings of a certain online social networking service are such, that messages posted by a user are visible to their author and to all of her/his contacts, as well as to the contacts of each of her/his contacts. In the set X of users of the service, introduce the relation $\bowtie$ defined as

$$x \bowtie y \Leftrightarrow x \text{ can read the posts written by } y,$$

for any x and y in X. Is $\bowtie$ an equivalence relation? If it is, what is the quotient space $X/\bowtie$? How can the default privacy settings of the networking service be modified, in order to make it (or make it remain) an equivalence relation? In that case, what is the quotient space $X/\bowtie$?

2.20   Consider the following family $\mathcal{F} = \{f_i\}$ of mappings of complex numbers with strictly positive imaginary part:

$$f_i : \{z = x + iy \in \mathbb{C}, y > 0\} \rightarrow \text{Im}(f_i) \subseteq \mathbb{C}, \quad f_i : z \rightarrow \frac{az + b}{cz + d},$$

where the coefficients a, b, c and d are integer numbers.

- Work out the general constraint on a, b, c and d that has to be satisfied, so that the mappings f_i are well-defined and invertible functions.
- Work out a more restrictive constraint $\mathcal{C}$ (in the form of an equality relating the a, b, c and d coefficients) enforcing that the f_i mappings are well-defined and invertible functions, and that, in particular, at least one of them is the identity mapping $i : z \rightarrow z$. In the subset of mappings satisfying $\mathcal{C}$, identify the sets of coefficients (a, b, c, d) that correspond to the same mapping.
- Let $G \subseteq \mathcal{F}$ be the subset of $\mathcal{F}$ for which the constraint discussed in the previous point holds; work out an appropriate "multiplication" operation in G, which makes G a non-Abelian group.

2.21   Given a group G, a left G-space X, and an element $x \in X$, prove that the isotropy group of x is a subgroup of G.

2.22   Consider the left action of $SO(3)$ on the sphere $S^2 \subset \mathbb{R}^3$ defined by the matrix-times-column-vector multiplication. Parameterize the isotropy group of

$$x = \begin{pmatrix} -9/39 \\ -60/65 \\ 4/13 \end{pmatrix}.$$

2.23 Given the set $X = \{1, 2, 3, 4, 5\}$, construct the inverse of each of the following elements of $\text{Perm}(X)$:

- $(24)(35)$,
- (14235),
- $(1)(2)(34)$,
- (253),
- $(15)(24) \cdot (154)(23)$.

2.24 Let f be a function defined as

$$f : \mathbb{R} \to \mathbb{R}, \quad f : x \to \max \left(\{ n \in \mathbb{Z}, \text{ with the property that: } n \leq x \} \right),$$

where $\max(X)$ denotes the largest element of X. Consider two fixed, linearly independent elements $\vec{v}_1$ and $\vec{v}_2$ of the two-dimensional real vector space, and the following relation $\sim$ among elements $\vec{a}$ and $\vec{b}$ of $\mathbb{R}^2$:

$$\vec{a} \sim \vec{b} \Leftrightarrow \left\{ \left[a^1 - f(a^1) = b^1 - f(b^1) \right] \text{ AND } \left[a^2 - f(a^2) = b^2 - f(b^2) \right] \right\},$$

where a^1, a^2, b^1, and b^2 are defined by

$$\vec{a} = a^1 \vec{v}_1 + a^2 \vec{v}_2, \quad \vec{b} = b^1 \vec{v}_1 + b^2 \vec{v}_2.$$

Prove that $\sim$ is an equivalence relation. Given the quotient space defined as $\mathcal{T} = \mathbb{R}^2 / \sim$, define two (not trivially related to each other, and both different from multiplication by a fixed scalar factor) transformations of $\vec{v}_1$ and $\vec{v}_2$

$$\aleph : \mathbb{R}^2 \times \mathbb{R}^2 \to \mathbb{R}^2 \times \mathbb{R}^2, \quad \aleph : (\vec{v}_1, \vec{v}_2) \to (\vec{u}_1, \vec{u}_2),$$
$$\beth : \mathbb{R}^2 \times \mathbb{R}^2 \to \mathbb{R}^2 \times \mathbb{R}^2, \quad \beth : (\vec{v}_1, \vec{v}_2) \to (\vec{w}_1, \vec{w}_2)$$

such that, by repeating the same construction as above, but using either $(\vec{u}_1, \vec{u}_2)$ or $(\vec{w}_1, \vec{w}_2)$ instead of $(\vec{v}_1, \vec{v}_2)$, one obtains the same quotient space $\mathcal{T}$. Finally, let $\mathcal{B}$ be the group of transformations mapping $(\vec{v}_1, \vec{v}_2)$ to a generic pair $(\vec{w}_1, \vec{w}_2)$ satisfying the requirement that the quotient space obtained applying the construction remains the same. Discuss the properties of $\mathcal{B}$.

3 Representation Theory of Groups

In Chapter 2 we discussed the action of a group on a set. We also listed some examples of Lie groups, their elements being $n \times n$ matrices. For example, the elements of the orthogonal group $O(n, \mathbb{R})$ corresponded to rotations and reflections of vectors in $\mathbb{R}^n$. In this chapter, we will continue along these lines and consider the action of a generic group on a (complex) vector space, so that we can represent the elements of the group by matrices. However, a vector space is more than just a set, so in defining the action of a group on it, we will have to ensure that it respects the vector space structure.

3.1 Complex Vector Spaces and Representations

A *complex vector space* V can be thought of as an Abelian group, where the group product is denoted by + and called "sum," and the neutral element is denoted as $\vec{0}$, and in which an additional operation, called *scalar multiplication* by a complex number $\mu \in \mathbb{C}$ has been defined, such that the following conditions are satisfied:

VS1 $\mu(\vec{v}_1 + \vec{v}_2) = \mu\vec{v}_1 + \mu\vec{v}_2$,
VS2 $(\mu_1 + \mu_2)\vec{v} = \mu_1\vec{v} + \mu_2\vec{v}$,
VS3 $\mu_1(\mu_2\vec{v}) = (\mu_1\mu_2)\vec{v}$,
VS4 $1\,\vec{v} = \vec{v}$,
VS5 $0\,\vec{v} = \vec{0}$.

We usually suppress the multiplication symbol and write $\mu\vec{v}$ instead of $\mu \cdot \vec{v}$.

Note that we could have replaced complex numbers by real numbers, to define a *real vector space*. More generally, the set of scalars could have been replaced by any "field," i.e., by any set of "numbers" in which the operations of addition, subtraction, multiplication, and division can be defined. We focus on complex vector spaces, as they are particularly relevant for quantum mechanics. A comment on notation: We denote vectors with arrows, $\vec{v}$, but many textbooks denote them in boldface, **v**. If it is clear from the context whether one means a vector or its component, then one may also simply use the notation v for a vector.

Consider a *linear combination* of k vectors $\vec{v}_1, \ldots, \vec{v}_k \in V$ of the form $\sum_{i=1}^{k} \mu_i\vec{v}_i$. The vectors $\vec{v}_1, \ldots, \vec{v}_k$ are said to be *linearly independent* if $\sum_{i=1}^{k} \mu_i\vec{v}_i = \vec{0}$ only when all coefficients μ_i are simultaneously zero. The *dimension* of a vector space V is the maximum number of vectors of V that are linearly independent; the dimension of V is often denoted as dim V. If dim $V = n$, then a set $\{\vec{e}^1, \ldots, \vec{e}^n\}$ of linearly independent vectors of V is called a *basis* of the vector space. Given a basis, any vector $\vec{v}$ can be written in a form $\vec{v} = \sum_{i=1}^{n} v_i \vec{e}^i$, where the *components* v_i of the vector are fixed uniquely.

Given two vector spaces V_1 and V_2, a map $L : V_1 \to V_2$ is said to be a *linear map* if it satisfies

$$L(\mu_1 \vec{v}_1 + \mu_2 \vec{v}_2) = \mu_1 L(\vec{v}_1) + \mu_2 L(\vec{v}_2) \tag{3.1}$$

for all μ_1 and μ_2 in $\mathbb{C}$, and for all $\vec{v}_1, \vec{v}_2 \in V_1$. A linear map is also called a *linear transformation* or, especially in a physics context, a *linear operator*. If a linear map is also a bijection, then it is called an *isomorphism*, and the vector spaces V_1 and V_2 are said to be *isomorphic* to each other, $V_1 \cong V_2$. It then follows that $\dim V_1 = \dim V_2$. Moreover, all n-dimensional vector spaces are isomorphic. An isomorphism from V to itself is called an *automorphism*. The set of automorphisms of V is denoted $\mathrm{Aut}(V)$; it is a group, with composition of mappings $L \circ L'$ as the group product. The neutral element in $\mathrm{Aut}(V)$ is the identity automorphism (which maps every vector of V to itself); the existence of the inverse of every automorphism is guaranteed, since automorphisms, being isomorphisms, are bijections.

The *image* of a linear transformation is

$$\mathrm{Im}\, L = f(V_1) = \{L(\vec{v}_1)|\ \vec{v}_1 \in V_1\} \subset V_2, \tag{3.2}$$

while its *kernel* is the set of vectors of V_1 that map to the null vector $\vec{0}_2$ of V_2:

$$\mathrm{Ker}\, L = \{\vec{v}_1 \in V_1|\ L(\vec{v}_1) = \vec{0}_2\} \subset V_1. \tag{3.3}$$

It is an easy exercise (see Problem 3.1) to show that both the image and the kernel are vector spaces. We also mention the following two theorems, without proving them.

Theorem 3.1 $\dim V_1 = \dim \mathrm{Ker}\, L + \dim \mathrm{Im}\, L$.

Theorem 3.2   *A linear map $L : V \to V$ is an automorphism if and only if $\mathrm{Ker}\, L = \{\vec{0}\}$.*

Every linear map is defined uniquely by its action on the basis vectors:

$$L(\vec{v}) = L\left(\sum_{i=1}^{n} v_i \vec{e}^i\right) = \sum_{i} v_i L\left(\vec{e}^i\right). \tag{3.4}$$

One can then expand the vectors $L(\vec{e}^i)$ in the basis $\{\vec{e}^j\}$, denoting the components by L_{ji}:

$$L\left(\vec{e}^i\right) = \sum_{j} L_{ji} \vec{e}^j. \tag{3.5}$$

Now

$$L(\vec{v}) = \sum_{i}\sum_{j} v_i L_{ji} \vec{e}^j = \sum_{j}\left(\sum_{i} L_{ji} v_i\right)\vec{e}^j, \tag{3.6}$$

so the image vector $L(\vec{v})$ has components $L(\vec{v})_j = \sum_i L_{ji} v_i$. Let dim $V_1 =$ dim $V_2 = n$. Then, in the familiar matrix language:

$$\begin{pmatrix} L(\vec{v})_1 \\ L(\vec{v})_2 \\ \vdots \\ L(\vec{v})_n \end{pmatrix} = \begin{pmatrix} L_{11} & L_{12} & \cdots & L_{1n} \\ L_{21} & L_{22} & \cdots & L_{2n} \\ \vdots & \vdots & \ddots & \vdots \\ L_{n1} & \cdots & \cdots & L_{nn} \end{pmatrix} \begin{pmatrix} v_1 \\ v_2 \\ \vdots \\ v_n \end{pmatrix}. \tag{3.7}$$

We will often shorten the notation for linear maps and write $L\vec{v}$ instead of $L(\vec{v})$, and $L_1 L_2 \vec{v}$ instead of $L_1(L_2(\vec{v}))$. From the above it should also be clear that the group of automorphisms of V is isomorphic to the group of invertible $n \times n$ complex matrices

$$\mathrm{Aut}(V) = \{L : V \to V | L \text{ is an automorphism}\} \cong \mathrm{GL}(n, \mathbb{C}). \tag{3.8}$$

The group multiplication laws are the composition of maps and matrix multiplication.

Now we have the tools to give a definition of a representation of a group. The idea is that we define the action of a group G on a vector space V. If V were just a set, we would associate with every group element $g \in G$ a permutation $L_g \in \mathrm{Perm}(V)$. However, we have to preserve the vector space structure of V. So we define the action just as before, but replace the group $\mathrm{Perm}(V)$ of permutations of V by the group $\mathrm{Aut}(V)$ of automorphisms of V.

A *linear representation D* of a group G in a vector space V is a homomorphism

$$D : G \to \mathrm{Aut}(V),$$
$$g \to D(g). \tag{3.9}$$

The *dimension of the representation* is the dimension of the vector space dim V. Note the following:

1. D is a homomorphism: $D(g_1 g_2) = D(g_1) D(g_2)$.
2. $D(g^{-1}) = (D(g))^{-1}$.

Example Let $G = \mathbb{Z}_4 = \{e, c, c^2, c^3\}$ and $V = \mathbb{R}^2$. One possible representation of G is $D : G \to \mathrm{Aut}(V)$, with

$$D(c) = \begin{pmatrix} 0 & -1 \\ 1 & 0 \end{pmatrix},$$

$$D(c^2) = (D(c))^2 = \begin{pmatrix} -1 & 0 \\ 0 & -1 \end{pmatrix} = -\mathbb{1},$$

$$D(c^3) = (D(c))^3 = \begin{pmatrix} 0 & 1 \\ -1 & 0 \end{pmatrix},$$

$$D(e) = D(c^4) = (D(c))^4 = \begin{pmatrix} 1 & 0 \\ 0 & 1 \end{pmatrix} = \mathbb{1}. \tag{3.10}$$

Note that the matrix $D(c)$ corresponds to a $\pi/2$ rotation about the origin in the $\mathbb{R}^2$ plane.

A representation D is said to be *faithful* when it is an injective map. Then $g_1 \neq g_2 \Rightarrow D(g_1) \neq D(g_2)$, and the inverse of D exists. A representation is faithful if and only if $\mathrm{Ker}\, D = \{e\}$, where $\mathrm{Ker}\, D$ is to be understood as the kernel of the group homomorphism defined in Eq. (2.106), not as the kernel of a linear transformation introduced in Eq. (3.3).

Whatever $\mathrm{Ker}\, D$ is, D is always a faithful representation of the quotient group $G/\mathrm{Ker}\, D$.

We would next like to find and classify all possible representations of a group. First we need a criterion to determine when two representations are the same, or, more precisely, are *equivalent*.

Let D_1 and D_2 be two representations of a group G in two vector spaces V_1 and V_2. An *intertwining operator* is a linear map $A : V_1 \to V_2$ such that $D_2(g)A = AD_1(g)$ for all $g \in G$. If A is an isomorphism (which is possible when $\dim V_1 = \dim V_2$), the representations D_1 and D_2 are *equivalent*. In other words, there then exists a *similarity transformation* $D_2(g) = AD_1(g)A^{-1}$ for all $g \in G$.

Example Let $\dim V_1 = n$, $V_2 = \mathbb{C}^n$. Any n-dimensional representation is equivalent to a representation of G by invertible complex matrices, the homomorphism $D_2 : G \to \mathrm{GL}(n, \mathbb{C})$.

A *scalar product* in a vector space V is a map $V \times V \to \mathbb{C}$, $(\vec{v}_1, \vec{v}_2) \mapsto \langle \vec{v}_1 | \vec{v}_2 \rangle \in \mathbb{C}$ which satisfies the following properties:

(i) $\langle \vec{v} | \mu_1 \vec{v}_1 + \mu_2 \vec{v}_2 \rangle = \mu_1 \langle \vec{v} | \vec{v}_1 \rangle + \mu_2 \langle \vec{v} | \vec{v}_2 \rangle$,
(ii) $\langle \vec{v} | \vec{w} \rangle = \langle \vec{w} | \vec{v} \rangle^\star$,
(iii) $\forall \vec{v} \in V$, one has $\langle \vec{v} | \vec{v} \rangle = 0$ if and only if $\vec{v} = \vec{0}$.

Note that, by combining (i) and (ii), it follows that $\langle \mu_1 \vec{v}_1 + \mu_2 \vec{v}_2 | \vec{v} \rangle = \mu_1^\star \langle \vec{v}_1 | \vec{v} \rangle + \mu_2^\star \langle \vec{v}_2 | \vec{v} \rangle$.

Given a scalar product, it is always possible to make the basis vectors orthogonal to each other and normalized to unity, such that $\langle \vec{e}^i | \vec{e}^j \rangle = \delta^{ij}$.

A way to generate an orthonormalized basis of vectors, starting from a generic basis of $\{\vec{u}^1, \vec{u}^2, \ldots \vec{u}^n\}$ is by the *Gram–Schmidt method*, according to the following algorithm:

• The first vector of the orthonormalized basis is simply defined by normalizing $\vec{u}^1$ to unity

$$\vec{e}^1 = \frac{\vec{u}^1}{\sqrt{\langle \vec{u}^1 | \vec{u}^1 \rangle}}; \tag{3.11}$$

• each of the other vectors $\vec{e}^i$, for $1 < i \le n$, are obtained by subtracting from each $\vec{u}^i$ the components along the directions of the $\vec{e}^j$ already defined (for $1 \le j < i$)

$$\vec{w}^i = \vec{u}^i - \sum_{j=1}^{i-1} \langle \vec{v}^j | \vec{w}^i \rangle \vec{v}^j, \tag{3.12}$$

and then normalizing them to unity

$$\vec{e}^i = \frac{\vec{w}^i}{\sqrt{\langle \vec{w}^i | \vec{w}^i \rangle}}. \tag{3.13}$$

The *adjoint* $A^\dagger$ of an operator (or of a linear map) $A : V \to V$ is the one which satisfies $\langle \vec{v} | A^\dagger \vec{w} \rangle = \langle A\vec{v} | \vec{w} \rangle$ for all $\vec{v}, \vec{w} \in V$.

An operator (linear map) $U : V \to V$ is *unitary* if $\langle \vec{v} | \vec{w} \rangle = \langle U\vec{v} | U\vec{w} \rangle$ for all $\vec{v}, \vec{w} \in V$. Equivalently, a unitary operator must be such that $U^\dagger U$ is the identity operator in V. It follows that the corresponding $n \times n$ matrix must be unitary, i.e., an element of $U(n)$. Unitary operators form a subgroup $\mathrm{Unit}(V) \cong U(n)$ of $\mathrm{Aut}(V) \cong GL(n, \mathbb{C})$.

A *unitary representation* of a group G is a homomorphism $D : G \to \mathrm{Unit}(V)$.

If U_1 and U_2 are unitary representations of G in V_1 and V_2, and there exists an intertwining isomorphism $A : V_1 \to V_2$ that preserves the scalar product, $\langle A\vec{v} | A\vec{w} \rangle_{V_2} = \langle \vec{v} | \vec{w} \rangle_{V_1}$ for all $\vec{v}, \vec{w} \in V_1$, then the representations are said to be *unitarily equivalent*.

Example Every n-dimensional unitary representation is unitarily equivalent to a representation by unitary matrices, a homomorphism $G \to U(n)$.

The basic problem in group representation theory is to classify all unitary representations of a group, up to unitary equivalences. Before proceeding to this task, however, in the following subsection we discuss the relevance of symmetry transformations and representations in quantum mechanics.

3.1.1 Symmetry Transformations in Quantum Mechanics

In physics, a major field of application of unitary representations in complex vector spaces is quantum mechanics.

In quantum mechanics, the set of all possible states of a physical system is the Hilbert space $\mathcal{H}$, a complex vector space endowed with a scalar product, such that with a norm and distance defined by the scalar product it is *complete*.[1] In quantum mechanics, a state vector is usually denoted by $|\psi\rangle$ (as opposed to our previous notation $\vec{v}$) and called a *ket*, and the scalar product of two vectors $|\psi\rangle$ and $|\chi\rangle$ is written as $\langle \psi | \chi \rangle$. The norm of a state vector, its scalar product with itself $\langle \psi | \psi \rangle$, is the probability density of finding the physical system in that particular state, and, hence, is suitably normalized. More precisely, state vectors differing by a global phase factor correspond to the same physical state of the system, and are thus equivalent, $e^{i\theta} |\psi\rangle \sim |\psi\rangle$. Thus a (pure) physical state is represented as a *ray* $e^{i\theta} |\psi\rangle$ in the Hilbert space: this is called the *ray representation* of physical states.

Note that often the Hilbert space is an infinite-dimensional vector space, whereas in our discussion of representation theory we have been focusing on finite-dimensional vector spaces. While the infinite dimension of a vector space entails some subtleties, those are not of concern for the present discussion. In fact, in many cases finite-dimensional representations are relevant for quantum mechanics.

According to quantum mechanics, the time evolution of a state is controlled by the *Schrödinger equation*

$$i\hbar \frac{d}{dt} |\psi\rangle = H|\psi\rangle, \tag{3.14}$$

where H is the *Hamiltonian operator* (or Hamiltonian), which plays the rôle of the time-evolution operator of the system.

---

[1] This means that every Cauchy sequence of points in $\mathcal{H}$ also has a limit in $\mathcal{H}$.

A system is said to have a symmetry under a group G of transformations, when its properties remain invariant under transformations belonging to the group G. In order to describe this symmetry, one has to specify how it acts on the state vectors of the system, i.e., to find its representation in the vector space of the states, the Hilbert space.

If the system is invariant under the group of symmetries G, then the (density of) probability of finding the system in a state described by a certain state vector should remain invariant under application of a transformation in G. This means that every transformation in G must leave $\langle\psi|\psi\rangle$ invariant, for all vector states $|\psi\rangle$. This is the case for all unitary representations of G, as they preserve the scalar product of the Hilbert space.

Moreover, in a closed system, the probability should be preserved under the time evolution. Thus, unitarity of the representations must also be preserved under the time evolution. This can be summarized in a more formal way: If $g \mapsto U_g$ is a faithful unitary representation of a group G in the Hilbert space of a quantum mechanical system, such that for all $g \in G$

$$U_g H U_g^{-1} = H, \tag{3.15}$$

where H is the Hamilton operator of the system, then the group G is a *symmetry group* of the system. Note that, defining the *commutator* of two operators A and B as $[A, B] = AB - BA$, Eq. (3.15) can also be written as $HU_g = U_g H$, i.e., as $[H, U_g] = 0$.

The condition defined by Eq. (3.15) arises as follows. Suppose that a state vector $|\psi\rangle$ is a solution of the Schrödinger equation. Upon performing a symmetry transformation on the system, the state vector is mapped to a new vector $U_g|\psi\rangle$. But if the system is invariant under this symmetry, then the new state $U_g|\psi\rangle$ must also be a solution of the Schrödinger equation: $i\hbar(d/dt)U_g|\psi\rangle = HU_g|\psi\rangle$. From this, it follows that $i\hbar(d/dt)|\psi\rangle = i\hbar(d/dt)U_g^{-1}U_g|\psi\rangle = U_g^{-1}HU_g|\psi\rangle = H|\psi\rangle$. Since this must hold for a generic state vector $|\psi\rangle$ that is a solution of the Schrödinger equation, one can deduce that $U_g^{-1}HU_g = H$.

Consider the eigenstates of the Hamiltonian, or energy eigenstates, $|\phi_n\rangle$, associated with the energy levels E_n; they are defined by the condition

$$H|\phi_n\rangle = E_n|\phi_n\rangle. \tag{3.16}$$

In general, energy levels may be degenerate, say with k linearly independent energy eigenstates $\{|\phi_{n1}\rangle, \ldots, |\phi_{nk}\rangle\}$. The set of linear combinations of these eigenstates $\mathcal{H}_n$ is called an *eigenspace*: It is a subspace of the full Hilbert space, and a k-dimensional vector space. If the system has a symmetry group G, then

$$HU_g|\phi_n\rangle = U_g H|\phi_n\rangle = E_n U_g|\phi_n\rangle, \tag{3.17}$$

so all states $U_g|\phi_n\rangle$ are eigenstates associated with the same energy level E_n. Thus the representation $D : g \to U_g$ maps the eigenspace $\mathcal{H}_n$ to itself. In other words, the representation can be thought of as a k-dimensional representation of G acting in $\mathcal{H}_n$.

Conversely, if a physical quantum system has a symmetry group G, then its representations determine the possible degeneracies of the energy levels of the system.

3.2 Reducibility of Representations

Irreducible representations are a particularly important type of representations, because, as will be discussed below, a generic representation can always be decomposed into irreducible representations. Before introducing the notion of irreducible representation at a formal level, we start with some definitions.

A subset W of a vector space V is called a *vector subspace* if it contains all possible linear combinations of its elements; If $\vec{v} \in W$ and $\vec{w} \in W$, then $\lambda\vec{v} + \mu\vec{w} \in W$ for every $\lambda \in \mathbb{C}$ and for every $\mu \in \mathbb{C}$.

The eigenspace $\mathcal{H}_n$ associated with an energy level E_n of a quantum-mechanical Hamiltonian H, that was introduced in Section 3.1.1, is an example of a subspace of the Hilbert space $\mathcal{H}$ of the system. Indeed, $\mathcal{H}_n$ is a subset of $\mathcal{H}$, and, by its definition as the linear span of the eigenstates of H with eigenvalue E_n, it contains all possible linear combinations of its elements.

Let D be a representation of a group G in a vector space V. The representation space V is also called a *G-module*.

A subspace W of V is said to be a *submodule* if it is closed under the action of the group G, namely if $\vec{w} \in W \Rightarrow D(g)\vec{w} \in W$ for all $g \in G$. If W is a submodule, then the restriction of $D(g)$ to W is an automorphism $D(g)_W : W \to W$.

Note that, for every group G, every representation D, and every vector space V, $\{\vec{0}\}$ and V itself are always submodules. However, they are called *trivial* submodules.

A representation $D : G \to \text{Aut}(V)$ is said to be *irreducible* if the only submodules are the trivial ones, $\{\vec{0}\}$ and V. Otherwise the representation is said to be *reducible*.

Example Choose a basis $\{\vec{e}^i\}$ in V, and let $\dim V = n$. Suppose that, for every $g \in G$, the matrix $D(g)$ with elements $D(g)_{ij} = \langle \vec{e}^i | D(g) | \vec{e}^j \rangle$ has the form

$$D(g) = \begin{pmatrix} M(g) & S(g) \\ 0 & T(g) \end{pmatrix}, \tag{3.18}$$

where $M(g)$ is an $n_1 \times n_1$ matrix, $T(g)$ is an $n_2 \times n_2$ matrix, $n_1 + n_2 = n$, and $S(g)$ is an $n_1 \times n_2$ matrix. Then the representation is reducible, since the subspace of V defined as

$$W = \left\{ \begin{pmatrix} \vec{v} \\ \vec{0} \end{pmatrix} \mid \vec{v} = \begin{pmatrix} v_1 \\ \vdots \\ v_{n_1} \end{pmatrix}, v_1 \in \mathbb{C}, v_2 \in \mathbb{C}, \dots, v_{n_1} \in \mathbb{C} \right\} \tag{3.19}$$

is a submodule:

$$D(g) \begin{pmatrix} \vec{v} \\ \vec{0} \end{pmatrix} = \begin{pmatrix} M(g)\vec{v} + S(g)\vec{0} \\ T(g)\vec{0} \end{pmatrix} = \begin{pmatrix} M(g)\vec{v} \\ \vec{0} \end{pmatrix} \in W. \tag{3.20}$$

If, in addition, $S(g) = 0$ for all $g \in G$, then the representation is manifestly built up by simply combining two representations $M(g)$ and $T(g)$, and is then an example of a class of representations that will be defined shortly, i.e., completely reducible representations.

Given two vector spaces V_1 and V_2, the *direct sum of vector spaces*, denoted as $V_1 \oplus V_2$, is the vector space having as its elements all pairs (v_1, v_2), with $v_1 \in V_1$, $v_2 \in V_2$, and with the addition of vectors and multiplication by a scalar respectively defined as

$$(v_1, v_2) + (v_1', v_2') = (v_1 + v_1', v_2 + v_2') \tag{3.21}$$

$$\lambda(v_1, v_2) = (\lambda v_1, \lambda v_2). \tag{3.22}$$

It is a simple exercise (see Problem 3.2) to show that $\dim(V_1 \oplus V_2) = \dim V_1 + \dim V_2$.

If a scalar product has been defined in V_1 and V_2, one can define a scalar product in $V_1 \oplus V_2$ by

$$\langle (v_1, v_2) | (v_1', v_2') \rangle = \langle v_1 | v_1' \rangle + \langle v_2 | v_2' \rangle. \tag{3.23}$$

Suppose that D_1 is a representation of a group G in V_1, and D_2 is a representations of G in V_2. Then, the *direct-sum representation* $D_1 \oplus D_2$ is defined as the representation acting on $V_1 \oplus V_2$ as

$$(D_1 \oplus D_2)(g)(v_1, v_2) = (D_1(g)v_1, D_2(g)v_2). \tag{3.24}$$

In this case it is useful to adopt the notation

$$V_1 = \left\{ \begin{pmatrix} \vec{v}_1 \\ \vec{0} \end{pmatrix} \right\}; \quad V_2 = \left\{ \begin{pmatrix} \vec{0} \\ \vec{v}_2 \end{pmatrix} \right\} \tag{3.25}$$

so that

$$V_1 \oplus V_2 = \left\{ \begin{pmatrix} \vec{v}_1 \\ \vec{v}_2 \end{pmatrix} \right\}. \tag{3.26}$$

Then, the matrices associated with group elements in the direct-sum representation are in block-diagonal form:

$$(D_1 \oplus D_2)(g) = \begin{pmatrix} D_1(g) & 0 \\ 0 & D_2(g) \end{pmatrix}. \tag{3.27}$$

A representation D in a vector space V is said to be *completely reducible* if for every nontrivial submodule $W \subset V$ (with $W \neq V$ and $W \neq \{\vec{0}\}$) there exists a *complementary submodule* W', such that $V = W \oplus W'$ and $D \cong D_W \oplus D_{W'}$.

Note that, according to the definition, to show that a representation D is completely reducible, one has to prove that D is equivalent to the direct-sum representation $D_W \oplus D_{W'}$. For the matrices of the representation, this means that there must be a similarity transformation which makes all the $D(g)$ matrices block-diagonal:

$$AD(g)A^{-1} = \begin{pmatrix} D_W(g) & 0 \\ 0 & D_{W'}(g) \end{pmatrix}. \tag{3.28}$$

As an aside, the *direct sum of matrices* is an operation that can be used to describe the case of two transformations acting separately on two independent vectors. Let A be an $N \times N$ matrix and B an $M \times M$ matrix. Their direct sum $A \oplus B$ is a matrix of size $(N + M) \times (N + M)$ that is defined as follows:

$$A \oplus B = \begin{pmatrix} A & 0 \\ 0 & B \end{pmatrix} = \begin{pmatrix} a_{11} & \dots & a_{1N} & 0 & \dots & 0 \\ a_{21} & \dots & a_{2N} & 0 & \dots & 0 \\ \dots & \dots & \dots & \dots & \dots & \dots \\ a_{N1} & \dots & a_{NN} & 0 & \dots & 0 \\ 0 & \dots & 0 & b_{11} & \dots & b_{1M} \\ 0 & \dots & 0 & b_{21} & \dots & b_{2M} \\ \dots & \dots & \dots & \dots & \dots & \dots \\ 0 & \dots & 0 & b_{M1} & \dots & b_{MM} \end{pmatrix}. \tag{3.29}$$

Note that the direct sum of two matrices is not commutative: $A \oplus B \neq B \oplus A$, but the two resulting matrices can be mapped into each other by a permutation of rows and columns. It is obvious that $\operatorname{tr}(A \oplus B) = \operatorname{tr} A + \operatorname{tr} B$ and $\det(A \oplus B) = (\det A) \cdot (\det B)$.

Given a completely reducible representation D, its *reduction* is defined as its decomposition into the direct sum of irreducible representations $D_1, D_2, \dots$, such that

$$D \cong D_1 \oplus D_2 \oplus D_3 \oplus \cdots \tag{3.30}$$

(then it also follows that $\dim D = \sum_i \dim D_i$). This is possible only if D is completely reducible.

It turns out that the representations of a particular type – which, as we mentioned in Section 3.1.1, is especially relevant for quantum mechanics – are always completely reducible.

Theorem 3.3 *Unitary representations are completely reducible.*

Proof   Since we are discussing unitary representations, it is implied that the representation space V has a scalar product. Let W be a submodule, and let $W_\perp = \{\vec{v} \in V | \langle \vec{v} | \vec{w} \rangle = 0 \ \forall \vec{w} \in W\}$ be its *orthogonal complement*. We leave it as an exercise (see Problem 3.3) to show that $V \cong W \oplus W_\perp$. We then show that $W_\perp$ is also a submodule, i.e., that it is closed under the action of G. Let $\vec{v} \in W_\perp$, and denote the unitary representation by U. For all $\vec{w} \in W$ and for all $g \in G$, one has

$$\begin{aligned} \langle U(g)\vec{v} | \vec{w} \rangle &= \langle U(g)\vec{v} | U(g)U^{-1}(g)\vec{w} \rangle \\ &= \langle \vec{v} | U^\dagger(g)U(g)U^{-1}(g)\vec{w} \rangle \\ &= \langle \vec{v} | U^{-1}(g)\vec{w} \rangle, \end{aligned} \tag{3.31}$$

where the last equality follows since U is unitary. Then

$$\langle U(g)\vec{v} | \vec{w} \rangle = \langle \vec{v} | U^{-1}(g)\vec{w} \rangle = \langle \vec{v} | \vec{w}' \rangle, \quad \text{with } \vec{w}' \in W \tag{3.32}$$

having used the fact that W is a G-module), and finally

$$\langle U(g)\vec{v} | \vec{w} \rangle = \langle \vec{v} | \vec{w}' \rangle = 0, \tag{3.33}$$

because $\vec{v} \in W_\perp$. Thus $U(g)\vec{v} \in W_\perp$, so $W_\perp$ is closed under the action of G. □

If G is a finite group, one can say more.

Theorem 3.4   *Let D be a finite-dimensional representation of a finite group G in a vector space V. Then there exists a scalar product in V, such that D is a unitary representation.*

Proof In a finite-dimensional vector space it is always possible to define a scalar product, e.g. by choosing a basis and defining $\langle \vec{v} | \vec{w} \rangle = \sum_{i=1}^{n} v_i^\star w_i$, where v_i and w_i are the components of the vectors $\vec{v}$ and $\vec{w}$, respectively, in the chosen basis. Given the scalar product $\langle \ldots | \ldots \rangle$, one can introduce the "group-averaged" scalar product $\langle\langle \ldots | \ldots \rangle\rangle$ defined as

$$\langle\langle \vec{v} | \vec{w} \rangle\rangle = \frac{1}{|G|} \sum_{g' \in G} \langle D(g') \vec{v} | D(g') \vec{w} \rangle. \tag{3.34}$$

It is straightforward to show that $\langle\langle \ldots | \ldots \rangle\rangle$ satisfies the requirements of a scalar product. Furthermore,

$$\langle\langle D(g) \vec{v} | D(g) \vec{w} \rangle\rangle = \frac{1}{|G|} \sum_{g' \in G} \langle D(g') D(g) \vec{v} | D(g') D(g) \vec{w} \rangle$$

$$= \frac{1}{|G|} \sum_{g' \in G} \langle D(g'g) \vec{v} | D(g'g) \vec{w} \rangle$$

$$= \frac{1}{|G|} \sum_{g'' \in G} \langle D(g'') \vec{v} | D(g'') \vec{w} \rangle$$

$$= \langle\langle \vec{v} | \vec{w} \rangle\rangle; \tag{3.35}$$

in other words, D is unitary with respect to the scalar product $\langle\langle \ldots | \ldots \rangle\rangle$. □

It is then easy to prove the following theorem, due to Heinrich Maschke.

Theorem 3.5 (Maschke's theorem) *Every finite-dimensional representation of a finite group is completely reducible.*

Proof According to Theorem 3.4, every finite-dimensional representation D of a finite group G is unitary. From Theorem 3.3, it then follows that D is completely reducible. □

3.3 Irreducible Representations

As shown in the previous section, many representations of interest are completely reducible, and can be decomposed into a direct sum of irreducible representations. The goal of this section is to develop methods to identify *inequivalent* irreducible representations, that allow one to classify the latter. For this purpose, we first discuss some general theorems, starting from a lemma due to Issai Schur.

Theorem 3.6 (Schur's lemma) *Given two irreducible representations D_1 and D_2 of a group G, every intertwining operator between them is either a null map or an isomorphism. In the latter case the representations are equivalent, $D_1 \cong D_2$.*

Proof Let us consider an intertwining operator, to be denoted as A, between the representations: $D_2(g)A = AD_1(g)$ for all $g \in G$. Let us first examine if A can be an injection. Note first that, if $\text{Ker}\, A \equiv \{\vec{v} \in V_1 | A\vec{v} = \vec{0}_2\} = \{\vec{0}_1\}$, then A is an injection, because if $A\vec{v} = A\vec{w}$ then $A(\vec{v} - \vec{w}) = 0$, which implies $\vec{v} - \vec{w} \in \text{Ker}\, A = \{\vec{0}_1\}$, namely

$\vec{v} = \vec{w}$. So what is Ker A? Recall that Ker A is a subspace of V_1. In addition, Ker A is also a submodule, namely it is closed under the action of G: if $\vec{v}$ is an arbitrary element of Ker A, then, for all $g \in G$, one has $AD_1(g)\vec{v} = D_2(g)A\vec{v}$ because A is an intertwining operator, and $D_2(g)A\vec{v} = \vec{0}_2$, because $\vec{v} \in$ Ker A. Hence $D_1(g)\vec{v} \in$ Ker A, i.e., Ker A is a submodule. But since D_1 is an irreducible representation, it does not admit nontrivial submodules; as a consequence, either Ker $A = V_1$ or Ker $A = \{\vec{0}_1\}$. In the former case all vectors of V_1 map to the null vector of V_2, so A is a null map $A = 0$. In the latter case, A is an injection.

By similar reasoning, one can examine whether A is also a surjection. Given an arbitrary $\vec{v}_2 \in$ Im A, by definition of Im A there exists $\vec{v}_1 \in V_1$ such that $\vec{v}_2 = A\vec{v}_1$. Then $D_2(g)\vec{v}_2 = D_2(g)A\vec{v}_1$ and, since A is an intertwiner, the latter expression equals $AD_1(g)\vec{v}_1$. Hence, $D_2(g)\vec{v}_2 = AD_1(g)\vec{v}_1$, which means that also $D_2(g)\vec{v}_2 \in$ Im A. Thus, Im A is a submodule of V_2. But since D_2 is irreducible, either Im $A = \{\vec{0}_2\}$, i.e., $A = 0$, or Im $A = V_2$ namely A is a surjection.

We deduce that either A is the null map, or A is a linear operator which is a bijection, that is an isomorphism.   □

Corollary (Corollary of Schur's lemma)   *If D is an irreducible representation of a group G in a complex vector space V, then the only operator which commutes with all $D(g)$ is a multiple of the identity operator.*

Proof   If $AD(g) = D(g)A$ for all $g \in G$, then for every complex number μ also $(A - \mu\mathbb{1})D(g) = D(g)(A - \mu\mathbb{1})$. According to Schur's lemma 3.6, either $(A - \mu\mathbb{1})$ is an isomorphism, so that $(A - \mu\mathbb{1})^{-1}$ exists for all $\mu \in \mathbb{C}$, or $A - \mu\mathbb{1} = 0$. However, if A has finite dimension, then the fundamental theorem of algebra implies that the polynomial equation $\det(A - \mu\mathbb{1}) = 0$ has at least one complex solution for μ, thus $(A - \mu\mathbb{1})$ cannot be invertible, hence it is not an isomorphism. (The same holds also if A has infinite dimension, although the proof is more subtle.) Hence it must be $A - \mu\mathbb{1} = 0$, namely $A = \mu\mathbb{1}$.   □

We will next discuss a sequence of theorems, starting from the rather abstract *fundamental orthogonality theorem*.

Theorem 3.7 (Fundamental orthogonality theorem)   *Given a finite group G, let U_α denote a generic unitary irreducible representation (acting on a vector space V_α), assuming that α labels inequivalent irreducible representations uniquely, i.e., that, given two unitary irreducible representations U_α and U_β, $\alpha = \beta$ if and only if U_α and U_β are the same representation, up to isomorphisms. Let U_α act on the vector space V_α and let U_β act on the vector space V_β. Then, for all $\vec{v}, \vec{w} \in V_\alpha$ and for all $\vec{u}, \vec{t} \in V_\beta$:*

$$\sum_{g \in G} \left(\langle \vec{w} | U_\alpha(g) \vec{v} \rangle_{V_\alpha} \right)^\star \langle \vec{u} | U_\beta(g) \vec{t} \rangle_{V_\beta} = \frac{|G|}{\dim V_\alpha} \delta_{\alpha\beta} \left(\langle \vec{w} | \vec{t} \rangle \right)^\star \langle \vec{v} | \vec{u} \rangle. \qquad (3.36)$$

In the latter expression, $\dim V_\alpha$ *is finite.*

Proof   Note that $\left(\langle \vec{w} | U_\alpha(g) \vec{v} \rangle_{V_\alpha} \right)^\star = \langle \vec{v} | U_\alpha(g)^\dagger \vec{w} \rangle_{V_\alpha}$. As U_α is a unitary representation, the latter expression can be rewritten as $\langle \vec{v} | U_\alpha(g^{-1}) \vec{w} \rangle_{V_\alpha}$.

Let v_i and w_j denote the components of $\vec{v}$ and $\vec{w}$ in an orthonormalized basis $\{\vec{e}^i\}$ in V_α. Similarly, let u_l and t_k denote the components of $\vec{u}$ and $\vec{t}$ in an orthonormalized basis $\{\vec{f}^l\}$ in V_β. Then, Eq. (3.36) can be rewritten as

$$\sum_{i,j,k,l}\sum_{g\in G} v_i^\star w_j t_k^\star u_l \left(U_\alpha(g^{-1})\right)_{ij} \left(U_\beta(g)\right)_{kl} = \sum_{i,j,k,l} t_k^\star w_j v_i^\star u_l \frac{|G|}{\dim V_\alpha}\delta_{\alpha\beta}\delta_{il}\delta_{kj}, \quad (3.37)$$

where $\left(U_\alpha(g^{-1})\right)_{ij}$ and $\left(U_\beta(g)\right)_{kl}$ are defined as

$$\left(U_\alpha(g^{-1})\right)_{ij} = \langle \vec{e}^i|U_\alpha(g^{-1})\vec{e}^j\rangle_{V_\alpha}, \qquad \left(U_\beta(g)\right)_{kl} = \langle \vec{f}^k|U_\beta(g)\vec{f}^l\rangle_{V_\beta}. \quad (3.38)$$

As Eq. (3.37) must be true for all $\vec{v}, \vec{w} \in V_\alpha$ and for all $\vec{u}, \vec{t} \in V_\beta$, it follows that Eq. (3.36) holds if and only if

$$\sum_{g\in G} \left(U_\alpha(g^{-1})\right)_{ij} \left(U_\beta(g)\right)_{kl} = \frac{|G|}{\dim V_\alpha}\delta_{\alpha\beta}\delta_{il}\delta_{jk}. \quad (3.39)$$

Consider, then, an arbitrary linear operator from $L : V_\beta \to V_\alpha$ and define

$$A = \sum_{g\in G} U_\alpha(g^{-1})LU_\beta(g). \quad (3.40)$$

Let h be an arbitrary element of G; then

$$\begin{aligned}
U_\alpha(h)A &= \sum_{g\in G} U_\alpha(h)U_\alpha(g^{-1})LU_\beta(g) \\
&= \sum_{g\in G} U_\alpha(hg^{-1})LU_\beta(g) \\
&= \sum_{g'\in G} U_\alpha((g')^{-1})LU_\beta\,(g'h) \\
&= \sum_{g'\in G} U_\alpha((g')^{-1})LU_\beta\,(g')\,U_\beta(h), \quad (3.41)
\end{aligned}$$

having defined $(g')^{-1} = hg^{-1}$, from which it follows that $g = g'h$, having used the fact that U_α and U_β are representations, and that $\sum_{g\in G} = \sum_{g'\in G}$. In the last expression, one recognizes the definition of A, so

$$U_\alpha(h)A = AU_\beta(h). \quad (3.42)$$

As Eq. (3.42) holds for all $h \in G$, A is an intertwining operator between the representations U_α and U_β. As a consequence of Schur's lemma 3.6, A must either be the null map, or an isomorphism between the representations U_α and U_β. In the latter case, the representations are then equivalent to each other, and the corollary to Schur's lemma implies that A is proportional to the identity operator

$$A = \sum_{g\in G} U_\alpha(g^{-1})LU_\beta(g) = \mu_L \delta_{\alpha\beta}\mathbb{1}. \quad (3.43)$$

Note that this must hold for every linear operator L, and the complex number μ_L depends on L. Choosing L to be a matrix whose entries are all zero, except for one element $L_{jk} = 1$, for fixed j and k, i.e., $L_{pq} = \delta_{pj}\delta_{qk}$, and denoting the corresponding number μ_L as μ_{jk}, the (i, l) entry of A reads

$$A_{il} = \sum_{g \in G} \left(U_\alpha(g^{-1})\right)_{ij} \left(U_\beta(g)\right)_{kl}. \tag{3.44}$$

Thus, when $\alpha = \beta$ the trace of A reads

$$\operatorname{tr} A = \sum_{g \in G} \sum_i \left(U_\alpha(g^{-1})\right)_{ij} \left(U_\alpha(g)\right)_{ki}, \tag{3.45}$$

and Eq. (3.43) implies that this quantity must be equal to $\mu_{jk} \dim V_\alpha$. This leads to

$$
\begin{aligned}
\mu_{jk} &= \frac{1}{\dim V_\alpha} \sum_{g \in G} \sum_i \left(U_\alpha(g^{-1})\right)_{ij} \left(U_\alpha(g)\right)_{ki} \\
&= \frac{1}{\dim V_\alpha} \sum_{g \in G} \sum_i \left(U_\alpha(g)\right)_{ki} \left(U_\alpha(g^{-1})\right)_{ij} \\
&= \frac{1}{\dim V_\alpha} \sum_{g \in G} \left(U_\alpha(gg^{-1})\right)_{kj} \\
&= \frac{1}{\dim V_\alpha} \sum_{g \in G} \left(U_\alpha(e)\right)_{kj} \\
&= \frac{1}{\dim V_\alpha} \sum_{g \in G} \delta_{kj} \\
&= \frac{|G|}{\dim V_\alpha} \delta_{jk}. \tag{3.46}
\end{aligned}
$$

Thus, for this particular choice of L, the i, l entry of A in Eq. (3.43) is

$$\sum_{g \in G} \left(U_\alpha(g^{-1})\right)_{ij} \left(U_\beta(g)\right)_{kl} = \frac{|G|}{\dim V_\alpha} \delta_{\alpha\beta} \delta_{jk} \delta_{il}. \tag{3.47}$$

This proves Eq. (3.39), which is equivalent to Eq. (3.36).   □

Note that the entries of $U_\alpha(g^{-1})$ and $U_\beta(g)$ appearing in Eq. (3.39) are basis-dependent quantities defined as in Eqs. (3.38). Having defined the $\{\vec{e}^i\}$ basis in V_α, each $(U_\alpha(g))_{ij}$ can be interpreted as the component labeled by g of a vector specified by the ordered (i, j) pair and by the representation label α. Then, Eq. (3.39) can be interpreted as an *orthogonality relation* among these vectors; This orthogonality relation is based on the scalar product defined as $\langle \vec{a} | \vec{b} \rangle = \sum_{i=1}^{|G|} a_i^\star b_i$. Note that in a $|G|$-dimensional vector space there can be at most $|G|$ mutually orthogonal vectors. As the ordered index pair (i, j) takes $(\dim V_\alpha)^2$ possible values, the upper bound on the total number of mutually orthogonal vectors is

$$\sum_\alpha (\dim V_\alpha)^2 \leq |G|, \tag{3.48}$$

where the sum is taken over all possible unitary inequivalent representations, labeled by α. In fact, it turns out that the bound given by Eq. (3.48) is saturated, i.e., that this non-strict inequality is actually an equality. This is the statement of the following theorem.

Theorem 3.8 (Burnside's theorem)   *Given a finite group G, its inequivalent irreducible representations U_α, acting on vector spaces V_α, satisfy*

$$\sum_\alpha (\dim V_\alpha)^2 = |G|. \tag{3.49}$$

Burnside's theorem restricts the possible dimensions of irreducible representations.

Example   As an example, consider the group of permutations of three elements, S_3, whose order is $|S_3| = 6$. According to Eq. (3.49), there are only two possibilities for the dimensions of the inequivalent unitary irreducible representations of S_3:

- Either there exists a representation of dimension 2 and two inequivalent irreducible representations of dimension 1,
- or there are six inequivalent irreducible representations, all of them of dimension 1.

As will be shown later, it turns out that S_3 has only three inequivalent unitary irreducible representations, so the inequivalent irreducible representations of this group have dimensions 2, 1, and 1.

3.4 Characters

Characters provide a convenient tool to classify inequivalent irreducible representations. Let $\{\vec{e}^1, \ldots, \vec{e}^n\}$ be an orthonormal basis in an n-dimensional vector space V with respect to the scalar product $\langle \ldots | \ldots \rangle$. The *trace of a linear operator A* is defined as

$$\operatorname{tr} A \equiv \sum_{i=1}^n \langle \vec{e}^i | A \vec{e}^i \rangle. \tag{3.50}$$

Note that the trace is well defined, since it is independent of the chosen basis; if $\{\vec{u}^1, \ldots, \vec{u}^n\}$ is another basis, then

$$\operatorname{tr} A = \sum_i \langle \vec{e}^i | A \vec{e}^i \rangle = \sum_{ij} \langle \vec{e}^i | \vec{u}^j \rangle \langle \vec{u}^j | A \vec{e}^i \rangle = \sum_{ij} \langle A^\dagger \vec{u}^j | \vec{e}^i \rangle \langle \vec{e}^i | \vec{u}^j \rangle \tag{3.51}$$

$$= \sum_{ij} \langle A^\dagger \vec{u}^j | \vec{u}^j \rangle = \sum_j \langle \vec{u}^j | A \vec{u}^j \rangle.$$

As the operator A is associated with the $n \times n$ matrix with components $A_{ij} = \langle \vec{e}^i | A \vec{e}^j \rangle$, $\operatorname{tr} A$ is equal to the trace of the matrix.

Let $U_\alpha(g)$ be a unitary representation in V of a finite group G; the *character of the representation U_α* is defined as the map

$$\chi_\alpha : G \to \mathbb{C},$$

$$g \to \chi_\alpha(g) = \operatorname{tr} U_\alpha(g). \tag{3.52}$$

Note that equivalent representations have the same characters; if the unitary representation W is equivalent to U_α, then there exists an isomorphism A such that

$W = AU_\alpha A^{-1}$. Then, the character of W maps every element $g \in G$ to $\operatorname{tr} W(g) = \operatorname{tr}(AU_\alpha(g)A^{-1}) = \operatorname{tr}(U_\alpha(g)A^{-1}A) = \operatorname{tr} U_\alpha(g)$, where we used the fact that the trace of any product of matrices is invariant under cyclic permutations of the factors.

Recall that in Section 2.9 we introduced conjugation as a way a group can act on itself. More precisely, the conjugation of a group G was defined as a group homomorphism $L : G \to \operatorname{Perm}(G)$, mapping each element $g \in G$ to a permutation defined as $L_g : G \to G, g_0 \to L_g(g_0) = gg_0g^{-1}$. The orbits $\{gg_0g^{-1} \mid g \in G\}$ defined by conjugation were called conjugacy classes. Then, the invariance of the trace of products of matrices under cyclic permutations of its factors implies that group elements related to each other by conjugation have the same characters. Hence, characters can be interpreted as mappings

$$\chi_\alpha : \{\text{conjugacy classes of } G\} \to \mathbb{C}. \tag{3.53}$$

Note also that the character of the identity element equals the dimension of the representation $\chi_\alpha(e) = \operatorname{tr} U_\alpha(e) = \operatorname{tr} \mathbb{1}_{V_\alpha} = \dim V_\alpha$, which is the dimension of the representation U_α.

Multiplying Eq. (3.39) by $\delta_{ij}\delta_{kl}$, and summing over i, j, k, and l, one obtains the following *orthogonality relation for characters*

$$\sum_{g \in G} \chi_\alpha^\star(g)\chi_\beta(g) = |G|\,\delta_{\alpha\beta}, \tag{3.54}$$

which can be used for the reduction of a representation into its irreducible components.

If an irreducible representation U_α appears multiple times in the direct sum that expresses a reducible representation U, e.g.,

$$U = U_1 \oplus U_1 \oplus U_1 \oplus U_2 \oplus U_3 \oplus \cdots , \tag{3.55}$$

then one can shorten the notation and multiply each irreducible representation by an integer n_α (called the *multiplicity* of the representation U_α in the decomposition of U) to account for how many times U_α appears

$$U = 3U_1 \oplus U_2 \oplus U_3 \oplus \cdots = \bigoplus_\alpha n_\alpha U_\alpha, \tag{3.56}$$

where the $\bigoplus_\alpha$ symbol is the analogue of $\sum_\alpha$ for direct sums, i.e., denotes direct summation over the α index.

Since the trace is a linear operation, the character χ of the representation satisfies

$$\chi = \sum_\alpha n_\alpha \chi_\alpha \tag{3.57}$$

with the same coefficients n_α. If one knows the character χ of the reducible representation U and the characters χ_α of all irreducible representations, then the multiplicities of each irreducible representation in the decomposition of U can be computed using Eq. (3.54):

$$n_\alpha = \frac{1}{|G|}\sum_{g \in G} \chi_\alpha^\star(g)\chi(g). \tag{3.58}$$

Thus, knowing all the multiplicities, the decomposition of the representation U is known. Note that if G has k conjugacy classes, denoted as $C_1, C_2, \ldots, C_k$, and $|C_i|$ is the number of group elements in C_i, then Eq. (3.58) can be rewritten as

$$n_\alpha = \frac{1}{|G|} \sum_{i=1}^{k} |C_i| \chi_\alpha^\star(C_i) \chi(C_i). \qquad (3.59)$$

In practice, characters of finite groups can be looked up from character tables.

The orthogonality of characters of the irreducible representations of finite groups can be interpreted as a useful orthogonality relation for certain vectors, called *character vectors*. Given the conjugacy classes $C_1, C_2, \ldots, C_k$, Eq. (3.54) implies

$$\sum_{i=1}^{k} |C_i| \chi_\alpha^\star(C_i) \chi_\beta(C_i) = |G|\, \delta_{\alpha\beta}. \qquad (3.60)$$

Consider the character vectors defined as

$$\vec{\chi}_\alpha = \left(\sqrt{|C_1|}\chi_\alpha(C_1),\ \sqrt{|C_2|}\chi_\alpha(C_2),\ \ldots,\ \sqrt{|C_k|}\chi_\alpha(C_k) \right) \in \mathbb{C}^k. \qquad (3.61)$$

The left-hand side of Eq. (3.60) is then the complex scalar product $\langle \vec{\chi}_\alpha | \vec{\chi}_\beta \rangle$ in $\mathbb{C}^k$. The number of vectors $\vec{\chi}_\alpha$ is the same as the number of irreducible representations. Equation (3.60) means that these vectors are mutually orthogonal, hence their number cannot be larger than the dimension of the vector space k, the number of conjugacy classes. Again, it can be shown that the numbers are actually the same.

Theorem 3.9 *The number of unitary irreducible representations of a finite group is the same as the number of its conjugacy classes.*

Then, it is possible to show the following theorem.

Theorem 3.10 *If G is a finite Abelian group, then all of its unitary irreducible representations have dimension one.*

Proof If the group is Abelian, then the conjugacy class of each element contains only the element itself: This is so because, for every g_0 and for every g, the fact that the group is Abelian implies $g g_0 g^{-1} = g_0 g g^{-1} = g_0$. Hence the number of conjugacy classes of a finite Abelian group is the same as the order $|G|$ of the group, and, according to Theorem 3.9, this is then also the number of unitary irreducible representations. Furthermore, Burnside's theorem 3.8 implies

$$\sum_{\alpha=1}^{|G|} (\dim U_\alpha)^2 = |G|. \qquad (3.62)$$

Since the number of addends on the left-hand side of Eq. (3.62) is $|G|$, and each of them, being the dimension of a representation, is a positive number, it must be $\dim U_\alpha = 1$ for every α. $\qquad \qquad \square$

In fact, it is possible to show that Theorem 3.10 holds for continuous Abelian groups, too (hence the word "finite" in the statement of Theorem 3.10 could be dropped).

Example Let us discuss how to use characters in practice, by constructing the character table for the S_3 group. In this character table, different rows correspond to different irreducible representations, and different columns correspond to different conjugacy classes. At their crossings, we insert the values of the characters for each conjugacy class in each irreducible representation.

Conjugacy classes of the permutation groups consist of permutations of the same cycle type. For S_3, they are represented by the following.

- The identity permutation e, which can be described by three cycles of length one as $(1)(2)(3)$.
- Transpositions (for example, $(12)(3)$, which interchanges the first two elements leaving the third unchanged) described by one cycle of length two and one cycle of length one.
- Cyclic permutations (for instance, (123), which maps the first element to the second, the second to the third, and the third to the first) described by one cycle of length three.

The only element in the first conjugacy class is the identity permutation, while the second conjugacy class contains three elements (the transpositions interchanging the three possible pairs out of three elements), and the last conjugacy class contains two elements (i.e., (123) and (132)). We label the three conjugacy classes as $1e$, $3(12)$, and $2(123)$, with the numbers in front denoting the number of elements in the conjugacy class.

According to Theorem 3.9, S_3 has then three irreducible representations. From our discussion in the example of application of Burnside's theorem 3.8, we know that S_3 could either have three irreducible representations (of which two of dimension 1, and one of dimension 2) or six irreducible representations, all of dimension 1. We can now rule out the latter possibility, and deduce that the dimensions of the irreducible representations must be 1, 1 and 2; let us denote them as **1**, **1′**, and **2**.

In general, we know that, for every group G, one of the one-dimensional representations is always the trivial representation, whereby all group elements are represented by the number 1. Also, the identity element of every group is always represented by the identity matrix of the corresponding dimension. For $G = S_3$, this information allows one to fill the first row and the first column of the character table: This is shown in Table 3.1, where the remaining entries are denoted by a, b, c, and d.

Let us then introduce the character vectors, which, according to Eq. (3.61), are defined starting from the rows of the character table, and multiplying each entry by the square root of the number of group elements in the corresponding conjugacy class:

$$\vec{\chi}_1 = (1, \sqrt{3}, \sqrt{2}),$$
$$\vec{\chi}_{1'} = (1, \sqrt{3}a, \sqrt{2}b),$$
$$\vec{\chi}_2 = (2, \sqrt{3}c, \sqrt{2}d).$$

Table 3.1. The character table of S_3 so far

	$1e$	$3(12)$	$2(123)$
1	1	1	1
1′	1	a	b
2	2	c	d

Table 3.2.	The character table of S_3		
	$1e$	$3(12)$	$2(123)$
1	1	1	1
1′	1	−1	1
2	2	0	−1

Then, the orthogonality relation for characters expressed in Eq. (3.60) yields the equations

$$\vec{\chi}_1 \cdot \vec{\chi}_{1'} = 1 + 3a + 2b = 0,$$
$$\vec{\chi}_1 \cdot \vec{\chi}_2 = 2 + 3c + 2d = 0,$$
$$\vec{\chi}_{1'} \cdot \vec{\chi}_2 = 2 + 3ac + 2bd = 0.$$

In order to solve this system of equations in four unknown (a priori complex) numbers, one would need further independent equations – for example, by imposing the normalization of the character vectors to $|G|$, according to Eq. (3.60). Instead, we note that for the S_3 group, the nontrivial one-dimensional representation **1′** can be defined as the one assigning $±1$ to each permutation, according to the parity of the permutation. Of the tree conjugacy classes, $3(12)$ is the one containing the odd permutations, while the other ones have the even permutations. Thus we set $a = -1$ and $b = 1$, which is a solution of the first equation above. The remaining two equations become

$$2 + 3c + 2d = 0,$$
$$2 - 3c + 2d = 0,$$

with solution $c = 0$, $d = -1$. So we arrive at the end result for the character table, shown in Table 3.2.

It is easy to verify that the character vectors are correctly normalized, according to Eq. (3.60):

$$\vec{\chi}_1 \cdot \vec{\chi}_1 = 1^2 + \left(\sqrt{3}\right)^2 + \left(\sqrt{2}\right)^2 = 6,$$
$$\vec{\chi}_{1'} \cdot \vec{\chi}_{1'} = 1^2 + \left(-\sqrt{3}\right)^2 + \left(\sqrt{2}\right)^2 = 6,$$
$$\vec{\chi}_2 \cdot \vec{\chi}_2 = 2^2 + 0^2 + \left(-\sqrt{2}\right)^2 = 6.$$

Finally, we show how to use the characters to solve the reduction problem for a representation. Consider the following three-dimensional representation U of S_3:

$$U(e) = \begin{pmatrix} 1 & 0 & 0 \\ 0 & 1 & 0 \\ 0 & 0 & 1 \end{pmatrix}, \quad U(12) = \begin{pmatrix} 0 & 1 & 0 \\ 1 & 0 & 0 \\ 0 & 0 & 1 \end{pmatrix},$$

$$U(13) = \begin{pmatrix} 0 & 0 & 1 \\ 0 & 1 & 0 \\ 1 & 0 & 0 \end{pmatrix}, \quad U(23) = \begin{pmatrix} 1 & 0 & 0 \\ 0 & 0 & 1 \\ 0 & 1 & 0 \end{pmatrix},$$

$$U(132) = \begin{pmatrix} 0 & 1 & 0 \\ 0 & 0 & 1 \\ 1 & 0 & 0 \end{pmatrix}, \quad U(123) = \begin{pmatrix} 0 & 0 & 1 \\ 1 & 0 & 0 \\ 0 & 1 & 0 \end{pmatrix}.$$

This representation is unitary, hence completely reducible. On the other hand, since it has dimension three, it cannot be irreducible.

One way to find the irreducible representations into which it can be decomposed, is by searching for a basis in which the matrices become (block) diagonal. This could be tedious. However, the reduction problem can be dramatically simplified by means of characters. Calculating the traces of a representative for each of the three conjugacy classes yields the character vector

$$\vec{\chi}_U = \left(3, \sqrt{3}, 0\right), \tag{3.63}$$

in which the components are in the same order as before (i.e., e, (12), (123)), and the $\sqrt{C_i}$ factor for each component has been taken into account, according to Eq. (3.61).

The decomposition of U into fundamental irreducible representations can then be obtained by means of Eq. (3.59), which yields

$$n_1 = \frac{1}{6} \sum_{i=1}^{3} |C_i| \chi_1^\star(C_i) \chi_U(C_i) = 1,$$

$$n_{1'} = \frac{1}{6} \sum_{i=1}^{3} |C_i| \chi_{1'}^\star(C_i) \chi_U(C_i) = 0,$$

$$n_2 = \frac{1}{6} \sum_{i=1}^{3} |C_i| \chi_2^\star(C_i) \chi_U(C_i) = 1,$$

or, equivalently, by noting that

$$\vec{\chi}_U = \left(3, \sqrt{3}, 0\right) = \left(1, \sqrt{3}, \sqrt{2}\right) + \left(2, 0, -\sqrt{2}\right) = \vec{\chi}_1 + \vec{\chi}_2, \tag{3.64}$$

so that the representation U decomposes into the direct sum of the trivial and the two-dimensional irreducible representation:

$$U = \mathbf{1} \oplus \mathbf{2}. \tag{3.65}$$

3.5 The Regular Representation

Given a group G, a particularly interesting representation is the *regular representation*. It is defined as the linear representation of the group that is induced by the action of the group on itself by a left or right translation. In the former case, one talks about a *left-regular representation*, while the latter defines the *right-regular representation*.

Note that, since the results of the multiplication of a group element g on the left (gh) or on the right (hg) of another arbitrary $h \in G$ are generally different (unless the group is Abelian), the left-regular and right-regular representations are generally different. However, as we will discuss below, they turn out to be closely related to each other.

Let us consider a finite group G of order $|G|$, and let $V \cong \mathbb{C}^{|G|}$ be a $|G|$-dimensional complex vector space, on which the regular representation will act. Given a basis in V, one can label the vectors of a basis (which can be chosen to be orthonormalized) by the elements of G, i.e., $\{\vec{e}_g \,|g \in G\}$. Thus every element of G is unambiguously associated with a basis vector of V. Now suppose that the group G acts on itself by the left translation $L : G \to \mathrm{Perm}(G)$; every element g is associated with a permutation L_g, acting on G as

$$L_g : G \mapsto G, \qquad h \mapsto L_g(h) = gh. \tag{3.66}$$

Now we define a $|G|$-dimensional representation in V. A group element g is represented by an automorphism $D(g) \in \mathrm{Aut}(V)$ defined first by its action on the basis vectors $\vec{e}_h$ labeled by the elements $h \in G$

$$D(g) : V \mapsto V, \qquad \vec{e}_h \mapsto D(g)\vec{e}_h = \vec{e}_{gh}, \tag{3.67}$$

and then extending linearly, so that

$$D(g)(\vec{v}) = D(g)\left(\sum_{h \in G} v_h \vec{e}_h\right) = \sum_{h \in G} v_h D(g)\vec{e}_h = \sum_{h \in G} v_h \vec{e}_{gh}. \tag{3.68}$$

From Eq. (3.67), one can find the $|G| \times |G|$ matrix $D(g)$ that represents the element g. Using the fact that the entry of $D(g)$ in the row labeled by g' and in the column labeled by h can be written as

$$\begin{aligned} D(g)_{g',h} &= \left(\vec{e}_{g'}\right)^{\mathrm{T}} D(g)\vec{e}_h \\ &= \left(\vec{e}_{g'}\right)^{\mathrm{T}} \vec{e}_{gh}, \end{aligned} \tag{3.69}$$

and using the fact that the $\vec{e}_i$ vectors are orthonormalized, we obtain

$$D(g)_{g',h} = \begin{cases} 1, & \text{if } g' = gh \\ 0, & \text{if } g' \neq gh \end{cases}. \tag{3.70}$$

Thus, in the regular representation, the elements g are represented by matrices where every element is 1 or 0. Note also that each row of the $D(g)$ matrix has exactly one nonvanishing entry, which is equal to 1. When $g = e$, the unit element of the group, the only entries of $D(e)$ that are equal to 1 are those for $g' = h$, i.e., the diagonal ones. The unit element e is thus represented by the $|G|$-dimensional identity matrix, whose trace is $|G|$.

On the other hand, when g is not the unit element of G, $g' = gh$ is *always* different from h, otherwise, by multiplying both sides of the equality $gh = h$ by h^{-1}, one would have $g = ghh^{-1} = hh^{-1} = e$, in contradiction with the assumption, and therefore in this case all diagonal elements of $D(g)$ vanish, hence $D(g)$ is traceless. Thus the character $\chi(g) = \mathrm{tr}\, D(g)$ of the regular representation has the property

$$\chi(g) = |G|\delta_{g,e}. \tag{3.71}$$

It is easy to check (see Problem 3.4) that the matrices of the left-regular representation obey the group multiplication law, i.e., that $D(g)D(g') = D(gg')$.

Example   Consider the cyclic groups $\mathbb{Z}_n = \{e, a, a^2, \ldots, a^{n-1}\}$. We have $a^{i+1} = aa^i$, so $D(a)_{i,j} = \delta_{i,j+1}$, where it is assumed that the matrix entry is nonvanishing when the $i = j + 1$ relation holds modulo n, i.e., that $D(a)_{i+1,i} = 1$ for every i from 0 to $n - 2$, and also $D(a)_{0,n-1} = 1$. Thus the generating element is represented by the $n \times n$ matrix

$$D(a) = \begin{pmatrix} 0 & 0 & 0 & \cdots & 0 & 0 & 1 \\ 1 & 0 & 0 & \cdots & 0 & 0 & 0 \\ 0 & 1 & 0 & \cdots & 0 & 0 & 0 \\ \cdots & \cdots & \cdots & \cdots & \cdots & \cdots & \cdots \\ 0 & 0 & 0 & \cdots & 1 & 0 & 0 \\ 0 & 0 & 0 & \cdots & 0 & 1 & 0 \end{pmatrix} \tag{3.72}$$

and $D(a^n) = (D(a))^n$.

The right-regular representation is defined similarly. We start with right transla-tions, but acting from the right with *the inverse element*

$$h \mapsto R_g(h) = hg^{-1}. \tag{3.73}$$

The reason for the presence of g^{-1} (instead of g) on the right-hand side of Eq. (3.73) is given shortly. Denoting the right-regular representation with D', we define its action in V by

$$\vec{e}_h \mapsto D'(g)\vec{e}_h = \vec{e}_{hg^{-1}}, \tag{3.74}$$

so that Eq. (3.70) is replaced by

$$D'(g)_{g'h} = \begin{cases} 1 \, , \; \text{if } g' = hg^{-1} \\ 0 \, , \; \text{if } g' \neq hg^{-1}. \end{cases} \tag{3.75}$$

The reason for the inverse of g appearing in Eq. (3.73) is that the right-regular representation preserves the order of group multiplication, $D'(g)D'(g') = D'(gg')$. We leave the proof as an exercise, in Problem 3.5.

One can prove (see Problem 3.6) that the left-regular and right-regular representa-tions are isomorphic to each other, therefore instead of speaking of the left-regular and right-regular representations separately, one usually speaks about the regular representation, denoting it as D.

The regular representation D is in general reducible, and it is particularly instructive to study its decomposition into its irreducible components, which we denote as D_α:

$$D = \bigoplus_\alpha n_\alpha D_\alpha, \tag{3.76}$$

where, as in Eq. (3.56), the multiplicity coefficient n_α counts how many times the corresponding irreducible representation D_α appears in the decomposition. The multiplicities n_α can be evaluated by means of Eq. (3.58):

$$n_\alpha = \frac{1}{|G|} \sum_{g \in G} \chi_\alpha^\star(g)\chi(g),$$

where $\chi_\alpha^\star(g)$ denotes the complex conjugate of the character of g in the representa-tion D_α, while $\chi(g)$ is the character of g in the regular representation D. As discussed above, however, the characters in the regular representation are all vanishing, except for the character of the unit element of G, which is equal to $|G|$. From this one obtains

$$n_\alpha = \frac{1}{|G|}|G|\chi_\alpha^\star(e) = \chi_\alpha^\star(e) = \dim D_\alpha,$$

having used the fact that unit element of the group is always represented by the identity matrix, and hence its character equals the dimension of the representation.

Then, Eq. (3.76) implies that $\chi(g) = \sum_\alpha n_\alpha \chi_\alpha(g)$, and evaluating this identity for $g = e$ we finally obtain

$$|G| = \sum_\alpha (\dim D_\alpha)^2 , \tag{3.77}$$

in agreement with Burnside's theorem 3.8.

The idea behind regular representations can be extended to define *permutation representations*. Let G be a a finite group acting on a finite set X,

$$X \ni x \mapsto gx, \tag{3.78}$$

where $g \in G$. We define a $|X|$-dimensional representation of G in $\mathbb{C}^{|X|}$ as follows. We pick an orthonormal basis in $\mathbb{C}^{|X|}$ and label the basis vectors by the elements of X: $\{\vec{e}_x \,|\, x \in X\}$. The *permutation representation* P of G is defined by its action on basis vectors

$$P(g) : \mathbb{C}^{|X|} \mapsto \mathbb{C}^{|X|},$$
$$\vec{e}_x \mapsto P(g)\vec{e}_x = \vec{e}_{gx}, \tag{3.79}$$

and extending linearly for all vectors $\vec{v} \in \mathbb{C}^{|X|}$, as before. The matrix elements $P(g)_{yx}$ are thus given by $P(g)_{yx} = \delta_{y,gx}$.

We conclude this chapter by continuing our discussion of vector spaces. In the next section we introduce dual vectors and tensors, which are a generalization of the concept of a vector, and discuss their properties. They are very important in quantum mechanics, where they are usually presented using the notation introduced by Paul Dirac.

3.6 Dual Vectors and Tensors

Given a complex vector space V of dimension n, the *dual vector space* $V^\star$ is defined as the set of linear functions from V to $\mathbb{C}$:

$$V^\star = \left\{ f : V \to \mathbb{C} \,|\, \forall \vec{v}, \vec{w} \in V, \ \forall \alpha, \beta \in \mathbb{C} : f\left(\alpha\vec{v} + \beta\vec{w}\right) = \alpha f(\vec{v}) + \beta f(\vec{w}) \right\}. \tag{3.80}$$

$V^\star$ is itself a vector space, with the sum given by the sum of functions, and the multiplication by a complex number simply defined as the multiplication of a function by a complex number. The null vector of $V^\star$ is the null function:

- $(f_1 + f_2)(\vec{v}) = f_1(\vec{v}) + f_2(\vec{v})$, for all f_1 and f_2 in $V^\star$ and for all $v \in V$,
- $(\mu \cdot f)(\vec{v}) = \mu \cdot (f(\vec{v}))$, for all f in $V^\star$, for all $\mu \in \mathbb{C}$, and for all $\vec{v}$,
- $0_{V^\star}(\vec{v}) = 0$, for all $\vec{v} \in V$.

Each element of $V^\star$ is a *dual vector*.

Example   Dual vectors play a very important rôle in quantum mechanics. Let us consider a Hilbert space V (which, for simplicity, we assume to be finite-dimensional, although this is not an essential restriction). In particular, it is a complex vector space with a scalar product. In Dirac's notation, the vectors are represented by the *ket* notation, namely, instead of writing $v \in V$ we write $|\psi\rangle \in V$. For every ket vector $|\psi\rangle$ there is a corresponding *bra* vector $\langle\psi|$. The bra vector is an element of the dual vector space $V^\star$; indeed, using the scalar product of the Hilbert space, every bra vector $\langle\psi|$ can be interpreted as a linear map acting as

$$\langle\psi| : V \mapsto \mathbb{C},$$
$$|\phi\rangle \mapsto \langle\psi|\phi\rangle. \tag{3.81}$$

The fact that Eq. (3.81) defines a linear function follows from the linearity of the scalar product in its second argument:

$$\forall \alpha_1, \alpha_2 \in \mathbb{C}, \ \forall |\phi_1\rangle, |\phi_2\rangle \in V: \ \langle\psi|\alpha_1\phi_1 + \alpha_2\phi_2\rangle = \alpha_1\langle\psi|\phi_1\rangle + \alpha_2\langle\psi|\phi_2\rangle, \tag{3.82}$$

which holds for every $\langle\psi| \in V^\star$.

Given a basis $\{\vec{e}_1, \ldots, \vec{e}_n\}$ of V, each vector $\vec{v}$ of V can be written as $\vec{v} = v^i\vec{e}_i$. We can then define a *dual vector basis* as the set $\{e^{\star 1}, e^{\star 2}, \ldots, e^{\star n}\}$ of linear functions from V to $\mathbb{C}$ such that

$$e^{\star i}\left(\vec{e}_j\right) = \delta^i{}_j. \tag{3.83}$$

Lemma 3.11 *The vectors $\left\{e^{\star 1}, \ldots, e^{\star n}\right\}$ are linearly independent.*

Proof The lemma can be proven by *reductio ad absurdum*. Assume that the $\left\{e^{\star 1}, \ldots, e^{\star n}\right\}$ vectors are not linearly independent: Then there exist coefficients $\mu_1, \ldots, \mu_n$ such that

$$\sum_{j=1}^{n} \mu_j e^{\star j} = 0 \tag{3.84}$$

and at least one of the coefficients μ_j is different from zero. However,

$$\sum_{j=1}^{n} \mu_j e^{\star j}(\vec{e}_i) = \mu_i = 0 \tag{3.85}$$

for all $i = 1, \ldots, n$, which is a contradiction. $\qquad\square$

Lemma 3.12 *Every dual vector $f \in V^\star$ can be expanded in the $\left\{e^{\star 1}, \ldots, e^{\star n}\right\}$ basis as*

$$f = \sum_{j=1}^{n} f_j e^{\star j} \tag{3.86}$$

with uniquely defined coefficients

$$f_j = f(e_j). \tag{3.87}$$

Proof Consider an arbitrary linear function $f : V \to \mathbb{C}$ and an arbitrary vector $\vec{v} = \sum_j v^j\vec{e}_j$; then the linearity of f implies

$$f(\vec{v}) = f\left(\sum_j v^j\vec{e}_j\right) = \sum_j v^j f\left(\vec{e}_j\right) = \sum_j v_j f_j, \tag{3.88}$$

where

$$f\left(\vec{e}_j\right) = f_j. \tag{3.89}$$

Then Eq. (3.88) can be rewritten as

$$f(\vec{v}) = \sum_{j=1}^n v^j f_j = \sum_{j=1}^n \sum_{i=1}^n v_j f_i \delta^i_{\ j} = \sum_{i=1}^n \sum_{j=1}^n v_j f_i e^{\star i}\left(\vec{e}_j\right)$$

$$= \sum_{i=1}^n f_i e^{\star i} \left(\sum_{j=1}^n v_j \vec{e}_j \right) = \sum_{i=1}^n f_i e^{\star i}\left(\vec{v}\right), \tag{3.90}$$

using Eq. (3.83) and the linearity of the $e^{\star i}$ functions. As Eq. (3.90) holds for every $\vec{v} \in V$, it is an equality among the functions:

$$f = \sum_{i=1}^n f_i e^{\star i}. \tag{3.91}$$

Since $e^{\star 1}, \ldots, e^{\star n}$ are linearly independent, the linear combination (3.91) is unique.

□

The above lemmas establish that $\left\{e^{\star 1}, e^{\star 2}, \ldots, e^{\star n}\right\}$ is a basis for $V^\star$, called the *dual basis*, and that

$$\dim V^\star = \dim V. \tag{3.92}$$

Example   An electron is a spin $j = 1/2$ particle, with spin quantum numbers $m = \pm\, 1/2$ ("spin up" and "spin down"). In general, the state of an electron can be a superposition of the two basis states; in Dirac ket notation we write the quantum state as

$$|\psi\rangle = a\,|j = 1/2, m = 1/2\rangle + b\,|j = 1/2, m = -1/2\rangle, \tag{3.93}$$

where a and b are complex coefficients satisfying the unit normalization condition $|a|^2 + |b|^2 = 1$. Except for the normalization condition, the state is thus a vector in the two-dimensional complex vector space $V = \mathbb{C}^2$. Instead of the Dirac notation, one can write the state as a vector $\vec{v} = (v^1, v^2) = (a, b) \in \mathbb{C}^2$, and denote the basis vectors as

$$e_1 \equiv |j = 1/2, m = 1/2\rangle, \qquad e_2 \equiv |j = 1/2, m = -1/2\rangle, \tag{3.94}$$

thus

$$|\psi\rangle \equiv \vec{v} = a e_1 + b e_2 \in \mathbb{C}^2. \tag{3.95}$$

Instead of an electron, we could consider an arbitrary quantum mechanical system with a two-dimensional Hilbert space $\mathbb{C}^2$ (dropping the unit normalization for the moment), so that all vectors (states) are of the above form.

In the context of quantum information theory, each unit normalized vector is called a *qubit*, which is short for "quantum bit." A classical bit is a state (for example, in a computer memory) that has two possible values, 0 or 1. If we introduce yet another notation for the two basis vectors of $\mathbb{C}^2$,

$$|0\rangle \equiv e_1, \qquad |1\rangle \equiv e_2, \tag{3.96}$$

inspired by the two states of a classical bit, then a qubit state is a linear combination

$$|\psi\rangle \equiv \vec{v} = a|0\rangle + b|1\rangle \tag{3.97}$$

with infinite possibilities for the components a and b satisfying the normalization $|a|^2 + |b|^2 = 1$. The dual basis vectors of $(\mathbb{C}^2)^\star$ are the bra states

$$e^{\star 1} = \langle 0|, \qquad e^{\star 2} = \langle 1|, \tag{3.98}$$

and the dual vector $\langle\psi|$, the bra state counterpart of the ket $|\psi\rangle$, has the expansion

$$\langle\psi| = a^\star e^{\star 1} + b^\star e^{\star 2} = a^\star\langle 0| + b^\star\langle 1|. \tag{3.99}$$

The unit normalization condition is then

$$1 = \langle\psi|\psi\rangle = a^\star a \underbrace{e^{\star 1}(e_1)}_{=1} + a^\star b \underbrace{e^{\star 1}(e_2)}_{=0} + b^\star a \underbrace{e^{\star 1}(e_2)}_{=0} + b^\star b \underbrace{e^{\star 2}(e_2)}_{=1}$$

$$= |a|^2 + |b|^2 = (\mathrm{Re}\, a)^2 + (\mathrm{Im}\, a)^2 + (\mathrm{Re}\, b)^2 + (\mathrm{Im}\, b)^2. \tag{3.100}$$

Two complex numbers correspond to four real parameters (the real and imaginary parts). The normalization condition reduces the number of the independent real parameters down to three, parameterizing a three-sphere S^3. One possible convention to parameterize the normalized unit vector is

$$|\psi\rangle \equiv \vec{v} = e^{i\gamma}\left[\cos\left(\frac{\theta}{2}\right)|0\rangle + e^{i\phi}\sin\left(\frac{\theta}{2}\right)|1\rangle\right], \tag{3.101}$$

where γ, θ, and ϕ are real numbers. However, it is possible to use an even more economical parameterization, since the overall phase of a physical state in quantum mechanics is not observable, and all states $e^{i\gamma}|\psi\rangle$ are equivalent descriptions of the same physical state. Therefore, the overall phase factor $\exp(i\gamma)$ appearing on the right-hand side of Eq. (3.101) can be ignored, and the normalized state can be parameterized as

$$|\psi\rangle \equiv \vec{v} = \cos\left(\frac{\theta}{2}\right)|0\rangle + e^{i\phi}\sin\left(\frac{\theta}{2}\right)|1\rangle. \tag{3.102}$$

So far we have not specified the range of the parameters. It turns out that $\theta \in [0, \pi], \phi \in [0, 2\pi)$, so in other words they are the two angles that customarily are used to parameterize the two-sphere S^2. In this context the two-sphere is called the *Bloch sphere*, named after the Swiss-American physicist Felix Bloch, and shown in Fig. 3.1. If we think of S^2 as the unit sphere in $\mathbb{R}^3$ with $x^2 + y^2 + z^2 = 1$ using the standard Cartesian coordinates x, y, z, a point on the sphere is specified by the vector $(x, y, z) = (\sin\theta\cos\phi, \sin\theta\sin\phi, \cos\theta)$, called the *Bloch vector* in this context. In passing, we also mention that the choice of coordinates in Eq. (3.101) and the reduction from the three-sphere S^3 to the two-sphere S^2 is also related to something called the *Hopf fibration*, which we shall discuss in Section 4.2. Parameterizing the states in terms of the Bloch sphere defined in Eq. (3.102) is convenient, since one can show that opposite points on the Bloch sphere correspond to orthogonal states; for example, the North pole $\theta = 0$ is $|0\rangle$ and the South pole $\theta = \pi$ is $|1\rangle$. More generally, given a generic qubit state $|\alpha\rangle$, described by parameters θ and ϕ, the

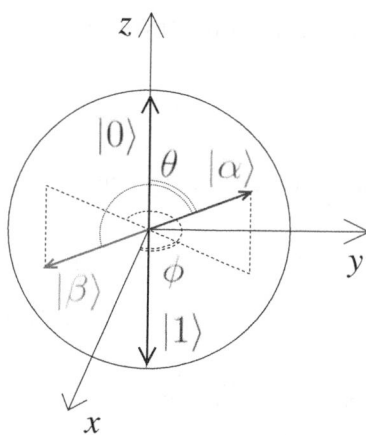

Fig. 3.1   The Bloch sphere.

antipodal point of the corresponding Bloch vector corresponds to angles $(\pi - \theta)$ and $(\phi + \pi)$, respectively, and is associated with the qubit

$$|\beta\rangle = \cos\left(\frac{\pi - \theta}{2}\right)|0\rangle + \exp[i(\phi + \pi)]\sin\left(\frac{\pi - \theta}{2}\right)|1\rangle$$

$$= \sin\left(\frac{\theta}{2}\right)|0\rangle - \exp(i\phi)\cos\left(\frac{\theta}{2}\right)|1\rangle, \tag{3.103}$$

and it is then straightforward to show that

$$\langle\beta|\alpha\rangle = \left[\sin\left(\frac{\theta}{2}\right)\langle0| - \exp(-i\phi)\cos\left(\frac{\theta}{2}\right)\langle1|\right]\left[\cos\left(\frac{\theta}{2}\right)|0\rangle + \exp(i\phi)\sin\left(\frac{\theta}{2}\right)|1\rangle\right]$$

$$= 0. \tag{3.104}$$

Thus moving from one orthonormal basis to another orthonormal basis corresponds to a rotation on the Bloch sphere.

The expression

$$f(\vec{v}) = \sum_{i,j} f_i e^{\star i}(v^j \vec{e}_j) = \sum_{i,j} f_i v^j e^{\star i}(\vec{e}_j) = \sum_i f_i v^i \tag{3.105}$$

can be interpreted as an *inner product*:

$$\langle\,\cdot\,,\cdot\,\rangle:\ V^{\star} \times V \mapsto \mathbb{C},$$

$$(f, \vec{v}) \mapsto \langle f, \vec{v}\rangle = f(\vec{v}) = \sum_i f_i v^i. \tag{3.106}$$

Note that the inner product is formally different from the scalar product $\langle\cdot|\cdot\rangle : V \times V \to \mathbb{C}$. Whereas the scalar product receives as its argument an ordered pair of vectors (v, w), the arguments of the inner product are a dual vector f and a vector v.

Example In quantum mechanics, the scalar product is constructed via an inner product. Consider two state vectors $\vec{v} \equiv |\psi\rangle$ and $\vec{w} \equiv |\chi\rangle$. For the latter, the corresponding dual vector is the bra $f_{\vec{w}} \equiv \langle\chi|$. The inner product $f_{\vec{w}}(v) \equiv \langle\chi|\psi\rangle$ is the same as the scalar product of the two vectors. Let $e_i \equiv |i\rangle$, for $i = 1, \ldots, n,$

be the basis vectors of the Hilbert space, assuming here a finite dimension n, and expand the state vectors

$$\vec{v} \equiv |\psi\rangle = \sum_i \psi_i |i\rangle, \quad \vec{w} = |\chi\rangle = \sum_i \chi_i |i\rangle. \tag{3.107}$$

In component notation, we can think of these as column vectors, e.g.,

$$|\psi\rangle = \begin{pmatrix} \psi_1 \\ \psi_2 \\ \vdots \\ \psi_n \end{pmatrix}. \tag{3.108}$$

The dual vector of $|\chi\rangle$, the bra $f_{\vec{w}} = \langle\chi|$, is the adjoint row vector $\vec{w}^\dagger = |\chi\rangle^\dagger$,

$$f_{\vec{w}} = \langle\chi| = \vec{w}^\dagger = (\chi_1^\star, \chi_2^\star, \ldots, \chi_n^\star), \tag{3.109}$$

and we can see explicitly that the inner product can be identified with the complex scalar product

$$\langle\chi|\psi\rangle = \vec{w}^\dagger \vec{v} = (\chi_1^\star, \chi_2^\star, \ldots, \chi_n^\star) \begin{pmatrix} \psi_1 \\ \psi_2 \\ \vdots \\ \psi_n \end{pmatrix} = \sum_{i=1}^n \chi_i^\star \psi_i. \tag{3.110}$$

Consider two vector spaces V and W, and the linear map $f : V \to W$; furthermore, let g be a linear function from W to $\mathbb{C}$, i.e., $g \in W^\star$. Then, also the composition map $g \circ f : V \to \mathbb{C}$ (where the notation means that f is applied first, then g: $(g \circ f)(\vec{v}) = g(f(\vec{v}))$ for every $\vec{v} \in V$) is linear:

$$g \circ f \in V^\star. \tag{3.111}$$

This can be rephrased by saying that the linear map $f : V \to W$ induces a map $f^\star$ from the dual vector spaces of W and V that is defined as

$$f^\star : W^\star \mapsto V^\star,$$
$$g \mapsto g \circ f, \tag{3.112}$$

where $f^\star(g)$ is called the *pullback* of g.

Note that, for every $\vec{v} \in V$, one can use the inner product $\langle\cdot, \cdot\rangle$ to define a linear function

$$\omega_{\vec{v}} : V^\star \mapsto \mathbb{C},$$
$$f \mapsto \omega_{\vec{v}}(f) = \langle f, v \rangle. \tag{3.113}$$

The bilinearity of the $\langle\cdot, \cdot\rangle$ inner product implies that $\omega_{\vec{v}}$ is an element of $(V^\star)^\star$, the *dual of the dual* of the vector space V. In fact, it is an instructive exercise (see Problem 3.7) to prove that *every* element of $(V^\star)^\star$, i.e., every linear function from $V^\star$ to $\mathbb{C}$ can be constructed in this way, i.e., associated with a vector $\vec{v} \in V$. Thus $(V^\star)^\star$, the dual of the dual of the vector space V, can be identified with V itself: $(V^\star)^\star = V$. We can then define an action of a vector $\vec{v}$ on a dual vector f by Eq. (3.113)

$$\vec{v} : V^\star \mapsto \mathbb{C},$$
$$f \mapsto \vec{v}(f) = \langle f, \vec{v} \rangle = f(\vec{v}). \tag{3.114}$$

Example   In quantum mechanics, we saw that dual vector bra states are obtained from vector ket states by the adjoint, $\langle\psi| = |\psi\rangle^\dagger$. It is easy to see that the dual of the dual is the vector itself, $(|\psi\rangle^\dagger)^\dagger = \langle\psi|^\dagger = |\psi\rangle$.

Given a complex vector space V, one can generalize the notions of vector and dual vector and introduce a *tensor of type* (p, q), also called a (p, q) tensor. We begin with formal definitions, and then proceed to some more transparent ways to construct tensors. First, we define a tensor of type (p, q) as a function of p dual vectors and q vectors that is multilinear, i.e., linear in each of its arguments,

$$T : \overbrace{V^\star \times \cdots \times V^\star}^{p} \times \overbrace{V \times \cdots \times V}^{q} \mapsto \mathbb{C}. \tag{3.115}$$

Every linear combination $aT + bU$ of (p, q) tensors T and U is also a (p, q) tensor, hence tensors of type (p, q) form a vector space; this follows from the fact that linear combinations preserve the multilinearity of (3.115).

Example   Let us discuss some examples of tensors.

- A tensor T of type $(0, 1)$ is a linear function from V to $\mathbb{C}$, i.e., a dual vector: $T \in V^\star$.
- A tensor T of type $(1, 0)$ is a linear function from $V^\star$ to $\mathbb{C}$, i.e., an element of the dual of the dual of V. As discussed above, $(V^\star)^\star$ can be identified with V itself, so T can be interpreted as an element of V itself.
- A tensor T of type $(1, 2)$ is a linear function $R : V^\star \times V \times V \to \mathbb{C}$. Choosing the basis $\{\vec{e}_i\}$ in V and the dual basis $\{e^{\star i}\}$ in $V^\star$, one can then write

$$T(f, \vec{v}, \vec{w}) = T\left(f_i e^{\star i}, v^j \vec{e}_j, w^k \vec{e}_k\right) = T^i_{jk} f_i v^j w^k, \tag{3.116}$$

where T^i_{jk} are the *components of the tensor* in the chosen basis: They are defined as

$$T^i_{jk} = T(e^{\star i}, \vec{e}_j, \vec{e}_k) \tag{3.117}$$

and are basis-dependent quantities. For a given basis, once all of the tensor components are specified, the tensor is uniquely determined.

Note the positioning of the indices in Eq. (3.117): In general, the components of tensors of type (p, q) are defined as

$$T^{i_1 \cdots i_p}{}_{j_1 \cdots j_q} = T\left(e^{\star i_1}, \ldots, e^{\star i_p}, \vec{e}_{j_1}, \ldots \vec{e}_{j_q}\right) \tag{3.118}$$

and have p upper indices and q lower indices. Note also that the components of a tensor form a multidimensional array, generalizing the two-dimensional array of a matrix.

Tensors with more indices can be constructed from tensors with fewer indices through the *tensor product* operation: Let R be a tensor of type (p, q) and let S be a tensor of type (l, m). Their tensor product $T = R \otimes S$ is defined as the tensor of type $(p + l, q + m)$ such that

$$\begin{aligned}
&T(f_1, \ldots, f_p, f_{p+1}, \ldots, f_{p+l}, \vec{v}_1, \ldots, \vec{v}_q, \vec{v}_{q+1}, \ldots, \vec{v}_{q+m}) \\
&= R(f_1, \ldots, f_p, \vec{v}_1, \ldots, \vec{v}_q) S(f_{p+1}, \ldots, f_{p+l}, \vec{v}_{q+1}, \ldots, \vec{v}_{q+m}). \tag{3.119}
\end{aligned}$$

Note that the tensor product is a bilinear operation:

$$(a_1 R_1 + a_2 R_2) \otimes (b_1 S_1 + b_2 S_2) \tag{3.120}$$
$$= a_1 b_1 R_1 \otimes S_1 + a_1 b_2 R_1 \otimes S_2 + a_2 b_1 R_2 \otimes S_1 + a_2 b_2 R_2 \otimes S_2,$$

for all complex coefficients a_1, a_2, b_1 and b_2.

In terms of components, Eq. (3.119) can be written as:

$$T^{i_1 \ldots i_p i_{p+1} \ldots i_{p+l}}{}_{j_1 \ldots j_q j_{q+1} \ldots j_{q+m}} = R^{i_1 \ldots i_p}{}_{j_1 \ldots j_q} S^{i_{p+1} \ldots i_{p+l}}{}_{j_{q+1} \ldots j_{q+m}}. \tag{3.121}$$

The definition looks complicated, so let us consider some simple examples.

Example The basis vectors e_i are $(1,0)$ tensors. Writing $e_i = \delta_i^j e_j$ we can see that the components of e_i are all 0, except for the ith component, which is 1. In other words, the components of e_i are δ_i^j. Likewise, the dual basis vectors $e^{\star k}$ are $(0,1)$ tensors, with components δ_l^k. One can construct a $(1,1)$ tensor by constructing a tensor product $e_i \otimes e^{\star k}$. According to the definition (3.119), it acts on f and $\vec{v}$ as

$$e_i \otimes e^{\star k}(f, \vec{v}) = e_i(f) e^{\star k}(\vec{v}) = f(e_i) e^{\star k}(\vec{v}) = f_i v^k, \tag{3.122}$$

having used Eq. (3.114) in the second equality. On the other hand, Eq. (3.121) implies that the components of the product tensor $e_i \otimes e^{\star k}$ are $\delta_i^j \delta_l^k$.

Example Moving to quantum mechanics and Dirac notation, $|i\rangle = e_i$, while $\langle k| = e^{\star k}$, the $(1,1)$ tensors introduced above can be written as

$$e_i \otimes e^{\star k} \equiv |i\rangle \otimes \langle k| \equiv |i\rangle\langle k|. \tag{3.123}$$

Their action on a dual vector $\langle \chi|$ and on a vector $|\psi\rangle$ is

$$\langle \chi|, |\psi\rangle \mapsto \langle \chi|i\rangle\langle k|\psi\rangle = \chi_k^\star \psi_i, \tag{3.124}$$

i.e., it projects out the product of their kth and ith components. In quantum-mechanical notation, it is common to drop the tensor product symbol $\otimes$.

Example Next, we consider a $(1,0)$ tensor $\vec{w}$ with components w^i in a given basis, and a $(0,1)$ tensor g with components g_k in the corresponding dual basis, and form a $(1,1)$ tensor $\vec{w} \otimes g$. According to Eq. (3.119), and using the bilinearity expressed in Eq. (3.120), we obtain

$$\vec{w} \otimes g(f, \vec{v}) = \vec{w}(f) g(\vec{v}) = w^i e_i(f) g_k e^{\star k}(\vec{v}) = w^i f(e_i) g_k e^{\star k}(\vec{v}) = w^i f_i g_k v^k, \tag{3.125}$$

showing that the components of $\vec{w} \otimes g$ are $w^i g_k$.

Example The previous examples suggest that $e_i \otimes e^{\star k}$ form a basis for the vector space of $(1,1)$ tensors. In other words, a $(1,1)$ tensor T can be expanded as

$$T = \sum_{ik} T^i{}_k e_i \otimes e^{\star k}, \tag{3.126}$$

where $T^i{}_k$ are its components. The same expression can be rewritten using Dirac notation:

$$T = \sum_{ik} T^i{}_k |i\rangle\langle k|. \tag{3.127}$$

Let us check the consistency of this expansion

$$T(e^{\star i}, e_k) = \sum_{jl} T^j{}_l e_j \otimes e^{\star l}(e^{\star i}, e_k) = \sum_{jl} T^j{}_l e_j(e^{\star i}) e^{\star l}(e_k) = T^i{}_k, \tag{3.128}$$

so we recover the components. Using Dirac notation,

$$T(e^{\star i}, e_k) = \langle i|T|k\rangle = \sum_{jl} T^j{}_l \langle i|j\rangle\langle l|k\rangle = T^i{}_k. \tag{3.129}$$

Also,

$$T(f, \vec{v}) = T^j{}_l e_j \otimes e^{\star l}(f, \vec{v}) = T^j{}_l f_j v^l, \tag{3.130}$$

as expected. Once again, using Dirac notation with $f = \langle \chi|$, $\vec{v} = |\psi\rangle$,

$$T(f, \vec{v}) = \langle \chi|T|\psi\rangle = \sum_{jl} T^j{}_l \langle \chi|j\rangle\langle l|\psi\rangle = \sum_{jl} T^j{}_l \chi_j^\star \psi_l. \tag{3.131}$$

We can form repeated tensor products; the tensor product is associative:

$$R \otimes S \otimes T = (R \otimes S) \otimes T = R \otimes (S \otimes T). \tag{3.132}$$

Finally, we define the operation of *contraction* of a tensor: This is an operation that, when applied to a tensor of type (p, q), produces a tensor of type $(p-1, q-1)$

$$\underbrace{T}_{(p,q)} \mapsto \underbrace{T_{c(ij)}}_{(p-1,q-1)}, \tag{3.133}$$

where $T_{c(ij)}$ is a tensor of type $(p-1, q-1)$ and is defined as

$$T_{c(ij)}(f_1, \ldots, f_{p-1}; \vec{v}_1, \ldots, \vec{v}_{q-1})$$

$$= \sum_k T(f_1, \ldots, \overbrace{e^{\star k}}^{i\text{th}}, \ldots, f_{p-1}; \vec{v}_1, \ldots, \overbrace{\vec{e}_k}^{j\text{th}}, \ldots, \vec{v}_{q-1}). \tag{3.134}$$

Note the sum over k on the right-hand side of Eq. (3.134). This equation can be rewritten in component form as

$$T_{c(ij)}{}^{l_1 \ldots l_{p-1}}{}_{m_1 \ldots m_{q-1}} = \sum_k T^{l_1 \ldots l_{i-1} k l_i \ldots l_{p-1}}{}_{m_1 \ldots m_{j-1} k m_j \ldots m_{q-1}}. \tag{3.135}$$

Thus, in the contraction one chooses the ith upper index and the jth lower index to be the same label k, which is then summed over.

3.6.1 Visualizing Contractions by Tensor Diagrams

Besides contracting indices of a single tensor, one can also consider products of various types of tensors and contract indices among them. Then it becomes helpful to visualize the contracted tensors with *tensor diagrams*, also known as *Penrose graphical notation*, proposed by Roger Penrose. The starting point is to represent a particular tensor with a specific shape, with lines sticking out representing the indices of its components. For example, a $(2, 1)$ tensor M with components $M^{\lambda \alpha}{}_\beta$ could be represented by, say, a square, with three lines sticking out representing the indices.

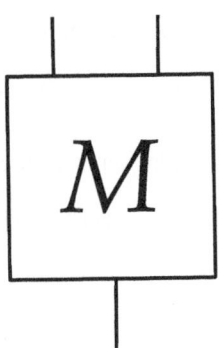

Fig. 3.2 A diagram representing the components $M^{\lambda\alpha}{}_{\beta}$.

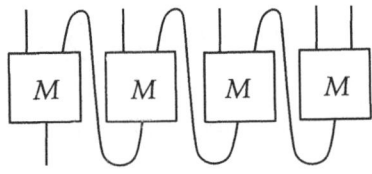

Fig. 3.3 A diagram representing the product $M^{\lambda_1\alpha_1}{}_{\beta}M^{\lambda_2\alpha_2}{}_{\alpha_1}M^{\lambda_3\alpha_3}{}_{\alpha_2}M^{\lambda_4\alpha}{}_{\alpha_3}$.

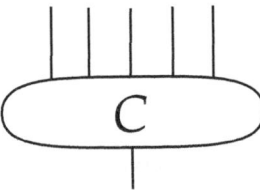

Fig. 3.4 A diagram representing $C^{\lambda_1\lambda_2\lambda_3\lambda_4\alpha}{}_{\beta}$.

To make a distinction between upper and lower indices, two lines representing λ and α could point upwards, and one line representing β could point downwards, as shown in Fig. 3.2.

Consider a product C of tensors M with contracted indices as follows:

$$C^{\lambda_1\lambda_2\lambda_3\lambda_4\alpha}{}_{\beta} = M^{\lambda_1\alpha_1}{}_{\beta}M^{\lambda_2\alpha_2}{}_{\alpha_1}M^{\lambda_3\alpha_3}{}_{\alpha_2}M^{\lambda\alpha}{}_{\alpha_3}. \tag{3.136}$$

Staring from the right-hand side, one can notice which indices have been contracted. This becomes clearer when we represent it with a diagram, using squares for the M. The result is shown in Fig. 3.3.

If one wishes, one can invent a new shape for the tensor C on the left-hand side of Eq. (3.136), for example an oval with six lines sticking out of it, as in Fig. 3.4.

We could then represent Eq. (3.136) diagrammatically showing the above diagrams together with an equal sign between them. Products of contracted tensors can of course include tensors of various types. Consider as an example the product $U^{\mu_1\nu_1}{}_{\alpha_1\alpha_2}W^{\lambda}{}_{\nu_1\alpha_3\nu_2}U^{\nu_2\mu_2}{}_{\alpha_4\alpha_4}$. A diagrammatic representation is shown in Fig. 3.5.

It should be clear that such products and diagrams can be used as building blocks for more complicated diagrams, by combining them together as desired. In this way a single diagram can give a nice visual representation of a very complicated

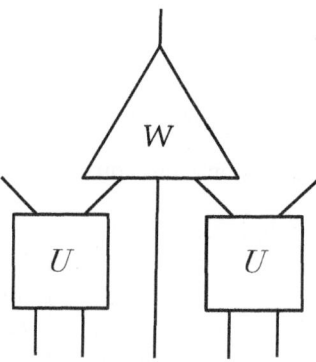

Fig. 3.5 A diagram representing $U^{\mu_1\nu_1}{}_{\alpha_1\alpha_2} W^{\lambda}{}_{\nu_1\alpha_3\nu_2} U^{\nu_2\mu_2}{}_{\alpha_4\alpha_4}$.

structure involving many tensors and contracted indices. This technique is central in the methods of *tensor networks*, which have many uses in quantum physics, machine learning, and biology, for example.

3.6.2 Tensor Products of Vector Spaces and Dual Vector Spaces

As we stated above, (p, q) tensors form a vector space. Let us arrive at this from another point of view. Let V and W be vector spaces, with dim $V = n$ and dim $W = m$. The *tensor product of vector spaces* V and W, denoted $V \otimes W$, is the set of all finite linear combinations of tensor products $v \otimes w$ with $v \in V$ and $w \in W$. If $\{e_i\}$ is a basis of V and $\{e'_j\}$ is a basis of W, then

$$V \otimes W = \sum_{i=1}^{n} \sum_{j=1}^{m} \lambda^{ij} e_i \otimes e'_j, \tag{3.137}$$

where the λ^{ij} are complex coefficients. From Eq. (3.137) one can see that

$$\dim(V \otimes W) = nm. \tag{3.138}$$

An element of $V \otimes W$ that can be written as $v \otimes w$ for some $v \in V$ and $w \in W$ is called a *pure tensor* or a *simple tensor*. Note that not all elements of $V \otimes W$ are simple.

Example   In Dirac notation, pure tensors in $V \otimes W$ have the form $|\psi\rangle|\chi\rangle$, while generic tensors have the form

$$|\Psi\rangle = \sum_{ij} \lambda^{ij} |i\rangle|j\rangle', \tag{3.139}$$

where $\{|i\rangle\}$ is a basis of V and $\{|j\rangle'\}$ is a basis of W.

Example   The state

$$|\Psi\rangle = \frac{1}{\sqrt{2}}|0\rangle|0\rangle + \frac{1}{\sqrt{2}}|1\rangle|1\rangle \tag{3.140}$$

is not a pure tensor, since it cannot be written as a tensor product

$$(a|0\rangle + b|1\rangle)(c|0\rangle + d|1\rangle) \tag{3.141}$$

for any choice of complex coefficients a, b, c, d. Indeed this tensor product can be rewritten as

$$ac|0\rangle|0\rangle + ad|0\rangle|1\rangle + bc|1\rangle|0\rangle + bd|1\rangle|1\rangle \qquad (3.142)$$

and, if this expression were equal to (3.140), then the following conditions should hold:

$$\begin{cases} ac = \frac{1}{\sqrt{2}} \\ ad = 0 \\ bc = 0 \\ bd = \frac{1}{\sqrt{2}} \end{cases}. \qquad (3.143)$$

It is easy to see that there exists no solution to this system of equations: The first equation implies that neither a nor c can be zero; since a is different from zero, the second equation implies $d = 0$, but then the fourth equation does not have any solution.

Note that, in general, $V \otimes W \neq W \otimes V$. However, they are isomorphic vector spaces. This follows from both having the same dimension, or by constructing an explicit isomorphism. In a similar fashion, one can construct the tensor products of dual vector spaces $V^\star \otimes W^\star$ and of a vector space and a dual vector space $V \otimes W^\star$. This is easiest using a basis and a dual basis:

$$V^\star \otimes W^\star = \sum_{i=1}^{n} \sum_{j=1}^{m} \lambda_{ij} e^{\star i} \otimes e'^{\star j}, \qquad (3.144)$$

$$V \otimes W^\star = \sum_{i=1}^{n} \sum_{j=1}^{m} \lambda_j^i e_i \otimes e'^{\star j}, \qquad (3.145)$$

where λ_{ij} and λ_j^i are complex coefficients.

Note that elements $L = \sum_{i=1}^{n} \sum_{j=1}^{m} \lambda_j^i e_i \otimes e'^{\star j}$ of $V \otimes W^\star$ are *linear operators* $L : W \to V$. This can be seen using the Dirac notation. We denote

$$L = \sum_{i=1}^{n} \sum_{j=1}^{m} \lambda_j^i |i\rangle\langle j| \qquad (3.146)$$

and expand a vector $|\chi\rangle \in W$ in the basis $|j\rangle$,

$$|\chi\rangle = \sum_{k=1}^{m} \chi_k |k\rangle, \qquad (3.147)$$

then L acts on it by

$$L|\chi\rangle = \sum_{i=1}^{n} \sum_{j=1}^{m} \sum_{k=1}^{n} \lambda_j^i \chi_k |i\rangle \underbrace{\langle j|k\rangle}_{=\delta_{jk}} = \sum_{i=1}^{n} \sum_{j=1}^{m} \lambda_j^i \chi_j |i\rangle, \qquad (3.148)$$

so $L|\chi\rangle$ is a vector in V with components $\sum_{j=1}^{m} \lambda_j^i \chi_j$, indexed by the free index i.

Since the tensor product space is a vector space, one can form a tensor product of that with another vector space. The tensor product is associative, in the sense that there is a natural isomorphism

$$(V_1 \otimes V_2) \otimes V_3 \cong V_1 \otimes (V_2 \otimes V_3) \qquad (3.149)$$

so that one can write $V_1 \otimes V_2 \otimes V_3$ for short. Now we can define (p, q) tensors in an alternative way. We define first a vector space

$$T_q^p(V) = \overbrace{V \otimes \cdots \otimes V}^{p} \otimes \overbrace{V^\star \otimes \cdots \otimes V^\star}^{q}. \tag{3.150}$$

Using the basis, elements $T \in T_q^p(V)$ can be expanded as

$$T = \sum_{i_1 \ldots i_p, j_1 \ldots j_q} T^{i_1 \ldots i_p}{}_{j_1 \ldots j_q} e_{i_1} \otimes \cdots \otimes e_{i_p} \otimes e^{\star j_1} \otimes \cdots \otimes e^{\star j_q}. \tag{3.151}$$

We recognize that elements of $T \in T_q^p(V)$ are (p, q) tensors; this can be taken as their alternative definition. Acting with T on the basis vectors projects out its components $T^{i_1 \ldots i_p}{}_{j_1 \ldots j_q}$. Let us check this with a $(1, 2)$ tensor; we first expand

$$T = \sum_{lmn} T^l{}_{mn} e_l \otimes e^{\star m} \otimes e^{\star n}. \tag{3.152}$$

Now, acting on the basis vectors projects out the components

$$T(e^{\star i}, \vec{e}_j, \vec{e}_k) = \sum_{lmn} T^l{}_{mn} e_l(e^{\star i}) e^{\star m}(e_j) e^{\star n}(e_k) = \sum_{lmn} T^l{}_{mn} \delta^i_l \delta^m_j \delta^n_k = T^i{}_{jk}, \tag{3.153}$$

as expected.

Example   Let us consider n distinguishable copies of a quantum system. "Distinguishable" means there is a label that identifies the copies. A concrete realization of this could be having n electrons in a one-dimensional lattice; this is called a *spin chain*. Then the lattice sites, enumerated by an index $k = 1, \ldots, n$, identify the copies. Another important example is in quantum computation, where a central concept is a *quantum register* of n qubits. Quantum computation in the circuit model is performed by a *quantum circuit*, a sequence of unitary operators, called gates, acting on the qubits. The mathematical description is the same for both examples; the unnormalized quantum state space (the Hilbert space) V of the full system consisting of all the copies is a tensor product of the two-dimensional vector spaces $\mathbb{C}^2$. Note that we will impose the unit normalization in the end; thus we start with

$$V = \overbrace{\mathbb{C}^2 \otimes \mathbb{C}^2 \otimes \cdots \mathbb{C}^2}^{n \text{ factors}} \cong \mathbb{C}^{2^n}. \tag{3.154}$$

A basis of this 2^n-dimensional complex vector space is given by the tensor products of individual normalized basis vectors

$$e_{i_1} \otimes e_{i_2} \otimes \cdots \otimes e_{i_n} \equiv |i_1\rangle \otimes |i_2\rangle \otimes \cdots \otimes |i_n\rangle, \tag{3.155}$$

where $i_1, i_2, \ldots, i_n \in \{0, 1\}$. In physics it is customary to write the basis states in a condensed notation, suppressing the $\otimes$ symbols and writing a basis vector as a single ket:

$$|i_1 i_2 \cdots i_n\rangle \equiv |i_1\rangle |i_2\rangle \cdots |i_n\rangle \equiv |i_1\rangle \otimes |i_2\rangle \otimes \cdots \otimes |i_n\rangle. \tag{3.156}$$

An unnormalized quantum state is then an $(N, 0)$ tensor in the tensor product space,

$$|\psi\rangle = \sum_{i_1, i_2, \ldots, i_n = 0, 1} \psi^{i_1 i_2 \ldots i_n} |i_1 i_2 \cdots i_n\rangle, \tag{3.157}$$

and the normalization condition $\langle \psi | \psi \rangle = 1$ introduces a nontrivial constraint on the allowed values of the components. The 2^n complex coefficients $\psi^{i_1 i_2 \cdots i_N}$ (or the 2^{n+1} real parameters) are first restricted to be the components of a unit vector, i.e., a point on the $(2^{n+1} - 1)$-dimensional unit sphere $S^{2^{n+1}-1}$. Then, remembering that the ray representation of quantum states (the equivalence up to a global phase $e^{i\varphi}$) requires an identification of points under the action of S^1, one finds that the $\psi^{i_1 i_2 \cdots i_N}$ parameters can be identified with elements of a *complex projective space*. In general, the latter is defined as

$$\mathbb{C}P^N = S^{2N+1}/S^1, \tag{3.158}$$

and, in particular, in the $N = 1$ case, the $\mathbb{C}P^1$ space is $\mathbb{C}P^1 = S^3/S^1$, which is S^2 by the Hopf fibration. Thus, the $\psi^{i_1 i_2 \cdots i_N}$ coefficients can be identified with elements of the complex projective space $\mathbb{C}P^{2^n - 1}$.

For the dual basis vectors, the Dirac notation is similar to the above example, using the bra notation:

$$\langle j_1 j_2 \cdots j_N | \equiv \langle j_1 | \langle j_2 | \cdots \langle j_N | \equiv \langle j_1 | \otimes \langle j_2 | \otimes \cdots \otimes \langle j_N |. \tag{3.159}$$

In Dirac notation, a (p, q) tensor would then have the expansion

$$T = \sum_{i_1 \ldots i_p j_1 \ldots j_q} T^{i_1 \ldots i_p}{}_{j_1 \ldots j_q} | i_1 \rangle \cdots | i_p \rangle \langle j_1 | \cdots \langle j_q |. \tag{3.160}$$

At this point we will adopt the *Einstein summation convention*: Repeated indices are understood to be summed over, and the explicit summation symbols $\sum$ are not shown. Thus, Eq. (3.151) would be rewritten as

$$T = T^{i_1 \ldots i_p}{}_{j_1 \ldots j_q} e_{i_1} \otimes \cdots \otimes e_{i_p} \otimes e^{\star j_1} \otimes \cdots \otimes e^{\star j_q}. \tag{3.161}$$

The tensor components satisfy a characteristic transformation rule under a change of basis. For notational brevity, we only consider a $(1, 2)$ tensor. Let

$$e'_k = R^l_k e_l, \quad e'^{\star i} = S^i_j e^{\star j} \tag{3.162}$$

be the transformation between two orthonormal bases $\{e_i\}$ and $\{e'_i\}$. Since $e'^{\star i}(e'_k) = e^{\star i}(e_k) = \delta^i_k$, one must have

$$\delta^i_k = e'^{\star i}(e'_k) = S^i_j R^l_k e^{\star j}(e_l) = S^i_j R^l_k \delta^j_l = S^i_j R^j_k,$$

which means that S is the inverse of R:

$$S^i_j = (R^{-1})^i_j.$$

Let us denote the components of a $(1, 2)$ tensor in the two bases as $T^i{}_{jk}$ and $T'^i{}_{jk}$. We get

$$T = T'^i{}_{jk} e'_i \otimes e'^{\star j} \otimes e'^{\star k} = T'^i{}_{jk} R^l_i (R^{-1})^j_m (R^{-1})^k_n e_l \otimes e^{\star m} \otimes e^{\star n}$$

$$= T^l{}_{mn} e_l \otimes e^{\star m} \otimes e^{\star n} \tag{3.163}$$

so that the components transform as

$$T^l{}_{mn} = R^l_i (R^{-1})^j_m (R^{-1})^k_n T'^i{}_{jk}. \tag{3.164}$$

Transformation rules for other (p, q) tensors can be derived in a similar manner.

As a side remark, we note that in repeated tensor products of a vector space, it is convenient to introduce a *tensor power notation*; given a nonnegative integer n, we write

$$V^{\otimes n} \equiv \overbrace{V \otimes \cdots \otimes V}^{n}.\tag{3.165}$$

With this notation, $T_q^p(V) = V^{\otimes p} \otimes V^{\star \otimes q}$.

Let us now consider the tensor product of tensors defined in Eq. (3.119). It should correspond to a tensor product map

$$T_q^p(V) \otimes T_n^m(V) \to T_{q+n}^{p+m}(V).\tag{3.166}$$

There is one subtlety in making contact with Eq. (3.119). Consider, for example, the tensor product of a $(1, 1)$ tensor R with a $(1, 0)$ tensor S. At first we might attempt

$$R \otimes S = (R_j^i e_i \otimes e^{\star j}) \otimes (S^k e_k) = R_j^i S^k e_i \otimes e^{\star j} \otimes e_k.\tag{3.167}$$

However, the expansion of a $(2, 1)$ tensor involves the basis vectors $e_i \otimes e_j \otimes e^{\star k}$ rather than $e_i \otimes e^{\star j} \otimes e_k$, and the tensor product is not commutative. One has to define the tensor product (3.167) in such a way that all basis vectors of V and dual basis vectors of $V^\star$ are grouped together. That is, instead of (3.167) we define

$$R \otimes S = R_k^i S^j e_i \otimes e_j \otimes e^{\star k}.\tag{3.168}$$

Now, consider again a (p, q) tensor R and a (m, n) tensor S. Expanding both,

$$R = R^{i_1 \ldots i_p}{}_{j_1 \ldots j_q} e_{i_1} \otimes \cdots \otimes e_{i_p} \otimes e^{\star j_1} \otimes \cdots \otimes e^{\star j_q}$$
$$S = S^{k_1 \ldots k_m}{}_{l_1 \ldots l_n} e_{k_1} \otimes \cdots \otimes e_{k_m} \otimes e^{\star l_1} \otimes \cdots \otimes e^{\star l_n}\tag{3.169}$$

the tensor product $T = R \otimes S$ can be built as

$$R \otimes S = R^{i_1 \ldots i_p}{}_{j_1 \ldots j_q} S^{k_1 \ldots k_m}{}_{l_1 \ldots l_n}$$
$$e_{i_1} \otimes \cdots \otimes e_{i_p} \otimes e_{k_1} \otimes \cdots \otimes e_{k_m} \otimes e^{\star j_1} \otimes \cdots \otimes e^{\star j_q} \otimes e^{\star l_1} \otimes \cdots \otimes e^{\star l_n}$$
$$= R^{i_1 \ldots i_p}{}_{j_1 \ldots j_q} S^{i_{p+1} \ldots i_{p+m}}{}_{j_{q+1} \ldots j_{q+n}}$$
$$e_{i_1} \otimes \cdots \otimes e_{i_p} \otimes e_{i_{p+1}} \otimes \cdots \otimes e_{i_{p+m}} \otimes e^{\star j_1} \otimes \cdots \otimes e^{\star j_q} \otimes e^{\star j_{q+1}} \otimes \cdots \otimes e^{\star j_{q+n}},\tag{3.170}$$

in agreement with Eq. (3.121).

3.6.3 Tensor Products of Linear Operators

We begin by recalling that elements $L \in W \otimes V^\star$ are linear operators $L : V \to W$. Conversely, one can represent every linear operator from V to W in a tensor product basis. Let $\{e_i\}$ be an orthonormal basis in V and let $\{e_k'\}$ be an orthonormal basis in W. Every linear operator $A : V \to W$ is then uniquely determined by its components with respect to the two bases. One can then identify the linear operator with the matrix having these components as its entries, $A = (A_j^i)$. One way to explicitly extract the components is to use the scalar product in W and let A act on the basis vectors of V,

$$A_j^i = \langle e_i' | A e_j \rangle.\tag{3.171}$$

Another way to extract the components of A is to use the dual basis $\{e'^{\star i}\}$ of $W^{\star}$ and compute the inner products

$$A^i_j = \langle e'^{\star i}, Ae_j \rangle. \tag{3.172}$$

Knowing the components, we can alternatively represent the operator A as a $(1,1)$ tensor:

$$A = A^i_j e'_i \otimes e^{\star j}. \tag{3.173}$$

When acting on a vector $v = v^k e_k \in V$, it gives

$$Av = A^i_j e'_i e^{\star j}(v^k e_k) = A^i_j v^k e'_i e^{\star j}(e_k) = A^i_j v^k \delta^j_k e'_i = A^i_j v^j e'_i, \tag{3.174}$$

which is a vector of W with components $A^i_j v^j$, as expected. Note the product of the matrix (A^i_j) and the vector (v^j). Recall also that in Dirac notation one would write

$$A = A^i_j |i\rangle\langle j| \tag{3.175}$$

with components

$$A^i_j = \langle i|A|j \rangle. \tag{3.176}$$

For example, for an operator acting in a two-dimensional vector space, $i,j = 0, 1$, we obtain the corresponding 2×2 matrix

$$A = (A^i{}_j) = \begin{pmatrix} \langle 0|A|0 \rangle & \langle 0|A|1 \rangle \\ \langle 1|A|0 \rangle & \langle 1|A|1 \rangle \end{pmatrix}. \tag{3.177}$$

In quantum mechanics it is often very useful to construct tensor products of linear operators. Let

$$A : V_1 \rightarrow V_2, \quad B : W_1 \rightarrow W_2 \tag{3.178}$$

be two linear operators. We define the *tensor product of linear operators* by defining first its action on the simple tensors of $V_1 \otimes W_1$:

$$(A \otimes B)(v \otimes w) = A(v) \otimes B(w) \quad \forall v \otimes w \in V_1 \otimes W_1. \tag{3.179}$$

Next, choosing a basis $\{e_i\}$ in V_1 and a basis $\{e'_j\}$ in W_1, the action of $A \otimes B$ on a generic $(2,0)$ tensor $T = T^{ij} e_i \otimes e'_j$ in $V_1 \otimes W_1$ can be defined as

$$A \otimes B : V_1 \otimes W_1 \mapsto V_2 \otimes W_2,$$
$$T \mapsto (A \otimes B)(T) = T^{ij} A(e_i) \otimes B(e'_j). \tag{3.180}$$

Expanding the linear operators in a basis, we get the expansion for their tensor product

$$A \otimes B = A^i_j B^k_l e_i \otimes e_k \otimes e'^{\star j} \otimes e'^{\star l}. \tag{3.181}$$

In Dirac notation, this would be

$$A \otimes B = A^i_j B^k_l |i\rangle \otimes |k\rangle \otimes \langle j| \otimes \langle l| \equiv A^i_j B^k_l |i, k\rangle\langle j, l|, \tag{3.182}$$

where the last expression involves the common shorthand notation for the tensor product basis states, in which the tensor product symbols are omitted. As before, the next step consists in generalizing this definition to multiple tensor products of linear operators; let $T \in V_1 \otimes \cdots \otimes V_n$. Then

$$(A_1 \otimes A_2 \otimes \cdots \otimes A_n)(T) = T^{i_1 \cdots i_n} A_1(e_{i_1}) \otimes A_2(e_{i_2}) \otimes \cdots A_n(e_{i_n}), \qquad (3.183)$$

where A_k are linear operators $V_k \to W_k$. In particular, in tensor products of a linear operator A with itself, the tensor power notation is frequently used:

$$A^{\otimes n} \equiv \overbrace{A \otimes \cdots \otimes A}^{n}. \qquad (3.184)$$

Generalizing further, we can then form linear combinations of tensor products of linear operators, to construct more general linear operators acting on $V_1 \otimes \cdots \otimes V_n$. Let $\{A_k^{(r)}\}$, with $r = 1, \ldots, s$, be a collection of linear operators from V_k to W_k. Then, the operator A defined as

$$A = \sum_{r=1}^{s} c_r (A_1^{(r)} \otimes A_2^{(r)} \otimes \cdots \otimes A_n^{(r)}) \qquad (3.185)$$

with coefficients $c_r \in \mathbb{C}$ is a linear operator

$$A : V_1 \otimes \cdots \otimes V_n \to W_1 \otimes \cdots \otimes W_n. \qquad (3.186)$$

In component form, using the shorthand version of the Dirac notation, Eq. (3.185) reads

$$A = \sum_r A_{j_1}^{(r)i_1} A_{j_2}^{(r)i_2} \cdots A_{j_n}^{(r)i_n} |i_1, i_2, \ldots, i_n\rangle\langle j_1, j_2, \ldots, j_n|. \qquad (3.187)$$

For a generic operator (3.186) acting on a tensor product of vector spaces, there is no reason to try to break up its components into a sum of products; instead we use the generic tensor component notation and write

$$A = A_{j_1 j_2 \cdots j_n}^{i_1 i_2 \cdots i_n} |i_1, i_2, \ldots, i_n\rangle\langle j_1, j_2, \ldots, j_n|. \qquad (3.188)$$

3.6.4 Kronecker Product and Outer Product

The tensor product is closely related to the *Kronecker product* of matrices and to the *outer product* of a vector and a dual vector.

Let A be an $N \times N$ matrix with components a_{ij} and let B an $M \times M$ matrix with components b_{ij}. The Kronecker product $A \otimes B$ of the matrices is denoted with the same symbol $\otimes$ as the tensor product, and is defined to be an $(NM) \times (NM)$ matrix with the structure

$$
\begin{aligned}
A \otimes B &= \begin{pmatrix} a_{11} & \cdots & a_{1N} \\ \cdots & \cdots & \cdots \\ a_{N1} & \cdots & a_{NN} \end{pmatrix} \otimes \begin{pmatrix} b_{11} & \cdots & b_{1M} \\ \cdots & \cdots & \cdots \\ b_{M1} & \cdots & b_{MM} \end{pmatrix} \qquad (3.189) \\
&= \begin{pmatrix} a_{11}B & a_{12}B & \cdots & a_{1N}B \\ \cdots & \cdots & \cdots & \cdots \\ a_{N1}B & a_{N2}B & \cdots & a_{NN}B \end{pmatrix} \\
&= \begin{pmatrix} a_{11}b_{11} & a_{11}b_{12} & \cdots & a_{11}b_{1M} & a_{12}b_{11} & \cdots & a_{1N}b_{1M} \\ a_{11}b_{21} & a_{11}b_{22} & \cdots & a_{11}b_{2M} & a_{12}b_{21} & \cdots & a_{1N}b_{2M} \\ \cdots & \cdots & \cdots & \cdots & \cdots & \cdots & \cdots \\ a_{N1}b_{M1} & a_{N1}b_{M2} & \cdots & a_{N1}b_{MM} & a_{N2}b_{M1} & \cdots & a_{NN}b_{MM} \end{pmatrix}.
\end{aligned}
$$

The Kronecker product of two matrices is not commutative: $A \otimes B \neq B \otimes A$, but the two resulting matrices can be mapped into each other by a permutation of rows and columns.

In particular, let us consider Kronecker products of vectors and dual vectors. We first represent the basis vectors e_i as column vectors or $N \times 1$ matrices,

$$e_1 = \begin{pmatrix} 1 \\ 0 \\ 0 \\ \vdots \\ 0 \end{pmatrix}, \quad e_2 = \begin{pmatrix} 0 \\ 1 \\ 0 \\ \vdots \\ 0 \end{pmatrix}, \quad \ldots, \quad e_N = \begin{pmatrix} 0 \\ 0 \\ 0 \\ \vdots \\ 1 \end{pmatrix}, \tag{3.190}$$

so that every vector $\vec{v}$ is represented by a column vector with its components as entries:

$$\vec{v} = v^1 e_1 + \cdots + v^N e_N = \begin{pmatrix} v^1 \\ v^2 \\ v^3 \\ \vdots \\ v^N \end{pmatrix}. \tag{3.191}$$

We then represent the dual basis vectors as row vectors, or $1 \times N$ matrices,

$$e^{\star 1} = (1,0,0,\ldots,0), \quad e^{\star 2} = (0,1,0,\ldots,0), \quad \ldots, \quad e^{\star N} = (0,0,0,\ldots,1), \tag{3.192}$$

so that every dual vector f can be expressed as a row vector

$$f = f_1 e^{\star 1} + \cdots f_N e^{\star N} = (f_1, f_2, f_3, \ldots, f_N). \tag{3.193}$$

Then the Kronecker product $\vec{v} \otimes f$ between a vector and a dual vector yields an $N \times N$ matrix

$$\vec{v} \otimes f = \begin{pmatrix} v^1 \\ v^2 \\ v^3 \\ \vdots \\ v^N \end{pmatrix} \otimes (f_1, f_2, f_3, \ldots, f_N) \tag{3.194}$$

$$= \begin{pmatrix} v^1 f_1 & v^1 f_2 & v^1 f_3 & \cdots & v^1 f_N \\ v^2 f_1 & v^2 f_2 & v^2 f_3 & \cdots & v^2 f_N \\ \vdots & \vdots & & & \vdots \\ v^N f_1 & v^N f_2 & & & v^N f_N \end{pmatrix}.$$

Note that we arrive at the same result if we first expand $\vec{v}$ and f and compose the outer products of the basis and dual basis vectors:

$$\vec{v} \otimes f = (v^i e_i) \otimes (f_j e^{\star j}) = v^i f_j (e_i \otimes e^{\star j}) \tag{3.195}$$

$$= \begin{pmatrix} v^1 f_1 & v^1 f_2 & v^1 f_3 & \cdots & v^1 f_N \\ v^2 f_1 & v^2 f_2 & v^2 f_3 & \cdots & v^2 f_N \\ \vdots & \vdots & & & \vdots \\ v^N f_1 & v^N f_2 & & & v^N f_N \end{pmatrix},$$

since $e_i \otimes e^{\star j}$ can be represented by a matrix where every entry is 0, except the one on the ith row and jth column, which is 1.

The Kronecker product of a vector $\vec{v}$ with N components and a vector $\vec{w}$ with M components is a vector with NM components,

$$\vec{v} \otimes \vec{w} = \begin{pmatrix} v^1 \\ v^2 \\ v^3 \\ \vdots \\ v^N \end{pmatrix} \otimes \begin{pmatrix} w^1 \\ w^2 \\ w^3 \\ \vdots \\ w^M \end{pmatrix} = \begin{pmatrix} v^1 \vec{w} \\ v^2 \vec{w} \\ v^3 \vec{w} \\ \vdots \\ v^N \vec{w} \end{pmatrix} = \begin{pmatrix} v^1 w^1 \\ v^1 w^2 \\ \vdots \\ v^1 w^M \\ v^2 w^1 \\ \vdots \\ v^2 w^M \\ \vdots \\ v^N w^1 \\ \vdots \\ v^N w^M \end{pmatrix}. \tag{3.196}$$

Likewise, a product of a dual (row) vector f with N components and a dual (row) vector g vector with M components is a dual (row) vector with NM components.

Example Let us consider a two-dimensional quantum state space (of qubits, for example), with the labeling for the basis vectors

$$e_0 = |0\rangle = \begin{pmatrix} 1 \\ 0 \end{pmatrix}, \quad e_1 = |1\rangle = \begin{pmatrix} 0 \\ 1 \end{pmatrix}. \tag{3.197}$$

We then consider the basis vectors

$$|i_1 i_2 \cdots i_N\rangle = e_{i_1} \otimes e_{i_2} \otimes \cdots \otimes e_{i_N} \tag{3.198}$$

for the 2^N-dimensional tensor product space $(\mathbb{C}^2)^{\otimes N}$. It is instructive to use the Kronecker product of vectors and write these in the 2^N-component column vector notation. As an example, we consider the case $N = 3$:

$$|000\rangle = \begin{pmatrix} 1 \\ 0 \end{pmatrix} \otimes \begin{pmatrix} 1 \\ 0 \end{pmatrix} \otimes \begin{pmatrix} 1 \\ 0 \end{pmatrix} = \begin{pmatrix} 1 \\ 0 \end{pmatrix} \otimes \begin{pmatrix} 1 \\ 0 \\ 0 \\ 0 \end{pmatrix} = \begin{pmatrix} 1 \\ 0 \\ 0 \\ 0 \\ \vdots \\ 0 \end{pmatrix}$$

$$|001\rangle = \begin{pmatrix} 1 \\ 0 \end{pmatrix} \otimes \begin{pmatrix} 1 \\ 0 \end{pmatrix} \otimes \begin{pmatrix} 0 \\ 1 \end{pmatrix} = \begin{pmatrix} 1 \\ 0 \end{pmatrix} \otimes \begin{pmatrix} 0 \\ 1 \\ 0 \\ 0 \end{pmatrix} = \begin{pmatrix} 0 \\ 1 \\ 0 \\ 0 \\ \vdots \\ 0 \end{pmatrix}$$

$$|010\rangle = \begin{pmatrix} 1 \\ 0 \end{pmatrix} \otimes \begin{pmatrix} 0 \\ 1 \end{pmatrix} \otimes \begin{pmatrix} 1 \\ 0 \end{pmatrix} = \begin{pmatrix} 1 \\ 0 \end{pmatrix} \otimes \begin{pmatrix} 0 \\ 0 \\ 1 \\ 0 \end{pmatrix} = \begin{pmatrix} 0 \\ 0 \\ 1 \\ 0 \\ \vdots \\ 0 \end{pmatrix}$$

$$\vdots$$

$$|111\rangle = \begin{pmatrix} 0 \\ 1 \end{pmatrix} \otimes \begin{pmatrix} 0 \\ 1 \end{pmatrix} \otimes \begin{pmatrix} 0 \\ 1 \end{pmatrix} = \begin{pmatrix} 0 \\ 1 \end{pmatrix} \otimes \begin{pmatrix} 0 \\ 0 \\ 0 \\ 1 \end{pmatrix} = \begin{pmatrix} 0 \\ 0 \\ 0 \\ 0 \\ \vdots \\ 1 \end{pmatrix}. \tag{3.199}$$

Note that the basis vectors are labeled by $n = 0, \ldots, 7$ written in base 2: 000, 001, 010, $\ldots$, 111. In other words, given $n = j_2 2^2 + j_1 2^1 + j_0 2^0$, the corresponding basis state is $|j_2 j_1 j_0\rangle$. For a general N, one can associate the $|j_{N-1} j_{N-2} \cdots j_0\rangle \equiv |i_1 i_2 \cdots i_N\rangle$ basis state, with $i_k = j_{N-k}$ for every $1 \le k \le N$, with the label defined as

$$n = \sum_{a=0}^{N-1} j_a 2^a. \tag{3.200}$$

In the context of quantum computing, this basis is called the *computational basis*.

For two (column) complex vectors $\vec{v}$ and $\vec{w}$ one may alternatively define an *outer product*. There one first maps the vector $\vec{w}$ to the complex conjugate of its transpose row vector, which we denote as $\vec{w}^\dagger$:

$$\vec{w}^\dagger = (w_1^\star, w_2^\star, \ldots, w_N^\star). \tag{3.201}$$

The outer product is then the Kronecker product of $\vec{v}$ and $\vec{w}^\dagger$, using Eq. (3.194) for $\vec{v} \otimes \vec{w}^\dagger$ to obtain the $N \times N$ matrix. When dealing with real vector spaces, the adjoint vector reduces to the transpose vector, i.e., the outer product is then $\vec{v} \otimes \vec{w}^T$. In Dirac notation, suppose the ket vectors $|\psi\rangle = \psi_i |i\rangle$ and $|\chi\rangle = \chi_j |j\rangle$ are expanded in the basis of V. The adjoint of $|\chi\rangle$ is, then, the bra vector $\langle\chi| = |\chi\rangle^\dagger$ and the outer product is the operator

$$|\psi\rangle\langle\chi| = \psi_i \chi_j^\star |i\rangle\langle j|, \tag{3.202}$$

where $(\psi_i \chi_j^\star)$ denote the entries of the $N \times N$ matrix of coefficients.

These concepts can be used, for example, in the context of tensor products of linear operators. As a simple example, suppose we have an operator $A \otimes B$ acting on a pure tensor $\vec{v} \otimes \vec{w}$ in $V \otimes W$. The image $T \equiv A(\vec{v}) \otimes B(\vec{w})$ has components

$$T^{ik} = A^i_j B^k_l v^j w^l. \tag{3.203}$$

We can rephrase this using Kronecker products. The Kronecker product of the vectors is

$$\vec{v} \otimes \vec{w} = \begin{pmatrix} v^1 \vec{w} \\ v^2 \vec{w} \\ v^3 \vec{w} \\ \vdots \\ v^N \vec{w} \end{pmatrix} \tag{3.204}$$

and the Kronecker product of the matrices $A = (A^i{}_j)$ and $B = (B^k{}_l)$ is

$$A \otimes B = \begin{pmatrix} A^1{}_1 B & A^1{}_2 B & \cdots & A^1{}_N B \\ \cdots & \cdots & \cdots & \cdots \\ A^N{}_1 B & A^N{}_2 B & \cdots & A^N{}_N B \end{pmatrix}. \tag{3.205}$$

Thus one obtains

$$
T = \begin{pmatrix} A^1{}_1 B & A^1{}_2 B & \cdots & A^1{}_N B \\ \cdots & \cdots & \cdots & \cdots \\ A^N{}_1 B & A^N{}_2 B & \cdots & A^N{}_N B \end{pmatrix} \begin{pmatrix} v^1 \vec{w} \\ v^2 \vec{w} \\ v^3 \vec{w} \\ \vdots \\ v^N \vec{w} \end{pmatrix}
$$

$$
= \begin{pmatrix} (A^1{}_j v^j) B \vec{w} \\ (A^2{}_j v^j) B \vec{w} \\ (A^3{}_j v^j) B \vec{w} \\ \vdots \\ (A^N{}_j v^j) B \vec{w} \end{pmatrix} = \begin{pmatrix} (A\vec{v})^1 B \vec{w} \\ \vdots \\ (A\vec{v})^N B \vec{w} \end{pmatrix}. \tag{3.206}
$$

3.6.5 Traces, Partial Traces, and Determinants

Let A be a linear operator on V with dim $V = N$ and B a linear operator on W with dim $W = M$. Given a basis in V and a basis in W, the operators can be identified with the matrices of their coefficients. As we discussed earlier, their tensor product $A \otimes B$ becomes the Kronecker product of the matrices, which is an $NM \times NM$ matrix. The trace of this matrix using the Dirac notation is taken over the tensor product basis $\{|i\rangle|j\rangle\}$, with $1 \le i \le N$ and $1 \le j \le M$, of $V \otimes M$:

$$\mathrm{tr}(A \otimes B) = \sum_{i=1}^{N} \sum_{j=1}^{M} \langle j| \langle i|(A \otimes B)|i\rangle|j\rangle = \sum_{i=1}^{N} \sum_{j=1}^{M} \langle i|A|i\rangle \langle j|B|j\rangle$$

$$= \sum_{i=1}^{N} \langle i|A|i\rangle \sum_{j=1}^{M} \langle j|B|j\rangle, \tag{3.207}$$

so we see that

$$\mathrm{tr}(A \otimes B) = (\mathrm{tr}\, A) \cdot (\mathrm{tr}\, B). \tag{3.208}$$

We may also need to calculate the determinant of the Kronecker product matrix $A \otimes B$. We leave it as an exercise (see Problem 3.8) to prove that

$$\det(A \otimes B) = (\det A)^M \cdot (\det B)^N. \tag{3.209}$$

Note that the tensor product operators $A \otimes B$ form just a subset of linear operators $V \otimes W \rightarrow V \otimes W$. Using Dirac notation, and a basis $\{|i\rangle|j\rangle\}$ of $V \otimes W$, a general linear operator has the form

$$L = L_{ij,kl}|i\rangle|j\rangle\langle k|\langle l|, \tag{3.210}$$

which reduces to the simple product tensor if the coefficients can be decomposed into a product form

$$L_{ij,kl} = A_{ik}B_{jl} \tag{3.211}$$

in a suitably chosen basis. This is not always possible, so then the trace of L is just

$$\text{tr}(L) = L_{ij,ij}, \tag{3.212}$$

having assumed the Einstein summation convention, whereby repeated indices are summed over.

In quantum mechanics, it is important to consider also the notion of *partial trace*. Suppose that we have a composite quantum system consisting of two subsystems with Hilbert spaces (state spaces) V and W, of dimensions N and M, respectively. Let L be an operator in the composite system, acting on $V \otimes W$. If the observers have no access, say, to the subsystem W, then they are led to average the operator L over the subsystem W, which amounts to a partial trace of L over W. (The detailed justification of this statement is given in courses of quantum mechanics.) Let us denote the partial traces over the subsystems V and W as tr_V and tr_W, respectively, and the trace over the full composite system $V \otimes W$ as tr. From Eq. (3.207), one can see that the trace over the full composite system is given by the product of the subsystem traces; formally:

$$\text{tr} = \text{tr}_V \cdot \text{tr}_W. \tag{3.213}$$

Furthermore, Eq. (3.207) also implies that, if one takes the partial trace over W of a tensor product operator $A \otimes B$, one obtains

$$\text{tr}_W(A \otimes B) = A\,\text{tr}_W(B) = \text{tr}_W(B)A, \tag{3.214}$$

which is an $N \times N$ matrix, or an operator acting on the subsystem V. Likewise, if one instead takes the partial trace over the subsystem V, one obtains

$$\text{tr}_V(A \otimes B) = \text{tr}_V(A)B, \tag{3.215}$$

an operator acting in W. Consider, then, a generic operator L acting in $V \otimes W$; using the Dirac notation, it has the expansion

$$L = L_{ij,kl}|i\rangle_V|j\rangle_{WW}\langle k|_V\langle l|, \tag{3.216}$$

where we included subscripts to emphasize which vector space the basis vectors belong to. Then, by computing the partial trace over W one obtains

$$\begin{aligned}
\text{tr}_W L &= \sum_m L_{ij,kl}|i\rangle_{VW}\langle m|j\rangle_{WW}\langle k|m\rangle_{WV}\langle l| \\
&= L_{ij,kl}\;_W\langle k|j\rangle_W|i\rangle_{VV}\langle k| \\
&= \left(\sum_j L_{ij,jl}\right)|i\rangle_{VV}\langle l|, \tag{3.217}
\end{aligned}$$

which is an operator in V. Note that, in the expression above, we have used the Einstein summation convention, but we added an explicit summation symbol for the repeated index j for the sake of clarity. Likewise, computing the partial trace over V gives

$$\text{tr}_V L = \sum_m L_{ij,kl} \,_V\langle m|i\rangle_V |j\rangle_W \,_W\langle k|\,_V\langle l|m\rangle_V$$

$$= L_{ij,kl} \,_V\langle l|i\rangle_V |j\rangle_W \,_W\langle k|$$

$$= \left(\sum_i L_{ij,ki}\right)|j\rangle_W \,_W\langle k|. \tag{3.218}$$

Example Let us consider the state vector of a coupled two-qubit system, labeling the two basis vectors $|0\rangle$ and $|1\rangle$ with subscripts 1 and 2, respectively, for the two single-qubit subsystems:

$$|\psi\rangle = \frac{1}{\sqrt{2}}\left(|0\rangle_1|0\rangle_2 + |1\rangle_1|1\rangle_2\right). \tag{3.219}$$

Then, consider the pure tensor operator

$$L = |\psi\rangle\langle\psi|$$
$$= \frac{1}{2}\left(|0\rangle_1|0\rangle_2\,_2\langle0|\,_1\langle0| + |1\rangle_1|1\rangle_2\,_2\langle0|\,_1\langle0| + |0\rangle_1|0\rangle_2\,_2\langle1|\,_1\langle1| + |1\rangle_1|1\rangle_2\,_2\langle1|\,_1\langle1|\right). \tag{3.220}$$

This is an operator with nonzero coefficients

$$L_{00,00} = L_{11,00} = L_{00,11} = L_{11,11} = \frac{1}{2}. \tag{3.221}$$

Then, let us compute the partial trace over the subsystem 2. We get

$$\text{tr}_2 L = \frac{1}{2}\left(_2\langle0|0\rangle_2\,|0\rangle_1\,_1\langle0| + _2\langle1|0\rangle_2\,|1\rangle_1\,_1\langle0| + _2\langle0|1\rangle_2\,|0\rangle_1\,_1\langle1| + _2\langle1|1\rangle_2\,|0\rangle_1\,_1\langle0|\right)$$

$$= \frac{1}{2}\left(|0\rangle_1\,_1\langle0| + |1\rangle_1\,_1\langle1|\right). \tag{3.222}$$

Noting that $|0\rangle_1\,_1\langle0| + |1\rangle_1\,_1\langle1|$ is the identity operator $\mathbb{1}_1$ in the subsystem 1, we obtain

$$\text{tr}_2 L = \frac{1}{2}\mathbb{1}_1, \tag{3.223}$$

with the matrix of coefficients being the 2×2 diagonal matrix $\frac{1}{2}\mathbb{1}_2$.

3.7 A Spin-Chain Example

We continue by briefly discussing a *quantum Heisenberg spin chain*, which is an elementary model of magnets. A ferromagnetic material can be thought of as consisting of tiny elementary bar magnets, particles with a magnetic moment. When the alignment of the elementary magnets has a preferred direction, which is observable on a macroscopic scale, the material appears to be magnetic. At the microscopic level, the existence of magnetic moments can be interpreted in terms of the spin of particles; for example, an electron acts like a tiny bar magnet with a dipole magnetic field, associated with a magnetic moment

$$\vec{\mu}_s = -g_S \mu_B m \vec{e}, \tag{3.224}$$

where g_S is called the spin g-factor (for an electron, $g_S \approx 2$), μ_B is the Bohr magneton, $m = \pm 1/2$ is the spin quantum number, and $\vec{e}$ is a unit vector in the direction of the spin vector.

Quantum Heisenberg spin chains are models of one-dimensional lattices of elementary magnetic moments, with nearest-neighbor interactions that tend to make the neighboring magnetic moments aligned and parallel, or aligned and oppositely oriented, resulting in a net magnetization. We start by considering linear operators of the form

$$\mathbb{1}_2 \otimes \mathbb{1}_2 \otimes \cdots \mathbb{1}_2 \otimes \sigma_k^a \otimes \sigma_{k+1}^b \otimes \mathbb{1}_2 \otimes \cdots \otimes \mathbb{1}_2 \tag{3.225}$$

where the subscripts k and $k + 1$ have been added to the Pauli matrices to emphasize that they act on the electrons at the kth site and at the $(k + 1)$th site, respectively. Acting on a basis state

$$|i_1 \cdots i_{k-1} i_k i_{k+1} i_{k+2} \cdots i_N\rangle = |i_1\rangle \otimes \cdots |i_{k-1}\rangle \otimes |i_k\rangle \otimes |i_{k+1}\rangle \otimes |i_{k+2}\rangle \otimes \cdots |i_N\rangle \tag{3.226}$$

the operator defined in (3.225) produces the state

$$|i_1\rangle \otimes \cdots |i_{k-1}\rangle \otimes \sigma^a |i_k\rangle \otimes \sigma^b |i_{k+1}\rangle \otimes |i_{k+2}\rangle \otimes \cdots |i_N\rangle. \tag{3.227}$$

It is customary to adopt a shorter notation $\sigma_k^a \otimes \sigma_{k+1}^b$ for the operator defined in (3.225), suppressing the unit matrices and presenting only the nontrivial action. With this shorter notation, one can define a set of local operators, labeled by k, of the form

$$H_{k,k+1} = \frac{\lambda}{2} \left[\mathbb{1}_k \otimes \mathbb{1}_{k+1} - \sum_{a=1}^{3} \sigma_k^a \otimes \sigma_{k+1}^a \right], \tag{3.228}$$

where λ is a real parameter controlling the strength and alignment of the nearest-neighbor interactions. Finally, we sum over the sites to construct the Hamiltonian operator

$$H = \sum_{k=1}^{N-1} H_{k,k+1}. \tag{3.229}$$

This is the *Heisenberg XXX quantum spin chain*. Note that there exist other variants of Heisenberg spin chains, with different Hamiltonian operators. Also, one can either assume the sites to be arranged in a periodic chain, identifying the first and the Nth site, and letting the sum run up to N, which yields a closed spin chain, or having an infinite lattice with the sum running over all integers, which corresponds to an infinite chain. The Hamiltonian has a discrete set of degenerate eigenvalues E_n and eigenstates $|\psi_n\rangle$,

$$H|\psi_n\rangle = E_n|\psi_n\rangle. \tag{3.230}$$

When $\lambda > 0$, the interaction defined in Eq. (3.228) tends to align the neighboring spins: In that case, the interaction and the resulting ground state, with energy E_0, are said to be *ferromagnetic*. Conversely, for $\lambda < 0$ the interaction and the ground state are called *antiferromagnetic*.

The model has an SU(2) symmetry which helps in finding and organizing the eigenstates. Defining

$$\mathbf{J}^a = \frac{1}{2} \sum_{k=1}^{N} \sigma_k^a, \tag{3.231}$$

it is easy to see that the three $\mathbf{J}^a$ operators satisfy the commutation relations of the generators of the SU(2) Lie group

$$[\mathbf{J}^a, \mathbf{J}^b] = i\epsilon^{abc} \mathbf{J}^c, \tag{3.232}$$

where ϵ^{abc} is totally antisymmetric under permutation of any pair of indices, and $\epsilon^{123} = 1$. Moreover, one can show that they commute with the Hamiltonian:

$$[\mathbf{J}^a, H] = 0 \quad \forall a \in \{1, 2, 3\}. \tag{3.233}$$

This means that the Heisenberg XXX spin chain has an SU(2) symmetry. Then the energy eigenstates for every E_n correspond to irreducible spin-j representations of the SU(2) group. Note that we started with the smallest irreducible two-dimensional representation for a single electron, but as a result of taking N copies we ended up with a large, 2^N-dimensional vector space V. The action of SU(2) on V is reducible. Since every state can be expressed in the basis of energy eigenstates, which belong to irreducible representations, studying the energy eigenstates of this system corresponds to solving the reduction problem.

Problems

3.1   Show that the image and the kernel of a linear transformation are vector spaces.

3.2   Given two vector spaces V_1 and V_2, prove that the dimension of their direct sum is $\dim(V_1 \oplus V_2) = \dim V_1 + \dim V_2$.

3.3   Given a unitary representation of a group on the vector space V, in which a scalar product $\langle \dots | \dots \rangle$ is defined, and given a submodule W, and its orthogonal complement $W_\perp = \{\vec{v} \in V | \langle \vec{v} | \vec{w} \rangle = 0 \ \forall \vec{w} \in W\}$, show that $V \cong W \oplus W_\perp$.

3.4   Given a finite group G, prove that the matrices of its left-regular representation, with elements defined by Eq. (3.70), satisfy the group multiplication law, i.e., that $L(g)L(g') = L(gg')$.

3.5   Given a finite group G, prove that the matrices of its right-regular representation, defined according to Eq. (3.73), obey the group multiplication law, i.e., $R(g)R(g') = R(gg')$. Prove also that this would not be true, if in Eq. (3.73) one had defined $\rho_g(h) = hg$ instead of $\rho_g(h) = hg^{-1}$.

3.6   Given a finite group G, prove that its left-regular and right-regular representations are isomorphic to each other.

3.7   Given a vector space V, prove that every $\omega \in (V^\star)^\star$ can be uniquely associated with a vector $\vec{v} \in V$, such that $\omega(f) = \langle f, \vec{v} \rangle$.

3.8   Consider a linear operator A acting on a vector space V of finite dimension N and a linear operator B acting on a vector space W of finite dimension M. Assuming that orthonormal bases exist in V and in W, and that the operators are represented in terms of the matrices of their coefficients in these bases, prove that the determinant of the tensor product $A \otimes B$ is given by $\det(A \otimes B) = (\det A)^M \cdot (\det B)^N$.

4

Differentiable Manifolds

The topic of this chapter is differential topology, which provides a concept of "shape" for sets. We begin by introducing basic topological concepts, such as topology (the notion of open sets, which are then used to define continuous mappings), homeomorphism, and topological invariants. We shall omit detailed proofs when they are not essential for moving further. Next, we introduce differential structure on manifolds, to extend calculus from Euclidean spaces $\mathbb{R}^n$ to the more general setting of differentiable manifolds. Finally, in Chapter 5 we will equip differentiable manifolds with a metric and introduce Riemannian manifolds. In all, we will introduce step by step a hierarchy of structures, illustrated in Fig. 4.1.

4.1 Topology

4.1.1 Topological Spaces

Defining a *topology* for a set X amounts to defining which of its subsets are *open sets*. These must satisfy some consistency conditions, extending the properties of open sets familiar from calculus in Euclidean spaces. We begin with the formal definition.

Let X be a set, and let $\tau = \{X_\alpha\}_{\alpha \in I}$ be a (finite or infinite) collection of subsets of X. Then (X, τ) is a *topological space*, if

(i) $\emptyset \in \tau$ and $X \in \tau$;
(ii) all possible unions of X_αs belong to τ : $\bigcup_{\alpha \in I'} X_\alpha \in \tau, \forall I' \subseteq I$;
(iii) all intersections of any *finite number* of X_αs belong to τ: $\bigcap_{i=1}^{n} X_{\alpha_i} \in \tau$.

Each X_α is called an *open set* of X in the topology τ, and τ is said to give a *topology* to X.

Within the same set X there are several possible definitions of topologies; below are some common examples.

Examples of Topologies

(i) $\tau = \{\emptyset, X\}$. This is the smallest possible choice, called the *trivial topology*.
(ii) $\tau = \{$all subsets of $X\}$. This is the maximal choice, called the *discrete topology*. The collection of all subsets of a set X is often denoted $\mathcal{P}(X)$ and is called the *power set* of X.
(iii) Let $X = \mathbb{R}$, $\tau = \{$open intervals $]a, b[$ and their unions$\}$. This is known as the *usual topology*.

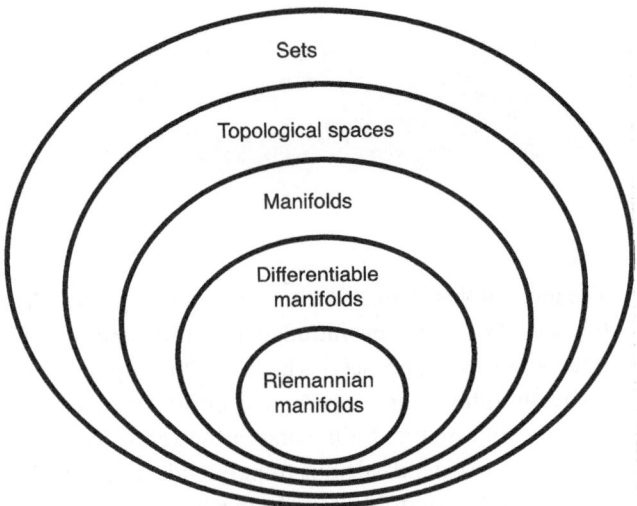

Fig. 4.1 A hierarchy of concepts to be introduced in Chapters 4 and 5.

(iv) $X = \mathbb{R}^n$, $\tau = \{\,]a_1, b_1[\times \cdots \times]a_n, b_n[$ and their unions$\}$ is the usual topology in higher dimensions.

(v) If (X, τ) and (Y, σ) are topological spaces, then the product space defined as $X \times Y = \{(x, y) | x \in X, y \in Y\}$ is a topological space, with the natural topology being the *product topology*: $\tau \times \sigma = \{(X_\alpha, Y_\beta) | X_\alpha \in \tau, Y_\beta \in \sigma\}$. An example of these are the usual topologies of the n-dimensional Euclidean spaces $\mathbb{R}^n$.

The first two examples above lead one to introduce a comparison in size between topologies. Let τ_1 and τ_2 be two topologies of X. If every element of τ_1 is also an element of τ_2, we denote $\tau_1 \subset \tau_2$ and say that the topology τ_1 is *coarser* (or *weaker* or *smaller*) than τ_2. Respectively, τ_2 is a *finer* topology than τ_1 (alternatively *stronger* or *larger* topology). Thus the trivial topology is the coarsest, and the discrete topology is the finest topology on X.

The next two topologies use open intervals and open cubes as a basis to define a topology on $\mathbb{R}^n$. We say that a collection B of subsets of X is a *basis* if it satisfies the following two properties.

(i) $B \subset \tau$.
(ii) Every non-empty $U \in \tau$ can be expressed as a union of sets in B.

The above conditions then guarantee that we can use the basis B to define a topology, as we did with the open intervals (a basis of $\mathbb{R}$) in the above example. In this way we obtain a topology *generated* by the basis B.

So far there is no notion of a distance between two points in a set. We define it next, by introducing a metric in a set. We will then proceed to use the metric to define a natural topology.

A *metric* on X is a function $d : X \times X \to \mathbb{R}$ such that

(i) $d(x, y) = d(y, x)$ (*symmetry*);
(ii) $d(x, y) \geq 0$, and $d(x, y) = 0$ if and only if $x = y$ (*positive definiteness*);
(iii) $d(x, y) + d(y, z) \geq d(x, z)$ (*triangle inequality*).

A space with a metric (X, d) is called a *metric space*.

Example   Let $X = \mathbb{R}^n$,

$$d_p(x, y) = \left(\sum_{i=1}^{n} |x_i - y_i|^p \right)^{1/p}, \text{ where } p \geq 1. \tag{4.1}$$

For $p = 2$, this defines the *Euclidean metric*. We write $d_p(x, y) = ||x - y||_p$ and call $|| \ldots ||_p$ the *p-norm* (or L^p-*norm*) in $\mathbb{R}^n$.

Example   Consider now a slightly more exotic example: Let $X \neq \emptyset$ be a set and define $d(x, x) = 0 \ \forall x \in X$ and $d(x, y) = 1 \ \forall x, y \in X, x \neq y$. This defines the *discrete metric*.

If X has a metric, then we can use it to define the *metric topology* on X, by choosing as the basis B the open balls

$$U_\epsilon(x) = \{ y \in X | \ d(x, y) < \epsilon \}$$

for all $x \in X$, and for every $\epsilon > 0$, and defining a topology τ as the unions of any number of elements of B, and the intersections of any finite number of elements of B: This topology is called the *metric topology*. One can see that in $\mathbb{R}$ with the Euclidean metric the open balls are the open intervals, so the metric topology is the same as the usual topology. More generally, one can show that the metric topology of $\mathbb{R}^n$ with metric d_p is the same as the usual topology. Another name for this topology is the *standard topology*. In a set X, the metric topology associated with the discrete metric is the same as the discrete topology.

Let (X, τ) be a topological space, and let $A \subset X$ be a subset. The topology τ induces the *relative topology* or the *subspace topology* τ' in A,

$$\tau' = \{ U_i \cap A \mid U_i \in \tau \}.$$

This allows one to obtain a topology for all subsets of $\mathbb{R}^n$ (such as the spheres S^m, $1 \leq m < n$).

Having defined the topology, i.e., the open sets, we can use this definition to introduce a notion of *continuous* functions.

Let (X, τ) and (Y, σ) be topological spaces. A map $f : X \rightarrow Y$ is said to be *continuous* if and only if the inverse image of every open set $V \in \sigma$, $f^{-1}(V) = \{ x \in X | f(x) \in V \}$, is an open set in X: $f^{-1}(V) \in \tau$.

Importantly, continuous bijections give us a way to decide which topological spaces are "the same," or *homeomorphic*.

A function $f : X \rightarrow Y$ is a *homeomorphism* if f is continuous, and has an inverse $f^{-1} : Y \rightarrow X$ which is also continuous.

Example   There exist continuous functions with a noncontinuous inverse. As an example, take a set X and equip it with two topologies τ_1 and τ_2 so that τ_2 is a proper subset of τ_1. The identity map id : $(X, \tau_1) \rightarrow (X, \tau_2), x \rightarrow x$ is obviously invertible and continuous, but the inverse map is not continuous.

If there exists a homeomorphism $f : X \rightarrow Y$, then we say that X is *homeomorphic* to Y, and vice versa. We denote this by $X \approx Y$.

Intuitively speaking, X and Y are homeomorphic if it is possible to continuously deform X to Y (without cutting or pasting). The usual way to describe this to a layperson is to explain that a coffee cup (of the usual design with a handle) can be continuously deformed into a donut shape, thus the two are homeomorphic.

We leave it as an exercise (see Problem 4.4) to show that two spaces being homeomorphic is an equivalence relation. The collection of all possible topological spaces is thus partitioned into equivalence classes of homeomorphic spaces. Next we would like to classify, or at least to describe different non-homeomorphic spaces. We need some new concepts, properties that can be used for the classification purpose. To this end, we will next define various *topological invariants*, i.e., properties that are invariant under homeomorphisms. Conversely, if we find that a topological invariant for X_1 is not equal to the same topological invariant for X_2, then X_1 is not homeomorphic to X_2.

A topological space X is said to be *connected* when it cannot be written as $X = X_1 \bigcup X_2$, with X_1 and X_2 both open, non-empty and disjoint, i.e., $X_1 \bigcap X_2 = \emptyset$.

If $X \neq \emptyset$ is not connected, its maximal connected subsets are called its *connected components*.

Example The plane with the x-axis removed, $\{(x,y) \in \mathbb{R}^2 | y \neq 0\}$, is not connected. The connected components are the upper half-plane $y > 0$ and the lower half-plane $y < 0$.

Example The *punctured plane* $\mathbb{R}^2 \setminus \{(0,0)\}$ is connected. It is homeomorphic to the infinite cylinder $S^1 \times \mathbb{R}$, which is also connected.

The *neighborhood* N of a point $x \in X$ is a subset $N \subset X$ such that there exists an open set $U \in \tau$, such that $x \in U$ and $U \subset N$, where N does not have to be an open set, as shown in Fig. 4.2.

Let S be a subset of a topological space X. A point $x \in S$ is a *limit point* of S if every neighborhood of x contains at least one point $y \in S$ with $y \neq x$.

Limit points are sometimes also called *cluster points* or *accumulation points*.

The *closure* $\bar{S}$ of S is the union of S and all of its limit points.

It can be shown that if S is connected, then $\bar{S}$ is also connected. Furthermore, all subsets S' between S and $\bar{S}$, $S \subset S' \subset \bar{S}$, are also connected.

Example The limit points of the open interval $S =]0,1[\subset \mathbb{R}$ are 0 and 1, the closure $\bar{S}$ is the closed interval $[0,1]$. The half-open interval $]0,1]$ is an example of a subset S'.

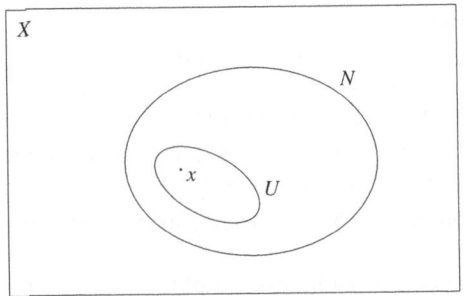

Fig. 4.2   A neighborhood of a point.

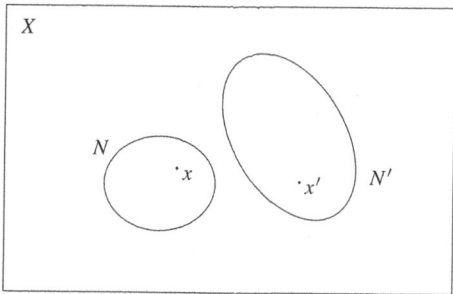

Fig. 4.3 A sketch of the property characterizing a Hausdorff space.

(X, τ) is a *Hausdorff space* if, for an arbitrary pair $x, x' \in X$, $x \neq x'$, there always exist neighborhoods N and N' such that $x \in N$, $x' \in N'$ and $N \cap N' = \emptyset$, as shown in Fig. 4.3.

In physics, it is rare to consider spaces that are not Hausdorff spaces.

Example $\mathbb{R}^n$ with the usual topology is a Hausdorff space. In fact, all spaces X with a metric topology are Hausdorff spaces.

Example Conversely, a topological space X with the trivial topology is a simple example of a topological space that is not a Hausdorff space.

A subset $A \subset X$ is a *closed set* if its complement $X \setminus A = \{x \in X \mid x \notin A \}$ is open.

Note that the notions of open and closed set are not mutually exclusive; in particular, X and $\emptyset$ are both open and closed.

One can show that the closure $\bar{A}$ of a subset A is closed.

A collection $\{A_i\}$ of subsets $A_i \subset X$ is called a *cover* of X if $\bigcup_i A_i = X$.

If all subsets A_i are open sets in the topology τ of X, then $\{A_i\}$ is called an *open cover*.

Let the collections $U = \{A_i\}_{i \in I}$ and $V = \{B_j\}_{j \in J}$ be two covers of X. The cover V is a *refinement* of the cover U, when for every $B_j, j \in J$ there exists an A_i in U such that $B_j \subset A_i$.

Example The collection $U = \{(-2, 1), (-1, 2), (0, 2)\}$ is an open cover of the open interval $X = (-2, 2) \subset \mathbb{R}$, and the collection $V = \{(-2, 0), (-1, 1), (0, 2)\}$ is a refinement of U.

A cover $U = \{A_i\}_{i \in I}$ of X is *locally finite* if, for every $x \in X$, there exists a neighborhood V_x of x such that only finitely many A_is have a non-empty intersection with V_x, as shown in Fig. 4.4.

A topological space (X, τ) is said to be *paracompact* when every open cover of X has a refinement that is a locally finite open cover.

Loosely speaking, paracompact spaces allow one to construct partitions of unity: collections of functions that are nonvanishing only locally, with their sum equal to one at every point.

We can now proceed to define topological manifolds.

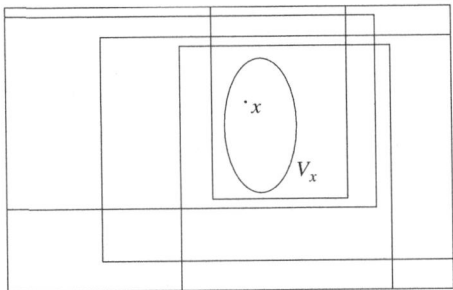

Fig. 4.4   A locally finite cover.

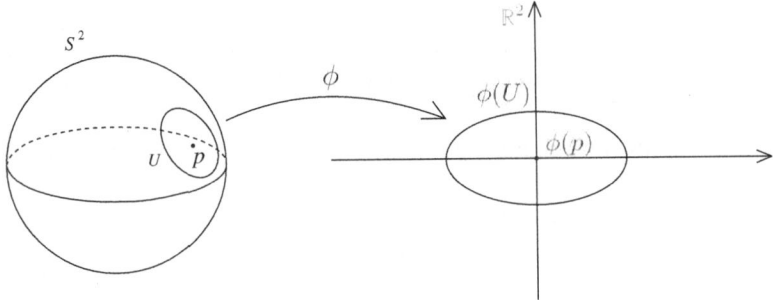

Fig. 4.5   A two-sphere is a manifold of dimension two.

4.1.2 Manifolds

Intuitively, a manifold is a topological space which locally "looks like" a piece of $\mathbb{R}^n$, i.e., is locally homeomorphic to an open subset of $\mathbb{R}^n$. More formally, a topological space (X, τ) is said to be a *manifold* if it is a paracompact Hausdorff space, and is locally homeomorphic to an open subset of $\mathbb{R}^n$. That is, every point $p \in X$ is contained in an open set U_i such that there exists a homeomorphism $\phi_i : U_i \to \phi_i(U_i) \subset \mathbb{R}^n$. The natural number n is the *dimension* of the manifold X, denoted as $\dim X = n$.

Examples

(i) A point is a zero-dimensional manifold.

(ii) $\mathbb{R}^n$ is a manifold of dimension n.

(iii) S^1 is a one-dimensional manifold; in the open segments U_i of S^1 the maps $\phi_i : U_i \to \mathbb{R}, \theta \to \theta$ are the homeomorphisms.

(iv) By similar arguments, n-spheres S^n are n-dimensional manifolds. See Fig. 4.5 for an example of a 2-sphere.

(v) Distinct intersecting lines do not define a manifold, because open sets containing an intersection point are not homeomorphic to any open subset of $\mathbb{R}$.

(vi) A sphere with a half-line sticking out (like the one shown in Fig. 4.6) is not a manifold.

(vii) If M and N are two manifolds of dimensions m and n, respectively, then the product $M \times N$ with the product topology is a manifold called the *product manifold* of dimension $m + n$. Simple examples of these product manifolds are

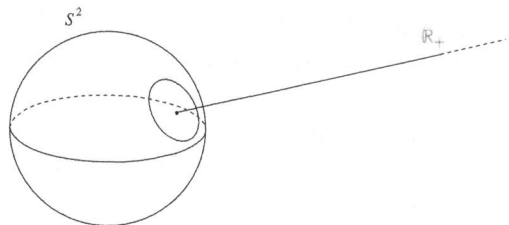

Fig. 4.6 A sphere with a half-line sticking out is not a manifold, because the open subset containing the intersection point is not homeomorphic to any open subset of $\mathbb{R}^1$ or $\mathbb{R}^2$.

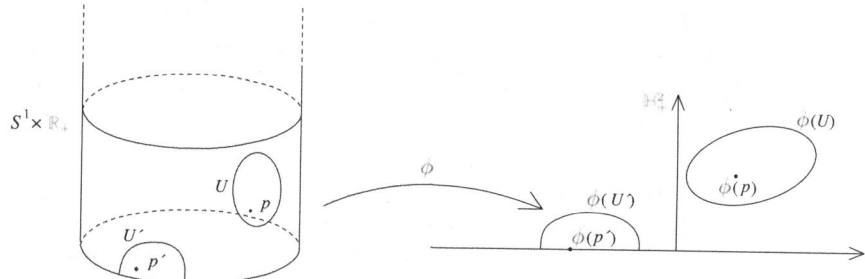

Fig. 4.7 An example of a manifold with a boundary.

the n-torus $T^n = S^1 \times \cdots \times S^1$ (with n factors) and the infinite two-dimensional cylinder $\mathbb{R} \times S^1$.

Let $\mathbb{H}^n_+$ denote the Euclidean (upper) half-space[1] $\{x = (x_1, \ldots, x_n) \in \mathbb{R}^n | x_n \geq 0\}$, with the relative topology induced by the usual topology of $\mathbb{R}^n$. The boundary of $\mathbb{H}^n_+$ is the $(n-1)$-dimensional plane $\partial \mathbb{H}^n_+ = \{x = (x_1, \ldots, x_n) \in \mathbb{R}^n | x_n = 0\}$, and the interior is $\text{int}(\mathbb{H}^n_+) = \{x = (x_1, \ldots, x_n) \in \mathbb{R}^n | x_n > 0\}$. Note that if U is an open set in $\mathbb{R}^n$ such that a part of it does not belong to the intersection $U \cap \mathbb{H}^n_+$, the intersection is by definition an open set in $\mathbb{H}^n_+$, but may not be open in $\mathbb{R}^n$.

A topological space (X, τ) is a *manifold with boundary* if there exists an n such that (X, τ) is a paracompact Hausdorff space locally homeomorphic to an open subset of $\mathbb{H}^n_+$. That is, every point $p \in X$ is contained in an open set U_i such that there exists a homeomorphism $\phi_i : U_i \to \phi_i(U_i) \subset \mathbb{H}^n_+$. Points of X that map to the boundary of $\mathbb{H}^n_+$ are called *boundary points*. The *boundary* of X, denoted as ∂X, is the set of all boundary points.

Examples of Manifolds with a Boundary

(i) The half-infinite cylinder $S^1 \times \mathbb{R}_+ = S^1 \times [0, \infty)$, shown in Fig. 4.7, is a manifold with a boundary.

(ii) The closed cube $I^n = [0, 1] \times \cdots \times [0, 1] = [0, 1]^n \subset \mathbb{R}^n$ is a manifold with boundary. Its boundary ∂I^n is a $(n-1)$-dimensional manifold homeomorphic to S^{n-1}.

---

[1] Later in this section, it will be more convenient to define $\mathbb{H}^n_+$ with $x_1 \geq 0$ instead of $x_n \geq 0$; this is just a relabeling of the coordinate axes.

(iii) More generally, if M is a n-dimensional manifold with boundary, ∂M is a $(n-1)$-dimensional manifold: We leave the proof as an exercise (see Problem 4.5).

(iv) If M is a $(n-1)$-dimensional manifold, then $M \times [0, 1]$ is a n-dimensional manifold with boundary.

(v) If M is a manifold with boundary ∂M, then $M \times \mathbb{R}$ is a manifold with boundary and $\partial(M \times \mathbb{R}) = \partial M \times \mathbb{R}$.

A topological space (X, τ) is *compact* when, for every open covering $\{U_i \mid i \in I\}$, there exists a *finite* subset $J \subset I$ such that $\{U_i \mid i \in J\}$ is also a covering of X, i.e., when every open covering has a finite subcovering.

This formal definition of compactness is not very intuitive. For metric spaces (in particular for subspaces of Euclidean spaces $\mathbb{R}^n$), however, the meaning of compactness can be clarified by using a theorem due to Heinrich Eduard Heine and Émile Borel, that we introduce below.

A metric space (X, d) is said to be *bounded* if there exist a positive real number R (called radius) and a point $x_0 \in X$ (called center), such that the inequality $d(x_0, x) < R$ holds for every $\forall x \in X$. Similarly, a subspace $S \subset X$ is said to be bounded if $\exists R > 0$ and $\exists x_0 \in S$ such that $d(x_0, x) < R$ for every $x \in S$. Note that a subspace $S \subset X$ can be bounded even if the space X is not.

The notion of boundedness helps in elucidating the notion of compact space, through the following theorem.

Theorem 4.1 (Heine–Borel theorem)   *A subspace S of an Euclidean space $\mathbb{R}^n$ is compact if and only if it is closed and bounded.*

Proof   Omitted. $\square$

In more general metric spaces, the situation is more complicated. Compactness always implies closedness and boundedness, but the converse is not necessarily true. However, if a metric space is such that its closedness and boundedness also imply its compactness, then the metric space is said to have the *Heine–Borel property*.

A useful topological invariant for compact two-dimensional manifolds X is the *Euler characteristic*, whose definition can be formulated by finding a polyhedron K homeomorphic to $X \subset \mathbb{R}^3$. For example, a two-sphere S^2 is homeomorphic to a cube. In general, every polyhedron K consists of vertices (points) which are connected by edges (line segments between two vertices) and faces (two-dimensional surfaces whose boundaries are composed of edges). Let $V(K)$ denote the number of vertices of K, $E(K)$ the number of its edges, and $F(K)$ the number of its faces. The Euler characteristic $\chi(X) = \chi(K)$ is then defined as the following combination of these three numbers:

$$\chi(X) = \chi(K) = V(K) - E(K) + F(K). \tag{4.2}$$

For example, $\chi(S^2) = \chi(\text{cube}) = 8 - 12 + 6 = 2$. We leave it to the reader to deduce that for a torus $\chi(T^2) = 16 - 32 + 16 = 0$. By construction, the Euler characteristic is invariant under homeomorphisms, because all manifolds homeomorphic to X are also homeomorphic to the polyhedron K, on which the counting is based.

We conclude this section by summarizing the topological invariants, i.e., quantities or properties that are invariant under homeomorphisms, that we have discussed so far:

1. connectedness,
2. compactness,
3. being a Hausdorff space, and
4. Euler characteristic.

4.2 Homotopy

4.2.1 Homotopy Groups

We will now use what we have learned from group theory to study the properties of loops (and their higher-dimensional counterparts) embedded in topological spaces, in order to define the *homotopy groups*. These turn out to be very useful topological invariants, with many important physics applications, for example for the classification of defects appearing in phase transitions in statistical mechanics and in quantum field theory. We begin by defining loops and paths.

Let X be a topological space, and $I = [0, 1] \subset \mathbb{R}$. A continuous map $\alpha : I \to X$ is a *path* in X. The path α shown in Fig. 4.8 starts at $\alpha_0 = \alpha(0)$ and ends at $\alpha_1 = \alpha(1)$.

If $\alpha_0 = \alpha_1 \equiv x_0$, then α is a *loop* with *base point* x_0, as shown in Fig. 4.9. Notice that a loop may cross itself arbitrarily many times, since the existence of crossings does not violate the continuity of the map α.

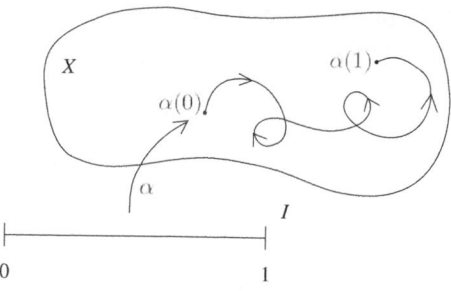

Fig. 4.8   A path.

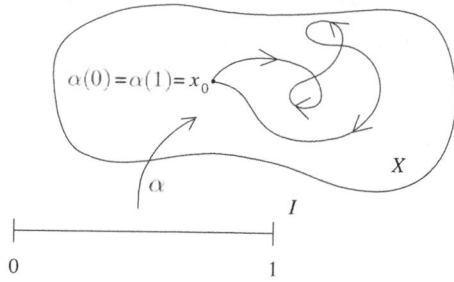

Fig. 4.9   A loop with base point x_0.

The *product* of two loops α and β with the same base point x_0, denoted by $\alpha * \beta$, is the loop

$$(\alpha * \beta)(t) = \begin{cases} \alpha(2t) & 0 \le t \le \frac{1}{2}, \\ \beta(2t-1) & \frac{1}{2} \le t \le 1. \end{cases}$$

Note that the product of loops is not a commutative operation, i.e., in general $\alpha * \beta \neq \beta * \alpha$.

In physics, t could be thought of as a time parameter keeping count on how a loop is traversed. In a product loop, the total time interval (of extent 1) is divided in half, with the first half spent moving along the loop α, and the second half moving along the loop β. An example of a product of two loops is shown in Fig. 4.10.

Homotopy

The concept of homotopy is based on the idea of transforming a continuous function f into another continuous function g (doing this through a map that is a continuous function itself). To this purpose, we introduce $s \in I = [0, 1]$, a parameter keeping track of the deformation starting at $s = 0$ and ending at $s = 1$, as follows. Let f and g be continuous functions $f, g : X \to Y$ between two topological spaces X and Y. A continuous function $F : X \times [0, 1] \to Y, (x, s) \to F(x, s)$ is called a *homotopy* between f and g if $F(x, 0) = f(x)$ and $F(x, 1) = g(x)$ for all $x \in X$. Figure 4.11 shows an example of homotopy.

Next we define the notion of continuously deforming a loop α into another loop β. Heuristically, one can think of this process as the deformation of a rubber band without cutting it (but loops, unlike rubber bands, can cross themselves).

Let α and β be two loops with base point x_0 in X. Then α and β are said to be *homotopic* (a relation denoted as $\alpha \sim \beta$) if there exists a homotopy between them, i.e., a continuous map $F : I \times I \to X, (t, s) \to F(t, s)$ such that

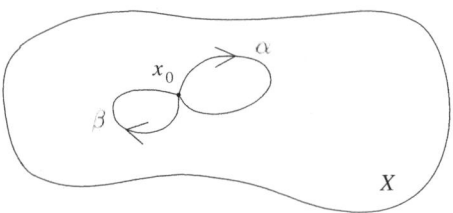

Fig. 4.10   A product of two loops.

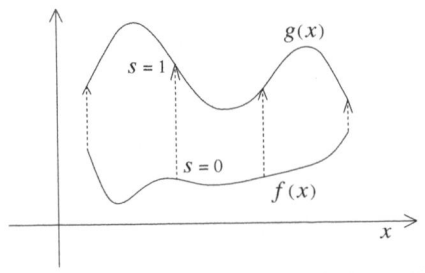

Fig. 4.11   An example of homotopy: As the parameter s is varied from 0 to 1 the f function is continuously deformed into the g function.

$$F(t,0) = \alpha(t) \qquad\qquad \forall t \in I$$
$$F(t,1) = \beta(t) \qquad\qquad \forall t \in I$$
$$F(0,s) = F(1,s) = x_0 \qquad \forall s \in I.$$

Note that sometimes (particularly in the mathematics literature) homotopy is defined without requiring the third condition above, i.e., $F(0,s) = F(1,s) = x_0 \ \forall s \in I$. Strictly speaking, including this requirement defines a homotopy *relative* to the endpoints $\{0,1\}$ of the interval I. We will not make the distinction here.

Homotopy is an equivalence relation.

(i) $\alpha \sim \alpha$: This can be proven by choosing $F(t,s) = \alpha(t) \ \forall t \in I$.
(ii) If $\alpha \sim \beta$ then also $\beta \sim \alpha$: To prove this, if $\alpha \sim \beta$ because there exists the homotopy $F(t,s)$, then $F(t,1-s)$ is the homotopy proving that $\beta \sim \alpha$.
(iii) If $\alpha \sim \beta$ and $\beta \sim \gamma$ then also $\alpha \sim \gamma$: This can be proven by noting that, if $\alpha \sim \beta$, then there exists a homotopy $F(t,s)$ such that $F(t,0) = \alpha(t)$ and $F(t,1) = \beta(t)$; similarly, since $\beta \sim \gamma$, there is a homotopy $G(t,s)$ such that $G(t,0) = \beta(t)$ and $G(t,1) = \gamma(t)$. Then the map

$$H(t,s) = \begin{cases} F(t,2s) & 0 \le s \le \frac{1}{2} \\ G(t,2s-1) & \frac{1}{2} \le s \le 1 \end{cases}$$

is a homotopy between α and γ, so $\alpha \sim \gamma$.

The equivalence class $[\alpha] = \{$loops homotopic with $\alpha\}$ is called the *homotopy class* of α.

Next, we introduce a product between homotopy classes. To this purpose, we introduce the following lemma.

Lemma 4.2   *If $\alpha \sim \alpha'$ and $\beta \sim \beta'$, then $(\alpha * \beta) \sim (\alpha' * \beta')$.*

Proof   Let $F(t,s)$ be a homotopy between α and α' and let $G(t,s)$ be a homotopy between β and β'. Then

$$H(t,s) = \begin{cases} F(2t,s) & 0 \le t \le \frac{1}{2} \\ G(2t-1,s) & \frac{1}{2} \le t \le 1 \end{cases}$$

is a homotopy between $\alpha * \beta$ and $\alpha' * \beta'$. $\square$

One can then define the product of homotopy classes as $[\alpha] * [\beta] \equiv [\alpha * \beta]$, and by the lemma this definition is sensible, i.e., the product is independent of the choice of representatives in $[\alpha]$ and in $[\beta]$.

The set of homotopy classes of loops at $x_0 \in X$ is called the *fundamental group* (or the *first homotopy group*) of X at x_0 and is denoted by $\pi_1(X,x_0)$. As the following theorem shows, calling $\pi_1(X,x_0)$ a *group* is not a misnomer.

Theorem 4.3   *Given a topological space X and a point $x \in X$, the first homotopy group of X at x_0 is a group, with the product between homotopy classes as the group multiplication.*

Proof   To prove that $\pi_1(X,x_0)$ is a group, we show that the product between homotopy classes defined above satisfies the requirements of a group multiplication.

(i) The product operation is closed in $\pi_1(X, x_0)$: for all $[\alpha], [\beta] \in \pi_1(X, x_0)$ we
have $[\alpha] * [\beta] = [\alpha * \beta] \in \pi_1(X, x_0)$, since $\alpha * \beta$ is also a loop at x_0.

(ii) To prove that the product is associative, one can show that $\big((\alpha * \beta) * \gamma\big) \sim$
$\big(\alpha * (\beta * \gamma)\big)$ by considering the homotopy defined as

$$F(t, s) = \begin{cases} \alpha\left(\frac{4t}{1+s}\right) & 0 \le t \le \frac{1+s}{4} \\ \beta(4t - s - 1) & \frac{1+s}{4} \le t \le \frac{2+s}{4} \\ \gamma\left(\frac{4t-s-2}{2-s}\right) & \frac{2+s}{4} \le t \le 1 \end{cases}.$$

From this, it follows that $[(\alpha * \beta) * \gamma] = [\alpha * (\beta * \gamma)] \equiv [\alpha * \beta * \gamma]$.

(iii) The product has the unit element $e = [C_{x_0}]$, where C_{x_0} is the constant path
defined as $C_{x_0}(t) = x_0$ for every $t \in I$. This follows from the existence of the
homotopies

$$F(t, s) = \begin{cases} \alpha\left(\frac{2t}{1+s}\right) & 0 \le t \le \frac{1+s}{2} \\ x_0 & \frac{1+s}{2} \le t \le 1 \end{cases},$$

which shows that $(\alpha * C_{x_0}) \sim \alpha$, and

$$G(t, s) = \begin{cases} x_0 & 0 \le t \le \frac{1-s}{2} \\ \alpha\left(\frac{2t-1+s}{1+s}\right) & \frac{1-s}{2} \le t \le 1 \end{cases},$$

which shows that $(C_{x_0} * \alpha) \sim \alpha$. It follows that $[\alpha * C_{x_0}] = [C_{x_0} * \alpha] = [\alpha]$.

(iv) Given an arbitrary element $[\alpha] \in \pi_1(X, x_0)$, the inverse element $[\alpha]^{-1}$ can be
defined as the homotopy class of $\alpha(1-t)$. To show that $[\alpha]^{-1}$ is really the inverse
of $[\alpha]$, i.e., that $[\alpha * \alpha^{-1}] = [\alpha^{-1} * \alpha] = [C_{x_0}]$, one can define

$$F(t, s) = \begin{cases} \alpha(2t(1 - s)) & 0 \le t \le \frac{1}{2} \\ \alpha(2(1 - t)(1 - s)) & \frac{1}{2} \le t \le 1 \end{cases}.$$

Now we have $F(t, 0) = \alpha * \alpha^{-1}$ and $F(t, 1) = C_{x_0}$, so $\alpha * \alpha^{-1} \sim C_{x_0}$. Similarly,
$\alpha^{-1} * \alpha \sim C_{x_0}$, so we have proven that $[\alpha^{-1} * \alpha] = [\alpha * \alpha^{-1}] = [C_{x_0}]$. □

4.2.2 Properties of the Fundamental Group

So far we have defined the fundamental group with reference to the base point x_0 of
the loops. Often one can remove this restriction. We first need some definitions.

If in a topological space X all pairs of points x_0 and x_1 can be connected with a
path, we say that X is *path-connected*, or *pathwise connected*.

Theorem 4.4 *A path-connected topological space is always connected.*

Proof In this proof we use the fact that $[0, 1]$ is connected. Let X be path-connected.
Suppose X is not connected, so that it has two non-empty open subsets X_1 and X_2
with $X_1 \cap X_2 = \emptyset$, such that $X = X_1 \cup X_2$. Since X is path-connected, one can choose
two points $x_1 \in X_1$ and $x_2 \in X_2$ and connect them with a path $\alpha : [0, 1] \to X$ such that
$\alpha(0) = x_1$ and $\alpha(1) = x_2$. Since α is a continuous function, $U_1 = \alpha^{-1}(X_1)$ and $U_2 =$
$\alpha^{-1}(X_2)$ are disjoint open subsets of $[0, 1]$. One can also verify that $U_1 \cup U_2 = [0, 1]$.
Moreover, $0 \in U_1$ and $1 \in U_2$. It then follows that $[0, 1]$ is not connected, which is a
contradiction. □

Note that the converse statement is not necessarily true, i.e., a connected topological space is not necessarily path-connected. However, it can be shown that a connected *metric* space is also path-connected.

It may not be obvious what a connected, but not path-connected, space could look like. One example is given by "the topologist's sine curve" S, a subset of $\mathbb{R}^2$ that is defined as $S = \{(x,y) \in \mathbb{R}^2 | y = \sin(1/x), x > 0\} \cup (0,0)$, i.e., the points on a curve that oscillates more and more densely as x approaches zero from the positive side, with the origin added. The subset $\{(x,y) \in \mathbb{R}^2 | y = \sin(1/x), x > 0\}$ is connected, and $(0,0)$ is its limit point. Thus S is also connected. To prove that S is not path-connected, a good strategy would be to establish that there are no paths that would connect the origin $(0,0)$ to the rest of the sine curve points $y = \sin(1/x)$ with $x > 0$. We leave the rigorous proof of this as an exercise (see Problem 4.6).

There is another related notion of connectedness, which refers to *arcs*.

An *arc* is a path α which is also a homeomorphism between I and its image $\alpha(I) \subset X$.

A path can cross itself, but an arc cannot; otherwise the inverse map to I would not exist, because it would not be single-valued, as a crossing means that two different points $t_1 \neq t_2 \in I$ map to the same point $\alpha(t_1) = \alpha(t_2)$.

If all pairs of points x_0 and x_1 in X can be connected with an arc, we say that X is *arc-connected* (or *arcwise connected*).

Since an arc is also a path, arc-connectedness implies path-connectedness. The converse is not necessarily true. It can be shown that if X is a Hausdorff space, then path-connectedness implies arc-connectedness. Thus, for Hausdorff spaces the two concepts are equivalent. We will use path-connectedness in what follows.

Now we list some facts, omitting the proofs.

(i) If x_0 and x_1 can be connected by a path, then $\pi_1(X, x_0) \cong \pi_1(X, x_1)$. If X is path-connected, then the fundamental group is independent of the choice of x_0, up to an isomorphism. We can then drop x_0 from the notation and denote the fundamental group simply as $\pi_1(X)$. This statement can be proven using the following argument: Let η be a path from x_1 to x_0. If α is a loop at x_0, then joining it with the path η and its inverse η^{-1} gives a loop $\eta^{-1} * \alpha * \eta$, which is a loop at x_1. Then construct a map $i : \pi_1(X, x_0) \to \pi_1(X, x_1)$, $[\alpha] \to i([\alpha]) = [\eta^{-1} * \alpha * \eta]$, and show that it is well defined and an isomorphism.

(ii) $\pi_1(X)$ is a topological invariant: If $X \approx Y$, then $\pi_1(X) \cong \pi_1(Y)$.

4.2.3 Examples of Fundamental Groups

(i) $\pi_1(\mathbb{R}^2) = \{e\} = \mathbb{Z}_1$, the trivial group.
(ii) $\pi_1(\mathbb{R}^2 \setminus \{\text{a point}\}) = \mathbb{Z}$; see Fig. 4.12.
(iii) $\pi_1(S^1) = \mathbb{Z}$.
(iv) $\pi_1(T^2) = \pi_1(S^1 \times S^1) = \mathbb{Z} \times \mathbb{Z}$.

One can show that $\pi_1(X \times Y) = \pi_1(X) \times \pi_1(Y)$ for path-connected spaces X and Y. This was used above in the example (iv). Note that in the physics literature the trivial group is sometimes denoted by 0, e.g., $\pi_1(\mathbb{R}^2) = 0$. Another somewhat confusing convention in the physics literature is to use the symbol $\oplus$ for the direct product group: $\pi_1(X) \oplus \pi_1(Y)$ instead of $\pi_1(X) \times \pi_1(Y)$.

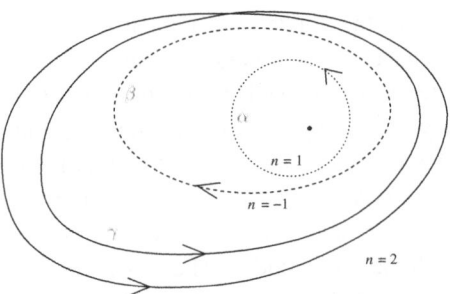

Fig. 4.12 Loops in the punctured plane $\mathbb{R}^2 \setminus \{$ a point$\}$. The black dot denotes the puncture, i.e. the point that is removed from the plane. Three loops, α, β, and γ, are shown. They cannot be deformed across the dot without cutting and gluing back together, hence each of them can be associated with an integer $n \in \mathbb{Z}$, indicating how many times the loop travels around the puncture, and n is called the *winding number*. Note that loops are oriented, as indicated by the arrows. Counterclockwise orientation counts as positive, while clockwise orientation is associated with negative winding. In this example, the winding numbers of the loops α, β, and γ are $+1$, -1, and $+2$, respectively. Loops with the same winding number n can be continuously deformed to each other (i.e., are homotopic), and form a homotopy class $[n]$ characterized by its winding number. In products of loops (first deformed to share a common base point for the product), clockwise winding cancels counterclockwise winding. Thus, multiplying a loop φ with winding number n and a loop ψ with winding number m results into a product loop
$\varphi * \psi$ with winding number $n + m$. The product of homotopy classes then satisfies
$[n] * [m] = [\varphi] * [\psi] = [\varphi * \psi] = [n + m]$. This establishes an isomorphism between the group $\pi(\mathbb{R}^2 \setminus \{$ a point$\})$ and the group $(\mathbb{Z}, +)$.

A topological space X is said to be *simply connected* when it is path-connected and its fundamental group is trivial, $\pi_1(X) = \{e\} \equiv \mathbb{Z}_1$.

For example, $\mathbb{R}^2$ is path-connected and its fundamental group is trivial, hence it is simply connected. $\mathbb{R}^2 \setminus \{$ a point$\}$, instead, is connected and path-connected, but not simply connected.

The *real projective space* is defined as $\mathbb{R}P^n = \{$lines through the origin in $\mathbb{R}^{n+1}\}$. If $x = (x^0, x^1, \ldots, x^n) \neq 0$, then x defines a line. Noting that all $y = \lambda x$ for some nonzero $\lambda \in \mathbb{R}$ are on the same line, we can define the equivalence relation: $y \sim x \Leftrightarrow y = \lambda x$, $\lambda \in \mathbb{R} \setminus \{0\} \Leftrightarrow$ (x and y are on the same line). The lines are the equivalence classes $[x]_\sim$.

Thus $\mathbb{R}P^n = \{[x]_\sim | x \in \mathbb{R}^{n+1} \setminus \{0\}\}$ with the above equivalence relation. One can also define the lines by surrounding the origin with an n-dimensional sphere of unit radius S^n and picking a point on a sphere. But, in this case, points that are related by a reflection through the origin define the same line. Let $\mathbb{Z}_2$ act on S^n by the unit map and the reflection through the origin. Then the lines in $\mathbb{R}^{n+1}$ are in one-to-one correspondence with the points in the quotient space $S^n/\mathbb{Z}_2$ (where opposite points are identified). Thus we can also identify $\mathbb{R}P^n = S^n/\mathbb{Z}_2$. Note that the $\mathbb{R}P^n$ spaces are path-connected.

Example $\mathbb{R}P^2 = S^2/\mathbb{Z}_2$.

To understand the construction of the topological space $S^2/\mathbb{Z}_2$, it is useful to think of S^2 as the set of points of $\mathbb{R}^3$ with coordinates (x, y, z) satisfying $x^2 + y^2 + z^2 = 1$.

Note that, if the point (x_0, y_0, z_0) belongs to S^2, so does the antipodal point $(-x_0, -y_0, -z_0)$, obtained from (x_0, y_0, z_0) by a reflection through the origin $(0, 0, 0)$. Consider then the two maps $e : S \rightarrow S$, $(x, y, z) \rightarrow (x, y, z)$ and $a : S \rightarrow S$, $(x, y, z) \rightarrow (-x, -y, -z)$. Note that e is the identity map in S^2, while a is an involution, i.e., $a^2 = e$. The set $\mathcal{R} = \{e, a\}$, with map composition as the group product, is a group isomorphic to $\mathbb{Z}_2$. Through this group, it is possible to introduce a relation among elements of S^2, that can be denoted as $\sim$ and defined as $P \sim Q$ when $\exists z \in \mathcal{R}$ such that $P = z(Q)$ (i.e., when the coordinates of P and the coordinates of Q are all equal, or all opposite). The $\sim$ relation is reflexive, symmetric, and transitive, hence it is an equivalence relation and it partitions the elements of S^2 into equivalence classes $[P]$. Each equivalence class contains exactly two distinct elements of S^2. Thus $S^2/\mathbb{Z}_2$ can be thought of as the set of these equivalence classes $\{[P]\}$. As the representative of each equivalence class, one can choose the point with $z > 0$ (if the two points in the equivalence class have $z \neq 0$, i.e., if they are not on the "equator" of the sphere), or the point with $y > 0$ (if the two points in the equivalence class have $z = 0$ and $y \neq 0$), or the point with $x = 1$ (for the equivalence class of the two points with $y = z = 0$). The set of points of $S^2/\mathbb{Z}_2$ can thus be represented by the points of the "northern hemisphere," and the points along the "equator," with the convention that all opposite points on the equator are identified. In turn, this set is homeomorphic to the unit disk with opposite boundary points being identified. $S^2/\mathbb{Z}_2$ is sometimes called a *crosscap* and is represented in Fig. 4.13.

Now let us consider loops in $S^2/\mathbb{Z}_2$. They fall into two homotopy classes. In the first class we have loops that cross the boundary of the disk an even number of times (including zero times). When a loop crosses the boundary, it continues on the other side, since the opposite points on the boundary are identified. These "even" loops are represented in Fig. 4.14. All such loops can be continuously contracted to a point, so the homotopy class of these loops $[C_{x_0}]$ is the unit element e of $\pi_1(\mathbb{RP}^2)$.

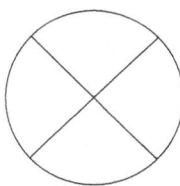

Fig. 4.13   A symbol for a crosscap. The diagonal lines represent the identification of opposite points on the boundary of the disk.

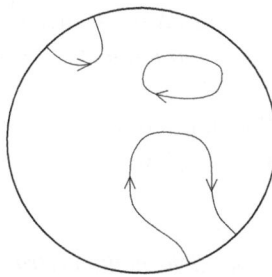

Fig. 4.14   "Even" loops in $S^2/\mathbb{Z}_2$. Note that a loop crossing the boundary continues on the opposite side of the disk.

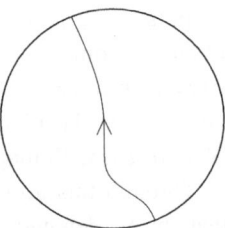

Fig. 4.15   An odd loop α in $S^2/\mathbb{Z}_2$.

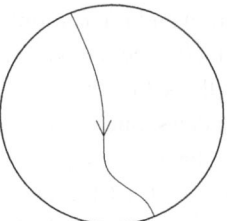

Fig. 4.16   The inverse loop α^{-1}.

The other homotopy class consists of loops that cross the boundary an odd number of times. Figure 4.15 shows a loop α that crosses the boundary of the disk only once. In the original picture of S^2 with antipodal points identified, this would be a path from a point to its antipodal point, but since the two points are identified, it is really a loop. If we rotate the loop by π, it becomes its inverse α^{-1}, depicted in Fig. 4.16. Thus $\alpha \sim \alpha^{-1}$, so $\alpha^2 = \alpha * \alpha \sim \alpha * \alpha^{-1} \sim C_{x_0}$. This means that the even loop α^2 can be contracted to a point. By induction, $\alpha^{2k+1} \sim \alpha$ and $\alpha^{2k} \sim C_{x_0}$. So there exist only two homotopy classes, $[C_{x_0}] \equiv e$ and $[\alpha]$. Since the group has two elements, it must be $\mathbb{Z}_2$: we conclude that

$$\pi_1(\mathbb{R}P^2) = \mathbb{Z}_2. \tag{4.3}$$

In fact, one can prove that $\pi_1(\mathbb{R}P^n) = \mathbb{Z}_2$ for all $n \geq 2$.

4.2.4 Retracts

Sometimes, in order to identify the homotopy group of a topological space, it is helpful to reduce its shape to an essential form. For that purpose, we introduce some new concepts. In the following definitions, let $A \neq \emptyset$ be a subspace of a topological space X.

A continuous map $f : X \to A$, such that the restriction of f to A is the identity map, $f|_A = \mathrm{id}_A$, is called a *retraction*. If a retraction exists, we say that A is a *retract* of X.

Example   The unit circle S^1 can be thought of as a retract of $\mathbb{R}^2 \setminus \{(0,0)\}$ and also as a retract of $\mathbb{R}^2 \setminus \{(0,0),(0,2)\}$. A possible retraction is $f(x) = \frac{x}{||x||}$, as shown in Fig. 4.17.

A subspace A is a *deformation retract* of X, if there exists a continuous map

$$F : X \times I \to X, \ (x,s) \to F(x,s)$$

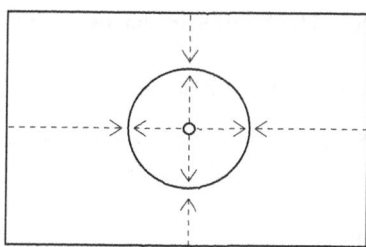

A retraction.

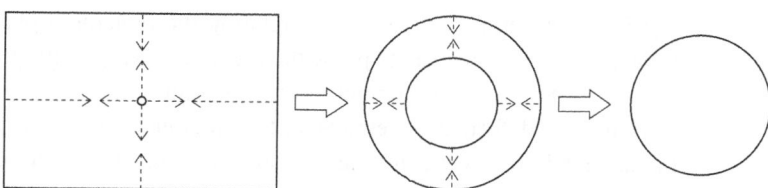

Deformation retraction.

that satisfies the following three properties:

(i) $F(x, 0) = x$ $\forall x \in X$ (at the beginning of the interval I, i.e., for $s = 0$, F is the identity map on X),

(ii) $F(y, 1) = y$ $\forall y \in A$ (at the end of the interval I, i.e., for $s = 1$, the restriction of F on A is the identity map in A), and

(iii) $F(x, 1) \in A$ $\forall x \in X$ (at the end of the interval I, the function F maps every point of X to A).

The properties (ii) and (iii) mean that the identity map in X is continuously deformed to a retraction of X to A (or, in other words, a retraction is homotopic to the identity map). Figure 4.18 shows an example of a deformation retraction.

Note that sometimes in the literature property (ii) is replaced with a stronger requirement, as follows.

$\forall s \in I : F(y, s) = y$ $\forall y \in A$ (for *all* values of s, the restriction of F to A is the identity map on A.)

This actually defines what is called a *strong deformation retract* in mathematics.

Example S^n is a (strong) deformation retract of $\mathbb{R}^{n+1} \setminus \{0\}$. This can be shown by considering, for instance, the map

$$F(x, s) = \left((1 - s) + \frac{s}{||x||} \right) x,$$

where $||x||$ is the norm (length) of x.

Example A unit circle S^1 is *not* a deformation retract of $\mathbb{R}^2 \setminus \{(0, 0), (0, 2)\}$. Thus, being a deformation retract is a stronger property than being a retract. For example, the mapping above no longer works: Since the image points have a hole at the image of the missing point $(0, 2)$, the mapping is not continuous for all values of s.

If a deformation retract of X can be a point ($A = \{x_0\}$), then X is said to be *contractible*.

Example The Euclidean spaces $\mathbb{R}^n$ are contractible.

The following theorem holds.

Theorem 4.5 *Let A be a path-connected deformation retract of X. Then $\pi_1(X, x_0) = \pi_1(A, x_0) = \pi_1(A)$ for all $x_0 \in A$.*

Proof Omitted. $\square$

This theorem often helps to identify fundamental groups. As a simple application, $\pi_1(\mathbb{R}^2 \setminus \{a\ point\}) = \pi_1(S^1) = \mathbb{Z}$, relating the punctured plane to its deformation retract S^1. This is one way to prove the previous example (iii) of fundamental groups, since S^1 is a deformation retract of $\mathbb{R}^2 \setminus \{a\ point\}$.

To find the deformation retracts of, for example, a plane with n punctures, and their fundamental groups, we introduce a new concept: The topological space obtained by gluing together n circles S^1 at a single point is called a *rose with n petals*, or an *n-bouquet*. As an example, Fig. 4.19 shows a rose with four petals.

Note that roses with $n > 1$ petals are not manifolds, because of the crossing point. A rose with one petal, S^1, is a deformation retract for a plane with a point removed. Similarly, one can show that a rose with n petals is a deformation retract for a plane with n points removed. It is a bit more complicated to see that a rose with two petals (a "figure eight") is also a deformation retract for a torus with one point removed, $T^2 \setminus \{a\ point\}$; this is shown in Fig. 4.20.

The fundamental groups of roses with n petals are free groups with n generators. To see this, we first observe that a loop can wind around a single point $k \in \mathbb{Z}$ times, where the sign of k corresponds to the orientation of the loop, just as in the case of a circle. Moreover, we also note that a loop cannot be continuously moved from one petal to another one. Therefore, for a rose with n petals there are n basic loops $\alpha_1, \ldots, \alpha_n$. Their products α_i^k give loops around the petal i with winding number k. It is less obvious that the product is noncommutative; for example, $\alpha_1 * \alpha_2$ is

Fig. 4.19   A rose with four petals.

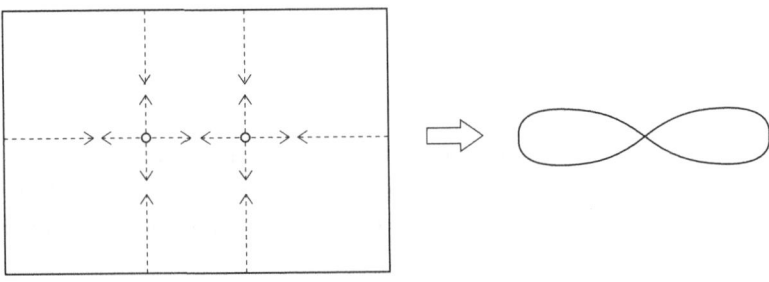

Fig. 4.20   Deformation retraction of a plane with two punctures to a rose with two petals.

a loop where first petal 1 is travelled around once, then petal 2, while the product $\alpha_2 * \alpha_1$ describes a different loop, traveling through the petals in the opposite order. Since the loops cannot be continuously moved to different petals, the two product loops are not homotopic. This observation generalizes to arbitrary products of loops. Thus, the fundamental group of a rose with n petals consists of the set of generators $X_n = \{[\alpha_1], \ldots, [\alpha_n]\}$, the homotopy classes of the loops through each petal (with the same orientation), and their arbitrary products (words of arbitrary length), i.e., is the free group with n generators $F(X_n)$.

4.2.5 Higher Homotopy Groups

As we discussed earlier, a loop was a continuous map of the closed one-dimensional interval I to a topological space, with the endpoints (boundary of the closed interval) mapping to the same point. An obvious generalization is to increase the dimensionality by starting with a closed n-dimensional interval $I^n = \{(t_1, \ldots, t_n)| \ 0 \leq t_i \leq 1, \ 1 \leq i \leq n\}$ with the boundary $\partial I^n = \{(t_1, \ldots, t_n)| \ \text{some } t_i = 0 \text{ or } 1\}$.

A continuous map $\alpha : I^n \rightarrow X$ that maps every point on ∂I^n to the same point $x_0 \in X$ is called an *n-loop* at $x_0 \in X$. Let α and β be n-loops at x_0. We say that α is homotopic to β, denoting this as $\alpha \sim \beta$, when there exists a continuous map $F : I^n \times I \rightarrow X$ such that

$$F(t_1, \ldots, t_n, 0) = \alpha(t_1, \ldots, t_n)$$
$$F(t_1, \ldots, t_n, 1) = \beta(t_1, \ldots, t_n)$$
$$F(t_1, \ldots, t_n, s) = x_0 \quad \forall s \in I \text{ when } (t_1, \ldots, t_n) \in \partial I^n.$$

This homotopy relation $\alpha \sim \beta$ is an equivalence relation that partitions the set of n-loops into homotopy classes $[\alpha]$. Again, strictly speaking the above defines a homotopy relative to ∂I^n.

Products $\alpha * \beta$ of n-loops are defined by splitting the interval of one of the parameters, say t_1:

$$\alpha * \beta(t_1, \ldots, t_n) = \begin{cases} \alpha(2t_1, t_2, \ldots, t_n) & 0 \leq t_1 \leq \frac{1}{2} \\ \beta(2t_1 - 1, t_2, \ldots, t_n). & \frac{1}{2} \leq t_1 \leq 1 \end{cases}$$

The inverse loop α^{-1} is then defined by

$$\alpha^{-1}(t_1, \ldots, t_n) = \alpha(1 - t_1, \ldots, t_n).$$

The trivial loop C_{x_0} is $C_{x_0}(t_1, \ldots, t_n) = x_0, \ \forall(t_1, \ldots, t_n) \in I^n$.

The set of homotopy classes of n-loops $\pi_n(X, x_0)$ with the product $[\alpha] * [\beta] = [\alpha * \beta]$ and with the unit element $[C_{x_0}]$ is again a group, called the *nth homotopy group* of X at x_0. If X is path-connected, then $\pi_n(X, x_0)$ is again independent of the base point x_0 and one can shorten the notation to $\pi_n(X)$. It classifies homotopically inequivalent continuous maps of S^n to X. In particular, $\pi_2(X)$ classifies homotopically inequivalent closed two-dimensional surfaces in X.

Some properties of the fundamental group generalize to the higher homotopy groups, for instance:

(i) If $A \subset X$ is a deformation retract of X, then $\pi_n(A, x_0) = \pi_n(X, x_0)$.

(ii) If X and Y are path-connected topological spaces, then $\pi_n(X \times Y) = \pi_n(X) \times \pi_n(Y)$ (again, in physics notation the $\oplus$ symbol can be used instead of $\times$).

However, the higher homotopy groups are simpler than the fundamental group. In particular, the following theorem holds.

Theorem 4.6 *The n-dimensional homotopy groups $\pi_n(X, x_0)$ are Abelian for $n > 1$.*

Proof Omitted.   □

Example $\pi_2(S^2) = \mathbb{Z}$. This homotopy group consists of the equivalence classes associated with the inequivalent ways to map a two-sphere to a two-sphere, just like $\pi_1(S^1)$ classifies the inequivalent ways of mapping a circle onto a circle. In general, for n-spheres S^n, one has $\pi_n(S^n) = \mathbb{Z}$. Likewise, one can argue that every k-loop with $k < n$ can be contracted to a point. Thus, $\pi_k(S^n) = \{e\}$ (the trivial group) for $k < n$.

It may not be immediately obvious that even if the dimension of X is strictly less than n, it is still possible to define n-loops α on X; this is possible because all that is required in the definition of an n-loop is that α is a continuous map of I^n to X. A trivial example, which we have already discussed, is the trivial loop C_{x_0}. Thus, it makes sense to study the nth homotopy groups of S^k even if $k < n$. Table 4.1 summarizes the results for $1 \le k, n \le 7$.

An important example of a nontrivial higher homotopy group is $\pi_3(S^2) = \mathbb{Z}$, classifying the inequivalent ways to wrap a three-sphere around a two-sphere. The key property here is the *Hopf fibration* of S^3, a construction whereby the three-sphere is decomposed into a two-sphere and circles at every point of the two-sphere in a nontrivial way. Then the two-sphere part of S^3 can be wrapped $n \in \mathbb{Z}$ times around the two-sphere, giving rise to the nontrivial homotopy.

The Hopf fibration is a surjective projection map $p : S^3 \to S^2$ such that the counter-image of every point $y \in S^2$, i.e., the set of all points $x \in S^3$ such that $p(x) = y$ (that one, with some slight abuse of notation, can indicate as $p^{-1}(y)$, even though p^{-1} is not defined as a function), is a circle. Consider the following embedding of S^3 into the two-dimensional complex space $\mathbb{C}^2$:

$$S^3 = \{(z_1, z_2) \in \mathbb{C}^2, \text{ with } |z_1|^2 + |z_2|^2 = 1\}. \tag{4.4}$$

Table 4.1. Table of homotopy groups for spheres

	S^1	S^2	S^3	S^4	S^5	S^6	S^7
π_1	$\mathbb{Z}$	$\mathbb{Z}_1$	$\mathbb{Z}_1$	$\mathbb{Z}_1$	$\mathbb{Z}_1$	$\mathbb{Z}_1$	$\mathbb{Z}_1$
π_2	$\mathbb{Z}_1$	$\mathbb{Z}$	$\mathbb{Z}_1$	$\mathbb{Z}_1$	$\mathbb{Z}_1$	$\mathbb{Z}_1$	$\mathbb{Z}_1$
π_3	$\mathbb{Z}_1$	$\mathbb{Z}$	$\mathbb{Z}$	$\mathbb{Z}_1$	$\mathbb{Z}_1$	$\mathbb{Z}_1$	$\mathbb{Z}_1$
π_4	$\mathbb{Z}_1$	$\mathbb{Z}_2$	$\mathbb{Z}_2$	$\mathbb{Z}$	$\mathbb{Z}_1$	$\mathbb{Z}_1$	$\mathbb{Z}_1$
π_5	$\mathbb{Z}_1$	$\mathbb{Z}_2$	$\mathbb{Z}_2$	$\mathbb{Z}_2$	$\mathbb{Z}$	$\mathbb{Z}_1$	$\mathbb{Z}_1$
π_6	$\mathbb{Z}_1$	$\mathbb{Z}_{12}$	$\mathbb{Z}_{12}$	$\mathbb{Z}_2$	$\mathbb{Z}_2$	$\mathbb{Z}$	$\mathbb{Z}_1$
π_7	$\mathbb{Z}_1$	$\mathbb{Z}_2$	$\mathbb{Z}_2$	$\mathbb{Z} \times \mathbb{Z}_{12}$	$\mathbb{Z}_2$	$\mathbb{Z}_2$	$\mathbb{Z}$

Then parameterize S^3 with three angles α_1, α_2, and β such that

$$z_1 = \exp\left(i\frac{\alpha_2 + \alpha_1}{2}\right) \sin\beta$$
$$z_2 = \exp\left(i\frac{\alpha_2 - \alpha_1}{2}\right) \cos\beta. \tag{4.5}$$

Next, we consider S^2 embedded in $\mathbb{C} \times \mathbb{R}$,

$$S^2 = \{(w,x) \in \mathbb{C} \times \mathbb{R}, \text{ with } |w|^2 + x^2 = 1\}, \tag{4.6}$$

and define the map

$$p : S^3 \rightarrow S^2,$$
$$(z_1, z_2) \rightarrow p(z_1, z_2) = (2z_1 z_2^\star, |z_2|^2 - |z_1|^2). \tag{4.7}$$

Since

$$2z_1 z_2^\star = 2e^{i\alpha_1} \sin\beta \cos\beta = e^{i\alpha_1} \sin(2\beta),$$
$$|z_2|^2 - |z_1|^2 = \cos^2\beta - \sin^2\beta = \cos(2\beta), \tag{4.8}$$

we can verify that p maps to S^2 as defined in Eq. (4.6), and is surjective. Moreover, if two points (z_1, z_2) and (z'_1, z'_2) map to the same point, they must satisfy

$$(z_1, z_2) = \lambda(z'_1, z'_2), \quad \text{with } |\lambda|^2 = 1, \tag{4.9}$$

so that $\lambda = e^{i\phi}$ with $\phi \in S^1$. Thus the counter-image $p^{-1}(y)$ for every $y \in S^2$ must be a circle. Note that, while this construction shows that S^3 locally "looks like" $S^2 \times S^1$, this is not true globally ($S^3 \neq S^2 \times S^1$). More precisely, S^3 with the Hopf fibration is an example of a *fiber bundle*, a manifold which locally can be decomposed into a *fiber* (in this case S^1) at every point of a *base space* (in this case S^2) in a nontrivial way. The visualization of how the circle fibers are linked with one another as one moves between the points on the two-sphere base space is quite complicated.

4.2.6 Homotopy Groups and Exact Sequences

The homotopy groups of N-dimensional spheres are interesting in physics also because of their relation to Lie groups. For example, S^3 is the group manifold of the group SU(2). Recall also that $S^N = \mathrm{O}(N+1)/\mathrm{O}(N) = \mathrm{SO}(N+1)/\mathrm{SO}(N)$. Furthermore, for spheres of odd dimension $N = 2k+1 \geq 3$, one also has $S^{2k+1} = \mathrm{U}(k+1)/\mathrm{U}(k)$. From Table 4.1 we see that

$$\pi_q(S^N) = \begin{cases} \mathbb{Z}_1 & \text{for } q < N \\ \mathbb{Z}, & \text{for } q = N \end{cases}$$

where $\mathbb{Z}_1 = \{e\}$ denotes the trivial group, containing only the identity element. For $q > N$, in general $\pi_q(S^N)$ can be nontrivial; in particular, the simplest nontrivial case is $\pi_3(S^2) = \mathbb{Z}$ (related to the Hopf fibration).

In order to compute the homotopy groups of various manifolds, it is often useful to resort to *exact sequences* of group homomorphisms.

A generic sequence of groups G_i and group homomorphisms $f_i : G_i \rightarrow G_{i+1}$

$$\cdots \longrightarrow G_i \xrightarrow{f_i} G_{i+1} \xrightarrow{f_{i+1}} G_{i+2} \longrightarrow \cdots$$

is said to be "exact" when, for every a, except, possibly, at the end of the sequence, one has $\mathrm{Im}\, f_a \cong \mathrm{Ker}\, f_{a+1}$, where the symbol $\cong$ denotes group isomorphism.

Recall that, due to the fundamental theorem of group homomorphisms (Theorem 2.6), given a group homomorphism f_i defined on a group G_i, one has

$$\mathrm{Im}\, f_i = G_i/\mathrm{Ker}\, f_i.$$

Given a Lie group G, and a compact Lie subgroup $H \subset G$, it is possible to prove that the following sequence:

$$\cdots \to \pi_q(H) \to \pi_q(G) \to \pi_q(G/H) \to \pi_{q-1}(H) \to \pi_{q-1}(G) \to \pi_{q-1}(G/H) \to \cdots,$$

constructed by repeating the basic block $\cdots \to \pi_m(H) \to \pi_m(G) \to \pi_m(G/H) \to \cdots$ for values of m which decrease by 1 every time, is exact.

Typically, the determination of nontrivial homotopy groups using exact sequences can be done by:

- considering a portion of the sequence starting from and ending in the trivial group $\mathbb{Z}_1$,
- using the definition of exact sequence,
- using the group homomorphisms' theorem.

As an example, consider the computation of $\pi_3\,(\mathrm{SO}(3))$. Since $\mathrm{SO}(3) \cong \mathrm{SU}(2)/\mathbb{Z}_2$, one can write

$$\pi_3(\mathbb{Z}_2) \xrightarrow{f} \pi_3\,(\mathrm{SU}(2)) \xrightarrow{g} \pi_3\,(\mathrm{SO}(3)) \xrightarrow{h} \pi_2(\mathbb{Z}_2).$$

First of all, we have $\pi_3(\mathbb{Z}_2) \cong \pi_2(\mathbb{Z}_2) \cong \mathbb{Z}_1$. Second, we note that $\pi_3\,(\mathrm{SU}(2)) \cong \mathbb{Z}$, because the group manifold of $\mathrm{SU}(2)$ is S^3. Next, one can observe that h necessarily maps every element of its domain $\pi_3\,(\mathrm{SO}(3))$ to the identity element of $\mathbb{Z}_1 \cong \pi_2(\mathbb{Z}_2)$. Hence, $\mathrm{Ker}\, h \cong \pi_3\,(\mathrm{SO}(3))$. Since the sequence is exact, one obtains $\mathrm{Im}\, g \cong \mathrm{Ker}\, h \cong \pi_3\,(\mathrm{SO}(3))$. From the homomorphism theorem, one also has $\mathrm{Im}\, g \cong \pi_3\,(\mathrm{SU}(2))/\mathrm{Ker}\, g$. Then, given that the sequence is exact, one has $\mathrm{Ker}\, g \cong \mathrm{Im}\, f$, but, since the domain of f is isomorphic to $\mathbb{Z}_1$, and f is a homomorphism, f necessarily maps the unique element of its domain to the identity element in $\pi_3\,(\mathrm{SU}(2))$, so $\mathrm{Im}\, f \cong \mathbb{Z}_1$. Thus $\mathrm{Ker}\, g = \mathbb{Z}_1$. To summarize, we find that

$$\pi_3\,(\mathrm{SO}(3)) \cong \mathrm{Ker}\, h \cong \mathrm{Im}\, g \cong \pi_3\,(\mathrm{SU}(2))/\mathrm{Ker}\, g \cong \pi_3\,(\mathrm{SU}(2))/\mathrm{Im}\, f$$

$$\cong \pi_3\left(S^3\right)/\mathrm{Im}\, f \cong \pi_3\left(S^3\right)/\mathbb{Z}_1 \cong \mathbb{Z}/\mathbb{Z}_1 \cong \mathbb{Z}.$$

4.3 Differentiable Manifolds

In this section we define additional structure for manifolds, enabling one to perform calculus on them. We start by introducing coordinates. A familiar setting, which can serve as a motivation, is the surface of the Earth, where geographic locations are specified by latitudes and longitudes. In schools, a traditional geography textbook was called the world atlas, a collection of maps (or *charts*) covering different geographical regions. On each chart there was a coordinate grid showing the latitudes and longitudes. Moving from one region to other regions was facilitated by the charts having some overlap, so that moving from one map to another one could identify the same location in an overlapping region, and then continue further. So if we (simplistically) think of the surface of the Earth as a two-sphere S^2, an atlas is simply a collection of overlapping regions covering the whole sphere, and for each region there are charts associating coordinates to all points on that region of the sphere. Now, we introduce a sequence of definitions to give mathematically rigorous meaning to such concepts.

Let M be a set (note that, so far, it does not necessarily have to be a topological manifold), and let U be a subset of M (a "region"). If there exists a bijection $\mathbf{x}$ from U to an open subset of $\mathbb{R}^n$, the pair $(U, \mathbf{x})$ is called a *chart*. We say that the chart is $\mathbb{R}^n$-valued.

Note that, since $\mathbf{x}$ takes values in $\mathbb{R}^n$ and is a bijection, it essentially "glues" the coordinates of $\mathbb{R}^n$ onto the region U of M, associating coordinates with each point of U and, depending on the details of the mapping, stretches and deforms the coordinate grids, as is necessary if M is, for example, the two-sphere. It is common to call U a *coordinate neighborhood* and $\mathbf{x}$ a *coordinate function*. When M is a manifold, we typically assume $\mathbf{x} : U \to \mathbf{x}(U)$ to be a homeomorphism.

We will also often use the inverse of the coordinate function to label a point $p \in U$ by the coordinates x that we are using in $\mathbb{R}^n$. In this case it is more convenient to use an alternative notation, e.g., ϕ^{-1} rather than $\mathbf{x}^{-1}$, the reason being that $U \ni p = \phi^{-1}(x)$ looks less confusing than $p = \mathbf{x}^{-1}(x)$. We illustrate this with the following example, which may appear to be overly pedantic at this point. Its purpose is to emphasize the distinction between a point p in M and the coordinates used to label it, which are borrowed from a patch of $\mathbb{R}^n$.

Example   Let $M = T^2 = S^1 \times S^1$, the two-dimensional torus. It would be natural to use two angles (φ_1, φ_2) to label a point on it. With the above definitions, the rigorous version of that convention is the following. Let us choose as a coordinate region U an open set on T^2. Then U is homeomorphic to the open rectangle $(-a, a) \times (-b, b) \subset \mathbb{R}^2$ for some $0 < a, b < \pi$, as shown in Fig. 4.21. In other words, there exists a homeomorphism (coordinate function) $\mathbf{x} : U \to (-a, a) \times (-b, b) \subset \mathbb{R}^2$, $\mathbf{x}(U) = (-a, a) \times (-b, b)$. Let us use the coordinates (φ_1, φ_2) for the points in $(-a, a) \times (-b, b)$, i.e., $-a < \varphi_1 < a$ and $-b < \varphi_2 < b$. A point p in U is then the inverse image of a point (φ_1, φ_2): $p = \mathbf{x}^{-1}(\varphi_1, \varphi_2)$. Conversely, the image of the point p on $\mathbb{R}^2$ can be labeled by (φ_1, φ_2), namely $\mathbf{x}(p) = (\varphi_1, \varphi_2)$. So, when we cavalierly write "$p = (\varphi_1, \varphi_2) \in T^2$," we are implicitly thinking of the coordinates as being "drawn

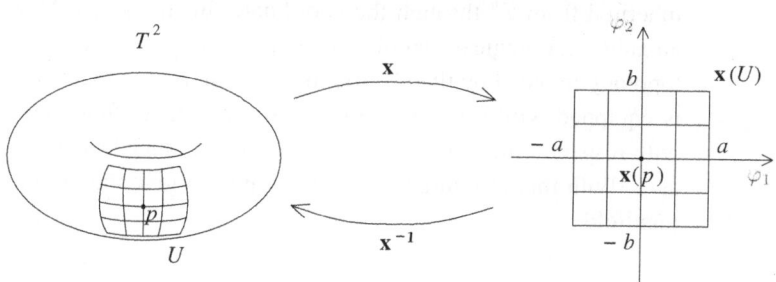

Fig. 4.21   A chart on a torus.

on the manifold." More precisely, the coordinates are "glued" by the coordinate function, and one should actually write $p = \mathbf{x}^{-1}(\varphi_1, \varphi_2) \in T^2$.

Let $(U, \mathbf{x})$ be a chart on M, and let p be a point in U. If $\mathbf{x}(p) = 0 \in \mathbb{R}^n$, the chart is *centered* at p.

This means that we choose the origin of the coordinate grid to be at the point p. As a geographical example, on the surface of the Earth, the origin of coordinates $(0,0)$, namely the point at $0°$ latitude North, $0°$ longitude East,[2] has been set to a fictional island called the Null Island in the Gulf of Guinea. The point $(10, 2)$ can then be found in Benin.

We are now ready to cover the whole set M with a collection $(U_\alpha, \mathbf{x}_\alpha)$ of overlapping charts, indexed by an index set $A \ni \alpha$, to define a "world atlas" of M. We can control how smoothly moving from one chart to another happens: The composite map $\mathbf{x}_\alpha \circ \mathbf{x}_\beta^{-1}$ can be defined in a non-empty intersection of two charts labeled by α and β, and this composite function is a mapping from a subset of $\mathbb{R}^n$ (the image of $U_\alpha \cap U_\beta$) to another subset of $\mathbb{R}^n$. Such maps are called *transition functions, overlap maps*, or *change of coordinate maps*.

Recall that a map $f \colon \mathbb{R}^n \supset U \to \mathbb{R}^n$ is said to be a *class-r map* (with $1 \leq r \leq \infty$), denoted as C^r, if it is continuous, and all of its partial derivatives up to order r exist and are continuous:

$$\frac{\partial^r f^l}{\partial (x^1)^{r_1} \ldots \partial (x^n)^{r_n}}, \quad f = (f^1, \ldots, f^n), \quad l = 1, \ldots, m, \quad r_1 + r_2 + \cdots + r_n = r.$$

A C^∞ map has derivatives up to arbitrary order and is also called a *smooth map*, or an *infinitely differentiable map*. This notion can be used to classify the smoothness of the composite maps $\mathbf{x}_\alpha \circ \mathbf{x}_\beta^{-1}$.

A collection $\mathcal{A} = \{(U_\alpha, \mathbf{x}_\alpha)\}_{\alpha \in A}$ of charts on M is an $\mathbb{R}^n$-valued *atlas of class r* (or a C^r atlas), if the following conditions are satisfied.

(i) The collection covers all of M, namely: $\cup_{\alpha \in A} U_\alpha = M$.
(ii) The intersections in $\mathbb{R}^n$ are open sets: $\mathbf{x}_\alpha(U_\alpha \cap U_\beta)$ are open in $\mathbb{R}^n$ for all α, β.
(iii) The overlap maps are C^r maps: Whenever $U_\alpha \cap U_\beta \neq \emptyset$, the function $\mathbf{x}_\alpha \circ \mathbf{x}_\beta^{-1}$ is a C^r diffeomorphism.

Two C^r atlases $\mathcal{A}_1$ and $\mathcal{A}_2$ are said to be *equivalent* if their union $\mathcal{A}_1 \cup \mathcal{A}_2$ is also a C^r atlas. As the name suggests, this relation is truly an equivalence relation among C^r atlases; each equivalence class is called a *differentiable structure*.

We will now short-cut some details. Note that so far M could be just a set. Given a C^r differentiable structure, one can use it to induce a topology on M (one inherited from $\mathbb{R}^n$ through the coordinate functions $\mathbf{x}_\alpha$). Now, suppose that M was already a topological manifold with a topology τ. Is this topology the same as the topology induced by the C^r structure? A criterion is that if the topological manifold is equipped with a C^r atlas, then if all coordinate functions are homeomorphisms with respect to the topology τ, the topology induced by the C^r structure is the same as τ. With this in mind, we are now going to give the definition of a differentiable manifold.

---

[2] Note that latitudes and longitudes are customarily expressed in arc degrees, not in radians.

A topological space M is an *n-dimensional differentiable manifold of class r* if

(i) M is provided with a collection $\mathcal{A} = \{(U_\alpha, \mathbf{x}_\alpha)\}$, where $\{U_\alpha\}$ is an open covering of M: $\bigcup_\alpha U_\alpha = M$, and every $\mathbf{x}_\alpha : U_\alpha \to U'_\alpha \subset \mathbb{R}^n$, U'_α open, is a homeomorphism, and

(ii) the collection $\mathcal{A}$ is a C^r atlas (all overlap maps $\psi_{\alpha\beta} \equiv \mathbf{x}_\alpha \circ \mathbf{x}_\beta^{-1}$ are C^r diffeomorphisms).

The number n is the *dimension of the manifold*; we denote dim $M = n$. In what follows, we will focus on C^∞ differentiable manifolds, often called *smooth manifolds* or, sometimes, *differentiable manifolds*.

Note that a given smooth manifold M can have several different differentiable structures; for example, S^7 has 28, while $\mathbb{R}^4$ has infinitely many differentiable structures.

Example   Let us realize S^n as a subset of $\mathbb{R}^{n+1}$:

$$S^n = \left\{ x \in \mathbb{R}^{n+1} \,\middle|\, \sum_{i=0}^{n} (x^i)^2 = 1 \right\}.$$

One possible atlas consists of the

• coordinate neighborhoods (the labels α are now called $i\pm$):

$$U_{i+} \equiv \{x \in S^n | x^i > 0\},$$
$$U_{i-} \equiv \{x \in S^n | x^i < 0\},$$

• coordinate functions:

$$\mathbf{x}_{i+}(x^0, \dots, x^n) = (x^0, \dots, x^{i-1}, x^{i+1}, \dots, x^n) \in \mathbb{R}^n,$$
$$\mathbf{x}_{i-}(x^0, \dots, x^n) = (x^0, \dots, x^{i-1}, x^{i+1}, \dots, x^n) \in \mathbb{R}^n,$$

which are projections onto the plane $x^i = 0$. See Fig. 4.22 for a sketch of the coordinate neighborhoods.

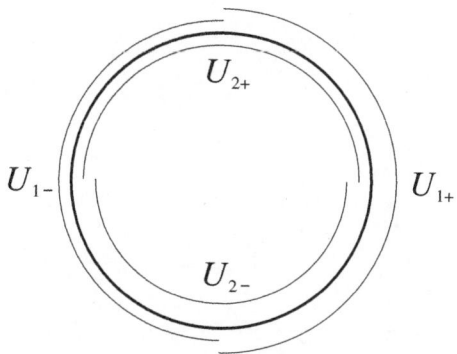

A sketch of the coordinate neighborhoods of S^1 discussed in the example in the text.

Assuming $i \neq j$, with $\alpha = \pm$, and $\beta = \pm$, the transition functions are

$$\psi_{i\alpha j\beta} = \mathbf{x}_{i\alpha} \circ \mathbf{x}_{j\beta}^{-1},$$
$$(x^0, \ldots, x^i, \ldots, x^{j-1}, x^{j+1}, \ldots, x^n)$$
$$\to \left(x^0, \ldots, x^{i-1}, x^{i+1}, \ldots, x^{j-1}, \beta\sqrt{1 - \sum_{k \neq j}(x^k)^2}, x^{j+1}, \ldots, x^n\right),$$

which are of class C^∞.

There are other compatible atlases, e.g., the stereographic projection.

As with topological manifolds, one can also define a *differential manifold with a boundary of class C^r* by restricting the coordinates into the half-space; one requires $\mathbf{x}_\alpha : U_\alpha \to U_i' \subset \mathbb{H}_+^n$, where U_i' is open in $\mathbb{H}_+^n$. Again, points with coordinate $x^m = 0$ belong to the *boundary* of M (denoted by ∂M). The transition functions $\mathbf{x}_{\alpha\beta} : \mathbf{x}_\beta(U_\alpha \cap U_\beta) \to \mathbf{x}_\alpha(U_\alpha \cap U_\beta)$ must now be C^r in an open set of $\mathbb{R}^n$ that contains $\mathbf{x}_\beta(U_\alpha \cap U_\beta)$.

4.3.1 Calculus on Manifolds

Let M and N be smooth manifolds with dimensions $\dim M = m$ and $\dim N = n$. Let f be a map $f : M \to N$, $p \to f(p)$. If for every chart $(V, \mathbf{y})$ with $f(p) \in V$ there exists a chart $(U, \mathbf{x})$ with $p \in U$ such that $f(U) \subset V$ and the composite map $\mathbf{y} \circ f \circ \mathbf{x}^{-1}$ shown in Fig. 4.23 is of class C^r, we say that the map f is of class C^r at p. If the map is of class C^r at every point $p \in M$, the map is said to be *of class C^r*.

Note that the definition is independent of the choice of charts, because by assumption the transition functions are smooth. If we choose another pair of charts $(U', \mathbf{x}'), (V', \mathbf{y}')$, we can decompose

$$\mathbf{y}' \circ f \circ \mathbf{x}'^{-1} = \overbrace{\mathbf{y}' \circ \mathbf{y}^{-1}}^{C^\infty} \overbrace{\mathbf{y} \circ f \circ \mathbf{x}^{-1}}^{C^r} \circ \overbrace{\mathbf{x} \circ \mathbf{x}'^{-1}}^{C^\infty}$$

so we see that $\mathbf{y}' \circ f \circ \mathbf{x}'^{-1}$ is also of class C^r.

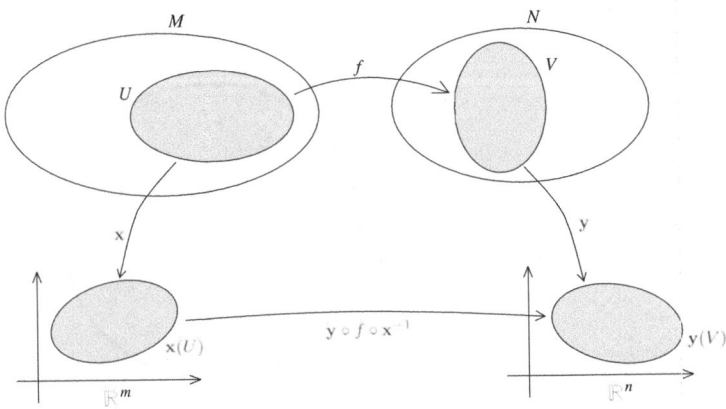

Fig. 4.23 The composite map $\mathbf{y} \circ f \circ \mathbf{x}^{-1}$.

The case when $N = \mathbb{R}$ is a somewhat special one. A function $f : M \to \mathbb{R}$ is of class C^r if it is of class C^r at every point $p \in M$ with a chart $(U, \mathbf{x})$ with $p \in U$. The set of C^∞ functions on M is denoted by $\mathcal{F}(M)$.

Let M and N be smooth manifolds. A homeomorphism $f : M \to N$ is called a C^r *diffeomorphism*, if f and f^{-1} are both class C^r maps. A C^∞ diffeomorphism is called simply a *diffeomorphism*. With the composition of maps, the set of all diffeomorphisms from M to itself is a group denoted $\mathrm{Diff}(M)$. Analogously, C^r diffeomorphisms form a group, denoted as $\mathrm{Diff}^r(M)$.

Example   Consider the two-sphere S^2 in $\mathbb{R}^3$ defined by $x^2 + y^2 + z^2 = 1$. Choose a $\varphi_0 \in (0, 2\pi)$. The map $r_{\varphi_0} : S^2 \to S^2$, $r_{\varphi_0}(\mathbf{x}) = (x \cos \varphi_0 - y \sin \varphi_0, x \sin \varphi_0 + y \cos \varphi_0, z)$ is a diffeomorphism. In polar coordinates, r_{φ_0} maps points by

$$(\cos \theta \cos \varphi, \cos \theta \sin \varphi, \sin \theta) \to (\cos \theta \cos(\varphi + \varphi_0), \cos \theta \sin(\varphi + \varphi_0), \sin \theta)),$$

so one can see that it corresponds to a rotation by an angle φ_0 about the z-axis.

Two smooth manifolds M and N are *diffeomorphic* when there exists a diffeomorphism $f : M \to N$.

Example   The open disk $B(0, 1) = \{(x, y) \in \mathbb{R}^2 | x^2 + y^2 < 1\}$ is diffeomorphic to the plane; for example, the map

$$f : B(0, 1) \to \mathbb{R}^2,$$

$$(x, y) \to f(x, y) = \left(\frac{x}{1 - \sqrt{1 - x^2 - y^2}}, \frac{y}{1 - \sqrt{1 - x^2 - y^2}} \right) \tag{4.10}$$

is a diffeomorphism.

The difference between a homeomorphism and a diffeomorphism is that homeomorphisms are continuous deformations, whereas diffeomorphisms are smooth deformations, which is a stronger requirement. For example, a plane is homeomorphic to a cone, but not diffeomorphic to it, as the mapping is not smooth at the apex of the cone. Similarly, the following map of a line to a V-shaped curve: $f(x) = |x|$ is not differentiable at the origin.

Note that manifolds being diffeomorphic is an equivalence relation, and the equivalence classes are called diffeomorphism classes. There exists also a local version of this relation, as follows.

A map $f : M \to N$ between two smooth manifolds of the same dimension is a *local diffeomorphism* when for every point $p \in M$ there exists an open subset $U_p, p \in U_p$, such that $f|_{U_p} : U_p \to f(U_p)$ is a diffeomorphism onto an open subset $f(U_p)$ of N.

Example   Consider $S^2/\mathbb{Z}_2$ where $\mathbb{Z}_2$ acts as a reflection through the origin, so that the antipodal points $\pm(x, y, z)$ on the sphere are identified. The map $S^2 \to S^2/\mathbb{Z}_2, (x, y, z) \to \pm(x, y, z)$ is a local diffeomorphism, but not a diffeomorphism since any two antipodal points on S^2 map to the same point.

Let $\tilde{M}$ and M be smooth manifolds. A surjective local diffeomorphism $\phi : \tilde{M} \to M$ is called a *smooth covering* if every point $p \in M$ has an open connected neighborhood U, such that each connected component $\tilde{U}_i$ of the inverse image $\phi^{-1}(U)$ is diffeomorphic to U via the restrictions $\phi_{\tilde{U}_i} : \tilde{U}_i \to U$. The space $\tilde{M}$ is called the *covering space* of M. If $\tilde{M}$ is simply connected (the fundamental group is trivial), then it is called a *universal covering space*.

Example Let us consider S^1 as the unit circle on the complex plane. The map $\mathbb{R} \to S^1$, $x \to \exp(ix)$ is a smooth covering, and $\mathbb{R}$ is a universal covering space of S^1. By a similar argument, $\mathbb{R}^n$ is an universal covering space of the n-torus T^n. The previous example shows that S^2 is a universal covering space of $S^2/\mathbb{Z}_2 = \mathbb{R}P^2$.

4.3.2 Tangent Vectors

Consider a pointlike body traveling in $\mathbb{R}^n$ along a path $\vec{x}(t) = (x^i(t))$ specifying the body's location as a function of time, in a region in which the temperature varies smoothly with the position and is specified by a function $T(\vec{x})$. Then, the local change of temperature as a function of time along the trajectory can be expressed using the chain rule,

$$\frac{dT(\vec{x}(t))}{dt} = \sum_i \frac{dx^i(t)}{dt} \frac{\partial T(\vec{x})}{\partial x^i}. \tag{4.11}$$

Recalling that the velocity vector $\vec{v}(t) = \left(v^i(t) \right) = \frac{d\vec{x}(t)}{dt}$ is tangential to the path, Eq. (4.11) is a directional derivative of the temperature function. In what follows, we adopt the Einstein summation convention, where repeated indices are understood to be summed over and the summation symbol is omitted. We have

$$\frac{dT(\vec{x}(t))}{dt} = v^i(t) \frac{\partial T(\vec{x})}{\partial x^i} = \vec{v}(t) \cdot \nabla T(\vec{x}) \equiv \nabla_{\vec{v}} T(\vec{x}) \tag{4.12}$$

with respect to the velocity vector $\vec{v}(t)$ along the path. So the tangent vector $\vec{v}$ along the curve, i.e., the velocity vector, gives rise to a directional derivative operator $\nabla_{\vec{v}}$ acting on functions.

More generally, suppose we have a vector-valued smooth function $\vec{V}(\vec{x}) = (V^i(\vec{x}))$ in $\mathbb{R}^n$ (each component V^i being a smooth function $\mathbb{R}^n \to \mathbb{R}$). Associated with it, we can compute the directional derivative of a smooth function $f : \mathbb{R}^n \to \mathbb{R}$ by acting on it with the directional derivative $\nabla_{\vec{V}}$, to obtain a smooth function $\nabla_{\vec{V}} f$. On the other hand, as is familiar from the theory of differential equations, solving the set of first-order equations (in component form)

$$\frac{dx^i(t)}{dt} = V^i\left(\vec{x}(t)\right) \tag{4.13}$$

gives rise (at least locally) to a family of integral curves, which are uniquely specified once an initial condition $\vec{x}(t_0) = \vec{x}_0$ is given. The vector $\vec{V}$ is then tangential to these integral curves at every point $\vec{x}$.

Tangent vectors on smooth manifolds are similarly defined using curves. In the following, we label components with Greek letters instead of Latin letters. Let M be a smooth manifold with local coordinates $x = (x^\mu(q)) = \phi(q)$ for points in a coordinate neighborhood U. Choose a point $p \in U$, and let $c : (a, b) \to M$ be a curve in U passing through the point $p = c(0)$ (we can assume $0 \in (a, b)$), as in Fig. 4.24.

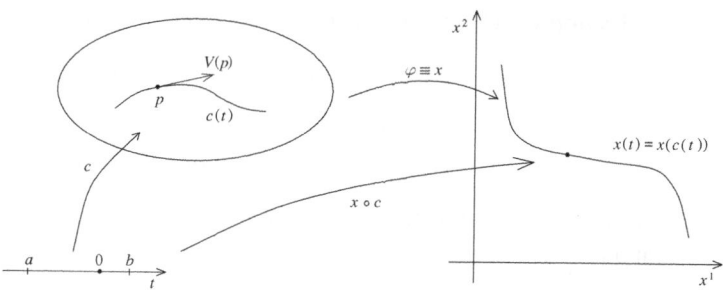

Fig. 4.24 A curve $c(t)$ and a tangent vector V_p at $p = c(0)$.

Then, let $f : M \to \mathbb{R}$ be a function. The rate of change of f along the curve c at the point p is

$$\left.\frac{df(c(t))}{dt}\right|_{t=0} = \left.\frac{\partial f}{\partial x^\mu}\frac{dx^\mu(c(t))}{dt}\right|_{t=0}, \qquad (4.14)$$

where $x^\mu(c(t)) = \varphi(c(t))$ are the curve points in local coordinates, and we used a shortened notation

$$\frac{\partial f}{\partial x^\mu} \equiv \frac{\partial \left(f \circ \varphi^{-1}(x)\right)}{\partial x^\mu}. \qquad (4.15)$$

In other words, $\frac{df(c(t))}{dt}$ is obtained by acting on the function f with the differential operator

$$V_p \equiv V(p)^\mu \left(\frac{\partial}{\partial x^\mu}\right)_p, \qquad (4.16)$$

where

$$V^\mu(p) = \left.\frac{dx^\mu(c(t))}{dt}\right|_{t=0}, \qquad (4.17)$$

so that

$$\left.\frac{df(c(t))}{dt}\right|_{t=0} \equiv \left. V_p[f]\right|_{t=0}. \qquad (4.18)$$

The operator V_p is called a *tangent vector* of M at p.

Comparing Eq. (4.18) and Eq. (4.17) with Eq. (4.12) and Eq. (4.13), one can see that $V_p[f]$ generalizes the concept of directional derivative of a function f along a curve, at a point p, from a curve in $\mathbb{R}^n$ to a curve in a smooth manifold M.

In general, $V_p[f]$ depends on the curve, but several curves can give rise to the same tangent vector V_p with components $V^\mu(p)$. We can see that two curves c_1 and c_2 give the same tangent vector V_p if and only if

(i) $c_1(0) = c_2(0) = p,$ $\qquad\qquad\qquad\qquad\qquad\qquad$ (4.19)

(ii) $\left.\dfrac{dx^\mu(c_1(t))}{dt}\right|_{t=0} = \left.\dfrac{dx^\mu(c_2(t))}{dt}\right|_{t=0}.$ $\qquad$ (4.20)

The property of having the same tangent vector at p defines an equivalence relation, which one can denote as $c_1 \sim c_2$, between curves passing through p. Thus equivalence classes $[c]$ can be identified with the tangent vectors V_p.

Example   Define three curves c_1, c_2, and c_3 from $\left(-\frac{\pi}{2}, \frac{\pi}{2}\right)$ to $\mathbb{R}^2$:

$$c_1(t) = (t+1, 1+\tanh t), \tag{4.21}$$

$$c_2(t) = (t+1, t+1), \tag{4.22}$$

$$c_3(t) = (t+\cos t, 1+\sin t). \tag{4.23}$$

All three curves pass through the point $p = (1, 1)$ at $t = 0$. The tangent vector to c_1 at p is

$$X_{c_1 p} = \frac{dx^1(c_1(t))}{dt}\bigg|_{t=0} \left(\frac{\partial}{\partial x^1}\right)_p + \frac{dx^2(c_1(t))}{dt}\bigg|_{t=0} \left(\frac{\partial}{\partial x^2}\right)_p$$

$$= (1)|_{t=0} \left(\frac{\partial}{\partial x^1}\right)_p + (\cosh^{-2} t)\big|_{t=0} \left(\frac{\partial}{\partial x^2}\right)_p$$

$$= \left(\frac{\partial}{\partial x^1}\right)_p + \left(\frac{\partial}{\partial x^2}\right)_p. \tag{4.24}$$

The tangent vectors to the other two curves are the same at p, hence all three curves belong to the same equivalence class.

The set of all tangent vectors at p is the *tangent space* $T_p M$ at p. It is a real vector space of dimension $\dim T_p M = m$. Linear combinations of elements of $T_p M$ are defined by

$$\lambda_1 X_{1p} + \lambda_2 X_{2p} = \left(\lambda_1 X_{1p}^\mu + \lambda_2 X_{2p}^\mu\right)\left(\frac{\partial}{\partial x^\mu}\right)_p, \tag{4.25}$$

where tangent vectors have been expanded in the basis

$$\{(e_\mu)_p\} = \left\{\left(\frac{\partial}{\partial x^\mu}\right)_p\right\}; \tag{4.26}$$

this basis is called the *coordinate basis*.

4.3.3 One-Forms, or Cotangent Vectors

Tangent vectors of a differentiable manifold M at a point p are elements of the vector space $T_p M$. Recalling the discussion in Section 3.6, we can now study the dual of this vector space. A dual vector, i.e., a linear function from $T_p M$ to $\mathbb{R}$ is called a *cotangent vector* or a *one-form*. The set of one-forms is the *dual vector space* and is denoted as $T_p^\star M$.

As before, one can define an inner product $\langle,\rangle : T_p^\star M \times T_p M \to \mathbb{R}$ such that

$$\langle w, v \rangle = w(v) \in \mathbb{R}, \tag{4.27}$$

for all pairs $w \in T_p^\star M$ and $v \in T_p M$, and such that it is linear in both of its arguments:

$$\langle w, \alpha_1 v_1 + \alpha_2 v_2 \rangle = w(\alpha_1 v_1 + \alpha_2 v_2) = \alpha_1 \langle w, v_1 \rangle + \alpha_2 \langle w, v_2 \rangle, \tag{4.28}$$

$$\langle \alpha_1 w_1 + \alpha_2 w_2, v \rangle = (\alpha_1 w_1 + \alpha_2 w_2)(v) = \alpha_1 \langle w_1, v \rangle + \alpha_2 \langle w_2, v \rangle. \tag{4.29}$$

Let $\{e_\mu\} = \{\frac{\partial}{\partial x^\mu}\}$ be a coordinate basis of $T_p M$. Note that the correct notation for the basis vectors would be $\{(\frac{\partial}{\partial x^\mu})_p\}$, but this is somewhat cumbersome, so we use the shortened form $\{\frac{\partial}{\partial x^\mu}\}$. The dual basis $\{e^{\star\mu}\}$ is denoted by $\{(dx^\mu)_p\}$ or, in a shortened form of notation, which we will use below, by dx^μ. By definition, the dual basis vectors satisfy

$$\left\langle dx^\mu, \frac{\partial}{\partial x^\nu} \right\rangle = dx^\mu\left(\frac{\partial}{\partial x^\nu}\right) = \delta^\mu{}_\nu. \tag{4.30}$$

Let us expand a one-form in the dual basis, $w = w_\mu dx^\mu$, and a tangent vector in the coordinate basis, $v = v^\nu \frac{\partial}{\partial x^\nu}$. Then

$$w(v) = \langle w, v \rangle = w_\mu v^\nu dx^\mu\left(\frac{\partial}{\partial x^\nu}\right) = w_\mu v^\mu. \tag{4.31}$$

Consider now a function $f \in \mathcal{F}(M)$, i.e., f is a smooth map from M to $\mathbb{R}$. Its *differential* $df \in T_p^\star M$ is defined as the map

$$df: v \rightarrow df(v) = \langle df, v \rangle \equiv v(f) = v^\mu \frac{\partial f}{\partial x^\mu} \tag{4.32}$$

for every $v \in T_pM$. Comparing with Eq. (4.31), we can read off that the components of df are $\frac{\partial f}{\partial x^\mu}$, so

$$df = \frac{\partial f}{\partial x^\mu} dx^\mu. \tag{4.33}$$

This resembles the concept of a differential df of a function on $U \subset \mathbb{R}^n$, familiar from calculus, except that, so far, the definition in Eq. (4.33) holds only at the point $p \in M$. Later on, this restriction will be lifted.

4.3.4 Tensors

Recall that a tensor of type (q, r) is a multilinear map from a product of dual vector spaces and vector spaces to real numbers. Now the vector space is taken to be the tangent vector space T_pM with the dual $T_p^\star M$. A (q, r)-tensor at p is then a multilinear map

$$T: \overbrace{T_p^\star M \times \cdots \times T_p^\star M}^{q} \times \overbrace{T_pM \times \cdots \times T_pM}^{r} \rightarrow \mathbb{R}. \tag{4.34}$$

We denote the vector space of (q, r)-tensors at $p \in M$ by $T_{r,p}^q(M)$. Note that $T_{0,p}^1 = (T_p^\star M)^\star = T_pM$ and $T_{1,p}^0(M) = T_p^\star M$.
The basis of $T_{r,p}^q(M)$ is the collection

$$\left\{ \frac{\partial}{\partial x^{\mu_1}} \otimes \cdots \otimes \frac{\partial}{\partial x^{\mu_q}} \otimes dx^{\nu_1} \otimes \cdots \otimes dx^{\nu_r} \right\} \tag{4.35}$$

with all possible index values. Being mappings of the form in Eq. (4.34), the basis vectors satisfy

$$\left(\frac{\partial}{\partial x^{\mu_1}} \otimes \cdots \otimes \frac{\partial}{\partial x^{\mu_q}} \otimes dx^{\nu_1} \otimes \cdots \otimes dx^{\nu_r} \right)\left(dx^{\alpha_1}, \ldots, dx^{\alpha_q}, \frac{\partial}{\partial x^{\beta_1}}, \ldots, \frac{\partial}{\partial x^{\beta_r}} \right)$$
$$= \delta^{\alpha_1}{}_{\mu_1} \cdots \delta^{\alpha_q}{}_{\mu_q} \delta^{\nu_1}{}_{\beta_1} \cdots \delta^{\nu_r}{}_{\beta_r}. \tag{4.36}$$

Note that in the latter expression we used

$$\frac{\partial}{\partial x^\mu}(dx^\alpha) \equiv \left\langle dx^\alpha, \frac{\partial}{\partial x^\mu} \right\rangle = \delta^\alpha{}_\mu. \tag{4.37}$$

On the left-hand side, $\frac{\partial}{\partial x^\mu}$ is interpreted as an element of $(T_p^\star M)^\star$: Recall that vectors can be identified with elements in the dual of the dual vector space.

Using the coordinate basis, one can expand a (q, r)-tensor as

$$T = T^{\mu_1 \cdots \mu_q}{}_{\nu_1 \cdots \nu_r} \frac{\partial}{\partial x^{\mu_1}} \otimes \cdots \otimes \frac{\partial}{\partial x^{\mu_q}} \otimes dx^{\nu_1} \otimes \cdots \otimes dx^{\nu_r} \tag{4.38}$$

so

$$T(w_1, \ldots, w_q; v_1, \ldots, v_r) = T^{\mu_1 \cdots \mu_q}{}_{\nu_1 \cdots \nu_r} w_{1\mu_1} \cdots w_{q\mu_q} v_1^{\nu_1} \cdots v_r^{\nu_r}, \tag{4.39}$$

where repeated upper and lower indices are summed over, according to Einstein's summation convention. The *tensor product* of a (q, r)-tensor $T \in T_{r,p}^q(M)$ and a (s, t)-tensor $U \in T_{t,p}^s(M)$ is the tensor $T \otimes U \in T_{r+t,p}^{q+s}(M)$ with

$$(T \otimes U)(w_1, \ldots, w_q, w_{q+1}, \ldots, w_{q+s}; v_1, \ldots, v_r, v_{r+1}, \ldots, v_{r+t})$$
$$= T(w_1, \ldots, w_q; v_1, \ldots, v_r)U(w_{q+1}, \ldots, w_{q+s}; v_{r+1}, \ldots, v_{r+t}).$$
$$= T^{\mu_1 \cdots \mu_q}{}_{\nu_1 \cdots \nu_r} w_{1\mu_1} \cdots w_{q\mu_q} v_1^{\nu_1} \cdots v_r^{\nu_r} U^{\alpha_1 \cdots \alpha_s}{}_{\beta_1 \cdots \beta_t} w_{(q+1)\alpha_1} \cdots w_{(q+s)\alpha_s} v_{r+1}^{\beta_1} \cdots v_{r+t}^{\beta_t}.$$

The meaning of this expression becomes more transparent by considering the components; the components of $T \otimes U$ are products of all components of T and U:

$$(T \otimes U)^{\mu_1 \cdots \mu_{(q+s)}}{}_{\nu_1 \cdots \nu_{(r+t)}} = T^{\mu_1 \cdots \mu_q}{}_{\nu_1 \cdots \nu_r} U^{\mu_{(q+1)} \cdots \mu_{(q+s)}}{}_{\nu_{(r+1)} \cdots \nu_{(r+s)}}. \tag{4.40}$$

Recall that a contraction $c(ij)$ maps a tensor $T \in T_{r,p}^q(M)$ to a tensor $T' \in T_{r-1,p}^{q-1}(M)$, whose components are defined as

$$T'^{\mu_1 \cdots \mu_{q-1}}{}_{\nu_1 \cdots \nu_{r-1}} = T^{\mu_1 \cdots \mu_{i-1} \rho \mu_i \cdots \mu_{q-1}}{}_{\nu_1 \cdots \nu_{j-1} \rho \nu_j \cdots \nu_{r-1}}, \tag{4.41}$$

where the ith upper index and the jth lower index are set to be the same (ρ) and summed over.

4.3.5 Vector Fields and Tensor Fields

We will now extend the definition of a tangent vector V_p at a point p by letting p vary over all of M in a smooth way; this will enable us to define a *vector field*.

First, we define the *tangent bundle* of M as the union of the tangent spaces at all points of M:

$$TM = \bigcup_{p \in M} T_p M.$$

Next, we choose a tangent vector V_p from every tangent space $T_p M$, for every point p in M. This corresponds to defining a map $V : M \to TM, p \to V_p$. This defines a *vector field* or, equivalently, a *smooth vector field*, if for every C^∞ function $f \in \mathcal{F}(M)$ the function $p \to V_p[f] : M \to \mathbb{R}$ is also a smooth function. We denote the $p \to V_p[f]$ mapping as $V[f]$. The set of smooth vector fields on M is denoted as $\chi(M)$. Figure 4.25 shows an example of a vector field on the two-dimensional sphere S^2.

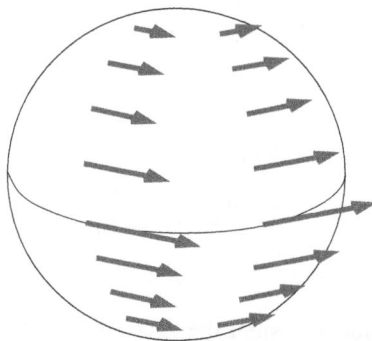

Fig. 4.25 A visualization of a vector field on a two-dimensional sphere. The strength of the vector field is visualized by the length of the arrows. For the sake of clarity, the vector field is shown at only 16 points on the surface of the sphere.

Let M be a smooth manifold with $\dim M = m$, and let $(U, \mathbf{x})$ be a chart on M with $\mathbf{x} = (x^\mu) = (x^1, \ldots, x^m)$. We define a vector field $\frac{\partial}{\partial x^\mu}$ by

$$\frac{\partial}{\partial x^\mu} : p \to \left(\frac{\partial}{\partial x^\mu}\right)_p, \quad \forall p \in U. \tag{4.42}$$

The set of fields $\{\frac{\partial}{\partial x^1}, \ldots, \frac{\partial}{\partial x^m}\}$ is called the *coordinate frame field*. Then, expanding a vector field V at a generic point p in U

$$V_p = V^\mu(p) \left(\frac{\partial}{\partial x^\mu}\right)_p \tag{4.43}$$

and letting p vary over U gives the expansion

$$V = V^\mu \frac{\partial}{\partial x^\mu} \tag{4.44}$$

in the coordinate neighborhood U, where the components V^μ are smooth functions in U.

The vectors are independent of a choice of coordinates, provided their components are transformed correctly. Let $x(p) = \varphi_i(p)$ and $y(p) = \varphi_j(p)$ be two coordinates with overlapping coordinate neighborhoods U_i and U_j. In the intersection $U_i \cap U_j$, the coordinate frame fields are related by the chain rule,

$$\frac{\partial}{\partial x^\mu} = \frac{\partial y^\nu}{\partial x^\mu} \frac{\partial}{\partial y^\nu}, \tag{4.45}$$

and the basis vector fields are simply related by the Jacobian matrix associated with the coordinate transformation, with components $\left(\frac{\partial y^\nu}{\partial x^\mu}\right)$. Then we can expand a smooth vector field X using either of the two sets of coordinates,

$$X = X^\mu \frac{\partial}{\partial x^\mu} = Y^\mu \frac{\partial}{\partial y^\mu}, \tag{4.46}$$

where we used different notations X^μ and Y^μ for the component functions, in order to emphasize that they are in general, different. By the chain rule expressed in Eq. (4.45), we have

$$X^\mu \frac{\partial}{\partial x^\mu} = X^\nu \frac{\partial y^\mu}{\partial x^\nu} \frac{\partial}{\partial y^\mu}, \tag{4.47}$$

thus we get the transformation rule for the components:

$$Y^\mu = X^\nu \frac{\partial y^\mu}{\partial x^\nu}. \tag{4.48}$$

Note the abuse of the notation:

$$X^\nu \frac{\partial y^\mu}{\partial x^\nu} \frac{\partial}{\partial x^\mu} \equiv X_p^\nu \frac{\partial(\varphi_j \circ \varphi_i^{-1})(x^\mu(p))}{\partial x^\nu(p)} \left(\frac{\partial}{\partial x^\mu}\right)_p. \tag{4.49}$$

Example Consider the vector field

$$V = \frac{\partial}{\partial\theta} + \frac{\partial}{\partial\varphi} \tag{4.50}$$

on S^2, using the polar angle coordinates (θ, φ). If we move to new coordinates $(x, y) = (\theta, \tan\varphi)$, then by the chain rule

$$V = \frac{\partial}{\partial\theta} + \frac{\partial}{\partial\varphi} = \frac{\partial}{\partial\theta} + \frac{\partial y}{\partial\varphi}\frac{\partial}{\partial y} = \frac{\partial}{\partial\theta} + (1 + \tan^2\varphi)\frac{\partial}{\partial y} = \frac{\partial}{\partial x} + (1 + y^2)\frac{\partial}{\partial y}. \tag{4.51}$$

Suppose that we act on a function f first by a vector field Y, then by a vector field X, where we recall that $Y[f]$ is a function. Examining $X[Y[f]]$, one can realize that the product XY is *not* a vector field:

$$XYf = X[Y[f]] = X^\mu \partial_\mu [Y^\nu \partial_\nu f] = \underbrace{X^\mu(\partial_\mu Y^\nu)\partial_\nu f}_{\text{a vector field}} + \underbrace{X^\mu Y^\nu \partial_\mu \partial_\nu f}_{\text{not a vector field}}. \tag{4.52}$$

However, the problematic second term does not appear if we replace XY with the *commutator* or *Lie bracket* $[X, Y] = XY - YX$:

$$[X, Y]f = X[Y[f]] - Y[X[f]] = X^\mu(\partial_\mu Y^\nu)\partial_\nu f - Y^\mu(\partial_\mu X^\nu)\partial_\nu f. \tag{4.53}$$

We can see that the commutator of vector fields $[X, Y]$ is again a vector field. In Problem 4.14 you will show the property

$$[X, fY] = (X[f])Y + f[X, Y], \tag{4.54}$$

where f is a function, and the *Jacobi identity*

$$[X, [Y, Z]] + [Y, [Z, X]] + [Z, [X, Y]] = 0, \tag{4.55}$$

for three vector fields X, Y, and Z. Note the cyclic permutations in Eq. (4.55).

Example Consider the vector fields $X = x^2\partial_x + y^2\partial_y$ and $Y = y\partial_x + x\partial_y$ on $\mathbb{R}^2$. We calculate the commutator $[X, Y]$. First we compute

$$\begin{aligned} XY &= x^2\partial_x(y\partial_x) + x^2\partial_x(x\partial_y) + y^2\partial_y(y\partial_x) + y^2\partial_y(x\partial_y) \\ &= x^2y\partial_x^2 + x^2\partial_y + x^3\partial_x\partial_y + y^2\partial_x + y^3\partial_x\partial_y + xy^2\partial_y^2 \end{aligned} \tag{4.56}$$

and

$$\begin{aligned} YX &= y\partial_x(x^2\partial_x) + y\partial_x(y^2\partial_y) + x\partial_y(x^2\partial_x) + x\partial_y(y^2\partial_y) \\ &= 2xy\partial_x + x^2y\partial_x^2 + y^3\partial_x\partial_y + x^3\partial_x\partial_y + 2xy\partial_y + xy^2\partial_y^2, \end{aligned} \tag{4.57}$$

then we compute the commutator as

$$[X, Y] = XY - YX = (y^2 - 2xy)\partial_x + (x^2 - 2xy)\partial_y. \tag{4.58}$$

Like vectors, one-forms can be organized to define a one-form field. First, we define the *cotangent bundle* as the union

$$T^\star M = \bigcup_{p \in M} T_p^\star M.$$

Then, we introduce a function $w : M \to T^\star M$, $p \to w_p$ mapping each point $p \in M$ to a one-form w_p; if for every smooth vector field $V \in \chi(M)$ the function

$$w[V] : M \to \mathbb{R},$$
$$p \to w[V](p) = w_p(V_p) \tag{4.59}$$

is smooth, then w is a *one-form field* (also called a *smooth one-form field*, or a *cotangent vector field*) on M. The set of one-form fields is denoted by $\Omega^1(M)$.

As with vector fields, let $\dim M = m$, and choose a chart $(U, \mathbf{x})$ on M with $\mathbf{x} = (x^\mu) = (x^1, \ldots, x^m)$. We define a one-form field dx^μ by

$$dx^\mu : p \to (dx^\mu)_p, \quad \forall p \in U; \tag{4.60}$$

then, the set of fields $\{dx^1, \ldots, dx^m\}$ is called the *coordinate coframe field* (or sometimes the *holonomic coframe field*). Note that

$$dx^\mu \left(\frac{\partial}{\partial x^\nu} \right) = \delta^\mu_\nu \tag{4.61}$$

everywhere in U.

Expanding a one-form field w at every point p,

$$w_p = w_\mu(p)(dx^\mu)_p, \tag{4.62}$$

and letting p vary over U, gives the expansion

$$w = w_\mu dx^\mu \tag{4.63}$$

in the coordinate neighborhood U. The components w_μ are smooth functions in U. A particular case of one-form fields are the differentials of functions df, with components $\frac{\partial f}{\partial x^\mu}$,

$$df = \frac{\partial f}{\partial x^\mu} dx^\mu \tag{4.64}$$

in U. Note that not all one-form fields are differentials of functions. We will discuss this in detail in Section 4.4.2; it will turn out that only the so-called *exact* one-forms w can be expressed as differentials, $w = df$, for some smooth function $f \in \mathcal{F}(M)$, everywhere on M. Others may be expressed as a differential only locally, in a neighborhood.

In U, one can also evaluate a one-form field w acting on a vector field v, using the coordinate expansions

$$w(v) = w_\mu v^\nu dx^\mu \left(\frac{\partial}{\partial x^\nu} \right) = w_\mu v^\mu; \tag{4.65}$$

this is a smooth function in U.

We now derive how the components of a one-form transform under a change of coordinates. Consider two coordinate patches U_i and U_j with $p \in U_i \cap U_j$, with coordinate functions $x = \varphi_i(p)$ and $y = \varphi_j(p)$, respectively. Then, in the intersection of U_i and U_j there exist two alternative expansions for a one-form field, $w = w_\mu dx^\mu = \tilde{w}_\nu dy^\nu$, and for a vector field, $v = v^\rho \frac{\partial}{\partial x^\rho} = \tilde{v}^\sigma \frac{\partial}{\partial y^\sigma}$. We already know that components of the vector field transform as $\tilde{v}^\nu = \frac{\partial y^\nu}{\partial x^\mu} v^\mu$. Using this to rewrite Eq. (4.65) in the two possible basis expansions gives

$$w(v) = w_\mu v^\mu = \tilde{w}_\nu \tilde{v}^\nu = \tilde{w}_\nu \frac{\partial y^\nu}{\partial x^\mu} v^\mu, \tag{4.66}$$

from which we find the transformed components

$$w_\mu = \tilde{w}_\nu \frac{\partial y^\nu}{\partial x^\mu} \quad \text{or} \quad \tilde{w}_\mu = w_\nu \frac{\partial x^\nu}{\partial y^\mu}. \tag{4.67}$$

The second equation appearing in (4.67) is obtained by swapping the rôles of the coordinates x and y. For a differential df, Eq. (4.67) implies

$$df = \frac{\partial f}{\partial x^\mu} dx^\mu = \frac{\partial f}{\partial x^\mu} \frac{\partial x^\mu}{\partial y^\nu} dy^\nu. \tag{4.68}$$

In particular the dual basis vectors transform then as

$$dx^\mu = \frac{\partial x^\mu}{\partial y^\nu} dy^\nu. \tag{4.69}$$

Equations (4.68) and (4.69) are easy to remember, and very useful in dealing with coordinate transformations. In particular, Eq. (4.69) gives an easy way to derive the more general rule encoded in Eq. (4.67):

$$w = w_\nu dx^\nu = w_\nu \frac{\partial x^\nu}{\partial y^\mu} dy^\mu = \tilde{w}_\mu dy^\mu, \tag{4.70}$$

from which one obtains

$$\tilde{w}_\mu = w_\nu \frac{\partial x^\nu}{\partial y^\mu}. \tag{4.71}$$

Example Let $M = \mathbb{R}^2$, and consider a one-form

$$w = ydx + xdy \tag{4.72}$$

using the Cartesian coordinates (x, y). It is possible to express w as a differential $w = df$, with $f(x, y) = xy$. We then introduce polar coordinates (r, φ), related to the Cartesian ones via the transformations

$$\begin{cases} x = r \cos \varphi \\ y = r \sin \varphi \end{cases}. \tag{4.73}$$

We can then express w in polar coordinates in two (and equivalent) alternative ways. One consists in calculating $dx = \cos \varphi dr - r \sin \varphi d\varphi$ and $dy = \sin \varphi dr + r \cos \varphi d\varphi$, and then substituting these and $x = r \cos \varphi$, $y = r \sin \varphi$ into Eq. (4.72). The other way, which is easier, is to first write $f = xy = \frac{1}{2} r^2 \sin 2\varphi$ and then compute

$$w = df = r \sin(2\varphi) \, dr + r^2 \cos(2\varphi) \, d\varphi, \tag{4.74}$$

which yields the same result as the one obtained in the first way.

Finally, the definition in Eq. (4.59) can be generalized to define tensor fields of arbitrary rank. To do this, we join together (q, r)-tensors $T(p) \in T^q_{r,p}(M)$ for all $p \in M$ to define a map, as follows. Given the smooth one-form fields $w_1, \ldots, w_q$ and the smooth vector fields $v_1, \ldots, v_r$, consider the map

$$T[w_1, \ldots, w_q; v_1, \ldots, v_r] : M \to \mathbb{R},$$

$$p \to T(p)(w_1(p), \ldots, w_q(p); v_1(p), \ldots, v_r(p)). \tag{4.75}$$

If the map (4.75) is smooth on M for all smooth one-form fields w_i and for all smooth vector fields v_j, then T is a *smooth (q, r)-tensor field* on M. In local coordinates, we can expand it as

$$T = T^{\mu_1 \mu_2 \cdots \mu_q}_{ v_1 v_2 \ldots v_r}(x) \left(\frac{\partial}{\partial x^{\mu_1}} \otimes \cdots \otimes \frac{\partial}{\partial x^{\mu_q}} \otimes dx^{v_1} \otimes \cdots \otimes dx^{v_r} \right). \tag{4.76}$$

The set of (q, r)-tensor fields $T^q_r(M)$ is a vector space. Special cases of (q, r)-tensor fields include $T^0_0(M) = \mathcal{F}(M)$ (smooth functions), $T^1_0(M) = \chi(M)$ (smooth vector fields), and $T^0_1(M) = \Omega_1(M)$ (smooth one-form fields).

Under a coordinate transformation, a tensor field of type (q, r) transforms as a product of q vector fields and r one-form fields. Note that the basis tensor fields

$$\frac{\partial}{\partial x^{\mu_1}} \otimes \cdots \otimes \frac{\partial}{\partial x^{\mu_q}} \otimes dx^{v_1} \otimes \cdots \otimes dx^{v_r}$$

are simple examples of a (q, r)-tensor. Thus, the components are multiplied by products of Jacobian matrices and their inverses. For example, considering $T \in T^1_2(M)$, a tensor field of type $(1, 2)$, which can be written as

$$T = T^{\alpha}_{\beta_1 \beta_2}(x) \frac{\partial}{\partial x^{\alpha}} \otimes dx^{\beta_1} \otimes dx^{\beta_2} = \tilde{T}^{\mu}_{v_1 v_2}(y) \frac{\partial}{\partial y^{\mu}} \otimes dy^{v_1} \otimes dy^{v_2}, \tag{4.77}$$

the transformation rule for the components is

$$\tilde{T}^{\mu}_{v_1 v_2}(y) = \frac{\partial y^{\mu}}{\partial x^{\alpha}} \frac{\partial x^{\beta_1}}{\partial y^{v_1}} \frac{\partial x^{\beta_2}}{\partial y^{v_2}} T^{\alpha}_{\beta_1 \beta_2}(x(y)). \tag{4.78}$$

4.3.6 Differential Map and Pullback

Let M and N be differentiable manifolds and let f be a smooth map $f : M \to N$. Let p be a point in M, with the image point $q = f(p) \in N$. Associated with f we can construct a map $f_{\star}$ from the tangent space $T_p M$ to $T_q N$, "pushing" tangent vectors on M to tangent vectors on N. The map $f_{\star}$ is called a *pushforward* (or a *differential map*) and is defined as follows.

If $g \in \mathcal{F}(N)$, i.e., $g : N \to \mathbb{R}$ is a smooth function, and $V \in T_p M$, then

$$(f_{\star} V)[g] = V[g \circ f]. \tag{4.79}$$

In other words, if V is a tangent vector of a curve $c(t)$, so that $V[g \circ f]$ characterizes the rate of change of a function $g \circ f$ along the curve $c(t)$, then $f_{\star} V$ characterizes the rate of change of the function g along the image curve $f(c(t))$.

Let us now discuss the explicit form of this map in local coordinates. Let x be local coordinates on M and let y be local coordinates on N. Instead of $q = f(p)$ we may use the local coordinates and write $y = y(x) \equiv f(x)$. Moreover, we can expand $V = V^\mu \frac{\partial}{\partial x^\mu}$ and $f_\star V = (f_\star V)^\nu \frac{\partial}{\partial y^\nu}$. We want to find the rule to compute the components $(f_\star V)^\nu$. Using the definition in Eq. (4.79), one can write

$$V[g \circ f] = V^\mu \frac{\partial(g(f(x)))}{\partial x^\mu} = V^\mu \frac{\partial(g(y(x)))}{\partial x^\mu} = V^\mu \frac{\partial g}{\partial y^\nu} \frac{\partial y^\nu}{\partial x^\mu}$$

$$= V^\mu \frac{\partial y^\nu}{\partial x^\mu} \frac{\partial g}{\partial y^\nu} \equiv (f_\star V)^\nu \frac{\partial g}{\partial y^\nu} = (f_\star V)[g], \tag{4.80}$$

so we can read off the rule for the components

$$(f_\star V)^\nu = V^\mu \frac{\partial y^\nu}{\partial x^\mu}, \tag{4.81}$$

where $y = y(x) = f(x)$. More precisely, $x^\mu = \varphi^\mu(p)$, $y^\nu = \psi^\nu(f(p))$ and

$$\frac{\partial y^\nu}{\partial x^\mu} = \frac{\partial(\psi \circ f \circ \varphi^{-1})^\nu}{\partial x^\mu}. \tag{4.82}$$

Example Let (x^1, x^2) and (y^1, y^2, y^3) be the coordinates in M and N, respectively, and let $V = a\frac{\partial}{\partial x^1} + b\frac{\partial}{\partial x^2}$ be a tangent vector at (x^1, x^2). Let $f : M \to N$ be a map whose coordinate presentation is

$$y = \left(x^1, x^2, \sqrt{1 - (x^1)^2 - (x^2)^2}\right).$$

Then

$$f_\star V = V^\mu \frac{\partial y^\alpha}{\partial x^\mu} \frac{\partial}{\partial y^\alpha} = a\frac{\partial}{\partial y^1} + b\frac{\partial}{\partial y^2} - \left(a\frac{y^1}{y^3} + b\frac{y^2}{y^3}\right)\frac{\partial}{\partial y^3}. \tag{4.83}$$

The function f also induces a map $f^\star : T^\star_{f(p)}N \to T^\star_p M$ of a one-form w on N "backwards" to a one-form $f^\star w$ on M, therefore the map is called a *pullback* (see Fig. 4.26). The definition of $f^\star w$ uses a pushforward of a tangent vector v,

$$(f^\star w)(V) = w(f_\star V), \tag{4.84}$$

where $V \in T_p M$ and $w \in T^\star_{f(p)}N$ are arbitrary.

Using the local coordinates x in M and y in N, expanding $w = w_\nu dy^\nu$, and again writing $y(x) \equiv f(x)$, we can derive a transformation rule for the components of $f^\star w$:

$$w(f_\star V) = w_\nu dy^\nu \left(V^\mu \frac{\partial y^\alpha}{\partial x^\mu} \frac{\partial}{\partial y^\alpha}\right) = w_\nu V^\mu \frac{\partial y^\nu}{\partial x^\mu} = (f^\star w)(V) = (f^\star w)_\mu V^\mu, \tag{4.85}$$

from which we get

$$(f^\star w)_\mu = w_\nu \frac{\partial y^\nu}{\partial x^\mu}. \tag{4.86}$$

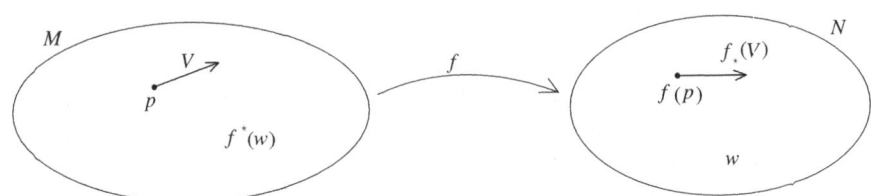

Fig. 4.26   Pushforward $f_\star(V)$ of a vector V and pullback $f^\star(w)$ of a one-form w.

Sacrificing some rigor, Eq. (4.86) can be derived more easily as follows:

$$w_\nu dy^\nu \rightarrow w_\nu dy^\nu(x) = w_\nu \frac{\partial y^\nu}{\partial x^\mu} dx^\mu \equiv (f^\star w)_\mu dx^\mu$$

$$\Rightarrow (f^\star w)_\mu = w_\nu \frac{\partial y^\nu}{\partial x^\mu}. \tag{4.87}$$

The pullback $f^\star$ can also be generalized to $(0, r)$-tensors; similarly, the differential map $f_\star$ can be generalized to $(q, 0)$-tensors.

4.3.7 Flow Generated by a Vector Field

Let X be a vector field on M. An *integral curve* $x(t)$ of X is a curve on M whose tangent vector at $x(t)$ is $X|_{x(t)}$. In local coordinates x, the vector field X is

$$X = X^\mu(x) \frac{\partial}{\partial x^\mu}. \tag{4.88}$$

Its integral curve $x(t)$ can be found as follows. In component form, the requirement of the tangent vector of the curve at a generic point $x(t)$ being equal to the value of the vector field X at $x(t)$ leads to the set of first-order differential equations

$$\frac{dx^\mu(t)}{dt} = X^\mu(x(t)). \tag{4.89}$$

The theorem of existence and uniqueness of solutions for ordinary differential equations guarantees that Eq. (4.89) has a unique solution, at least locally in some neighborhood of $t = 0$, once the initial condition $x^\mu(t = 0) = x_0^\mu$ has been specified. If M is compact, the solution exists for all t. As an example, Fig. 4.27 shows the integral curves of a vector field in the two-dimensional real vector space $\mathbb{R}^2$.

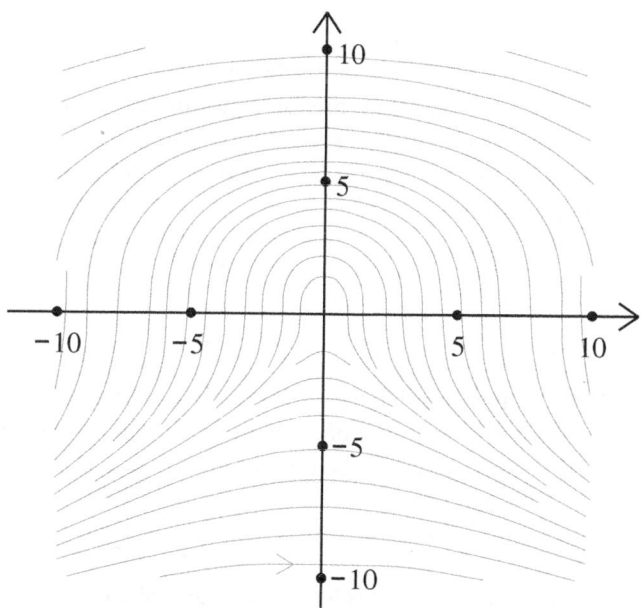

Fig. 4.27 Sketch of the integral curves for the vector field $X = y^2 \partial_x - 3x \partial_y$ on $\mathbb{R}^2$. The orientation of all curves is consistent with the one displayed for the curve at the bottom.

Let us denote the integral curve of X that goes through the point x_0 at $t = 0$ by $\sigma(t, x_0)$. Equation (4.89) can be rewritten as

$$\frac{d\sigma^\mu(t, x_0)}{dt} = X^\mu(\sigma(t, x_0)), \tag{4.90}$$

and the initial condition can be written as $\sigma^\mu(t = 0, x_0) = x_0^\mu$. The map $\sigma : I \times M \to M$ is called a *flow* generated by X, where I is an open interval around 0 in $\mathbb{R}$. Flows satisfy the property encoded in the following theorem.

Theorem 4.7   *A flow $\sigma : I \times M \to M$ satisfies*

$$\sigma(t, \sigma(s, x_0)) = \sigma(t + s, x_0)$$

for every value of t such that $t + s \in I$.

Proof   The left-hand and right-hand sides of Eq. (4.7) satisfy the same differential equation

$$\frac{d}{dt}\sigma^\mu(t, \sigma) = X^\mu(\sigma) = \frac{d}{dt}\sigma^\mu(t + s, \sigma)$$

and the same initial condition. Thus, by the theorem of existence and uniqueness of solutions for ordinary differential equations, they are the same map, as we wanted to prove.   □

Example   Let $M = \mathbb{R}^2$ and let $X((x, y)) = -y\frac{\partial}{\partial x} + x\frac{\partial}{\partial y}$ be a vector field in M. The flow generated by X is

$$\sigma(t, (x, y)) = (x \cos t - y \sin t, x \sin t + y \cos t). \tag{4.91}$$

Hence the flow through (x, y) is a circle centered at the origin.

In the example above, the flow associated with the vector field X looks like a rotation around the origin of the two-dimensional real vector plane. We can make this interpretation more precise and think of a vector field X as generating a one-dimensional group. For a fixed t, we can associate with $\sigma(t, x)$ a diffeomorphism $\sigma_t : M \to M$, $x \to \sigma(t, x)$. The family of diffeomorphisms $\{\sigma_t | t \in I\}$ is an Abelian group (when $I = \mathbb{R}$):

$$\sigma_t \cdot \sigma_s \equiv \sigma_t \circ \sigma_s = \sigma_{t+s}$$
$$\sigma_{-t} = (\sigma_t)^{-1} \tag{4.92}$$
$$\sigma_0 = \mathrm{id}_M.$$

This group is called a *one-parameter group of transformations*.

Conversely, from the one-parameter group of transformations we can extract the vector field X. Let $t = \epsilon$ be infinitesimally close to zero. Now,

$$\sigma_\epsilon^\mu(x) = \sigma^\mu(\epsilon, x) = \sigma^\mu(0, x) + \frac{d\sigma^\mu(t, x)}{dt}\bigg|_{t=0} \epsilon + O(\epsilon^2) \approx x^\mu + X^\mu(x)\epsilon. \tag{4.93}$$

In this context the vector field X is called the *infinitesimal generator* of the transformation σ_t. Given a vector field X, the corresponding flow is often denoted by

$$\sigma_t^\mu(x) = \sigma^\mu(t, x) = \exp(tX)x^\mu = (e^{tX})x^\mu \tag{4.94}$$

and is called the exponentiation of X. This is because, using the Taylor expansion of $\sigma_t^\mu(x)$ close to $t = 0$, one can write

$$
\begin{aligned}
\sigma_t^\mu(x) &= x^\mu + t \left. \frac{d\sigma^\mu(s,x)}{ds} \right|_{s=0} + \frac{1}{2!} t^2 \left. \frac{d^2\sigma^\mu(s,x)}{ds^2} \right|_{s=0} + \cdots \\
&= \left(1 + t\frac{d}{ds} + \frac{1}{2!} t^2 \frac{d^2}{ds^2} + \cdots \right) \sigma^\mu(s,x) \bigg|_{s=0} \\
&= \left. e^{t\frac{d}{ds}} \sigma^\mu(s,x) \right|_{s=0} \\
&= e^{tX} x^\mu .
\end{aligned}
$$

In physics, one is often interested in the infinitesimal transformation generated by a vector field X. The infinitesimal transformation can be written as

$$
x^\mu \to x^\mu + \delta x^\mu \tag{4.95}
$$

with

$$
\delta x^\mu = \epsilon X^\mu(x). \tag{4.96}
$$

4.3.8 Lie Derivative

Let $\sigma_t(x)$ be a flow on M generated by a vector field X. We would like to know how various tensor fields change under infinitesimal transformations of the form appearing in Eq. (4.95). The rate of change is given by an operation called the *Lie derivative*, denoted by $\mathcal{L}_X$. Let us begin with the simplest case, studying how the Lie derivative acts on a generic function f.

Lie Derivative of a Function

We started our introduction to tangent vectors and vector fields by asking how a function changes along a curve having X as its tangent vector, and found that the rate of change of a function was given by $X[f]$. We could also compare the values of the function at x and at $x^\mu + \delta x^\mu = x^\mu + \epsilon X^\mu(x)$, and denote the infinitesimal change in f as $\epsilon \delta f$, writing the value of the function f at the point $x^\mu + \epsilon X^\mu(x)$ as $f(x) \to f(x) + \epsilon \delta f(x)$. Thus,

$$
\begin{aligned}
\epsilon \delta f(x) &= f(x + \epsilon \delta x) - f(x) = f(x + \epsilon X) - f(x) \\
&= [f(x) + \epsilon X^\mu \partial_\mu f(x) + \cdots] - f(x) \approx \epsilon X^\mu \partial_\mu f(x),
\end{aligned} \tag{4.97}
$$

so that

$$
\delta f = X[f]. \tag{4.98}
$$

A more formal definition of the *Lie derivative of a function* $\mathcal{L}_X f$ can be given by using the flow to write $x^\mu + \epsilon X^\mu = \sigma_\epsilon^\mu(x)$ and defining

$$
\mathcal{L}_X f = \lim_{\epsilon \to 0} \frac{f(\sigma_\epsilon(x)) - f(x)}{\epsilon} = \lim_{\epsilon \to 0} \frac{f(x + \epsilon X) - f(x)}{\epsilon} = X[f], \tag{4.99}
$$

i.e., the Lie derivative of f coincides with $X[f]$.

Note that, since the commutator $[X, Y]$ of two vector fields is a vector field, we can also calculate the Lie derivative with respect to it. This yields

$$\mathcal{L}_{[X,Y]}f = [X, Y]f = (XY - YX)f$$
$$= \mathcal{L}_X(\mathcal{L}_Y f) - \mathcal{L}_Y(\mathcal{L}_X f) = ([\mathcal{L}_X, \mathcal{L}_Y])f. \tag{4.100}$$

Lie Derivative of a Vector Field

Let Y be another vector field on M. We now want to calculate the rate of change of Y along the curve $x^\mu(t) = \sigma_t^\mu(x)$. In order to evaluate the change, clearly one cannot subtract a vector at x from a vector at $\sigma_\epsilon(x)$, as the two vectors belong to generally different vector spaces. Hence, we need to map the latter vector back to the same point x, keeping track of its orientation. To this purpose, we use the pushforward map in Eq. (4.81) with the inverse flow back to x, namely $f_\star = (\sigma_{-\epsilon})_\star$. So we define the *Lie derivative of a vector field* Y by

$$\mathcal{L}_X Y = \lim_{\epsilon \to 0} \frac{1}{\epsilon} \left((\sigma_{-\epsilon})_\star Y|_{\sigma_\epsilon(x)} - Y|_x \right). \tag{4.101}$$

Let us rewrite this in a more transparent form. It is useful to think of the infinitesimal transformation $\sigma_\epsilon(x)$ as a coordinate transformation:

$$y \equiv \sigma_\epsilon^\mu(x) = x^\mu + \epsilon X^\mu(x) + O(\epsilon^2),$$

which can be inverted as

$$x^\mu = y^\mu - \epsilon X^\mu(y^\mu) + O(\epsilon^2), \tag{4.102}$$

and expand the two vectors

$$Y|_x = Y^\mu(x) \frac{\partial}{\partial x^\mu}$$
$$Y|_y = Y^\mu(y) \frac{\partial}{\partial y^\mu}. \tag{4.103}$$

A Taylor expansion of the components of the latter to first order in ϵ yields

$$Y|_y = (Y^\mu(x + \epsilon X)) \frac{\partial}{\partial y^\mu} \approx \left(Y^\mu(x) + \epsilon X^\nu(x) \frac{\partial Y^\mu(x)}{\partial x^\nu} \right) \frac{\partial}{\partial y^\mu}. \tag{4.104}$$

Then, by applying the differential map

$$\left((\sigma_{-\epsilon})_\star Y|_y \right)^\alpha = Y^\mu|_y \frac{\partial x^\alpha}{\partial y^\mu} = \left(Y^\mu(x) + \epsilon X^\nu(x) \frac{\partial Y^\mu(x)}{\partial x^\nu} \right) \left(\delta^\alpha{}_\mu - \epsilon \overbrace{\frac{\partial X^\alpha(y)}{\partial y^\mu}}^{\frac{\partial X^\alpha}{\partial x^\mu} + O(\epsilon)} \right)$$
$$= Y^\alpha(x) + \epsilon \left(X^\nu(x) \frac{\partial Y^\alpha}{\partial x^\nu} - Y^\mu(x) \frac{\partial X^\alpha}{\partial x^\mu} \right) + O(\epsilon^2) \tag{4.105}$$

we obtain

$$\mathcal{L}_X Y = \left(X^\nu \frac{\partial Y^\mu}{\partial x^\nu} - Y^\nu \frac{\partial X^\mu}{\partial x^\nu} \right) \frac{\partial}{\partial x^\mu} = [X, Y]. \tag{4.106}$$

In physics, one can interpret Eq. (4.106) by saying that the infinitesimal change of a vector field Y under an infinitesimal coordinate transformation generated by a vector field X is

$$\delta_X Y = \mathcal{L}_X Y = [X, Y]. \tag{4.107}$$

Lie Derivative of a One-Form Field

We now define the *Lie derivative of a one-form field*. Let $w \in \Omega^1(M)$ be a one-form field, i.e., a cotangent vector field. We define the Lie derivative of w along X as before, except that now we use the pullback $(\sigma_\epsilon)^\star$, whose expression in components is given in Eq. (4.86), to move the one-form to x in order to do the subtraction

$$\mathcal{L}_X w = \lim_{\epsilon \to 0} \frac{1}{\epsilon}\left(\sigma_\epsilon^\star w|_{\sigma_\epsilon(x)} - w|_x\right). \tag{4.108}$$

Using the infinitesimal coordinate transformation $y^\mu \equiv \sigma_\epsilon^\mu(x) \approx x^\mu + \epsilon X^\mu(x)$, it is possible to work out an explicit formula as above:

$$(\sigma_\epsilon^\star w)_\alpha = w_\beta(y)\frac{\partial y^\beta}{\partial x^\alpha} = w_\beta(x+\epsilon X)\frac{\partial}{\partial x^\alpha}(x^\beta + \epsilon X^\beta)$$
$$= (w_\beta(x) + \epsilon X^\mu \partial_\mu w_\beta(x))(\delta^\beta{}_\alpha + \epsilon \partial_\alpha X^\beta)$$
$$= w_\alpha + \epsilon(X^\mu \partial_\mu w_\alpha + w_\mu \partial_\alpha X^\mu).$$

Thus we obtain

$$\mathcal{L}_X w = \left(X^\mu \partial_\mu w_\alpha + w_\mu \partial_\alpha X^\mu\right)dx^\alpha$$
$$= X[w_\alpha]dx^\alpha + w_\mu dX^\mu. \tag{4.109}$$

Example Consider the vector field $X = y^3 \partial_x + x\partial_y$ and the one-form $w = x^2 dx + y^2 dy$ on $\mathbb{R}^2$. Using Eq. (4.109), the Lie derivative of w along the flow generated by X is found to be

$$\mathcal{L}_X w = X[w_\alpha]dx^\alpha + w_\mu dX^\mu$$
$$= X[x^2]dx + X[y^2]dy + x^2 d(y^3) + y^2 d(x)$$
$$= y^3 \partial_x(x^2)dx + x\partial_y(y^2) + 3x^2y^2 dy + y^2 dx$$
$$= (2xy^3 + y^2)dx + (2xy + 3x^2y^2)dy. \tag{4.110}$$

Lie Derivative of a Tensor Field

We next require that the Lie derivative satisfies the Leibniz rule by imposing that the condition

$$\mathcal{L}_X(T_1 \otimes T_2) = (\mathcal{L}_X T_1) \otimes T_2 + T_1 \otimes (\mathcal{L}_X T_2) \tag{4.111}$$

holds, even if pairs of indices have been contracted. We leave it to the reader (see Problem 4.15) to prove that the Leibniz rule is indeed satisfied when T_1 is a function, i.e., a $(0,0)$-tensor, and T_2 is a one-form or a vector field, or vice versa. Lie derivatives of tensors of higher rank can then be worked out by applying the rule and by induction.

Example Let us find the Lie derivative of a $(1,1)$-tensor

$$T = T_\mu{}^\nu dx^\mu \otimes e_\nu, \quad \text{where } e_\nu = \frac{\partial}{\partial x^\nu}. \tag{4.112}$$

Working step by step

$$\begin{aligned}
\mathcal{L}_X T &= (\mathcal{L}_X T)^\mu{}_\nu dx^\nu \otimes e_\mu + T^\mu{}_\nu (\mathcal{L}_X dx^\nu) \otimes e_\mu + T^\mu{}_\nu dx^\nu \otimes (\mathcal{L}_X e_\mu) \\
&= (X^\alpha \partial_\alpha T^\mu{}_\mu) dx^\nu \otimes e_\mu + T^\mu{}_\mu (\partial_\alpha X^\nu) dx^\alpha \otimes e_\mu - T^\mu{}_\nu dx^\nu \otimes (\partial_\mu X^\alpha) e_\alpha \\
&= (X^\alpha \partial_\alpha T^\mu{}_\nu + T^\mu{}_\alpha \partial_\nu X^\alpha - T^\alpha{}_\nu \partial_\alpha X^\mu) dx^\nu \otimes e_\mu, \tag{4.113}
\end{aligned}$$

hence

$$(\mathcal{L}_X T)^\mu{}_\nu = X^\alpha (\partial_\alpha T^\mu{}_\nu) + (\partial_\nu X^\alpha) T^\mu{}_\alpha - (\partial_\alpha X^\mu) T^\alpha{}_\nu. \tag{4.114}$$

Note that in the derivation we used the fact that $e_\mu = \frac{\partial}{\partial x^\mu}$, and the properties that $(e_\mu)^\alpha = \delta_\mu{}^\alpha$ and $(dx^\nu)_\alpha = \delta^\nu{}_\alpha$. Moreover, we also used Eq. (4.106) to obtain

$$(\mathcal{L}_X e_\mu)^\alpha = X^\nu \partial_\nu (e_\mu)^\alpha - (e_\mu)^\nu \partial_\nu X^\alpha = -\partial_\mu X^\alpha \tag{4.115}$$

as well as Eq. (4.109) to get

$$(\mathcal{L}_X dx^\nu)_\alpha = X^\mu \partial_\mu (dx^\nu)_\alpha + (dx^\nu)_\mu \partial_\alpha X^\mu = \partial_\alpha X^\nu. \tag{4.116}$$

For a generic (q,r)-tensor, a similar but tedious calculation gives for the components

$$\begin{aligned}
(\mathcal{L}_X T)^{\mu_1\mu_2\cdots\mu_q}{}_{\nu_1\nu_2\cdots\nu_r} &= X^\lambda \partial_\lambda T^{\mu_1\mu_2\cdots\mu_q}{}_{\nu_1\nu_2\cdots\nu_r} + \\
&+ (\partial_{\nu_1} X^\lambda) T^{\mu_1\cdots\mu_q}{}_{\lambda\nu_2\cdots\nu_r} + (\partial_{\nu_2} X^\lambda) T^{\mu_1\cdots\mu_q}{}_{\nu_1\lambda\nu_3\cdots\nu_r} + \cdots + (\partial_{\nu_r} X^\lambda) T^{\mu_1\cdots\mu_q}{}_{\nu_1\cdots\nu_{r-1}\lambda} - \\
&- (\partial_\lambda X^{\mu_1}) T^{\lambda\mu_2\cdots\mu_q}{}_{\nu_1\cdots\nu_r} - (\partial_\lambda X^{\mu_2}) T^{\mu_1\lambda\cdots\mu_q}{}_{\nu_1\cdots\nu_r} - \cdots - (\partial_\lambda X^{\mu_q}) T^{\mu_1\cdots\mu_{r-1}\lambda}{}_{\nu_1\cdots\nu_r}. \tag{4.117}
\end{aligned}$$

We conclude our discussion of the Lie derivative by noting some useful identities. The identity (4.100) generalizes to all tensor fields (see Problem 4.16):

$$\mathcal{L}_{[X,Y]} t = [\mathcal{L}_X, \mathcal{L}_Y] t. \tag{4.118}$$

Moreover, one can also show that for Lie derivatives of vector fields, the following property holds:

$$\mathcal{L}_X[Y, Z] = [\mathcal{L}_X Y, Z] + [Y, \mathcal{L}_X Z]. \tag{4.119}$$

Note that Eq. (4.119) can be rewritten as

$$[X, [Y, Z]] = [[X, Y], Z] + [Y, [X, Z]], \tag{4.120}$$

so that Eq. (4.119) reduces to the Jacobi identity

$$[X, [Y, Z]] + [Y, [Z, X]] + [Z, [X, Y]] = 0. \tag{4.121}$$

4.4 Differential Forms

A *differential form* of order r (or an r-form) is defined as a totally antisymmetric $(0, r)$-tensor:

$$p \in S_r : \quad w(v_{p(1)}, \ldots, v_{p(r)}) = \text{sgn}(p)\, w(v_1, \ldots, v_r), \tag{4.122}$$

where $\text{sgn}(p)$ is the sign of the permutation p:

$$\text{sgn}(p) = \begin{cases} +1 & \text{for an even permutation} \\ -1 & \text{for an odd permutation} \end{cases}.$$

In local coordinates (x^μ), total antisymmetry means that the components of the tensor $w_{\mu_1,\ldots,\mu_r}$ satisfy

$$w_{\mu_{p(1)},\ldots,\mu_{p(r)}} = \text{sgn}(p)\, w_{\mu_1,\ldots,\mu_r}. \tag{4.123}$$

Example Let ω be a two-form, i.e., an antisymmetric $(0,2)$-tensor. In local coordinates x^i, it can be written as

$$\omega = \omega_{\mu\nu} dx^\mu \otimes dx^\nu, \tag{4.124}$$

and by the antisymmetry $\omega_{\nu\mu} = -\omega_{\mu\nu}$ we can rewrite it as follows:

$$\omega = \frac{1}{2}\omega_{\mu\nu}(dx^\mu \otimes dx^\nu - dx^\nu \otimes dx^\mu). \tag{4.125}$$

Note that in this rewriting we have two terms corresponding to the two possible permutations of the two indices.

Example More generally, due to the total antisymmetry we can rewrite a generic r-form

$$w = w_{\mu_1,\ldots,\mu_r} dx^{\mu_1} \otimes \cdots \otimes dx^{\mu_r} \tag{4.126}$$

as

$$w = \frac{1}{r!} w_{\mu_1,\ldots,\mu_r} \sum_{p \in S_r} \text{sgn}(p) dx^{\mu_{p(1)}} \otimes \cdots \otimes dx^{\mu_{p(r)}}. \tag{4.127}$$

The example above leads us to define the *wedge product* of one-forms:

$$dx^{\mu_1} \wedge dx^{\mu_2} \wedge \cdots \wedge dx^{\mu_r} = \sum_{p \in S_r} \text{sgn}(p)\, dx^{\mu_{p(1)}} \otimes \cdots \otimes dx^{\mu_{p(r)}}. \tag{4.128}$$

Example Simple examples are

$$dx^\mu \wedge dx^\nu = dx^\mu \otimes dx^\nu - dx^\nu \otimes dx^\mu$$

and

$$dx^1 \wedge dx^2 \wedge dx^3 = dx^1 \otimes dx^2 \otimes dx^3 + dx^2 \otimes dx^3 \otimes dx^1 + dx^3 \otimes dx^1 \otimes dx^2$$
$$- dx^2 \otimes dx^1 \otimes dx^3 - dx^3 \otimes dx^2 \otimes dx^1 - dx^1 \otimes dx^3 \otimes dx^2.$$

We note two key properties of the wedge product:

(i) the wedge product vanishes, if the same index value appears at least twice;

(ii) a permutation of the factors changes the result of the wedge product multiplying it by the sign of the permutation:

$$dx^{\mu_1} \wedge \cdots \wedge dx^{\mu_r} = \text{sgn}(p) dx^{\mu_{p(1)}} \wedge \cdots \wedge dx^{\mu_{p(r)}}.$$

Example   For example,

$$dx^1 \wedge dx^2 \wedge dx^1 = 0 \qquad (4.129)$$

and

$$dx^4 \wedge dx^1 \wedge dx^2 \wedge dx^3 = -dx^1 \wedge dx^2 \wedge dx^3 \wedge dx^4. \qquad (4.130)$$

Since linear combinations of antisymmetric tensors of the same type are again antisymmetric, r-forms make up a vector space, to be denoted as $\Omega_p^r(M)$. Combining Eq. (4.127) and Eq. (4.128), we can see that r-forms can be expressed conveniently using the wedge products,

$$w = \frac{1}{r!} w_{\mu_1,\dots,\mu_r} dx^{\mu_1} \wedge dx^{\mu_2} \wedge \cdots \wedge dx^{\mu_r}. \qquad (4.131)$$

Thus $\{dx^{\mu_1} \wedge \cdots \wedge dx^{\mu_r} | \ \mu_1 < \mu_2 < \cdots < \mu_r\}$ forms the basis of the vector space of r-forms $\Omega_p^r(M)$.

We may assign a r-form smoothly at each point p on a manifold M, to obtain an r-form field. The r-form field will also be called an r-form for short. The number r is known as the *degree* of the form.

The vector spaces of r-forms (or of r-form fields) are denoted as $\Omega^r(M)$. Explicitly, for a manifold M of dimension $\dim M = n$,

$$\Omega^0(M) = \mathcal{F}(M) \quad \text{smooth functions on } M$$

$$\Omega^1(M) = T^\star(M) \quad \text{cotangent vectors, or one-form fields on } M$$

$$\Omega^2(M) = \text{span}\{dx^\mu \wedge dx^\nu \mid \mu < \nu\}$$

$$\vdots$$

$$\Omega^n M = \{f dx^1 \wedge dx^2 \wedge \dots \wedge dx^n | f \in \mathcal{F}(M)\}. \qquad (4.132)$$

The n-forms $\omega_n = f dx^1 \wedge dx^2 \wedge \cdots \wedge dx^n$ are known as *top forms* of the manifold M.

Example   Consider $M = \mathbb{R}^3$. We have the following r-forms ω_r. A zero-form is a smooth function,

$$\omega_0 = f(x, y, z), \qquad (4.133)$$

and a one-form is

$$\omega_1 = \omega_x(x, y, z) dx + \omega_y(x, y, z) dy + \omega_z(x, y, z) dz. \qquad (4.134)$$

For future use, note the components of the one-form ω_1 can be collected to a vector $\vec{\omega} \equiv (\omega_x, \omega_y, \omega_z)$, and one could write $\omega_1 = \vec{\omega} \cdot d\vec{r}$, where $d\vec{r} = (dx, dy, dz)$.

Using Eq. (4.131), we first write the two-form as

$$\omega_2 = \frac{1}{2} \Big[\omega_{xy}(x, y, z) dx \wedge dy + \omega_{yx}(x, y, z) dy \wedge dx$$

$$+ \omega_{yz} dy \wedge dz + \omega_{zy} dz \wedge dy + \omega_{zx} dz \wedge dx + \omega_{xz} dx \wedge dz \Big]. \qquad (4.135)$$

Using the antisymmetry of the components (e.g., $\omega_{yx} = -\omega_{yx}$) and the antisymmetry of the wedge products (e.g., $dy \wedge dx = -dx \wedge dy$), we can combine terms to arrive at

$$\omega_2 = \omega_{xy}(x, y, z) dx \wedge dy + \omega_{yz}(x, y, z) dy \wedge dz + \omega_{zx}(x, y, z) dz \wedge dx. \qquad (4.136)$$

Similarly, for future use, one can use the components of ω_2 to define a vector $\vec{\omega}' \equiv (\omega_{yz}, \omega_{zx}, \omega_{xy})$.

For the three-form we start with

$$\omega_3 = \frac{1}{3!} \Big[\omega_{xyz} dx \wedge dy \wedge dz + \omega_{yxz} dy \wedge dx \wedge dz + \omega_{xzy} dx \wedge dz \wedge dy$$

$$+ \omega_{zyx} dz \wedge dy \wedge dx + \omega_{yzx} dy \wedge dz \wedge dx + \omega_{zxy} dz \wedge dx \wedge dy \Big]. \quad (4.137)$$

Using the total antisymmetry of the components with respect to permutations of the indices, and the antisymmetry of the wedge product, one can prove that all of the six terms are equal, so Eq. (4.137) can be rewritten as

$$\omega_3 = \omega_{xyz} dx \wedge dy \wedge dz. \quad (4.138)$$

Finally, if one tries to construct a four-form, or a higher-degree form, by wedge products of dx, dy, or dz, it will involve at least two factors of the same type, and thus it will vanish.

Let $\dim M = m$. The $0 \leq r \leq m$ indices of the components of an r-form must each take different values out of the m possible values, so there are $\binom{m}{r}$ possible choices. Thus

$$\dim \Omega_p^r(M) = \binom{m}{r} = \frac{m!}{r!\,(m-r)!}. \quad (4.139)$$

Now we generalize the wedge product for the products of a q-form and an r-form: The generalized product will be called the *exterior product*. We begin with the formal definition.

The *exterior product* of a q-form ω and an r-form η is a $(q+r)$-form $\omega \wedge \eta$:

$$(\omega \wedge \eta)(v_1, \ldots, v_{q+r}) = \frac{1}{q!\,r!} \sum_{p \in S_{q+r}} \mathrm{sgn}(p)\omega(v_{p(1)}, \ldots, v_{p(q)}) \cdot \eta(v_{p(q+1)}, \ldots, v_{p(q+r)}).$$

$$(4.140)$$

When we introduce local coordinates, this formal definition reduces to the useful practical rule for constructing the exterior product:

$$\omega \wedge \eta = \frac{1}{q!\,r!} \omega_{\mu_1 \ldots \mu_q} \eta_{\nu_1 \ldots \nu_r} dx^{\mu_1} \wedge \cdots \wedge dx^{\mu_q} \wedge dx^{\nu_1} \wedge \cdots \wedge dx^{\nu_r}. \quad (4.141)$$

Note that, if $q + r > m = \dim(M)$, then $\omega \wedge \eta = 0$. Furthermore, the exterior product satisfies the following properties:

(i) $\omega \wedge \eta = (-1)^{qr} \eta \wedge \omega$,
(ii) if q is odd, then $\omega \wedge \omega = 0$,
(iii) $(\omega \wedge \eta) \wedge \xi = \omega \wedge (\eta \wedge \xi)$.

The proof is left to the reader (see Problem 4.18). Note that (i) implies (ii).

Example   In $\mathbb{R}^3$, consider the exterior product $\omega_1 \wedge \omega_2$. Explicitly,

$$
\begin{aligned}
\omega_1 \wedge \omega_2 &= \left(\omega_x(x,y,z)dx + \omega_y(x,y,z)dy + \omega_z(x,y,z)dz \right) \wedge \left(\omega_{xy}(x,y,z)dx \wedge dy + \right. \\
&\quad \left. + \omega_{yz}dy \wedge dz + \omega_{zx}dz \wedge dx \right) \\
&= \omega_x \omega_{yz} dx \wedge dy \wedge dz + \omega_y \omega_{zx} dy \wedge dz \wedge dx + \omega_z \omega_{xy} dz \wedge dx \wedge dy \\
&= \left(\omega_x \omega_{yz} + \omega_y \omega_{zx} + \omega_z \omega_{xy} \right) dx \wedge dy \wedge dz.
\end{aligned}
\tag{4.142}
$$

Note that if we use the vectors $\vec{\omega}$ and $\vec{\omega}'$ that we defined previously using the components of ω_1 and ω_2, the function inside the brackets of (4.142) is just the scalar product:

$$
\omega_1 \wedge \omega_2 = (\vec{\omega} \cdot \vec{\omega}') \, dx \wedge dy \wedge dz.
\tag{4.143}
$$

Example   Consider the two-dimensional real vector space $\mathbb{R}^2$. As will be discussed later, considering Cartesian coordinates (x, y), the two-form $dx \wedge dy$ can be related to the oriented infinitesimal area element $dxdy$. Let us now consider polar coordinates (r, θ), to which the Cartesian coordinates are related via $(x, y) = (r \cos\theta, r \sin\theta)$, and use them to express the differentials dx and dy and their exterior product:

$$
\begin{aligned}
dx \wedge dy &= (\cos\theta dr - r\sin\theta d\theta) \wedge (\sin\theta dr + r\cos\theta d\theta) \\
&= \cos\theta \sin\theta dr \wedge dr + r(\cos\theta)^2 dr \wedge d\theta \\
&\quad - r(\sin\theta)^2 d\theta \wedge dr - r^2 \sin\theta \cos\theta d\theta \wedge d\theta \\
&= rdr \wedge d\theta.
\end{aligned}
\tag{4.144}
$$

4.4.1 Exterior Derivative

Next we want to generalize the mapping from functions (zero-forms) f to differentials (one-forms) df to a mapping from r-forms ω to $(r+1)$-forms. We define the *exterior derivative d* as a linear map

$$
d : \Omega^r(M) \to \Omega^{r+1}(M)
\tag{4.145}
$$

such that

$$
d : \omega = \frac{1}{r!} \omega_{\mu_1 \ldots \mu_r} dx^{\mu_1} \wedge \cdots \wedge dx^{\mu_r} \to d\omega = \frac{1}{r!} \frac{\partial \omega_{\mu_1 \ldots \mu_r}}{\partial x^\nu} dx^\nu \wedge dx^{\mu_1} \wedge \cdots \wedge dx^{\mu_r}.
\tag{4.146}
$$

We may also write, in short,

$$
d = \frac{\partial}{\partial x^\nu} dx^\nu \wedge,
\tag{4.147}
$$

with the understanding that the partial derivative acts on the components of the r-form ω in $d\omega$.

The meaning of the definition of the exterior derivative for r-forms is easiest to understand with an example. We consider again the r-forms in $\mathbb{R}^3$ and construct their exterior derivatives. Starting from zero-forms (functions), we have

$$
d\omega_0 = df = \frac{\partial f}{\partial x} dx + \frac{\partial f}{\partial y} dy + \frac{\partial f}{\partial z} dz.
\tag{4.148}
$$

Thus the components of $d\omega_0$ are the components of ∇f.

Next, the exterior derivative of a one-form is

$$
\begin{aligned}
d\omega_1 &= \frac{\partial \omega_x}{\partial y} dy \wedge dx + \frac{\partial \omega_x}{\partial z} dz \wedge dx + \frac{\partial \omega_y}{\partial x} dx \wedge dy \\
&\quad + \frac{\partial \omega_y}{\partial z} dz \wedge dy + \frac{\partial \omega_z}{\partial x} dx \wedge dz + \frac{\partial \omega_z}{\partial y} dy \wedge dz \\
&= \left(\frac{\partial \omega_y}{\partial x} - \frac{\partial \omega_x}{\partial y} \right) dx \wedge dy + \left(\frac{\partial \omega_z}{\partial y} - \frac{\partial \omega_y}{\partial z} \right) dy \wedge dz + \left(\frac{\partial \omega_x}{\partial z} - \frac{\partial \omega_z}{\partial x} \right) dz \wedge dx.
\end{aligned}
\tag{4.149}
$$

Thus, recalling the definition of $\vec{\omega}$ as the vector built from the components of the one-form ω, i.e., $\vec{\omega} = (\omega_x, \omega_y, \omega_z)$, we observe that the components of $d\omega_1$ are the components of $\nabla \times \vec{\omega}$.

Then, the exterior derivative of a two-form is

$$
\begin{aligned}
d\omega_2 &= \frac{\partial \omega_{xy}}{\partial dz} dz \wedge dx \wedge dy + \frac{\partial \omega_{yz}}{\partial x} dx \wedge dy \wedge dz + \frac{\partial \omega_{zx}}{\partial y} dy \wedge dz \wedge dx \\
&= \left(\frac{\partial \omega_{yz}}{\partial x} + \frac{\partial \omega_{zx}}{\partial y} + \frac{\partial \omega_{xy}}{\partial z} \right) dx \wedge dy \wedge dz.
\end{aligned}
\tag{4.150}
$$

Thus, the components of $d\omega_2$ can be interpreted as the terms appearing in the divergence of the vector $\vec{\omega}' = (\omega_{yz}, \omega_{zx}, \omega_{xy})$, namely, $d\omega_2 = \nabla \cdot \vec{\omega}'$.

Thus we found that the exterior derivatives correspond to the gradient, curl and divergence.

Finally,

$$
d\omega_3 = 0.
\tag{4.151}
$$

Now, let us ask: What is $d(d\omega)$? Applying the definition twice, we notice that

$$
d(d\omega) = \frac{1}{r!} \left(\underbrace{\frac{\partial^2}{\partial x^\alpha \partial x^\beta}}_{\text{symmetric in } \alpha \text{ and } \beta} w_{\mu_1 \ldots \mu_r} \overbrace{dx^\alpha \wedge dx^\beta}^{\text{antisymmetric in } \alpha \text{ and } \beta} \wedge dx^{\mu_1} \wedge \cdots \wedge dx^{\mu_r} \right) = 0.
\tag{4.152}
$$

So $d^2 = 0$.

In particular, for $\dim M = 3$, the fact that the exterior derivative of the exterior derivative of a function is identically vanishing,

$$
d(df) = d(\partial_x f dx + \partial_y f dy + \partial_z f dz) = \left(\frac{\partial^2}{\partial x \partial y} - \frac{\partial^2}{\partial y \partial x} \right) dx \wedge dy + \cdots = 0,
$$

means that the curl of a gradient is identically vanishing $\nabla \times \nabla f = 0$. Similarly, the fact that the exterior derivative of the exterior derivative of a one-form is zero, $d(d\omega_1) = 0$, corresponds to the fact that the divergence of the curl of the vector $\vec{\omega}$ (whose components are the components of ω_1, and thus are arbitrary) is zero, $\nabla \cdot (\nabla \times \vec{\omega}) = 0$. Thus, one notices that the familiar rules of vector calculus stem from the single equation $d^2 = 0$.

In passing, we note that the exterior derivative of the pullback of a form is equal to the pullback of the exterior derivative,

$$
df^\star \omega = f^\star(d\omega),
\tag{4.153}
$$

and the proof is left to the reader in Problem 4.20.

4.4.2 Closed and Exact Differential Forms

Among differential forms, one can distinguish the following special cases. If

$$d\omega = 0, \tag{4.154}$$

then ω is said to be a *closed r-form*

If there exists an $(r-1)$-form ω_{r-1} such that

$$\omega_r = d\omega_{r-1}, \tag{4.155}$$

then ω_r is an *exact r*-form. Note that every exact form is also closed, since $d^2 = 0$.

Example The one-form

$$\alpha = ydx + xdy \tag{4.156}$$

is exact in $\mathbb{R}^2$ since $\alpha = d(xy)$. Conversely, the one-form

$$\beta = ydx \tag{4.157}$$

is not closed, so it cannot be exact either

Note that exactness is not necessarily true everywhere on the manifold or in the subset where ω_r is well defined. It may happen that Eq. (4.155) can be satisfied only in a subset $U \subset M$. Then we say that ω is exact in U. Note also that an exact form is always closed (in a subset U), but the opposite is not necessarily true. However, every closed form is exact at least locally. This is the statement of the following lemma, due to Henri Poincaré.

Lemma 4.8 (Poincaré's lemma) *Let B be an open ball in $\mathbb{R}^n$. Any smooth closed r form ω_r (with $1 \leq r \leq n$) is exact in B*

Proof Omitted.   □

One can extend the statement to a general manifold M and consider an open set U in a coordinate neighborhood V which is mapped by the coordinate function to an open set $\phi^{-1}(U)$ inside a ball B in $\mathbb{R}^n$. If a form ω is closed in U, its pullback will be closed in $\phi^{-1}(U) \subset B$, so it will also be exact there. Then ω must be exact in U. Thus, every closed form is locally exact. Another important fact is the following.

Theorem 4.1 *Let M be a contractible $(\pi_1(M) = \{e\})$ differentiable manifold. Then every closed form in M is exact in M.*

Proof Omitted.   □

Example Let $M = \mathbb{R}^2 \setminus \{(0,0)\}$. Consider the one-form

$$\omega = \frac{y}{x^2 + y^2}dx + \frac{x}{x^2 + y^2}dy. \tag{4.158}$$

This form is closed, but is it exact? It is illuminating to reexpress the Cartesian coordinates and the form in terms of polar coordinates, $(x,y) = (r\cos\varphi, r\sin\varphi)$. We find $x^2 + y^2 = r^2$,

$$dx = \cos\varphi dr - r\sin\varphi d\varphi, \qquad dy = \sin\varphi dr + r\cos\varphi d\varphi, \tag{4.159}$$

and finally

$$\omega = d\varphi, \tag{4.160}$$

so that ω is closed. It also looks like an exact form, but there is a subtlety. Recall that

$$\varphi = \arctan\left(\frac{y}{x}\right) \tag{4.161}$$

is a multivalued map, requiring a branch cut. In other words, φ is properly defined only, for example, in $U = \mathbb{R}^2 \setminus \{(x, 0)|x \geq 0\}$. So ω is exact and Eq. (4.160) is well defined in this region, which is a proper subset of M, and does not contain the full unit circle. Note also that U is contractible, while M is not.

It is interesting to ask when a closed form is not exact. This turns out to give one more way to classify differentiable manifolds. Let α and β be two closed r-forms. We say that α and β are *cohomologous*, denoting this relation as $\alpha \sim \beta$, when they differ by an exact form, $\alpha = \beta + d\gamma$. One can easily verify that this is an equivalence relation. Each of the equivalence classes $[\alpha]$ defined by the equivalence relation $\sim$ is a *cohomology class* of closed forms.

The set of cohomology classes can be given the structure of a group, by defining the group product of two cohomology classes as the cohomology class containing the sum of the representatives, i.e., $[\alpha] + [\beta] = [\alpha + \beta]$. The unit element is the set of exact forms $[0]$. In this way, the cohomology classes of closed r-forms form a group, called the rth *de Rham cohomology group* of the manifold M, denoted as $H^r_{\text{dR}}(M)$, or simply as $H^r(M)$ if it is clear from the context. The de Rham cohomology groups, which are named after Georges de Rham, give a way to classify differentiable manifolds. It is clear from the above theorem that if M is contractible, all of its de Rham cohomology groups are trivial, so $H^r_{\text{dR}}(M) = \{e\}$ for all $r = 1 \leq r \leq \dim(M)$.

Another way to think about the de Rham cohomology is the following: The exterior derivative induces the sequence of maps

$$0 \xrightarrow{i} \Omega^0 \xrightarrow{d_0} \Omega^1 \xrightarrow{d_1} \Omega^2 \xrightarrow{d_2} \cdots \xrightarrow{d_{m-2}} \Omega^{m-1} \xrightarrow{d_{m-1}} \Omega^m \xrightarrow{d_m} 0,$$

where $\Omega^r = \Omega^r(M)$, while i is the inclusion map $0 \hookrightarrow \Omega^0(M)$ and d_r denotes the map $d_r : \Omega^{r-1} \to \Omega^r$, $\omega \to d\omega$. Since $d^2 = 0$, we have

$$\underbrace{\text{Im}\, d_r}_{\text{exact } r\text{-forms}} \subset \underbrace{\text{Ker}\, d_{r+1}}_{\text{closed } (r+1)\text{-forms}} .$$

Such a sequence is called an *exact sequence*. This particular sequence is called the *de Rham complex*. The quotient space $\text{Ker}\, d_{r+1} / \text{Im}\, d_r$ is the rth de Rham cohomology group.

4.4.3 Interior Product, Exterior Derivative, and Lie Derivative

The exterior product maps r-forms to $(r + 1)$-forms. We can also define a map of r-forms to $(r - 1)$-form by using vector fields. We define an *interior product i* as a map $\chi(M) \times \Omega^r(M) \to \Omega^{r-1}(M)$ that maps a vector field X and an r-form ω to an $(r - 1)$-form $i_X\omega$,

$$i : (X, \omega) \to i_X\omega = \omega(X, \ldots, \ldots, \ldots), \tag{4.162}$$

where the ellipses denote $r-1$ arguments, so that when $i_X\omega$ acts on $r-1$ vector fields we have

$$i_X\omega(Y_1, \ldots, Y_{r-1}) = \omega(X, Y_1, \ldots, Y_{r-1}). \qquad (4.163)$$

This is more transparent in local coordinates. Let

$$X = X^\mu(x)\frac{\partial}{\partial x^\mu}, \qquad \omega = \frac{1}{r!}\omega_{\mu_1 \ldots \mu_r}(x)dx^{\mu_1} \wedge \cdots \wedge dx^{\mu_r}, \qquad (4.164)$$

then

$$i_X\omega = \frac{1}{(r-1)!}\omega_{\mu\mu_2 \ldots \mu_r}X^\nu \, dx^\mu \underbrace{\left(\frac{\partial}{\partial x^\nu}\right)}_{\delta^\mu_\nu} dx^{\mu_2} \wedge \cdots \wedge dx^{\mu_r} \qquad (4.165)$$

$$= \frac{1}{(r-1)!}X^\mu \omega_{\mu\mu_2 \ldots \mu_r}dx^{\mu_2} \wedge \cdots \wedge dx^{\mu_r}. \qquad (4.166)$$

For future use, we introduce the notation

$$dx^1 \wedge \cdots dx^{j-1} \wedge \widehat{dx^j} \wedge dx^{j+1} \wedge \cdots dx^n = dx^1 \wedge \cdots dx^{j-1} \wedge dx^{j+1} \wedge \cdots dx^n, \quad (4.167)$$

where $\widehat{dx^j}$ means that dx^j is omitted from the wedge products; Eq. (4.166) can then be rewritten as

$$i_X\omega = \frac{1}{r!}\sum_{p=1}^{r} X^{\mu_p}\omega_{\mu_1 \ldots \mu_p \ldots \mu_r}(-1)^{p-1}dx^{\mu_1} \wedge \cdots \wedge \widehat{dx^{\mu_p}} \wedge \cdots \wedge dx^{\mu_r}. \qquad (4.168)$$

Example   Let $X = x\partial_y - y\partial_x$ be a vector field, let $\eta = ydx + xdy$ be a one-form, and let $\omega = dx \wedge dy$ be a two-form on $\mathbb{R}^2$. One can show that

$$i_X\eta = -y^2 + x^2, \qquad (4.169)$$

is a function as expected. As for $i_X\omega$, on the other hand, one might mistakenly think that $i_X\omega = -ydy$, by combining X just with dx. However, this is not correct, because $dx \wedge dy$ denotes an *antisymmetrized* product. To get the correct result, one must first write ω as the antisymmetrized tensor

$$\omega = dx \otimes dy - dy \otimes dx, \qquad (4.170)$$

and then apply Eq. (4.162) to obtain

$$\begin{aligned}
i_X\omega &= dx(X) \otimes dy - dy(X) \otimes dx \\
&= x\underbrace{dx(\partial_y)}_{=0} \otimes dx - y\underbrace{dx(\partial_x)}_{=1} \otimes dy - x\underbrace{dy(\partial_y)}_{=1} \otimes dx + y\underbrace{dx(\partial_y)}_{=0} \otimes dx \\
&= -ydy - xdx. \qquad (4.171)
\end{aligned}$$

Equivalently, one can obtain the same result by first writing the two-form ω in terms of a sum over all (nonvanishing) wedge products, as in Eq. (4.131):

$$\omega = dx \wedge dy = \frac{dx \wedge dy - dy \wedge dx}{2} = \frac{1}{2!}(dx \wedge dy - dy \wedge dx), \qquad (4.172)$$

where we used the fact that $dx \wedge dy = -dy \wedge dx$, from which one obtains the coefficients of the two-form:

$$\omega_{12} = 1, \qquad \omega_{21} = -1. \qquad (4.173)$$

Then, applying Eq. (4.166):

$$i_X\omega = \frac{1}{(2-1)!}[-y \cdot 0 \, dx - y \cdot 1 \, dy + x \cdot (-1) \, dx + x \cdot 0 \, dy]$$
$$= -ydy - xdx, \tag{4.174}$$

in agreement with Eq. (4.171).

We note some properties of the interior product, leaving the proofs as Problem 4.24:

(i) $i_X i_Y \omega = -i_Y i_X \omega$. As a special case, this property implies the nilpotency of the interior product with respect to a vector field, $i_X^2 = 0$.

(ii) $i_X(\omega_r \wedge \eta_s) = (i_X\omega_r) \wedge \eta_s + (-1)^r \omega_r \wedge (i_X\eta_s)$.

Next we derive a useful formula, named after Élie Cartan, that relates the Lie derivative to the exterior derivative and the interior product.

Theorem 4.9 (Cartan's formula) *Given a vector field and an r-form field ω, one has*

$$\mathcal{L}_X\omega = (di_X + i_X d)\omega. \tag{4.175}$$

Proof We present a proof formulated in local coordinates. According to Eq. (4.117), the components of the Lie derivative of the r-form ω are

$$(\mathcal{L}_X\omega)_{\nu_1\ldots\nu_r} = \frac{1}{r!}X^\lambda \partial_\lambda \omega_{\nu_1\ldots\nu_r} + (\partial_{\nu_1}X^\lambda)\omega_{\lambda\nu_2\ldots\nu_r} + \cdots + (\partial_{\nu_r}X^\lambda)\omega_{\nu_1\ldots\nu_{r-1}\lambda}, \tag{4.176}$$

so that $\mathcal{L}_X\omega$ can be written as

$$\mathcal{L}_X\omega = \frac{1}{r!}X^\lambda \partial_\lambda \omega_{\mu_1\ldots\mu_r} dx^{\mu_1} \wedge \cdots \wedge dx^{\mu_r}$$
$$+ \frac{1}{r!}\sum_{p=1}^{r}(\partial_{\mu_p}X^\lambda)\omega_{\mu_1\ldots\mu_{p-1}\lambda\mu_{p+1}\ldots\mu_r} dx^{\mu_1} \wedge \cdots \wedge dx_r^\mu. \tag{4.177}$$

On the right-hand side of Eq. (4.175), using Eq. (4.168), the $di_X\omega$ term can be written as

$$di_X\omega = \frac{1}{r!}\sum_{p=1}^{r}(\partial_\lambda X^{\mu_p})\omega_{\mu_1\ldots\mu_p\ldots\mu_r}(-1)^{p-1}dx^\lambda \wedge dx^{\mu_1} \wedge \cdots \wedge \widehat{dx^{\mu_p}} \wedge \cdots \wedge dx^{\mu_r}$$
$$+ \frac{1}{r!}\sum_{p=1}^{r}X^{\mu_p}\partial_\lambda\omega_{\mu_1\ldots\mu_p\ldots\mu_r}(-1)^{p-1}dx^\lambda \wedge dx^{\mu_1} \wedge \cdots \wedge \widehat{dx^{\mu_p}} \wedge \cdots \wedge dx^{\mu_r}, \tag{4.178}$$

while the $i_X d\omega$ term is

$$i_X d\omega = \frac{1}{r!}X^\lambda \partial_\lambda \omega_{\mu_1\ldots\mu_r}dx^{\mu_1} \wedge \cdots \wedge dx^{\mu_r}$$
$$+ \frac{1}{r!}\sum_{p=1}^{r}X^{\mu_p}\partial_\lambda\omega_{\mu_1\ldots\mu_p\ldots\mu_r}(-1)^p dx^\lambda \wedge dx^{\mu_1} \wedge \cdots \wedge \widehat{dx^{\mu_p}} \wedge \cdots \wedge dx^{\mu_r}. \tag{4.179}$$

We note that the sum appearing on the second line of Eq. (4.178) and the one on the second line of Eq. (4.179) are equal, except for opposite factors $(-1)^{p-1}$ and $(-1)^p$, and thus will cancel against each other in the $d i_X \omega + i_X d\omega$ sum. On the other hand, in the first sum appearing on the right-hand side of Eq. (4.178), one can move the one-form dx^λ to the position of the missing dx^{μ_p}; this can be done by a sequence of $(p-1)$ pairwise permutations of dx^α terms, resulting in a factor $(-1)^{p-1}$ that cancels against the other factor of the same type. The dummy indices λ and μ_p can then be interchanged with each other. This leads to

$$(d i_X + i_X d)\omega = \frac{1}{r!} X^\lambda \partial_\lambda \omega_{\mu_1 \dots \mu_r} dx^{\mu_1} \wedge \dots \wedge dx^{\mu_r}$$
$$+ \frac{1}{r!} \sum_{p=1}^{r} (\partial_{\mu_p} X^\lambda) \omega_{\mu_1 \dots \mu_{p-1} \lambda \mu_{p+1} \dots \mu_r} dx^{\mu_1} \wedge \dots \wedge dx^{\mu_r}, \quad (4.180)$$

from which, using Eq. (4.177), one obtains

$$d i_X \omega + i_X d\omega = \mathcal{L}_X \omega, \quad (4.181)$$

namely Cartan's formula (4.175).   □

Example Let us consider the same vector field and two-form as in the previous example,

$$X = x\partial_y - y\partial_x, \quad \omega = dx \wedge dy. \quad (4.182)$$

The Lie derivative $\mathcal{L}_X \omega$ is associated with the infinitesimal change $\delta_X \omega$ of the two-form under a transformation generated by X. Let us compute

$$\delta_X \omega \equiv \mathcal{L}_X \omega = d i_X \omega + i_X d\omega = d i_X \omega, \quad (4.183)$$

having used the fact that ω is closed, i.e., $d\omega = 0$, thus

$$\delta_X \omega = d(-ydy - xdx) = -dy \wedge dy - dx \wedge dx = 0. \quad (4.184)$$

The result is not unexpected; $\omega = dx \wedge dy$ represents a volume element in $\mathbb{R}^2$, and the vector field X generates rotations which leave the volume element invariant.[3] Recall that, using polar coordinates that are related to the Cartesian ones via $(x,y) = (r\cos\phi, r\sin\phi)$, we have

$$\frac{\partial}{\partial \phi} = \frac{\partial x}{\partial \phi}\frac{\partial}{\partial x} + \frac{\partial y}{\partial \phi}\frac{\partial}{\partial y} = -r\sin\phi\partial_x + r\cos\phi\partial_y = -y\partial_x + x\partial_y = X, \quad (4.185)$$

so X generates shifts $\phi \to \phi + a$, which leave ω invariant; under X, the two-form $\omega = rdr \wedge d\phi$ is mapped to $rdr \wedge d(\phi + a) = rdr \wedge d\phi = \omega$.

4.4.4 Integration of Differential Forms

Previously we have focused on extending the concept of partial differentiation to manifolds. In this section we focus on integration. We begin by defining the line integral of a one-form.

[3] The interpretation as a volume element will be more clear shortly, when we discuss integrals of m-forms.

Line Integrals

Let M be a smooth manifold. A curve $c : [a, b] \to M$ is said to be *piecewise smooth* when there exist points $t_i \in [a, b]$ such that $a \equiv t_0 < t_1 < \cdots < t_n \equiv b$ and c is smooth in each interval $[t_i, t_{i+1}]$.

Let α be a one-form on M, and let c_i be the restriction of c on each interval $[t_i, t_{i+1}]$. We can then consider the pullback one-form $c_i^\star \alpha$. In local coordinates $x = (x^\mu)$, a segment of the curve is $c_i(t) = x_i(t) = (x_i^\mu(t))$, and the one-form can be expanded as $\alpha = \alpha_\mu(x) dx^\mu$. The pullback is then

$$c_i^\star \alpha = \alpha_\mu(\mathbf{x}_i(t)) \dot{x}^\mu(t) dt \equiv f_i(t) dt. \tag{4.186}$$

We define the *line integral* of α along the curve c as

$$\int_c \alpha = \sum_{i=0}^{n-1} \int_{t_i}^{t_{i+1}} c_i^\star \alpha = \sum_{i=0}^{n-1} \int_{t_i}^{t_{i+1}} f_i(t) dt. \tag{4.187}$$

Example Let us discuss an example with a curve consisting of just one smooth segment. Consider S^2 with polar coordinates (θ, φ), the curve $c : [0, 1] \to S^2$, $c(t) = (\theta(t), \varphi(t)) = (\pi t, 2\pi t)$, and the one-form $\alpha = d\theta + \sin \theta d\varphi$. The pullback is

$$c^\star \alpha = \dot{\theta}(t) dt + \sin \theta(t) \dot{\varphi}(t) dt = (\pi + 2\pi \sin(\pi t)) dt \tag{4.188}$$

and the line integral is

$$\int_c \alpha = \int_0^1 (\pi + 2\pi \sin(\pi t)) \, dt = \pi + 4. \tag{4.189}$$

Now suppose that the one-form α is exact on M, meaning that there exists a smooth function $g \in \mathcal{F}(M)$ so that α can be written as $\alpha = dg$. Denoting the endpoints of the curve as $p_0 = c(a)$ and $p_1 = c(b)$, the line integral simplifies to

$$\int_c \alpha = \int_c dg = g(p_1) - g(p_0). \tag{4.190}$$

In this case the line integral is also path-independent: If $\tilde{c}$ is any piecewise smooth curve with the same endpoints p_0 and p_1, then

$$\int_{\tilde{c}} \alpha = g(p_1) - g(p_0) = \int_c \alpha. \tag{4.191}$$

In particular, if c is any piecewise smooth *closed* curve, $c(a) = c(b) = p_0$, then

$$\oint_c \alpha = \oint_c dg = 0, \tag{4.192}$$

where we denoted a line integral over a closed curve by the $\oint$ symbol. Thus the exactness of a one-form implies that its line integral over every piecewise smooth and closed curve vanishes. The converse statement is encoded in the following theorem.

Theorem 4.10 *If*

$$\oint_c \alpha = 0 \tag{4.193}$$

for all piecewise smooth closed curves on a smooth manifold M, then α is exact on M.

Proof   Omitted.   □

Example   Let $M = \mathbb{R}^2 \setminus \{(0,0)\}$ and consider again the one-form

$$\omega = -\frac{y}{x^2 + y^2}dx + \frac{x}{x^2 + y^2}dy. \tag{4.194}$$

For the line integral on the circle of unit radius centered at the origin, we get

$$\oint_{r=1} \omega = \oint_{r=1} d\varphi = 2\pi. \tag{4.195}$$

Thus, the theorem above implies that ω cannot be exact everywhere on M. However, if we restrict ω to the proper subset $U = \mathbb{R}^2 \setminus \{(x,0)|x \geq 0\}$, which is the two-dimensional real plane with the nonnegative x semiaxis removed, we know from the discussion after Eq. (4.161) that ω is exact in U,

$$\omega = d\varphi, \tag{4.196}$$

with

$$\varphi = \arctan\left(\frac{y}{x}\right), \tag{4.197}$$

where the arctangent function is properly defined (single-valued) in U. Now

$$\oint_c \omega = \oint_c d\varphi = 0 \tag{4.198}$$

for all closed curves $c \subset U$.

Integrals of m-Forms

After the line integral, we move to integrate m-forms over an m-dimensional manifold. As a warm-up, we consider the integral of an m-form ω in a subset A of $\mathbb{R}^m$. Using Cartesian coordinates, the m-form can be written as

$$\omega = f(x)\, dx^1 \wedge dx^2 \wedge \cdots \wedge dx^m. \tag{4.199}$$

The expression $dx^1 \wedge dx^2 \wedge \cdots \wedge dx^m$ can be interpreted as the infinitesimal volume element $d^m x = dx^1 dx^2 \cdots dx^m$, the measure of the ordinary Riemann integral. Thus the integral of ω on A can be defined as

$$\int_A \omega = \int_A f(x)dx^1 \cdots dx^m. \tag{4.200}$$

Notice that the choice of coordinates also implies an implicit choice of orientation. For example, the usual coordinates x^1, x^2, and x^3 in the three-dimensional space $\mathbb{R}^3$ define what we usually call the right-handed frame. With a coordinate transformation

$$\begin{cases} y^1 = x^1 \\ y^2 = x^2 \\ y^3 = -x^3 \end{cases},$$

the new set of coordinates y^1, y^2, and y^3 defines a left-handed coordinate frame. Correspondingly, had one used the y^i coordinates in Eq. (4.199), then $dy^1 \wedge dy^2 \wedge dy^3$ would have implicitly induced a left-handed orientation into Eq. (4.200).

It turns out that the concept of orientation cannot be consistently extended over all manifolds. We need to define what manifolds are orientable. Moreover, one can define the integration over of an m-form over M only if M is an orientable manifold.

Let p be a point of M, with $p \in U_i \cap U_j$; denote the coordinates on U_i as x^μ, and those on U_j as y^μ. Then, the tangent space T_pM is spanned by $e_\mu = \frac{\partial}{\partial x^\mu}$ or by $\tilde{e}_\mu = \frac{\partial}{\partial y^\mu}$, and the two bases are related by the Jacobian matrix

$$\tilde{e}_\mu = \frac{\partial x^\nu}{\partial y^\mu} e_\nu, \tag{4.201}$$

whose determinant

$$J(p) = \det\left(\frac{\partial x^\mu}{\partial y^\nu}\right)_p. \tag{4.202}$$

is nonvanishing if the two bases are properly defined.

If $J(p) > 0$, we say that $\{e_\mu\}$ and $\{\tilde{e}_\mu\}$ define the *same orientation* at $p \in U_i \cap U_j$. Conversely, when $J(p)$ is negative, $\{e_\mu\}$ and $\{\tilde{e}_\mu\}$ define the *opposite orientation* at $p \in U_i \cap U_j$.

We say that $(M, \{U_i, x_i\})$, a manifold M with an atlas $\{U_i, x_i\}$, is *orientable* when, for every pair of overlapping charts U_i and U_j, the determinant $J = \det\left(\frac{\partial x_i^\mu}{\partial x_j^\nu}\right)$ is positive at every point $p \in M$. Note that i and j are fixed, while μ and ν denote the components of the matrix. In other words, the determinant is taken over μ and ν.

If M is orientable, then one can prove that there exists an m-form ω which is nonvanishing everywhere on M. This m-form ω is called a *volume element* and plays the rôle of an integration measure on M. Two volume elements ω and ω' are equivalent, if $\omega = h\omega'$, where $h \in \mathcal{F}$ is a smooth, positive function on M, i.e., $h(p) > 0$ for all $p \in M$; in that case, we denote $\omega \sim \omega'$, which is an equivalence relation.

If ω and ω'' are not equivalent, then $\omega = k\omega''$, where $k(p) < 0$ for all $p \in M$. So there are two equivalence classes for volume elements, corresponding to two inequivalent orientations. We call one of them a *right-handed orientation* and the other a *left-handed orientation*.

Integration of Forms

Consider a volume form ω in an m-dimensional orientable manifold M (recall that ω is a m-form). We first define the integral of ω in a coordinate neighborhood $U \subset M$, denoting the coordinate function by $\phi : U \to \phi(U) \subset \mathbb{R}^m$. We use the Cartesian coordinates x^μ of $\mathbb{R}^m$ to label the points in U by the inverse map $U \ni p = \phi^{-1}(x)$. We can now use the inverse of the coordinate function to pull back the volume form ω into $\phi(U)$, where the pullback $(\phi^{-1})^\star\omega$ can be written as

$$(\phi^{-1})^\star\omega = h(x)\, dx^1 \wedge \cdots \wedge dx^m. \tag{4.203}$$

The pullback $(\phi^{-1})^\star\omega$ can thus be integrated on the subset $\phi(U)$ using Eq. (4.200); the integral of ω over U is then

$$\int_U \omega = \int_{\phi(U)} (\phi^{-1})^\star\omega = \int_{\phi(U)} h(x)\, dx^1 \cdots dx^m. \tag{4.204}$$

Example   Let $M = S^2$. As in an earlier example, we choose U as the set of points of S^2 from which we remove a line from the north pole to the south pole. There exists a diffeomorphism ϕ such that $\phi(U) = (0, \pi) \times (0, 2\pi)$, and we denote a generic point $p \in U$ as $p = \phi^{-1}(\theta, \varphi)$ using the coordinates (θ, φ) in $(0, \pi) \times (0, 2\pi)$. The pullback of a volume form ω_2 on U is

$$(\phi^{-1})^\star \omega_2 = \sin\theta \, d\theta \wedge d\varphi, \qquad (4.205)$$

hence one can integrate

$$\int_U \omega_2 = \int_{\phi(U)} (\phi^{-1})^\star \omega_2 = \int_0^\pi \int_0^{2\pi} \sin\theta d\theta d\varphi = 4\pi. \qquad (4.206)$$

Next, we define an integral of a smooth function $f \in \mathcal{F}(M)$ over U with the volume element ω as the integration measure. Note that f is defined on M. In order to get a function to be integrated on $\phi(U)$, consider the composite function $f \circ \phi^{-1} = f(\phi^{-1})$. In other words, we know what is $f(p)$ and we combine this with $p = \phi^{-1}(x)$. We now define

$$\int_U f\omega = \int_{\phi(U)} f(\phi^{-1}) \, (\phi^{-1})^\star \omega = \int_{\phi(U)} f(\phi^{-1}(x))h(x) \, dx^1 \cdots dx^m. \qquad (4.207)$$

Recall that we defined a (smooth) manifold to be paracompact. This means that one can choose a locally finite open covering $\{U_i\}$ of M, so that every point $p \in M$ belongs to only a finite number of U_is. Let the coordinate function on each U_i be denoted as ϕ_i. We now define a *partition of unity*, with respect to the locally finite cover. It is a family of differentiable functions ϵ_i such that

(i) $0 \leq \epsilon_i(p) \leq 1$ for every $p \in M$,
(ii) $\epsilon_i(p) = 0$ for every $p \notin U_i$,
(iii) $\sum_i \epsilon_i(p) = 1$ for every $p \in M$.

Local finiteness of the cover guarantees that the sum in (iii) is a finite one, so that there are no convergence issues. We emphasize that the partition of unity $\{\epsilon_i\}$ depends on the choice of the cover $\{U_i\}$.

Now consider a function $f : M \to \mathbb{R}$. Associated with the partition of unity, we define new functions

$$f_i = f\epsilon_i \qquad (4.208)$$

so that

$$f(p) = f(p) \cdot 1 = f(p) \sum_i \epsilon_i(p) = \sum_i f_i(p). \qquad (4.209)$$

Since $f_i(p) = 0$ when $p \notin U_i$, we can use Eq. (4.207) to extend the integral over all M as follows

$$\int_M f\omega = \sum_i \int_{U_i} f_i\omega = \sum_i \int_{\phi_i(U_i)} f_i(\phi_i^{-1}(x))h(x) \, dx^1 \cdots dx^m. \qquad (4.210)$$

Again, the paracompactness condition ensures that the sum over i is finite and that there are no convergence problems for the sum. One can show that, although a different atlas $\{(V_i, \psi_i)\}$ gives different coordinates and a different partition of unity, the integral remains the same.

Example   Let $M = S^1$, and consider $U_1 = S^1 \setminus \{(1, 0)\}$ and $U_2 = S^1 \setminus \{(-1, 0)\}$. Further, let the inverse coordinate functions be

$$\varphi_1^{-1} : (0, 2\pi) \to U_1, \quad p = \varphi_1^{-1}(\theta_1) = (\cos \theta_1, \sin \theta_1),$$
$$\varphi_2^{-1} : (-\pi, \pi) \to U_2, \quad p = \varphi_2^{-1}(\theta_2) = (\cos \theta_2, \sin \theta_2).$$

With this cover, we define the following partition of unity:

$$\epsilon_1(p) = \epsilon_1(\varphi_1^{-1}(\theta_1)) = \sin^2\left(\frac{\theta_1}{2}\right),$$

$$\epsilon_2(p) = \epsilon_2(\varphi_2^{-1}(\theta_2)) = \cos^2\left(\frac{\theta_2}{2}\right). \tag{4.211}$$

Note that this satisfies the conditions (i), (ii), and (iii) defined above. Next we define the function $f : S^1 \to \mathbb{R}$ to be integrated as

$$f(p) = \begin{cases} f(\varphi_1^{-1}(\theta_1)) = \sin^2 \theta_1 \\ f(\varphi_2^{-1}(\theta_2)) = \sin^2 \theta_2 \end{cases} \tag{4.212}$$

and the volume form ω such that

$$(\varphi_i^{-1})^\star \omega = \begin{cases} 1 \cdot d\theta_1 & \text{on } \varphi_1(U_1) \\ 1 \cdot d\theta_2 & \text{on } \varphi_2(U_2) \end{cases}. \tag{4.213}$$

Thus $h(\theta_i) = 1$. Substituting these expressions into Eq. (4.210), we get

$$\int_{S^1} f\omega = \sum_{i=1}^2 \int_{U_i} f_i\omega = \sum_{i=1}^2 \int_{\varphi_i(U_i)} d\theta_i \, h(\theta_i) f(\varphi_i^{-1}(\theta_i)) \epsilon_i(\varphi_i^{-1}(\theta_i))$$

$$= \int_0^{2\pi} d\theta_1 \sin^2\left(\frac{\theta_1}{2}\right) \sin^2 \theta_1 + \int_{-\pi}^{\pi} d\theta_2 \cos^2\left(\frac{\theta_2}{2}\right) \sin^2 \theta_2$$

$$= \frac{\pi}{2} + \frac{\pi}{2} = \pi, \tag{4.214}$$

as expected.

    Although this construction may make integration on manifolds look complicated, one can often get away with not making an explicit distinction between coordinate neighborhoods and their images in $\mathbb{R}^n$, just thinking of the coordinates as being "drawn on the manifold." In the example above, one would just think of points on S^1 as θ, i.e., use just one patch, covering the whole manifold except for one point. There would then be no need for the partition of unity. The volume form would then be called $\omega = d\theta$, and the function to be integrated $f = \sin^2(\theta)$. (Note that, for a function f to be well defined on S^1, it must be periodic with period 2π.) Then one would simply compute

$$\int_{S^1} f\omega = \int_0^{2\pi} \sin^2 \theta \, d\theta = \pi. \tag{4.215}$$

Missing one point from S^1 with the single patch means removing a set of zero measure from the integration domain and it has no effect on the integral of a smooth function.

Stokes' Theorem

The power of differential forms becomes manifest in the generalization of the celebrated theorem formulated by George Gabriel Stokes for the integration of differential forms on manifolds. As we will see, it is an elegant way to capture many formulas from calculus into a single, easily memorizable equation, much like in our previous example with $d^2 = 0$ applied to r-forms in $\mathbb{R}^3$. We begin by recalling some familiar formulas.

The *fundamental theorem of calculus* states that the integral of a function $f(x)$ over an interval $[a, b]$ is

$$\int_a^b f(x)\, dx = F(b) - F(a), \tag{4.216}$$

where $F(x)$ is a *primitive of* $f(x)$, i.e., a function whose derivative is $F'(x) = f(x)$. As we discussed earlier in this subsection, the line integral on a smooth manifold of an exact one-form is the generalization of Eq. (4.216).

Then, recall that another well-known theorem, due to George Green, establishes a correspondence between a closed contour integral to a surface integral, with the contour C being the boundary of the surface S, $C = \partial S$. This theorem is often used in classical electromagnetism, where we use Green's theorem to relate the line integral of the electric field $\vec{E}$ along a closed path C to the surface integral of its curl on a surface S bounded by C:

$$\oint_C \vec{E} \cdot d\vec{x} = \int_S \left(\nabla \times \vec{E} \right) \cdot d\vec{a}, \tag{4.217}$$

where $d\vec{a} = \vec{n}\, da$, with da an infinitesimal area element and $\vec{n}$ a unit vector normal to the surface S, and oriented in such a way that C winds anticlockwise around $\vec{n}$. Furthermore, using *Faraday's law of induction*

$$\nabla \times \vec{E} = -\frac{\partial \vec{B}}{\partial t}, \tag{4.218}$$

one can rewrite Eq. (4.217) as

$$\oint_C \vec{E} \cdot d\vec{x} = -\int_S \left(\frac{\partial \vec{B}}{\partial t} \right) \cdot d\vec{a}. \tag{4.219}$$

Finally, *Gauss' theorem* (named after Carl Friedrich Gauss, and also known as the *divergence theorem*) relates a volume integral to a surface integral,

$$\int_V \left(\nabla \cdot \vec{E} \right) dv = \oint_{\partial V} \vec{E} \cdot d\vec{a}, \tag{4.220}$$

where V is a volume (with infinitesimal volume element dv) and ∂V denotes its boundary, which is a closed surface, and $d\vec{a} = \vec{n}\, da$, with da an infinitesimal area element and $\vec{n}$ a unit vector normal to the surface and oriented towards the exterior of the volume V. The right-hand side of Eq. (4.220) expresses the total flux of a vector quantity $\vec{E}$ through this surface. If $\vec{E}$ is an electric field, sourced by the charge density ρ, then

$$\nabla \cdot \vec{E} = \frac{\rho}{\varepsilon_0}, \tag{4.221}$$

where ε_0 is the absolute dielectric permittivity of the vacuum. In this case, Eq. (4.220) implies

$$\frac{Q}{\varepsilon_0} = \frac{1}{\varepsilon_0} \int_V \rho dv = \oint_{\partial V} \vec{E} \cdot d\vec{a} \equiv \Phi_E, \tag{4.222}$$

which relates the total electric charge Q enclosed in V to the electric flux Φ_E through ∂V.

We will now revisit, and generalize, these theorems, using the formalism of the integration of r-forms. We start by proving a theorem for integrals of forms in the lower half-space $\mathbb{H}^n_- = \{x = (x^1, \ldots, x^n) \in \mathbb{R}^n | x_1 \leq 0\}$. Note first that all $(n-1)$-forms can be expanded as

$$\omega = \sum_{j=1}^n f_j(x) dx^1 \wedge \cdots dx^{j-1} \wedge \widehat{dx^j} \wedge dx^{j+1} \wedge \cdots dx^n. \tag{4.223}$$

Theorem 4.11 *Let ω be an $(n-1)$-form with compact support on $\mathbb{H}^n_-$, including compact support on $\partial \mathbb{H}^n_-$. Then the following integral formula holds:*

$$\int_{\mathbb{H}^n_-} d\omega = \int_{\partial \mathbb{H}^n_-} \omega. \tag{4.224}$$

Proof Because of the expansion in Eq. (4.223) and linearity of Eq. (4.224), it is sufficient to consider the forms

$$\omega_j \equiv f_j(x) dx^1 \wedge \cdots dx^{j-1} \wedge \widehat{dx^j} \wedge dx^{j+1} \wedge \cdots dx^n. \tag{4.225}$$

Note that only one term in the expansion contributes to the right-hand side of Eq. (4.224), and it simplifies as follows

$$\int_{\partial \mathbb{H}^n_-} \omega = \int_{\partial \mathbb{H}^n_-} \omega_1 = \int_{\mathbb{R}^{n-1}} f_1(0, x^2, \ldots, x^n) dx^2 \wedge \cdots \wedge dx^n. \tag{4.226}$$

The fact that ω has compact support means that all functions f_j have compact support in $\mathbb{H}^n_-$ (including on the boundary). Then,

$$\int_{\mathbb{H}^n_-} d\omega_j = \int_{\mathbb{H}^n_-} d\left(f_j dx^1 \wedge \cdots dx^{j-1} \wedge \widehat{dx^j} \wedge dx^{j+1} \wedge \cdots dx^n \right)$$

$$= \int_{\mathbb{H}^n_-} \left(df_j \wedge dx^1 \wedge \cdots dx^{j-1} \wedge \widehat{dx^j} \wedge dx^{j+1} \wedge \cdots dx^n \right)$$

$$= \int_{\mathbb{H}^n_-} \left(\sum_k (\partial_k f_j) dx^k \wedge dx^1 \wedge \cdots dx^{j-1} \wedge \widehat{dx^j} \wedge dx^{j+1} \wedge \cdots dx^n \right)$$

$$= \int_{\mathbb{H}^n_-} \left(\partial_j f_j \right) dx^j \wedge dx^1 \wedge \cdots dx^{j-1} \wedge \widehat{dx^j} \wedge dx^{j+1} \wedge \cdots dx^n$$

$$= \int_{\mathbb{H}^n} \left(\partial_j f_j \right) dx^j dx^1 \cdots dx^{j-1} dx^{j+1} \cdots dx^n. \tag{4.227}$$

If $2 \leq j \leq n$, the final integral becomes (note that now there is no sum over j)

$$\int_{\mathbb{H}^n_-} d\omega_j = \int_{\mathbb{H}^{n-1}} \left(\int_{-\infty}^{\infty} \partial_j f_j dx^j \right) dx^1 \cdots dx^{j-1} dx^{j+1} \cdots dx^n = 0. \tag{4.228}$$

The integral of $\partial_j f_j$ over dx^j vanishes, because f_j has compact support and thus is zero at infinity.

If $j = 1$, we instead have an integral over a half-line,

$$
\int_{\mathbb{H}^n_-} d\omega_1 = \int_{\mathbb{R}^{n-1}} \left(\int_{-\infty}^0 \partial_1 f_1 dx^1 \right) dx^2 \cdots dx^n
$$

$$
= \int_{\mathbb{R}^{n-1}} dx^2 \dots dx^n \left(f_1(0, x^2, \dots, x^n) - 0 \right)
$$

$$
= \int_{\partial \mathbb{H}^n_-} \omega, \tag{4.229}
$$

in agreement with Eq. (4.224).   □

To move to consider a manifold (with a boundary) M, we need some preparation. First, a chart (U_i, ψ_i) on a smooth manifold (with a boundary) is said to be *nice* when, if $U_i \cap \partial M \neq \emptyset$, then ψ_i is a diffeomorphism onto $\mathbb{H}^n_-$, namely $\psi_i(U_i) = \mathbb{H}^n_-$, and if $U_i \cap \partial M = \emptyset$, then ψ_i is a diffeomorphism onto the interior $\text{int}(\mathbb{H}^n_-) = \{x \in \mathbb{H}^n_- | x^1 < 0\}$. So the essential feature of a nice chart is that it stretches a coordinate neighborhood to the lower half-space, or its interior. Note that $\psi_i(\partial U_i) = \partial \mathbb{H}^n_-$. We now introduce the following lemmas.

Lemma 4.12   *Let (U_i, ψ_i) be a nice chart on a manifold M of dimension $\dim M = n$, and let ω be a $(n-1)$-form. Then*

$$
\int_{U_i} d\omega = \int_{\partial U_i} \omega. \tag{4.230}
$$

Proof   Using the previous theorem and the definition of the integral,

$$
\int_{U_i} d\omega = \int_{\psi_i(U_i)} (\psi_i^{-1})^\star d\omega
$$

$$
= \int_{\mathbb{H}^n_-} (\psi_i^{-1})^\star d\omega = \int_{\mathbb{H}^n_-} d((\psi_i^{-1})^\star \omega)
$$

$$
= \int_{\partial \mathbb{H}^n_-} (\psi_i^{-1})^\star \omega = \int_{\partial U_i} \omega, \tag{4.231}
$$

in agreement with the statement of the theorem in Eq. (4.230).   □

Lemma 4.13   *Every (oriented) smooth manifold has an (oriented) atlas consisting of nice charts.*

Proof   Omitted.   □

Next we move to consider a manifold (with a boundary) M. We need to choose an orientable atlas $\{(U_i, \varphi_i)\}$ with nice charts such that $\{U_i\}$ is a locally finite cover, and a partition of unity $\{\epsilon_i\}$ subordinate to the cover. Note also that $\{(U_j \cap \partial M, \varphi_{i|\partial M})\}$ is an atlas of the boundary ∂M, and the restrictions $\{\epsilon_{i|\partial M}\}$ give the subordinate partition of unity.

Theorem 4.14 (Stokes' theorem)   *Let M be an oriented manifold, $\dim M = n$, with a nonvanishing or vanishing boundary ∂M, and ω a $(n-1)$-form on M. Then,*

$$
\int_M d\omega = \int_{\partial M} \omega. \tag{4.232}
$$

Proof Using the partition of unity and the definition of the integral, the quantity appearing on the left-hand side of Eq. (4.232) can be written as

$$
\int_M d\omega = \sum_i \int_{U_i} d(\epsilon_i \omega) = \sum_i \int_{\varphi_i(U_i)} (\varphi_i^{-1})^\star d(\epsilon_i \omega)
$$

$$
= \sum_i \int_{\varphi_i(U_i)} d((\varphi_i^{-1})^\star (\epsilon_i \omega)) = \sum_i \int_{\partial(\varphi_i(U_i))} (\varphi_i^{-1})^\star (\epsilon_i \omega)
$$

$$
= \sum_i \int_{\partial U_i} \epsilon_i \omega = \int_{\partial M} \omega, \tag{4.233}
$$

which proves Eq. (4.232). $\square$

4.5 Classical Mechanics and Symplectic Geometry

As one application of differential forms and vector fields, we consider a reformulation of classical mechanics with the help of differential forms, known as *symplectic geometry*. From the Hamiltonian formulation of classical mechanics, we know that the degrees of freedom of a classical system are associated with a *phase space* Γ: This is an even-dimensional space where a state of the system at time t is described by a point $x = (\vec{q}(t), \vec{p}(t))$, where $\vec{q} = (q^i)$ are the generalized canonical coordinates and $\vec{p} = (p^i)$ are the generalized canonical momenta. Now we want to think of the phase space Γ as a differentiable manifold of dimension $\dim \Gamma = 2n$. The dynamics of the system will be described by a function $H : \Gamma \to \mathbb{R}$ called the *Hamiltonian* (for simplicity, we only consider time-independent Hamiltonians). Let us consider generic coordinates $x = (x^\mu)$; a key property of the phase space is that there exists a closed two-form, called the *symplectic form*,

$$
\omega = \frac{1}{2}\omega_{\mu\nu}(x) dx^\mu \wedge dx^\nu, \qquad d\omega = 0, \tag{4.234}
$$

which is *nondegenerate*, meaning that $i_V \omega = 0$ if and only if the vector field $V \in \Gamma$ is identically vanishing. We then call the pair (Γ, ω) a *symplectic manifold*.

Note that components of ω form an antisymmetric matrix $(\omega_{\mu\nu})$, whose determinant is nonvanishing, due to the nondegeneracy condition. We can then consider its inverse matrix and denote its components as $\omega^{\mu\nu}$, thus

$$
\omega^{\mu\lambda}\omega_{\lambda\nu} = \delta^\mu_\nu. \tag{4.235}
$$

Using the inverse matrix $(\omega^{\mu\nu})$, we can define two important concepts. First, let f and g be two functions on the phase space. We define their *Poisson bracket* $\{f, g\}$ as

$$
\{f, g\} = \omega^{\mu\nu} \frac{\partial f}{\partial x^\mu} \frac{\partial g}{\partial x^\nu}. \tag{4.236}
$$

Second, associated with any function f we can define a vector field $X_f = X_f^\mu \frac{\partial}{\partial x^\mu}$, with components

$$
X_f^\mu = \omega^{\mu\nu} \frac{\partial f}{\partial x^\nu}. \tag{4.237}
$$

One can reformulate the definition in a coordinate-independent way, by defining a vector field Ωdf having the local coordinate expansion

$$\Omega df = \omega^{\mu\nu} \frac{\partial f}{\partial x^\nu} \frac{\partial}{\partial x^\mu}, \tag{4.238}$$

and rewriting Eq. (4.237) as

$$X_f = \Omega df. \tag{4.239}$$

In particular, if the function f is chosen to be the Hamiltonian H, we call the vector field X_H the *Hamiltonian vector field*. However, one often adopts the interpretation that any function f could define some dynamical system, and thus all X_f defined by Eq. (4.237) are called Hamiltonian vector fields.

So far, we have given the definitions with generic local coordinates of Γ, but they become more transparent if we again introduce canonical phase-space coordinates (q^i, p^i). This time they appear because of a theorem by Jean Gaston Darboux, known as *Darboux's theorem*. Without going into the detailed form of the theorem, it implies that locally at every point $p \in \Gamma$ there is a coordinate neighborhood U with local coordinates denoted as $(q^1, q^2, \ldots, q^n, p^1, p^2, \ldots, p^n)$ such that the symplectic form takes the form

$$\omega = dq^1 \wedge dp^1 + dq^2 \wedge dp^2 + \cdots + dq^n \wedge dp^n. \tag{4.240}$$

Since the symplectic form is locally exact, it is also possible to write

$$\omega = d\theta = d(q^1 dp^1 + \cdots + q^n dp^n), \tag{4.241}$$

where θ is called the *canonical one-form*. Then the system is invariant under transformations that map

$$\theta \to \theta + d\Lambda \tag{4.242}$$

where $\Lambda = \Lambda(q^1, \ldots, q^1, p^1, \ldots, p^n)$. The invariance of the system under transformations of the form in Eq. (4.242) is called *gauge invariance*.

The phase space can be covered by such charts. The coordinates (q^i, p^i) are the canonical coordinates, also called *Darboux coordinates*. Adopting a more uniform notation for the Darboux coordinates,

$$(y^1, \ldots, y^n, y^{n+1}, \ldots, y^{2n}) = (q^1, q^2, \ldots, q^n, p^1, p^2, \ldots, p^n), \tag{4.243}$$

Eq. (4.240) becomes

$$\omega = \frac{1}{2} \omega_{\mu\nu} dy^\mu \wedge dy^\nu \tag{4.244}$$

with the antisymmetric matrix of coefficients written (in terms of $n \times n$ blocks) as

$$\omega \equiv (\omega_{\mu\nu}) = \begin{pmatrix} \mathbb{0}_n & \mathbb{1}_n \\ -\mathbb{1}_n & \mathbb{0}_n \end{pmatrix}, \tag{4.245}$$

while the inverse matrix reads

$$\omega^{-1} \equiv (\omega^{\mu\nu}) = \begin{pmatrix} \mathbb{0}_n & -\mathbb{1}_n \\ \mathbb{1}_n & \mathbb{0}_n \end{pmatrix}. \tag{4.246}$$

In Darboux coordinates, the Poisson bracket in Eq. (4.236) takes the canonical form

$$\{f, g\} = \sum_{i,j=1}^{n} \left(\frac{\partial f}{\partial p^i} \frac{\partial g}{\partial q^i} - \frac{\partial f}{\partial q^i} \frac{\partial g}{\partial p^i} \right), \tag{4.247}$$

where, for the sake of clarity, we restored the summation symbol, and the Hamiltonian vector field X_H associated with the Hamiltonian $H(q,p)$ can be written as

$$X_H = \sum_{i=1}^{n} \left\{ \frac{\partial H}{\partial p^i} \frac{\partial}{\partial q^i} - \frac{\partial H}{\partial q^i} \frac{\partial}{\partial p^i} \right\}. \tag{4.248}$$

Note that the Darboux coordinates are not unique; it is possible to find a local coordinate transformation

$$(y^\mu) \rightarrow (Y^\mu(y)) = (Q^1, \ldots, Q^n, P^1, \ldots, P^n) \tag{4.249}$$

with

$$Q^i = Q^i(q^1, \ldots, q^n, p^1, \ldots, p^n), \qquad P^i = P^i(q^1, \ldots, q^n, p^1, \ldots, p^n) \tag{4.250}$$

that leave Eq. (4.240) invariant, i.e., that preserve the symplectic form:

$$\omega = dQ^1 \wedge dP^1 + dQ^2 \wedge dP^2 + \cdots + dQ^n \wedge dP^n. \tag{4.251}$$

Thus, they also preserve the Poisson bracket:

$$\{f, g\} = \sum_{i,j=1}^{n} \left(\frac{\partial f}{\partial P^i} \frac{\partial g}{\partial Q^i} - \frac{\partial f}{\partial Q^i} \frac{\partial g}{\partial P^i} \right). \tag{4.252}$$

Transformations of the type in Eq. (4.250) are called *canonical transformations*.

Let us study the flow associated with a Hamiltonian vector field,

$$\frac{dx^\mu(t)}{dt} = X_H^\mu(x(t)). \tag{4.253}$$

In Darboux coordinates, one can prove (see Problem 4.13), that the flow equation reduces to the Hamilton equations:

$$\frac{dq_i}{dt} = \frac{\partial H}{\partial p_i}, \qquad \frac{dp_i}{dt} = -\frac{\partial H}{\partial q_i}. \tag{4.254}$$

The integral curve thus represents the classical trajectory of the system in the phase space Γ under time evolution. We also expect the flow to be compatible with the symplectic form. Let us study what that means.

If we first invert Eq. (4.237) by $(\omega_{\mu\nu})$ and change notation from f to H, we obtain

$$\omega_{\mu\nu} X_H^\nu = \omega_{\mu\nu} \omega^{\nu\lambda} \partial_\lambda H = \partial_\mu H,$$

which means

$$dH = -i_{X_H}\omega. \tag{4.255}$$

We can then see that the flow generated by a Hamiltonian vector field preserves the symplectic form:

$$\mathcal{L}_{X_H}\omega = d i_{X_H}\omega = -ddH = 0. \tag{4.256}$$

One way to interpret this is that the Hamiltonian flow induces a gauge transformation, as in Eq. (4.242). Consider how the canonical one-form θ transforms under an infinitesimal flow:

$$\theta \to \theta + \delta_{X_H}\theta \equiv \theta + \mathcal{L}_{X_H}\theta, \tag{4.257}$$

where

$$\mathcal{L}_{X_H}\theta = (di_{X_H} + i_{X_H}d)\theta = di_{X_H}\theta + i_{X_H}\omega \tag{4.258}$$
$$= di_{X_H}\theta - dH = d(i_{X_H}\theta - H) \equiv d\Lambda_H,$$

so that Eq. (4.257) can be rewritten as a gauge transformation,

$$\theta \to \theta + d\Lambda_H \tag{4.259}$$

with

$$\Lambda_H = i_{X_H}\theta - H. \tag{4.260}$$

We can also make the converse statement of Eq. (4.256). Suppose the symplectic form is preserved by a transformation induced by a vector field X

$$\mathcal{L}_X\omega = 0; \tag{4.261}$$

then

$$di_X\omega = (di_X + i_Xd)\omega = \mathcal{L}_X\omega = 0, \tag{4.262}$$

so $i_X\omega$ is closed (and smooth) in some open ball $B \subset \Gamma$. We can then use Poincaré lemma 4.8, by which the one-form $i_X\omega$ must be exact in B; hence there exists a function F (at least locally defined in B) such that

$$i_X\omega = -dF. \tag{4.263}$$

Comparing Eq. (4.263) with Eq. (4.255), we can then interpret locally F as a Hamiltonian, for which $X \equiv X_F$ is the Hamiltonian vector field.

Problems

4.1 Show that $\mathbb{R}^n$ with the usual topology is a Hausdorff space.

4.2 Consider $\mathbb{R}^2$ equipped with the *discrete metric*

$$d(x,y) = \begin{cases} 1 & \text{if } x \neq y \\ 0 & \text{if } x = y \end{cases},$$

where x and y are elements of $\mathbb{R}^2$, and with the corresponding metric topology τ_d. Is $(\mathbb{R}^2, \tau_d)$ connected? Give a proof. Show also that $d(x,y)$ satisfies the definition of a metric.

4.3 Let $f : M \to N$ be a homeomorphism. Define a map $f_\star : \pi_1(M,x_0) \to \pi_1(N,f(x_0))$ such that $f_\star([\gamma]) = [f \circ \gamma]$. Show that $f_\star$ is an isomorphism, i.e., that $\pi_1(M,x_0) \cong \pi_1(N,f(x_0))$.

4.4 Show that the homeomorphism between topological spaces is an equivalence relation.

4.5 Prove that, if M is an n-dimensional manifold with boundary, ∂M is an $(n-1)$-dimensional manifold.

4.6 Given the subset of $\mathbb{R}^2$ defined as follows:

$$S = \{(x, y) \in \mathbb{R}^2 | y = \sin(1/x), x > 0\} \cup \{(0, 0)\},$$

prove that its subset $S_+ = \{(x, y) \in \mathbb{R}^2 | y = \sin(1/x), x > 0\}$ is path-connected and that S is connected, but not path-connected.

4.7 This problem focuses on the determination of homotopy groups.
   (i) Let $M = \mathbb{R}^3 \backslash \{$a point$\}$. Find $\pi_1(M)$ and $\pi_2(M)$.
   (ii) Let $M = \mathbb{R}^3 \backslash \{$a line$\}$. Compute $\pi_1(M)$.
   (iii) Let $M = \mathbb{R}^3 \backslash \{l_1, l_2\}$, where $l_1 \neq l_2$ are two parallel lines in $\mathbb{R}^3$. Determine $\pi_1(M)$.

4.8 Find the fundamental homotopy group of a torus with one point removed, $T^2 \backslash \{$a point$\}$. Hint: It may be useful to think of a torus as a rectangle with opposite sides identified.

4.9 Find an atlas and coordinates for a torus $T^2 = S^1 \times S^1$.

4.10 Let a differentiable manifold M_1 be the set of real numbers $\mathbb{R}$, with the coordinate $\phi_1(x) = x$, and M_2 also $\mathbb{R}$ but with the coordinate $\phi_2(x) = x^3$. Show that M_1 and M_2 are diffeomorphic. Hint: Find a suitable map $f : M_1 \rightarrow M_2$ and show that it is a diffeomorphism.

4.11 Derive the transformation rule for the components of the tensor

$$T = T^{\mu_1 \mu_2}{}_{\nu_1 \nu_2 \nu_3} \frac{\partial}{\partial x^{\mu_1}} \otimes \frac{\partial}{\partial x^{\mu_2}} \otimes dx^{\nu_1} \otimes dx^{\nu_2} \otimes dx^{\nu_3}$$

under the coordinate transformation $x \rightarrow y$.

4.12 Let the $(1, 0)$-tensor

$$R = \sum_{\mu=1}^{3} R^\mu \frac{\partial}{\partial x^\mu}$$

have the components

$$R^1 = a, \quad R^2 = a^2, \quad R^3 = a^3,$$

and let the $(0, 1)$-tensor

$$S = \sum_{\mu=1}^{3} S_\mu dx^\mu$$

have the components

$$S_1 = b, \quad S_2 = c, \quad S_3 = d.$$

Calculate all the components T^μ_ν of the $(1, 1)$-tensor $T = R \otimes S$.

4.13 **Hamilton's equations as a flow generated by a vector field.** In classical mechanics the equations of motion for a system with n degrees of freedom can be written as a set of first-order differential equations

$$\frac{dq_i}{dt} = \frac{\partial H}{\partial p_i}, \quad \frac{dp_i}{dt} = -\frac{\partial H}{\partial q_i}.$$

These equations are Hamilton's equations. We will next reformulate them in a different way. Let us define a vector field X_H in the phase space M,

$$X_H = \sum_{i=1}^{n} \left\{ \frac{\partial H}{\partial p_i} \frac{\partial}{\partial q_i} - \frac{\partial H}{\partial q_i} \frac{\partial}{\partial p_i} \right\}.$$

The vector field X_H gives rise to integral curves

$$x_H(t) = (q_1(t), \ldots, q_N(t), p_1(t), \ldots, p_N(t))$$

on the manifold M.

(i) Show that the equation defining the integral curves $x_H(t)$ is equivalent to Hamilton's equations.

(ii) Let $M = \mathbb{R}^2$, i.e., $n = 1$, and $H = \frac{1}{2}(p^2 + q^2)$. Find X_H and the generated flow $\sigma(t, x_0)$, where $x_0 = (q_0, p_0) = (1, 0)$. Illustrate it by a figure.

(iii) As before, but now with $H = \frac{1}{2}(p^2 - q^2)$ and $x_0 = (1, 1)$.

(iv) Now take $M = T^2$, with coordinates $q, p \in [0, 2\pi]$, and $H = \cos(p)$. Find the equation of the integral curve. Draw a figure of the curves on T^2.

4.14 Let X, Y, and Z be smooth vector fields on a differentiable manifold M and let f be a smooth function on M. Show that the following identities are true:

(i) $[X, fY] = (Xf)Y + f[X, Y]$,

(ii) $[X, [Y, Z]] + [Z, [X, Y]] + [Y, [Z, X]] = 0$.

4.15 Show that the Leibniz rule expressed by Eq. (4.111) is satisfied when $T_1 \equiv f$ is a function, and T_2 is a vector field X or a one-form ω.

4.16 Let T be a (p, q)-tensor field, and let X and Y be vector fields. Show that the Lie derivative satisfies $[\mathcal{L}_X, \mathcal{L}_Y]T = \mathcal{L}_{[X,Y]}T$. It is enough to consider the cases

(i) $(p, q) = (1, 0)$, i.e., T is a vector field, and

(ii) $(p, q) = (0, 1)$, i.e., T is a cotangent vector field.

The general case follows by applying the Leibniz rule.

4.17 Let $X = X^\mu(x)\frac{\partial}{\partial x^\mu}$ be a vector field and let $g = g_{\mu\nu}(x)dx^\mu \otimes dx^\nu$ be a $(0, 2)$-tensor. Compute the Lie derivative $\mathcal{L}_X g$.

4.18 Given the q-form ω, the r-form η, and the s-form ξ, show that the exterior product satisfies the properties:

(i) $\omega \wedge \omega = 0$, when the order of ω is odd,

(ii) $\omega \wedge \eta = (-1)^{qr} \eta \wedge \omega$,

(iii) $(\omega \wedge \eta) \wedge \xi = \omega \wedge (\eta \wedge \xi)$.

For the purposes of this problem, it is possible to use the following expansions

$$\omega = \frac{1}{q!}\omega_{\mu_1\mu_2\ldots\mu_q}dx^{\mu_1} \wedge dx^{\mu_2} \wedge \cdots \wedge dx^{\mu_q}, \qquad (4.264)$$

$$\eta = \frac{1}{r!}\eta_{\nu_1\nu_2\ldots\nu_r}dx^{\nu_1} \wedge dx^{\nu_2} \wedge \cdots \wedge dx^{\nu_r}, \qquad (4.265)$$

$$\xi = \frac{1}{s!}\eta_{\rho_1\rho_2\ldots\rho_s}dx^{\rho_1} \wedge dx^{\rho_2} \wedge \cdots \wedge dx^{\rho_s}, \qquad (4.266)$$

and direct computations.

4.19 Show that under a coordinate transformation $(x_1, \ldots, x_n) \rightarrow (y_1(\vec{x}), \ldots, y_n(\vec{x}))$ in $\mathbb{R}^n$, the n-form transforms as

$$dy^1 \wedge dy^2 \wedge \cdots \wedge dy^n = J(\vec{y}, \vec{x})dx^1 \wedge dx^2 \wedge \cdots \wedge dx^n,$$

where J is the Jacobian determinant

$$J(\vec{y}, \vec{x}) = \det\left(\frac{\partial y^i}{\partial x^j}\right).$$

Hint: Use the definition of the determinant of an $n \times n$ matrix.

4.20 Show that the exterior derivative of the pullback of a form equals the pullback of the exterior derivative: $df^\star\omega = f^\star(d\omega)$.

4.21 **Electromagnetic duality.** Let M be the four-dimensional Minkowski space, with coordinates x^0, x^1, x^2, and x^3. Let us define a linear operator $* : \Omega^r(M) \rightarrow \Omega^{4-r}(M)$, such that

$$r = 0 : \quad *1 = -dx^0 \wedge dx^1 \wedge dx^2 \wedge dx^3,$$

$$r = 1 : \quad *dx^i = -dx^j \wedge dx^k \wedge dx^0, \quad *dx^0 = -dx^1 \wedge dx^2 \wedge dx^3,$$

$$r = 2 : \quad *(dx^i \wedge dx^j) = dx^k \wedge dx^0, \quad *(dx^i \wedge dx^0) = -dx^j \wedge dx^k,$$

$$r = 3 : \quad *(dx^1 \wedge dx^2 \wedge dx^3) = -dx^0, \quad *(dx^i \wedge dx^j \wedge dx^0) = -dx^k,$$

$$r = 4 : \quad *(dx^0 \wedge dx^1 \wedge dx^2 \wedge dx^3) = 1,$$

where (i, j, k) is a symmetric permutation of $(1, 2, 3)$. The $*$ operator is called the *Hodge star operator*, mapping a differential form to its *Hodge dual*. We then define a one-form $A = A_\mu dx^\mu$, where $(A_\mu)_{\mu=0,\ldots,3} = (\phi, A)$ is the electromagnetic potential. The electromagnetic field-strength tensor is then the two-form $F = dA$, with the components

$$F_{\mu\nu} = \begin{pmatrix} 0 & -E_1 & -E_2 & -E_3 \\ E_1 & 0 & B_3 & -B_2 \\ E_2 & -B_3 & 0 & B_1 \\ E_3 & B_2 & -B_1 & 0 \end{pmatrix},$$

where

$$\vec{E} = -\nabla\phi - \frac{\partial}{\partial x^0}\vec{A}, \quad \vec{B} = \nabla \times \vec{A}.$$

Finally, we define a one-form $J = J_\mu dx^\mu = \rho dx^0 + j_k dx^k$, which corresponds to the electromagnetic current.

(i) Show that the identity $dF = d(dA) = 0$ corresponds to the Maxwell's equations

$$\nabla \cdot \vec{B} = 0, \quad \frac{\partial \vec{B}}{\partial x^0} + \nabla \times \vec{E} = 0.$$

(ii) Show that the equation $d * F = *J$ corresponds to the Maxwell equations

$$\nabla \cdot \vec{E} = \rho, \quad \nabla \times \vec{B} - \frac{\partial \vec{E}}{\partial x^0} = \vec{j}.$$

(iii) Show that the identity $0 = d(d * F) = d * J$ corresponds to the continuity equation of the current

$$\partial_\mu J^\mu = \frac{\partial \rho}{\partial x^0} + \nabla \cdot \vec{j} = 0.$$

4.22   Consider the following differential forms in $\mathbb{R}^3$:

$$\alpha = xdx + ydy + zdz, \quad \beta = zdx + xdy + ydz, \quad \gamma = xydz.$$

(i) Is α closed or exact? Is γ closed or exact?
(ii) Calculate $\alpha \wedge \beta$ and $(\alpha + \gamma) \wedge (\alpha + \gamma)$.

4.23   Examine if the following differential forms are closed, explaining why:
(i) $A \in \Omega^1(\mathbb{R}^2 \backslash \{0\})$, defined as

$$A = \frac{hc}{2\pi e} \frac{xdy - ydx}{x^2 + y^2},$$

where h, c, and e are real numbers;
(ii) $B \in \Omega^2(\mathbb{R}^3 \backslash \{0\})$, defined as

$$B = \frac{x_1 dx_2 \wedge dx_3 - x_2 dx_1 \wedge dx_3 + x_3 dx_1 \wedge dx_2}{|x|^3};$$

(iii) $\omega \in \Omega^3(\mathrm{GL}(n, \mathbb{C}))$, defined as

$$\omega = \mathrm{Tr}[(g^{-1}dg))^3] \equiv \sum_{i_1 \ldots i_6} (g_{i_1 i_2}^{-1} dg_{i_2 i_3} \wedge g_{i_3 i_4}^{-1} dg_{i_4 i_5} \wedge g_{i_5 i_6}^{-1} dg_{i_6 i_1}),$$

where g^{-1} is the inverse matrix of g, namely $g_{ij}^{-1} \equiv (g^{-1})_{ij}$.

4.24   Let X and Y be vector fields, let ω be a r-form, and let η be an s-form. Show that the following properties of the interior product are true:
(i) $i_X i_Y \omega = -i_Y i_X \omega$;
(ii) $i_X(\omega_r \wedge \eta_s) = (i_X \omega_r) \wedge \eta_s + (-1)^r \omega_r \wedge (i_X \eta_s)$.

4.25   Consider the vector field $X = x\partial_x + y\partial_y + z\partial_z$, the two-form $\alpha = 2zdx \wedge dy + 3ydx \wedge dz$, and the three-form $\omega = dx \wedge dy \wedge dz$ on $\mathbb{R}^3$. Calculate
(i) $i_X \alpha$,
(ii) $i_X \omega$,
(iii) $\mathcal{L}_X \alpha$.

4.26   Let $M = \mathbb{R}^3$, and $\alpha = \alpha_1 dx + \alpha_2 dy + \alpha_3 dz$. Show that Stokes' theorem implies

$$\int_S (\nabla \times \vec{\alpha}) \cdot d\vec{S} = \oint_{\partial S} \vec{\alpha} \cdot d\vec{s}, \tag{4.267}$$

where $\vec{\alpha} = (\alpha_1, \alpha_2, \alpha_3)$, while S is an arbitrary surface and ∂S denotes its boundary. In a similar vein, given the two-form $\phi = \frac{1}{2!}\phi_{\mu\nu}dx^{\mu} \wedge dx^{\nu}$, show that

$$\int_V \nabla \cdot \vec{\phi}\,dV = \int_{\partial V} \vec{\phi} \cdot d\vec{S}, \qquad (4.268)$$

where the components of the vector $\vec{\phi}$ are defined as $\phi^{\mu} = \varepsilon^{\mu\nu\rho}\phi_{\nu\rho}$, while V is an arbitrary volume and ∂V denotes its boundary. Discuss for what choice of orientations for the surface normal and for the line integral Eq. (4.267) and Eq. (4.268) hold. Hint: Using the definition of integral of an n-form, show how the integrals reduce to the usual integrals in $\mathbb{R}^3$ (and, for simplicity, for this exercise it is not necessary to consider several charts).

5 Riemannian Geometry

In this chapter, we equip differentiable manifolds with a metric and introduce concepts of differential geometry. This area of mathematics provides the language of the theory of general relativity. It gives a precise meaning to statements such as "the universe is expanding," "spacetime is curved," and "nothing can escape from a black hole." After learning the concepts of this chapter, you will have a mathematical background to study general relativity and learn more about the above-mentioned topics. However, the applications of differential geometry are not limited to general relativity alone, as its concepts and techniques are widely used in many other areas of physics, science, and engineering.

5.1 The Metric Tensor

We start this section by introducing the notions of a Riemannian metric and a pseudo-Riemannian metric, named after Bernhard Riemann.

Let M be a smooth manifold, and let p denote a generic point of M. A tensor field g is called a *Riemannian metric* on M when it is a $(0, 2)$-tensor field satisfying the following two properties:

(i) symmetry under exchange of its arguments: $g_p(U, V) = g_p(V, U)$ $\forall p \in M$ and for every U and $V \in T_pM$, the tangent space of M at the point p;
(ii) positive-definiteness: $g_p(U, U) \geq 0$ and $g_p(U, U) = 0$ if and only if $U = 0$.

If g satisfies (i), but, instead of (ii), g satisfies the following condition

(ii′) nondegeneracy: if $g_p(U, V) = 0$ for all $U \in T_pM$, then $V = 0$,

then g is called a *pseudo-Riemannian metric*.

In local coordinates in a neighborhood U, the metric is a symmetric tensor field

$$g = g_{\mu\nu}(x)dx^{\mu} \otimes dx^{\nu}. \tag{5.1}$$

In physics, the notation $ds^2 = g_{\mu\nu}(x)dx^{\mu}dx^{\nu}$ is common. We will motivate this notation shortly.

The components $g_{\mu\nu}(x) = g_{\nu\mu}(x)$ can be thought as the elements of a symmetric matrix $g(x)$, which can be diagonalized at every point x. If g is a Riemannian metric, then the positive-definiteness condition (ii) implies that no eigenvalue is zero, otherwise the corresponding eigenvector U_0 would be a nonzero vector satisfying $g_p(U_0, U_0) = 0$, in violation of the positive-definiteness condition. Similarly, if g is a pseudo-Riemannian metric, then from the nondegeneracy condition (ii′) follows that no eigenvalue is zero. In either case, it then follows that, if D is the dimension of M, then at every point $x \in M$ the matrix $g(x)$ has n_- negative and n_+ positive eigenvalues with $n_- + n_+ = D$. (Note that a theorem due to James Joseph Sylvester implies that the numbers of positive and negative eigenvalues for a symmetric matrix do not depend on the choice of basis.) Since $g(x)$ is a symmetric matrix with no zero eigenvalues, it can also be inverted. Let $g^{\mu\nu}(x)$ denote the components of the inverse matrix:

$$\left(g^{-1}(x)\right)_{\mu\nu} = g^{\mu\nu}(x). \tag{5.2}$$

Then, the $g^{\mu\nu}(x)$ define the components of the *inverse metric*.

Since $g_{\mu\nu}(x)$ are smooth functions in the coordinate neighborhood U, the eigenvalues of $g(x)$ cannot change sign in U, a continuous function cannot change sign without passing through zero, but zero eigenvalues are not allowed. Thus the numbers n_- and n_+ cannot vary from point to point and, using the smoothness of the transition functions, this observation can be extended to other charts. We call (n_-, n_+) the *signature* of the metric.

The pair (M, g) is called a *(pseudo)-Riemannian manifold* with signature (n_-, n_+). In particular, a Riemannian manifold has signature $(0, D)$. A pseudo-Riemannian manifold with signature $(1, D-1)$ is called a *Lorentzian manifold*, or a *spacetime*. In this case there are different conventions in physics. The definition above, that we use in this chapter, is often used in general relativity (and related areas), and is called the *mostly plus convention* for the signature. In contrast, in elementary particle physics the *mostly minus convention* $(D-1, 1)$ is usually preferred for spacetimes. We shall present an example of use of the mostly minus convention in Chapter 6. Note that in the mostly plus convention the determinant of the metric is always negative,

$$\det(g_{\mu\nu}(x)) < 0, \tag{5.3}$$

independently of the dimension D. The example below illustrates the two conventions.

Example The *Euclidean metric* δ on $\mathbb{R}^n$ is

$$\delta \equiv \delta_{ij} dx^i \otimes dx^j = dx^1 \otimes dx^1 + \cdots + dx^n \otimes dx^n. \tag{5.4}$$

Then $\delta(U, V) = \sum_{i=1}^n U^i V^i$. The Euclidean metric has signature $(0, n)$. The Riemannian manifold $(\mathbb{R}^n, \delta)$ is called the *n*-dimensional *Euclidean space*.

Next, we label the coordinates as x^μ with $\mu = 0, 1, \ldots, n-1$, and introduce the *Minkowski metric* η, named after Hermann Minkowski, in the mostly plus convention,

$$\eta = \eta_{\mu\nu} dx^\mu \otimes dx^\nu = -dx^0 \otimes dx^0 + dx^1 \otimes dx^1 + \cdots + dx^{n-1} \otimes dx^{n-1}, \tag{5.5}$$

with signature $(1, n-1)$.

The pseudo-Riemannian manifold $(\mathbb{R}^n, \eta)$ is called the *n*-dimensional *Minkowski spacetime*. If one wants to emphasize the convention, one can use the notation $\mathbb{R}^{1,n-1}$ to imply the mostly plus convention, and $\mathbb{R}^{n-1,1}$ for the mostly minus convention.

We will later also use metrics with a more general signature: Denoting the coordinates as x^M, with $M = 1, \ldots, n$, one can define

$$
\begin{aligned}
\eta_{p,q} &= \eta_{MN} dx^M \otimes dx^N \\
&= -dx^1 \otimes dx^1 - \cdots - dx^p \otimes dx^p + dx^{p+1} \otimes dx^{p+1} + \cdots + dx^n \otimes dx^n, \quad (5.6)
\end{aligned}
$$

where $p + q = n$. This metric has signature (p, q). We denote $(\mathbb{R}^n, \eta_{p,q})$ as $\mathbb{R}^{p,q}$.

On a Riemannian manifold one can define a length (or norm) of a vector X as

$$
||X|| = \sqrt{g(X, X)} = \sqrt{g_{\mu\nu} X^\mu X^\nu}, \quad (5.7)
$$

where the last expression involves components in local coordinates. The length of a vector field X can vary from point to point, since in general both $g_{\mu\nu}(x)$ and $X^\mu(x)$ are functions of x. Note that this definition of a metric refers only to local properties, while earlier we had defined a metric to give a distance between two points, through a definition that required the metric to satisfy a triangle inequality. We will clarify this apparent discrepancy shortly.

On a Riemannian manifold one can also define an angle ϕ between two vectors X and Y as

$$
\phi = \arccos\left(\frac{g(X, Y)}{||X|| \cdot ||Y||} \right), \quad (5.8)
$$

if both $||X||$ and $||Y||$ are positive.

The definition of the norm that we introduced above cannot be used in a pseudo-Riemannian manifold with dimension D and signature $(1, D - 1)$, because $g(X, X)$ can be negative or zero. In a pseudo-Riemannian manifold we make a distinction between three types of vectors. We say that a vector $X \in T_p M$ is

(i) *timelike*, if $g(X, X) < 0$,
(ii) *spacelike*, if $g(X, X) > 0$,
(iii) *lightlike* (or *null*), if $g(X, X) = 0$.

Note that, with this definition, a null vector is not necessarily the zero vector in $T_p M$ (although the zero vector *is* a null vector).

5.1.1 Lengths of Curves, Distance between Points

In physics, the metric is usually denoted $ds^2 = g_{\mu\nu} dx^\mu dx^\nu$. The reason for this is the following. Let $c(t) = x^\mu(t)$ be a curve on a manifold M with the metric g. The tangent vector of the curve in local coordinates has components $V^\mu = \dot{x}^\mu(t)$.

If $M = \mathbb{R}^3$ with the Euclidean metric $g_{\mu\nu} = \delta_{\mu\nu}$, the length of the curve between t_0 and t_1 (assuming $t_0 < t_1$) can be written as

$$
L_{\mathbb{R}^3}[c] = \int_{t_0}^{t_1} dt \sqrt{(\dot{x}^1)^2 + (\dot{x}^2)^2 + (\dot{x}^3)^2} = \int_{t_0}^{t_1} dt \sqrt{\delta_{\mu\nu} \dot{x}^\mu \dot{x}^\nu}. \quad (5.9)
$$

In a general Riemannian manifold, the length of the part of the curve between t_0 and t_1 is defined similarly, but now the components of the metric are $g_{\mu\nu}(x)$. Thus the length of a piecewise smooth curve $c : [t_0, t_1] \to M$ is

$$L[c] = \int_{t_0}^{t_1} dt \sqrt{g_{\mu\nu}(x(t))\dot{x}^\mu \dot{x}^\nu}. \tag{5.10}$$

Note that this definition is *reparameterization invariant*, i.e., invariant under a generic *reparameterization* of the curve c that can be expressed by a diffeomorphism $u : [t_0, t_1] \rightarrow [u_0, u_1], t \mapsto u(t)$. Indeed,

$$
\begin{aligned}
L[c] &= \int_{u_0}^{u_1} du \sqrt{g_{\mu\nu}(x(u))\frac{dx^\mu(u)}{du}\frac{dx^\nu(u)}{du}} \\
&= \int_{u_0}^{u_1} dt \frac{du}{dt} \sqrt{g_{\mu\nu}(x(u(t)))\frac{dx^\mu(u)}{du}\frac{dx^\nu(u)}{du}} \\
&= \int_{t_0}^{t_1} dt \sqrt{g_{\mu\nu}(x(u(t)))\frac{dx^\mu(u)}{du}\frac{du}{dt}\frac{dx^\nu(u)}{du}\frac{du}{dt}} \\
&= \int_{t_0}^{t_1} dt \sqrt{g_{\mu\nu}(x(t))\frac{dx^\mu(t)}{dt}\frac{dx^\nu(t)}{dt}}. \tag{5.11}
\end{aligned}
$$

If t_0 and t_1 are infinitesimally close, then writing $t_1 = t_0 + \Delta t$, one has

$$\Delta s \equiv L \approx \Delta t \sqrt{g_{\mu\nu}\dot{x}^\mu \dot{x}^\nu} \approx \Delta t \sqrt{g_{\mu\nu}\frac{\Delta x^\mu}{\Delta t}\frac{\Delta x^\nu}{\Delta t}} = \sqrt{g_{\mu\nu}\Delta x^\mu \Delta x^\nu}. \tag{5.12}$$

Changing the notation for the infinitesimal displacements, $\Delta s \rightarrow ds$ and $\Delta x^\mu \rightarrow dx^\mu$, one thus obtains

$$ds^2 = g_{\mu\nu}dx^\mu dx^\nu \tag{5.13}$$

for the square of an *infinitesimal length element ds*. One can then rewrite the length of the curve (5.10) as

$$L[c] = \int_c ds. \tag{5.14}$$

Using the definition (5.10), it is also possible to define a *distance $d(p, q)$* between two points p and q in M:

$$d(p, q) = \inf_{c \in \Gamma} \{L[c] \mid c : [t_0, t_1] \rightarrow M\}, \tag{5.15}$$

where the Γ denotes the set of all piecewise smooth curves $c(t)$ satisfying $c(t_0) = p$ and $c(t_1) = q$. This definition of the distance satisfies the properties of a metric in the sense of the earlier definition in Section 4.1.1.

Proof   To show that the definition of the distance (5.15) is symmetric, i.e., invariant under the interchange of the points, $d(p, q) = d(q, p)$, one can note that, if $c(t)$ denotes an arbitrary piecewise smooth curve from p to q, i.e., $c(t_0) = p$ and $c(t_1) = q$, then the curve defined as

$$\bar{c} : [t_0, t_1] \rightarrow M, t \rightarrow \bar{c}(t) = c(t_0 + t_1 - t) \tag{5.16}$$

is also piecewise smooth, satisfies $\bar{c}(t_0) = q$ and $\bar{c}(t_1) = p$, and has the same length as c. In particular, then, the inferior extremum of the lengths $L[\bar{c}]$ among all curves from q to p is also the same as the inferior extremum of the lengths $L[c]$ among all curves from p to q, i.e., $d(p, q) = d(q, p)$. Next, one can show that d satisfies the triangle inequality; given the two curves $c_1 : [t_0, t_1] \rightarrow M$ and $c_2 : [t'_0, t'_1] \rightarrow M$, such that the endpoint of c_1 is the same as the beginning point of c_2: $c_1(t_1) = c_2(t'_0)$, one can define the *concatenation* of the curves as the new curve $c_1 + c_2 : [0, t_1 - t_0 + t'_1 - t'_0] \rightarrow M$ such that

$$(c_1 + c_2)(t) = \begin{cases} c_1(t + t_0), & \text{for } t \in [0, t_1 - t_0] \\ c_2(t - (t_1 - t_0) + t'_0), & \text{for } t \in [t_1 - t_0, t'_1 - t'_0 + t_1 - t_0] \end{cases}.$$

(5.17)

With this definition, the length of the concatenation of the curves equals the sum of the lengths of the two curves: $L[c_1 + c_2] = L[c_1] + L[c_2]$. Next, consider three points p, q, and r in M. For any $\epsilon > 0$ there must exist a curve c_1 from p to q such that $d(p, q) \leq L[c_1] < d(p, q) + \epsilon$, otherwise the infimum would be strictly larger than $d(p, q)$. Likewise, there must exist a curve c_2 from q to r such that $d(q, r) \leq L[c_2] < d(q, r) + \epsilon$. Hence

$$d(p, r) \leq L[c_1 + c_2] = L[c_1] + L[c_2] < d(p, q) + d(q, r) + 2\epsilon, \qquad (5.18)$$

thus we recover the triangle inequality in the limit $\epsilon \rightarrow 0$. Finally, one can also show that $d(p, q) = 0 \Leftrightarrow p = q$. The fact that $p = q$ implies $d(p, q) = 0$ is easy to prove. For any $p \in M$, one always has $d(p, p) = 0$, because, according to the definition (5.10), the length of every curve (being the integral of a nonnegative quantity over a positively oriented interval) is always nonnegative, and such is also the inferior extremum appearing in the definition (5.10). Moreover, when $q = p$, one can easily show that there exists at least one piecewise smooth curve that saturates the inequality, i.e., that has exactly zero length, namely $c_{\text{const}}(t) = p$ for every $t \in [t_0, t_1]$, because in that case $\dot{x}^\mu(t) = 0$ for every μ and for all t. To prove that $d(p, q) = 0$ implies $p = q$, one can proceed with a *reductio ad absurdum*; if one assumes that $d(p, q) = 0$ but $p \neq q$, then one can choose a chart (U, ϕ), such that $p \in U$, and denote $x = \phi(p)$. Moreover, one can also choose a ball of radius $\delta > 0$ centered at x: $B_\delta(x) \subset \mathbb{R}^n$. The radius δ can be chosen such that $d(p, r) \geq \epsilon |x - z|$ for some ϵ for all points $r \in \phi^{-1}(B_\delta(x))$, $z = \phi(r)$. Thus, the point q cannot be in this region. If q is outside the region, then every curve that connects q to p must enter the region. But the segment of the curve contained in the region must have a length at least equal to $\epsilon\delta$. Thus one obtains $d(q, p) \geq \epsilon\delta > 0$, in contradiction with the assumption that $d(p, q) = 0$.   □

Defining the length of a curve on a Lorentzian manifold requires some additional specifications about the curve. To this purpose, we first introduce some definitions: A curve is said to be a *timelike curve* if its tangent vector is timelike at every point on the curve. Similarly, a *spacelike curve* is one for which the tangent vector is spacelike at every point, whereas a *lightlike curve* (or a *null curve*) is one whose tangent vector is lightlike at every point. The hypersurface spanned by all lightlike curves emanating from a point p in a spacetime M is called the *light cone* at point p. The lightcone is divided into two halves: the future light cone and the past light cone. All timelike curves passing through p stay inside the light cone, whereas all spacelike curves stay outside the light cone. The part of a timelike curve or a lightlike curve in the future light cone is called a *future-directed* timelike (or *future-directed* lightlike) curve, and the part in the past half of the light cone is called *past-directed*; see Fig. 5.1 for an illustration. A generic curve does not necessarily belong to one of these classes,

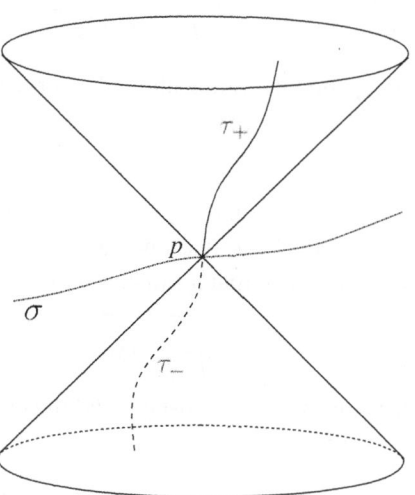

Fig. 5.1   A lightcone at p with various curves: The solid curve τ_+, lying in the future light cone (i.e., in the upper half of the cone), is a future-directed timelike curve, whereas the dashed curve τ_-, in the past light cone (the lower half of the cone) is a past-directed timelike curve. Finally, the dash-dotted curve σ, which lies outside the light cone, is a spacelike curve.

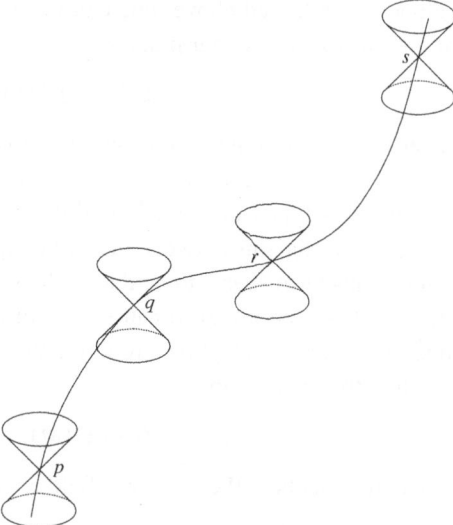

Fig. 5.2   Assuming that the time direction is represented as the vertical direction (as in Fig. 5.1), a generic curve can be timelike, lightlike, or spacelike in its different portions: As an example, the curve shown in this figure is timelike at the point p, lightlike at q, spacelike at r, and again timelike at s.

since its tangent vector can change from being spacelike to lightlike to timelike along various parts of the curve. See Fig. 5.2 for an example of a curve that is timelike, lightlike, spacelike, and again timelike in various parts of it.

If a curve is spacelike, its length is defined as in Eq. (5.10). This definition cannot be used for a timelike curve $(x^\mu(t))$, since $g_{\mu\nu}(x(t))\dot{x}^\mu(t)\dot{x}^\nu(t) < 0$ would make the square root appearing in Eq. (5.10) non-defined. For a timelike curve one instead defines a *proper time* τ by setting

$$d\tau = dt\sqrt{-g_{\mu\nu}(x(t))\dot{x}^\mu(t)\dot{x}^\nu(t)} \tag{5.19}$$

on the curve, and integrating along the curve to obtain

$$\tau = \int_{t_0}^{t_1} dt\sqrt{-g_{\mu\nu}(x(t))\dot{x}^\mu(t)\dot{x}^\nu(t)}, \tag{5.20}$$

setting $\tau = 0$ at the point $x(t_0)$. For a timelike curve, proper time thus replaces the concept of length. Note that one can also reparameterize the curve $t \to u(t)$ and search for a map such that u satisfies (compare with Eq. (5.11))

$$\sqrt{-g_{\mu\nu}(x(u))\frac{dx^\mu(u)}{du}\frac{dx^\nu(u)}{du}} = 1. \tag{5.21}$$

This means that $\tau = u + \text{constant}$, and the curve is now parameterized by its proper time, $c = (x^\mu(\tau))$.

5.1.2 Raising and Lowering Indices (A Musical Interlude)

In local coordinates, the components of the metric $g_{\mu\nu}(x)$ can be thought of as the entries of an $n \times n$ matrix

$$g = (g_{\mu\nu}(x)). \tag{5.22}$$

As we already mentioned above, one can invert the matrix, and denote the components of the inverse with upper indices,

$$g^{-1} = (g^{\mu\nu}(x)). \tag{5.23}$$

The reader may be familiar with manipulations to raise and lower indices, e.g., $A^\mu = g^{\mu\nu}A_\nu$, but a different placement of indices actually refers to components of tensors of different types and defined in different vector spaces. Behind these manipulations are isomorphisms between the vector spaces, called *musical isomorphisms*. The origin of this term seems to be due to the French mathematician Marcel Berger.

Let $\{e_\mu\} = \{\frac{\partial}{\partial x^\mu}\}$ be a (coordinate) basis in the space of vector fields $\chi(M)$ and let its dual basis $\{e^\mu\} = \{dx^\mu\}$ be a (coordinate) basis in the space of one-form fields $\Omega(M)$. Define the linear map

$$\flat : \chi(M) \to \Omega(M), \quad X \mapsto X^\flat, \tag{5.24}$$

where the components of the one-form field $X^\flat$ are related to those of the vector field X by

$$X^\flat_\mu(x) = g_{\mu\nu}(x)X^\nu(x). \tag{5.25}$$

Thus the map $\flat$ lowers the index. The notation with the $\flat$ symbol has its origin in music, where the $\flat$ symbol lowers the pitch of a note by a semitone (half-step). It is common to shorten the notation from $X^\flat_\mu$ to X_μ. The map $\flat$ is a linear map.

Then we define the other linear map

$$\sharp : \Omega(M) \to \chi(M), \quad \omega \mapsto \omega^\sharp, \tag{5.26}$$

where the components of the vector field $\omega^\sharp$ are related to those of the one-form field ω by

$$(\omega^\sharp)^\mu(x) = g^{\mu\nu}(x)\omega_\nu(x) \tag{5.27}$$

using the inverse metric. Also in this case, the origin of this notation is rooted in music, where $\sharp$ raises the pitch of a note by a semitone (half-step). We usually shorten the notation from $(\omega^{\sharp})^{\mu}$ to ω^{μ}. As in music, flat and sharp are inverses of one other, because $(\omega^{\sharp})^{\flat} = \omega$,

$$g_{\mu\nu}\left(g^{\nu\kappa}\omega_{\kappa}\right) = \omega_{\mu}, \tag{5.28}$$

and $(X^{\flat})^{\sharp} = X$,

$$g^{\mu\nu}\left(g_{\nu\kappa}X^{\kappa}\right) = X^{\mu}. \tag{5.29}$$

The space of vector fields $\chi(M)$ is the tangent bundle TM and the space of the one-form fields $\Omega(M)$ is the cotangent bundle T^*M, one can see that they are isomorphic by the maps $\flat$ and $\sharp$.

Raising and lowering of indices can be readily generalized for tensor fields of arbitrary type (p, q) to provide similar isomorphisms between the spaces $T_q^p M$.

5.2 The Induced Metric

Let (N, g_N) be a Riemannian manifold of dimension $\dim N = n$. We define an m-dimensional *submanifold* M of N as follows. First, let $f : M \rightarrow N$ be a smooth map such that the push $f_* : T_p M \rightarrow T_{f(p)} N$ is an injection. Then f is an *immersion*. If f itself is also an injection, such that M is diffeomorphic with its image $f(M)$, then f is an *embedding* of M in N and the image $f(M)$ is a *submanifold* of N. Since M and $f(M)$ are diffeomorphic, for simplicity one often says that M is a submanifold of N.

Example To understand the difference between an immersion and an embedding, for example a figure eight (a rose with two petals) is an immersion of S^1 in the plane $\mathbb{R}^2$, but not an embedding because of the intersection point. Note also that S^1 is not diffeomorphic with the figure eight.

Now the pullback f^* of f induces a natural metric g_M on M:

$$g_M = f^* g_N. \tag{5.30}$$

This equation becomes more intuitive and easier to remember if one writes the map f using local coordinates. Let y^{α} be local coordinates on N and let x^{μ} be local coordinates in M. The map $f : M \rightarrow N$ is

$$x = (x^1, \ldots, x^m) \mapsto f(x) \equiv y(x) = (y^1(x), \ldots, y^n(x)). \tag{5.31}$$

The metric g_N on N is

$$g_N = g_{N\alpha\beta}(y) dy^{\alpha} \otimes dy^{\beta}, \tag{5.32}$$

and the induced metric $g_M = f^* g_N$ is

$$g_M = g_{M\mu\nu}(x) dx^{\mu} \otimes dx^{\nu} \tag{5.33}$$

with the components

$$g_{M\mu\nu}(x) = g_{N\alpha\beta}(y(x))\frac{\partial y^\alpha}{\partial x^\mu}\frac{\partial y^\beta}{\partial x^\nu}. \tag{5.34}$$

In other words, this expression is analogous to the chain rule in derivation:

$$g_{N\alpha\beta}(y)dy^\alpha \otimes dy^\beta \rightarrow g_{N\alpha\beta}(y(x))\left(\frac{\partial y^\alpha}{\partial x^\mu}dx^\mu\right) \otimes \left(\frac{\partial y^\beta}{\partial x^\nu}dx^\nu\right)$$

$$= g_{N\alpha\beta}(y(x))\frac{\partial y^\alpha}{\partial x^\mu}\frac{\partial y^\beta}{\partial x^\nu}dx^\mu \otimes dx^\nu = g_{M\mu\nu}(x)dx^\mu \otimes dx^\nu. \tag{5.35}$$

Example Let (θ, φ) be the polar coordinates on the two-dimensional sphere S^2. We denote $x^1 = \theta$ and $x^2 = \varphi$. Consider the Euclidean metric $\delta_{\alpha\beta}$ on $\mathbb{R}^3$, and denote the coordinates as (y^1, y^2, y^3). Let $f : S^2 \rightarrow \mathbb{R}^3$ be the usual embedding: $f(\theta, \varphi) = (\sin\theta\cos\varphi, \sin\theta\sin\varphi, \cos\theta)$, i.e., $y(x) = (\sin x^1 \cos x^2, \sin x^1 \sin x^2, \cos x^1)$. One obtains the induced metric on S^2 as

$$g_{\mu\nu}dx^\mu \otimes dx^\nu = \delta_{\alpha\beta}\frac{\partial y^\alpha}{\partial x^\mu}\frac{\partial y^\beta}{\partial x^\nu}dx^\mu \otimes dx^\nu = d\theta \otimes d\theta + \sin^2\theta d\varphi \otimes d\varphi. \tag{5.36}$$

Thus the components of the induced metric on S^2 are $g_{11}(\theta, \varphi) = 1$, $g_{22}(\theta, \varphi) = \sin^2\theta$, and $g_{12}(\theta, \varphi) = g_{21}(\theta, \varphi) = 0$.

Instead of memorizing or deriving Eq. (5.34), a sometimes handy calculational strategy is to first compute exterior derivatives and then proceed straightforwardly (sacrificing some rigor with the notation) as follows:

$$dy^1 = \cos x^1 \cos x^2 dx^1 - \sin x^1 \sin x^2 dx^2$$
$$dy^2 = \cos x^1 \sin x^2 dx^1 + \sin x^1 \cos x^2 dx^2$$
$$dy^3 = -\sin x^1 dx^1, \tag{5.37}$$

then

$$dy^1 \otimes dy^1 = \cos^2 x^1 \cos^2 x^2 dx^1 \otimes dx^1 - \frac{1}{4}\sin(2x^1)\sin(2x^2)dx^1 \otimes dx^2$$

$$- \frac{1}{4}\sin(2x^1)\sin(2x^2)dx^2 \otimes dx^1 + \sin^2 x^1 \sin^2 x^2 dx^2 \otimes dx^2$$

$$dy^2 \otimes dy^2 = \cos^2 x^1 \sin^2 x^2 dx^1 \otimes dx^1 + \frac{1}{4}\sin(2x^1)\sin(2x^2)dx^1 \otimes dx^2$$

$$+ \frac{1}{4}\sin(2x^1)\sin(2x^2)dx^2 \otimes dx^1 + \sin^2 x^1 \cos^2 x^2 dx^2 \otimes dx^2$$

$$dy^3 \otimes dy^3 = \sin^2 x^1 dx^1 \otimes dx^1. \tag{5.38}$$

Summing up the three terms:

$$dy^1 \otimes dy^1 + dy^2 \otimes dy^2 = \cos^2 x^1 dx^1 \otimes dx^1 + \sin^2 x^1 dx^2 \otimes dx^2,$$

$$dy^1 \otimes dy^1 + dy^2 \otimes dy^2 + dy^3 \otimes dy^3 = dx^1 \otimes dx^1 + \sin^2 x^1 dx^2 \otimes dx^2,$$

$$\equiv d\theta \otimes d\theta + \sin^2\theta d\varphi \otimes d\varphi, \tag{5.39}$$

one finally obtains the same result as in Eq. (5.36).

5.3 Affine Connection

In Chapter 4 we discussed the Lie derivative as a way to generalize the notion of a directional derivative to that of tensor fields on smooth manifolds. In this chapter we introduce another generalization, the covariant derivative, which shares some properties with the Lie derivative, but also has some different features. Before defining the covariant derivative, it is convenient to introduce a slightly more abstract concept: the affine connection. To do so, recall that, given a manifold M, we defined $\chi(M)$ as the set of smooth vector fields on M and $\mathcal{F}(M)$ as the set of smooth functions on M. An *affine connection* ∇ is a map

$$\nabla : \chi(M) \times \chi(M) \mapsto \chi(M), \ (X, Y) \mapsto \nabla(X, Y) \equiv \nabla_X Y \qquad (5.40)$$

that satisfies the following properties for all $a, b \in \mathbb{R}$, and for all $f, g \in \mathcal{F}(M)$:

(i) linearity over $\mathcal{F}(M)$ in the first argument

$$\nabla_{(fX+gY)} Z = f \nabla_X Z + g \nabla_Y Z, \qquad (5.41)$$

(ii) linearity over $\mathbb{R}$ in the second argument

$$\nabla_X(aY + bZ) = a \nabla_X Y + b \nabla_X Z, \qquad (5.42)$$

(iii) the Leibniz rule

$$\nabla_X(fY) = X[f]Y + f \nabla_X Y. \qquad (5.43)$$

The vector field $\nabla_X Y$ is called the *covariant derivative* of the vector field Y with respect to the vector field X.

Let us compare $\nabla_X Y$ with the Lie derivative $\mathcal{L}_X Y$. Both satisfy the property of linearity over $\mathbb{R}$ in the second argument, since $\mathcal{L}_X(aY + bZ) = a\mathcal{L}_X Y + b\mathcal{L}_X Z$, and the Leibniz rule. However, for the Lie derivative, $\mathcal{L}_{fX} Y = [fX, Y] \neq f[X, Y] = f\mathcal{L}_X Y$, so the first property distinguishes the two.

Consider now a chart (U, φ) with coordinates $x = \varphi(p)$, and let $\{e_\nu = \frac{\partial}{\partial x^\nu}\}$ be the coordinate basis of $T_p M$. Since $\nabla_{e_\mu} e_\nu$ is a vector, we can expand it in the same basis,

$$\nabla_{e_\mu} e_\nu = \Gamma^\lambda_{\mu\nu} e_\lambda, \qquad (5.44)$$

to define the *connection coefficients* $\Gamma^\lambda_{\mu\nu}$. Their number is $(\dim M)^3$.

Example Consider the two-dimensional Euclidean plane $\mathbb{R}^2$ with the usual Cartesian coordinates, and denote the two orthonormal unit vectors as e_1 and e_2. Define an affine connection ∇ so that the covariant derivatives with respect to the e_i basis are

$$\nabla_{e_1} e_1 = \nabla_{e_1} e_2 = 0, \qquad \nabla_{e_2} e_1 = -e_2, \ \nabla_{e_2} e_2 = e_1. \qquad (5.45)$$

What are the connection coefficients Γ^k_{ij}? Comparing Eq. (5.45) with Eq. (5.44), one obtains

$$\Gamma^1_{11} = \Gamma^1_{12} = \Gamma^1_{21} = 0$$
$$\Gamma^1_{22} = 1 \qquad (5.46)$$

$$\Gamma^2_{11} = \Gamma^2_{22} = \Gamma^2_{12} = 0$$
$$\Gamma^2_{21} = -1.$$

One can express the covariant derivative $\nabla_X Y$ in the coordinate basis with the help of the connection coefficients. Let $X = X^\mu e_\mu$ and $Y = Y^\nu e_\nu$ be two vector fields. Denote $\nabla_\mu \equiv \nabla_{e_\mu}$. Now

$$\nabla_X Y = X^\mu \nabla_\mu (Y^\nu e_\nu)$$

$$= X^\mu e_\mu[Y^\nu]e_\nu + X^\mu Y^\nu \nabla_\mu e_\nu$$

$$= X^\mu \frac{\partial Y^\nu}{\partial x^\mu} e_\nu + X^\mu Y^\nu \Gamma^\lambda_{\mu\nu} e_\lambda$$

$$= X^\mu \left(\frac{\partial Y^\lambda}{\partial x^\mu} + \Gamma^\lambda_{\mu\nu} Y^\nu \right) e_\lambda$$

$$\equiv X^\mu (\nabla_\mu Y)^\lambda e_\lambda, \tag{5.47}$$

where in the first line we have used the linearity of ∇, in the second we have used the Leibniz rule, in the third we used the definition of the e_μ vector field and of the connection coefficients, while in the fourth we renamed the dummy summation index ν in the first term as λ, and finally we defined

$$(\nabla_\mu Y)^\lambda = \frac{\partial Y^\lambda}{\partial x^\mu} + \Gamma^\lambda_{\mu\nu} Y^\nu. \tag{5.48}$$

Note that $\nabla_X Y$ contains no derivatives of X, unlike $\mathcal{L}_X Y$.

An alternative way to think about the connection is as a map

$$\nabla : \chi(M) \to \chi(M) \otimes \Omega^1(M), \; Y \mapsto \nabla Y. \tag{5.49}$$

Then ∇Y is a linear combination of tensor products of vector fields and one-form fields. If we choose as Y one of the basis vector fields e_ν, we can expand

$$\nabla e_\nu = e_\lambda \otimes \omega^\lambda_\nu \equiv e_\lambda \omega^\lambda_\nu, \tag{5.50}$$

where ω^λ_ν are one-form fields, called *connection forms*. Note that we shortened the notation above by suppressing the tensor-product symbol. We can further expand ω^λ_ν in the dual basis $e^{*\kappa}$:

$$\omega^\lambda_\nu = \omega^\lambda_{\kappa\nu} e^{*\kappa}. \tag{5.51}$$

Let us use the same coordinate basis as in (5.44), so that $\{e^{*\mu}\} = \{dx^\mu\}$. Now we can evaluate the one-form fields ω^λ_ν in (5.50) at the basis vector field e_μ to get a vector field

$$(\nabla e_\nu)(e_\mu) = e_\lambda \omega^\lambda_\nu (e_\mu) = e_\lambda \omega^\lambda_{\kappa\nu} e^{*\kappa}(e_\mu) = e_\lambda \omega^\lambda_{\mu\nu}. \tag{5.52}$$

We want this vector field to be the same one as in Eq. (5.44), so

$$\nabla_{e_\mu} e_\nu = \Gamma^\lambda_{\mu\nu} e_\lambda = (\nabla e_\nu)(e_\mu) = \omega^\lambda_{\mu\nu} e_\lambda. \tag{5.53}$$

Thus, the components of the connection forms are the connection coefficients,

$$\omega^{\lambda}_{\mu\nu} = \Gamma^{\lambda}_{\mu\nu}, \tag{5.54}$$

so in the coordinate basis the connection forms take the simple form

$$\omega^{\lambda}_{\nu} = \Gamma^{\lambda}_{\mu\nu}dx^{\mu}. \tag{5.55}$$

Example   With the connection (5.45) in the previous example, the non-vanishing connection forms ω^i_j in $\mathbb{R}^2$ are

$$\omega^1_2 = \Gamma^1_{k2}dx^k = dx^2$$
$$\omega^2_1 = \Gamma^2_{k1}dx^k = -dx^2. \tag{5.56}$$

Using the components (5.48) we can express ∇Y in the coordinate basis,

$$\nabla Y = (\nabla_{\mu}Y)^{\lambda}\frac{\partial}{\partial x^{\lambda}}dx^{\mu} = \left(\frac{\partial Y^{\lambda}}{\partial x^{\mu}}dx^{\mu} + \Gamma^{\lambda}_{\mu\nu}dx^{\mu}Y^{\nu}\right)\frac{\partial}{\partial x^{\lambda}}$$
$$= (dY^{\lambda} + \omega^{\lambda}_{\nu}Y^{\nu})\frac{\partial}{\partial x^{\lambda}}. \tag{5.57}$$

Note that the affine connection was defined without any reference to a metric. In Section 5.7 we will take into account a metric of the manifold, and define metric-compatible connections, called *metric connections*. Among them, we will also define the so-called *torsionless connections*, which are completely determined by the metric and are called *Levi-Civita connections*. Levi-Civita connections are the most common connections in the theory of general relativity. Thus there will be a hierarchy, as depicted in Fig. 5.3.

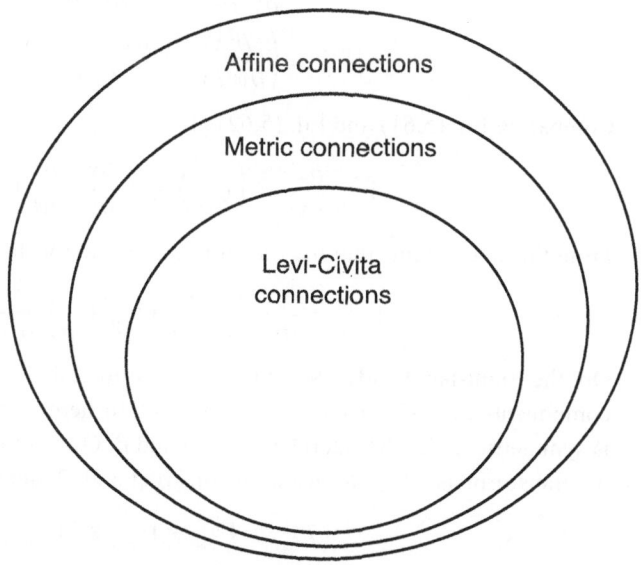

Fig. 5.3   A hierarchy of connections to be discussed in this chapter.

5.4 Coordinate Transformation Properties of Connection Coefficients

Let U and V be two overlapping charts, with coordinates x on U, with the basis of vector fields

$$e_\mu = \frac{\partial}{\partial x^\mu}, \tag{5.58}$$

and coordinates y on V with the basis

$$\tilde{e}_\nu = \frac{\partial}{\partial y^\nu} = \frac{\partial x^\mu}{\partial y^\nu} e_\mu. \tag{5.59}$$

We will later denote the components of the Jacobian matrix and its inverse by

$$J^\mu_\nu = \frac{\partial x^\mu}{\partial y^\nu}, \quad (J^{-1})^\nu_\mu = \frac{\partial y^\nu}{\partial x^\mu}. \tag{5.60}$$

We compare the connection coefficients in the intersection $U \cap V \neq \emptyset$ in the two coordinates. Using the y-coordinates,

$$\nabla_{\tilde{e}_\alpha} \tilde{e}_\beta = \tilde{\Gamma}^\gamma_{\ \alpha\beta} \tilde{e}_\gamma = \tilde{\Gamma}^\gamma_{\ \alpha\beta} \frac{\partial x^\nu}{\partial y^\gamma} e_\nu. \tag{5.61}$$

Alternatively, one can write

$$
\begin{aligned}
\nabla_{\tilde{e}_\alpha} \tilde{e}_\beta &= \nabla_{\tilde{e}_\alpha} \left(\frac{\partial x^\mu}{\partial y^\beta} e_\mu \right) \\
&= \left(\nabla_{\tilde{e}_\alpha} \frac{\partial x^\mu}{\partial y^\beta} \right) e_\mu + \frac{\partial x^\mu}{\partial y^\beta} \left(\nabla_{\tilde{e}_\alpha} e_\mu \right) \\
&= \frac{\partial^2 x^\mu}{\partial y^\alpha \partial y^\beta} e_\mu + \frac{\partial x^\mu}{\partial y^\beta} \frac{\partial x^\lambda}{\partial y^\alpha} \left(\nabla_{e_\lambda} e_\mu \right) \\
&= \left(\frac{\partial^2 x^\nu}{\partial y^\alpha \partial y^\beta} + \frac{\partial x^\lambda}{\partial y^\alpha} \frac{\partial x^\mu}{\partial y^\beta} \Gamma^\nu_{\ \lambda\mu} \right) e_\nu.
\end{aligned}
\tag{5.62}
$$

Comparing Eq. (5.61) and Eq. (5.62) gives

$$\tilde{\Gamma}^\gamma_{\ \alpha\beta} \frac{\partial x^\nu}{\partial y^\gamma} = \left(\frac{\partial^2 x^\nu}{\partial y^\alpha \partial y^\beta} + \frac{\partial x^\lambda}{\partial y^\alpha} \frac{\partial x^\mu}{\partial y^\beta} \Gamma^\nu_{\ \lambda\mu} \right). \tag{5.63}$$

From this one obtains that the transformation rule for the connection coefficients is

$$\tilde{\Gamma}^\gamma_{\ \alpha\beta} = \frac{\partial y^\gamma}{\partial x^\nu} \frac{\partial x^\lambda}{\partial y^\alpha} \frac{\partial x^\mu}{\partial y^\beta} \Gamma^\nu_{\ \lambda\mu} + \frac{\partial^2 x^\nu}{\partial y^\alpha \partial y^\beta} \frac{\partial y^\gamma}{\partial x^\nu}. \tag{5.64}$$

On the right-hand side, the first term is just the transformation rule for the components of a $(1,2)$-tensor, but in addition here one has a further term, which is symmetric under the interchange of α and β. Owing to the additional term, Γ does not transform as a $(1,2)$-tensor. To construct a $(1,2)$-tensor out of Γ, one can define

$$T^\gamma_{\ \alpha\beta} = \Gamma^\gamma_{\ \alpha\beta} - \Gamma^\gamma_{\ \beta\alpha} \equiv 2\Gamma^\gamma_{\ [\alpha\beta]}, \tag{5.65}$$

in which we introduced a notation for an antisymmetrization of indices,

$$T_{[\alpha\beta]} \equiv \frac{1}{2} (T_{\alpha\beta} - T_{\beta\alpha}). \tag{5.66}$$

The components $T^\gamma{}_{\alpha\beta}$ are the components of the *torsion tensor*, which will be discussed in more detail in Section 5.10. If the torsion tensor vanishes, $T^\gamma{}_{\alpha\beta} = 0$, the connection is said to be *torsion-free* or *symmetric*, since then

$$\Gamma^\gamma{}_{\alpha\beta} = \Gamma^\gamma{}_{\beta\alpha}. \tag{5.67}$$

This will be the case for the Levi-Civita connection used in general relativity.

Example   Continuing with the connection (5.45) in $\mathbb{R}^2$, we identify a torsion tensor with the nonvanishing coefficients

$$T^2{}_{12} = \Gamma^2{}_{12} - \Gamma^2{}_{21} = 0 - (-1) = 1, \quad T^2{}_{21} = \Gamma^2{}_{21} - \Gamma^2{}_{12} = -1, \tag{5.68}$$

while the other components vanish.

We can similarly derive a coordinate transformation law for the connection forms. Below we use the Jacobian matrix notation (5.60). First, we express the connection forms in the y-coordinates,

$$\nabla\tilde{e}_\beta = \tilde{e}_\kappa \tilde{\omega}^\kappa{}_\beta = e_\sigma J^\sigma_\kappa \tilde{\omega}^\kappa{}_\beta, \tag{5.69}$$

while, on the other hand,

$$\nabla\tilde{e}_\beta = \nabla(J^\sigma_\beta e_\sigma) = (dJ^\sigma_\beta)e_\sigma + J^\sigma_\beta \nabla e_\sigma = (dJ^\sigma_\beta)e_\sigma + J^\kappa_\beta \nabla e_\kappa$$
$$= e_\sigma(dJ^\sigma_\beta) + J^\kappa_\beta e_\sigma \omega^\sigma{}_\kappa = e_\sigma(dJ^\sigma_\beta + \omega^\sigma{}_\kappa J^\kappa_\beta). \tag{5.70}$$

Comparing Eq. (5.69) and Eq. (5.70), we obtain the transformation law for the connection forms:

$$\tilde{\omega}^\gamma{}_\beta = (J^{-1})^\gamma_\sigma \omega^\sigma{}_\kappa J^\kappa_\beta + (J^{-1})^\gamma_\sigma dJ^\sigma_\beta. \tag{5.71}$$

By combining the connection forms into a matrix $\omega = (\omega^\mu{}_\nu)$, Eq. (5.71) can be expressed in matrix form as

$$\tilde{\omega} = J^{-1}\omega J + J^{-1}dJ. \tag{5.72}$$

One can check that Eq. (5.71) reproduces the previous transformation rule (5.74). Extracting the connection coefficients using Eq. (5.55), one obtains

$$\tilde{\Gamma}_{\alpha\beta}dy^\alpha = \frac{\partial y^\gamma}{\partial x^\sigma}\Gamma^\sigma{}_{\lambda\kappa}\frac{\partial x^\kappa}{\partial y^\beta}dx^\lambda + \frac{\partial y^\gamma}{\partial x^\sigma}\frac{\partial}{\partial y^\alpha}\left(\frac{\partial x^\sigma}{\partial y^\beta}\right)dy^\alpha, \tag{5.73}$$

then transforming $dx^\lambda = \frac{\partial x^\lambda}{\partial y^\alpha}dy^\alpha$, extracting the components, and multiplying on the left by the inverse Jacobian, one obtains

$$\tilde{\Gamma}^\gamma{}_{\alpha\beta} = \frac{\partial y^\gamma}{\partial x^\nu}\frac{\partial x^\lambda}{\partial y^\alpha}\frac{\partial x^\mu}{\partial y^\beta}\Gamma^\nu{}_{\lambda\mu} + \frac{\partial^2 x^\nu}{\partial y^\alpha \partial y^\beta}\frac{\partial y^\gamma}{\partial x^\nu}, \tag{5.74}$$

in agreement with Eq. (5.74).

5.5 Parallel Transport and Holonomy

Through the covariant derivative, it is possible to define the concept of parallel transport of a vector field along a curve. Let $c : (t_0, t_1) \to M$ be a curve on M with coordinate representation $x^\mu = x^\mu(t)$. Its tangent vector is

$$V = V^\mu e_\mu \big|_{c(t)} = \frac{dx^\mu(t)}{dt} e_\mu \big|_{c(t)}. \tag{5.75}$$

If a vector field X satisfies

$$\nabla_V X = 0 \tag{5.76}$$

along the curve $c(t)$, then X is said to be *parallel-transported* along the curve $c(t)$. In component form, this is

$$\frac{dX^\mu}{dt} + \Gamma^\mu_{\nu\lambda} \frac{dx^\nu(t)}{dt} X^\lambda = 0. \tag{5.77}$$

Equation (5.77) can also be expressed by using the connection forms $\omega^\mu_{\ \lambda}$ appearing in Eq. (5.55): When evaluated at the tangent vector V, one obtains

$$\omega^\mu_{\ \lambda}(V) = \omega^\mu_{\nu\lambda} \frac{dx^\nu}{dt} = \Gamma^\mu_{\nu\lambda} \frac{dx^\nu}{dt}, \tag{5.78}$$

so that Eq. (5.77) takes the form

$$(\nabla_V X)^\mu = \frac{dX^\mu}{dt} + \omega^\mu_{\ \lambda}(V) X^\lambda = 0. \tag{5.79}$$

Note that, since the tangent vector V in Eq. (5.75) depends on the curve parameter t, so do the $\omega^\mu_{\ \lambda}(V)$. By denoting the ω components in a matrix notation

$$M^\mu_{\ \lambda}(t) \equiv \omega^\mu_{\ \lambda}(V), \tag{5.80}$$

Eq. (5.79) can be interpreted as a system of first-order ordinary differential equations,

$$\frac{dX^\mu}{dt} = -M^\mu_{\ \lambda}(t) X^\lambda. \tag{5.81}$$

Once the initial values $X^\mu(t_0)$ have been specified, the solution $X^\mu(t)$ will be unique.

Example   Let us consider again the connection on $\mathbb{R}^2$ defined in Eq. (5.45) that we used in our previous examples. Consider two straight lines parallel to the two coordinate axes

$$c_1(t) = (x^1(t), x^2(t)) = (t, x_0^2), \quad \text{tangent vector } U = (1, 0),$$
$$c_2(t) = (x^1(t), x^2(t)) = (x_0^1, t), \quad \text{tangent vector } V = (0, 1). \tag{5.82}$$

Let us first parallel-transport a vector $X = (X^1, X^2)$ along the line c_1. The parallel transport equation (5.77) becomes

$$\frac{dX^k}{dt} + \Gamma^k_{ij} U^i X^j = \frac{dX^k}{dt} + \Gamma^k_{1j} X^j = 0. \tag{5.83}$$

From Eq. (5.46), one can see that all connection components Γ^k_{1j} vanish. So in this case $\frac{dX^k}{dt} = 0$, which means that the vector X is unchanged when parallel-transported along a line parallel to the x^1 axis.

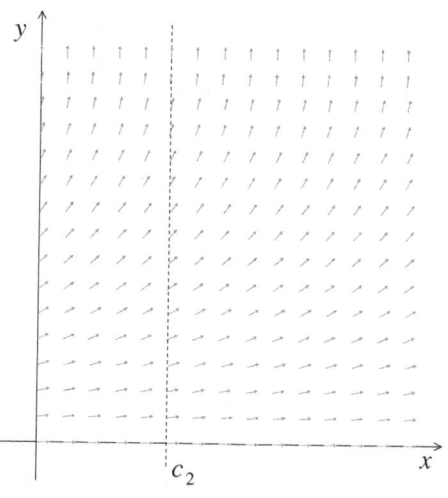

Fig. 5.4 Unit vectors $(1, 0)$ on the x-axis parallel transported with the torsion-including connection in Eq. (5.45) upwards along a straight line c_2 (shown in the figure as the dashed line) parallel to the y-axis. The effect of the torsion is that the vectors rotate anticlockwise during parallel transport.

Next we parallel-transport X along the line c_2. In this case Eq. (5.77) becomes

$$\frac{dX^k}{dt} + \Gamma^k_{ij} V^i X^j = \frac{dX^k}{dt} + \Gamma^k_{2j} X^j = 0. \tag{5.84}$$

In this case there are two nonvanishing connection components, $\Gamma^1_{22} = 1$ and $\Gamma^2_{21} = -1$, and one obtains a pair of coupled equations

$$\frac{dX^1}{dt} = -X^2, \tag{5.85}$$

$$\frac{dX^2}{dt} = X^1. \tag{5.86}$$

Assuming, as the initial condition on X, that at $t = 0$ the vector is $X(t = 0) = (X^1_0, X^2_0)$, the solution to Eq. (5.85) takes the form

$$X(t) = \begin{pmatrix} X^1(t) \\ X^2(t) \end{pmatrix} = \begin{pmatrix} \cos t & -\sin t \\ \sin t & \cos t \end{pmatrix} \begin{pmatrix} X^1_0 \\ X^2_0 \end{pmatrix}. \tag{5.87}$$

This means that parallel-transporting X along c_2 to (x^1_0, t) rotates it anticlockwise by an angle equal to t, as shown in Fig. 5.4.

5.5.1 Holonomy

An equivalent, but slightly more elegant, way to discuss the parallel transport of vector fields along curves consists in starting with an *Ansatz* for the parallel-transported vector $X^\mu(t)$:

$$X^\mu(t) = U^\mu_{\ \lambda}(t, t_0) X^\lambda(t_0). \tag{5.88}$$

Substituting this *Ansatz* into Eq. (5.81) yields

$$\frac{dU^{\mu}_{\nu}(t, t_0)}{dt} = -M^{\mu}_{\lambda}(t)U^{\lambda}_{\nu}(t, t_0) \tag{5.89}$$

with the initial condition

$$U^{\mu}_{\nu}(t_0) = \delta^{\mu}_{\nu}. \tag{5.90}$$

Interpreting $U^{\mu}_{\nu}(t_0)$ as the entries of a matrix $U = (U^{\mu}_{\nu})$, the solution to Eq. (5.89) can be written as

$$U(t, t_0) = \mathrm{P}\exp\left[-\int_{t_0}^{t} dt' M(t')\right], \tag{5.91}$$

where P exp is a *path-ordered exponential* (or *curve-ordered exponential*), defined as the infinite series

$$\mathrm{P}\exp\left[\int_{t_0}^{t} A(t)\right] \equiv 1 + \int_{t_0}^{t} dt'_1 A(t'_1) + \int_{t_0}^{t} dt'_1 \int_{t_0}^{t'_1} dt'_2 A(t'_1)A(t'_2) + \cdots . \tag{5.92}$$

Note that the ordering of factors appearing in the integrands on the right-hand side matters, since, in general, matrices $A(t'_1)$ and $A(t'_2)$ defined at different times do not necessarily commute with each other. However, if

$$[A(t), A(t')] = 0 \qquad \forall t,\ \forall t', \tag{5.93}$$

then the ordering does not matter and the path-ordered exponential simplifies to the usual exponential function,

$$\mathrm{P}\exp\left[\int dt' A(t')\right] = \exp\left[\int dt' A(t')\right]. \tag{5.94}$$

The solution (5.91) can then be used to transport vectors along the curve $c(t)$, and one obtains a map between tangent spaces

$$U_c : T_{c(t_0)}M \to T_{c(t)}M, \quad X(t_0) \mapsto X(t) = U_c(t, t_0)X(t_0), \tag{5.95}$$

where we expressed the *Ansatz* (5.88) in matrix notation and added a subscript c to emphasize that U_c depends on the curve. U_c is called the *holonomy operator*, or the *parallel-transport operator*: It is a linear operator and satisfies the multiplication properties

$$U_c(t_2, t_1)U_c(t_1, t_0) = U_c(t_2, t_0)$$
$$U_c(t_1, t_2)U_c(t_2, t_1) = \mathbb{1} \tag{5.96}$$

for arbitrary times $t_0 < t_1 < t_2$. Furthermore,

$$U_c(t_2, t_1)^{-1} = U_c(t_1, t_2). \tag{5.97}$$

Example In the previous example, transporting along the curve c_2 led to Eq. (5.85), from which one can read off the matrix

$$M(t) = M = \begin{pmatrix} 0 & 1 \\ -1 & 0 \end{pmatrix} \qquad (5.98)$$

or construct it by evaluating the nonvanishing connection forms $\omega^1{}_2 = dx^2$, $\omega^2{}_1 = -dx^2$ on the tangent vector $V = (0, 1)$,

$$M^1{}_2 = \omega^1{}_2(V) = 1, \ M^2{}_1 = \omega^2{}_1(V) = -1. \qquad (5.99)$$

The holonomy operator is now

$$U(t, 0) = \exp(-Mt) = \sum_{k=0}^{\infty} \frac{t^k}{k!}(-M)^k$$

$$= \sum_{k=0}^{\infty} \frac{t^{2k}}{(2k)!}(-M)^{2k} + \sum_{k=0}^{\infty} \frac{t^{2k+1}}{(2k+1)!}(-M)^{2k+1}$$

$$= \sum_{k=0}^{\infty} (-1)^k \frac{t^{2k}}{(2k)!} \mathbb{1} + \sum_{k=0}^{\infty} (-1)^k \frac{t^{2k+1}}{(2k+1)!}(-M) = \mathbb{1}\cos t - M\sin t$$

$$= \begin{pmatrix} \cos t & -\sin t \\ \sin t & \cos t \end{pmatrix}, \qquad (5.100)$$

which, when substituted into Eq. (5.88), reproduces Eq. (5.87). In the derivation above, we first separated the contributions from even and odd powers of M, and used the property that $M^2 = -\mathbb{1}$ to write $(-M)^{2k} = \mathbb{1}(-1)^k$ and $(-M)^{2k+1} = (-1)^k(-M)$.

Using parallel transport, it is possible to define a directional derivative of a vector field X defined in a neighborhood of the curve $c(t)$. We wish to evaluate the rate of change of X along the curve. Let us consider two different points $c(t_0 + \Delta t)$ and $c(t_0)$, and evaluate X at these points. Note that one cannot directly subtract $X(c(t_0))$ from $X(c(t_0 + \Delta t))$, since the two vectors belong to different tangent spaces. However, one can first parallel-transport $X(c(t_0 + \Delta t))$ along the curve to $c(t_0)$, and then do the subtraction there. One can then consider the limit in which the separation between the two points becomes infinitesimally small, and evaluate

$$\lim_{\Delta t \to 0} \frac{U(t_0, t_0 + \Delta t)X(c(t_0 + \Delta t)) - X(c(t_0))}{\Delta t} = \frac{d}{dt}[U(t_0, t)X(c(t))]_{|t=t_0}. \qquad (5.101)$$

Working in the coordinate basis, and extracting the components, the right-hand side becomes

$$\left[\frac{d}{dt} U^\mu{}_\nu(t_0, t)\Big|_{t=t_0}\right] X^\nu(c(t_0)) + U^\mu{}_\nu(t_0, t_0) \frac{dX^\nu(c(t))}{dt}\Big|_{t=t_0}$$

$$= M^\mu{}_\nu(t_0)X^\nu(c(t_0)) + \frac{dX^\mu}{dt}\Big|_{t=t_0} = \omega^\mu{}_\nu(V)X^\nu(c(t_0)) + \frac{dX^\mu}{dt}\Big|_{t=t_0}$$

$$= (\nabla_V X)^\mu, \qquad (5.102)$$

having used Eq. (5.79) in the last step. The definition (5.101) is thus an alternative way to define the covariant derivative. Note that in this derivation we also used

$$\frac{d}{dt}U(t_0, t) = \frac{d}{dt}U(t, t_0)^{-1}$$

$$= -U(t, t_0)^{-1}\left(\frac{d}{dt}U(t, t_0)\right)U(t, t_0)^{-1}$$

$$= -U(t, t_0)^{-1}\left(-M(t)U(t, t_0)\right)U(t, t_0)^{-1}$$

$$= U(t, t_0)^{-1}M(t)$$

$$= U(t_0, t)M(t)$$

$$= M(t_0)$$

$$= (\omega^\mu{}_\nu(V))_{t=t_0}, \tag{5.103}$$

where, in turn, we used the fact that $U(t, t_0)^{-1}U(t, t_0) = \mathbb{1}$, which implies

$$\frac{d}{dt}\left(U(t, t_0)^{-1}U(t, t_0)\right) = \frac{d}{dt}\left(U(t, t_0)^{-1}\right)U(t, t_0) + U(t, t_0)^{-1}\frac{d}{dt}U(t, t_0) = 0, \tag{5.104}$$

namely

$$\frac{d}{dt}\left(U(t, t_0)^{-1}\right)U(t, t_0) = -U(t, t_0)^{-1}\frac{d}{dt}U(t, t_0), \tag{5.105}$$

or, equivalently,

$$\frac{d}{dt}U(t, t_0)^{-1} = -U(t, t_0)^{-1}\left(\frac{d}{dt}U(t, t_0)\right)U(t, t_0)^{-1}. \tag{5.106}$$

5.5.2 The Geodesic Equation

If the tangent vector V itself is parallel transported along the curve $c(t)$,

$$\nabla_V V = 0, \tag{5.107}$$

then the curve $c(t)$ is called a *geodesic*. Equation (5.107) is the *geodesic equation*, which in component form reads

$$\frac{d^2 x^\mu}{dt^2} + \Gamma^\mu{}_{\nu\lambda}\frac{dx^\nu}{dt}\frac{dx^\lambda}{dt} = 0. \tag{5.108}$$

Geodesics can be interpreted as the straightest possible curves in a manifold with connection. In particular, when $M = \mathbb{R}^n$ and the connection vanishes, the geodesics are straight lines.

5.6 Covariant Derivative of Tensor Fields

As the next step, we generalize the action of the covariant derivative first to one-form fields and then to general tensor fields. As in the previous section, we will proceed in two alternative ways. We will first define the covariant derivative of a one-form field through some algebraic manipulations.

To start with, we define the covariant derivative of a function, $\nabla_X f$, to be the same as the directional derivative:

$$\nabla_X f = X[f]. \tag{5.109}$$

Then, we impose the condition (iii) in the definition of ∇, namely the Leibniz rule encoded in Eq. (5.43):

$$\nabla_X(fY) = (\nabla_X f)Y + f(\nabla_X Y), \tag{5.110}$$

to hold for any (tensor) product of tensor fields T_1 and T_2 of arbitrary types; this implies

$$\nabla_X(T_1 \otimes T_2) = (\nabla_X T_1) \otimes T_2 + T_1 \otimes (\nabla_X T_2). \tag{5.111}$$

We require that the rule must also be true when some of the indices of the components of the two tensors are contracted. Then the covariant derivative of a one-form can be defined as follows. Let ω be a one-form (i.e., a $(0,1)$-tensor field) and Y a vector field (a $(1,0)$-tensor field). Then $\langle \omega, Y \rangle \in \mathcal{F}(M)$ is a smooth function on M: Recall that $\langle \omega, Y \rangle \equiv \omega[Y] = \omega_\mu Y^\mu$. From Eq. (5.109) it follows

$$\nabla_X \langle \omega, Y \rangle = X(\omega[Y]) = X^\mu \frac{\partial}{\partial x^\mu}(\omega_\nu Y^\nu) = X^\mu \frac{\partial \omega_\nu}{\partial x^\mu} Y^\nu + X^\mu \omega_\nu \frac{\partial Y^\nu}{\partial x^\mu}. \tag{5.112}$$

On the other hand, one can think of $\langle \omega, Y \rangle$ as coming from the tensor product $T_1 \otimes T_2 = \omega \otimes Y$, with the μ index contracted. Then we can use Eq. (5.111) to write

$$\nabla_X \langle \omega, Y \rangle = \langle \nabla_X \omega, Y \rangle + \langle \omega, \nabla_X Y \rangle = (\nabla_X \omega)_\nu Y^\nu + \omega_\nu (\nabla_X Y)^\nu$$

$$= (\nabla_X \omega)_\nu Y^\nu + \omega_\nu X^\mu \frac{\partial Y^\nu}{\partial x^\mu} + \omega_\nu \Gamma^\nu_{\mu\alpha} X^\mu Y^\alpha. \tag{5.113}$$

Comparing Eq. (5.112) and Eq. (5.113) one finds

$$(\nabla_X \omega)_\nu = X^\mu \left(\frac{\partial \omega_\nu}{\partial x^\mu} - \Gamma^\alpha_{\mu\nu} \omega_\alpha \right). \tag{5.114}$$

In particular, for a coordinate-basis vector field $X = \frac{\partial}{\partial x^\mu}$, this reduces to

$$(\nabla_\mu \omega)_\nu = \frac{\partial \omega_\nu}{\partial x^\mu} - \Gamma^\alpha_{\mu\nu} \omega_\alpha. \tag{5.115}$$

Additionally, choosing a dual basis one-form $\omega = dx^\sigma$, one obtains

$$\nabla_\mu dx^\sigma = -\Gamma^\sigma_{\mu\nu} dx^\nu. \tag{5.116}$$

There exists an alternative, and equivalent, way to define the covariant derivative of a one-form, which is based on parallel transport, as in (5.101). Using (5.95), we define a parallel transport of one-forms

$$\tilde{U}_c : T_{c(t)} M \to T_{c(s)} M, \quad \omega_{c(t)} \mapsto \omega_{c(s)} = \tilde{U}_c(s,t) \omega_{c(t)} \tag{5.117}$$

by setting

$$(\tilde{U}_c(s,t) \omega_{c(t)})(X_{c(s)}) = \omega_{c(s)}(U_c(s,t) X_{c(t)}). \tag{5.118}$$

In particular, in the coordinate basis, for $X = \partial/\partial x^\nu$ and $\omega = dx^\mu$, Eq. (5.118) becomes

$$(\tilde{U}_c(s,t) dx^\mu) \left(\frac{\partial}{\partial x^\nu} \right) = dx^\mu \left(U_c(s,t) \left(\frac{\partial}{\partial x^\nu} \right) \right)$$

$$= dx^\mu \left(U^\lambda_\nu(s,t) \frac{\partial}{\partial x^\lambda} \right)$$

$$= U^\mu_\nu(s,t). \tag{5.119}$$

Thus $\tilde{U}(s,t) dx^\mu = U^\mu_\nu(t,s) dx^\nu$, and, analogously to Eq. (5.101), one can then evaluate

$$\frac{d}{dt} \left[\tilde{U}(t_0,t) \omega_{c(t)} \right]_{t=t_0} = \frac{d}{dt} \left[\omega_\mu U^\mu_\nu(t,t_0) dx^\nu \right]_{t=t_0}. \tag{5.120}$$

Proceeding as in (5.102) and extracting the components, one obtains

$$\frac{d\omega_\mu}{dt} - \omega^\nu{}_\mu(V)\omega_\nu = V^\alpha\left(\frac{\partial\omega_\mu}{\partial x^\alpha} - \Gamma^\nu_{\alpha\mu}\omega_\nu\right) = (\nabla_V\omega)_\mu, \qquad (5.121)$$

in agreement with Eq. (5.114).

Recall that $Y \mapsto \nabla Y$ can be regarded as a function that maps a $(1,0)$-tensor field to a $(1,1)$-tensor field with components

$$(\nabla_\mu Y)^\nu = \frac{\partial Y^\nu}{\partial x^\mu} + \Gamma^\nu_{\mu\lambda}Y^\lambda. \qquad (5.122)$$

Likewise, one can interpret $\omega \mapsto \nabla\omega$ as a map from $(0,1)$-tensor fields to $(0,2)$-tensor fields with components

$$(\nabla_\mu\omega)_\nu = \partial_\mu\omega_\nu - \Gamma^\lambda_{\mu\nu}\omega_\lambda. \qquad (5.123)$$

In physics literature, simplified notations

$$\nabla_\mu Y^\nu \equiv (\nabla_\mu Y)^\nu \qquad (5.124)$$

$$\nabla_\mu\omega_\nu \equiv (\nabla_\mu\omega)_\nu \qquad (5.125)$$

are used frequently. Another notation convention uses a comma to denote a partial derivative:

$$f_{,\mu} = \partial_\mu f = \frac{\partial f}{\partial x^\mu}, \qquad (5.126)$$

$$A^\mu{}_{,\nu} \equiv \partial_\nu A^\mu \qquad (5.127)$$

$$\omega_{\mu,\nu} \equiv \partial_\nu\omega_\mu \qquad (5.128)$$

and a semicolon to denote a covariant derivative:

$$Y^\nu{}_{;\mu} \equiv \nabla_\mu Y^\nu \qquad (5.129)$$

$$\omega_{\mu;\nu} \equiv \nabla_\nu\omega_\mu. \qquad (5.130)$$

Multiple derivatives are denoted by multiple indices after the comma or the semicolon, with the first index immediately after indicating the first derivative to be taken, followed by the next, and so on:

$$f_{,\mu\nu} \equiv \partial_\nu\partial_\mu f, \qquad (5.131)$$

$$Y^\lambda{}_{;\mu\nu} \equiv \nabla_\nu(\nabla_\mu Y^\lambda). \qquad (5.132)$$

Note that, while the partial derivatives of a differentiable function commute, e.g.,

$$f_{,\nu\mu} = \partial_\mu\partial_\nu f = \partial_\nu\partial_\mu f = f_{,\mu\nu}, \qquad (5.133)$$

this is not necessarily true for covariant derivatives, because $\nabla_\mu f = \partial_\mu f$ are the components of a one-form field df, and the covariant derivative of df has components in which the connection coefficients $\Gamma^\lambda_{\mu\nu}$ appear, as in Eq. (5.123). Thus,

$$f_{;\mu\nu} = \nabla_\nu\nabla_\mu f = \partial_\nu\partial_\mu f - \Gamma^\lambda_{\mu\nu}\partial_\lambda f = f_{,\mu\nu} - \Gamma^\lambda_{\mu\nu}f_{,\lambda}. \qquad (5.134)$$

The map ∇ generalizes to a map from (p, q)-tensor fields T to $(p, q + 1)$-tensor fields ∇T. For a generic tensor, the components of the covariant derivative ∇T can be shown to be

$$\nabla_\nu T^{\lambda_1 \dots \lambda_p}{}_{\mu_1 \dots \mu_q} = \partial_\nu T^{\lambda_1 \dots \lambda_p}{}_{\mu_1 \dots \mu_q} + \Gamma^{\lambda_1}{}_{\nu\rho} T^{\rho\lambda_2 \dots \lambda_p}{}_{\mu_1 \dots \mu_q} + \dots + \Gamma^{\lambda_p}{}_{\nu\rho} T^{\lambda_1 \dots \lambda_{p-1}\rho}{}_{\mu_1 \dots \mu_q}$$

$$- \Gamma^\rho{}_{\nu\mu_1} T^{\lambda_1 \dots \lambda_p}{}_{\rho\mu_2 \dots \mu_q} - \dots - \Gamma^\rho{}_{\nu\mu_q} T^{\lambda_1 \dots \lambda_p}{}_{\mu_1 \dots \mu_{q-1}\rho}. \tag{5.135}$$

Example Let g be a $(0, 2)$-tensor field (for example, a metric), then ∇g has the components

$$g_{\alpha\beta;\mu} = \nabla_\mu g_{\alpha\beta} = \partial_\mu g_{\alpha\beta} - \Gamma^\rho{}_{\mu\alpha} g_{\rho\beta} - \Gamma^\rho{}_{\mu\beta} g_{\alpha\rho}. \tag{5.136}$$

One more comment on notation. A convention, common in particular in general relativity textbooks, is to use brackets and square brackets to denote symmetrization and antisymmetrization of tensor indices. We denote a *symmetrization of indices* by inserting them in brackets, for example

$$T^{\lambda_1 \dots \lambda_p}{}_{(\mu_1\mu_2 \dots \mu_q)\nu} \equiv \frac{1}{q!} \sum_{\pi \in S_q} T^{\lambda_1 \dots \lambda_p}{}_{\mu_{\pi(1)}\mu_{\pi(2)} \dots \mu_{\pi(q)}\nu}, \tag{5.137}$$

where we sum over all permutations (re)ordering the first q lower indices. Likewise, the antisymmetrization, denoted by the square bracket, is similar, but with a minus sign for terms with odd permutations:

$$T^{\lambda_1 \dots \lambda_p}{}_{[\mu_1\mu_2 \dots \mu_q]\nu} \equiv \frac{1}{q!} \sum_{\pi \in S_q} \text{sgn}(\pi) T^{\lambda_1 \dots \lambda_p}{}_{\mu_{\pi(1)}\mu_{\pi(2)} \dots \mu_{\pi(q)}\nu}. \tag{5.138}$$

Example For example,

$$S_{(\mu\nu\lambda)} = \frac{1}{6} \left(S_{\mu\nu\lambda} + S_{\nu\lambda\mu} + S_{\lambda\mu\nu} + S_{\nu\mu\lambda} + S_{\mu\lambda\nu} + S_{\lambda\nu\mu} \right), \tag{5.139}$$

and

$$A_{[\mu\nu\lambda]} = \frac{1}{6} \left(A_{\mu\nu\lambda} + A_{\nu\lambda\mu} + A_{\lambda\mu\nu} - A_{\nu\mu\lambda} - A_{\mu\lambda\nu} - A_{\lambda\nu\mu} \right), \tag{5.140}$$

while

$$A_{[\mu\nu]\lambda} = \frac{1}{2} \left(A_{\mu\nu\lambda} - A_{\nu\mu\lambda} \right). \tag{5.141}$$

If one (anti)symmetrizes indices that are not next to each other, one can exclude the indices which are not part of the symmetrization, with a vertical bar. For example,

$$T_{[\mu|\nu|\lambda]} = \frac{1}{2} \left(T_{\mu\nu\lambda} - T_{\lambda\nu\mu} \right). \tag{5.142}$$

Indices can also come from partial or covariant derivatives. For example,

$$n_{[\mu} \nabla_\nu n_{\lambda]} = \frac{1}{6} \left(n_\mu \nabla_\nu n_\lambda + n_\nu \nabla_\lambda n_\mu + n_\lambda \nabla_\mu n_\nu - n_\nu \nabla_\mu n_\lambda - n_\mu \nabla_\lambda n_\nu - n_\lambda \nabla_\nu n_\mu \right). \tag{5.143}$$

5.6.1 Lie Derivative and Covariant Derivative

If $\Gamma^{\mu}_{\ \nu\lambda}$ is a *torsion-free* (symmetric) connection, then the Lie derivative can be rewritten in terms of the covariant derivative

$$(\mathcal{L}_X Y)^{\mu} = X^{\nu}\partial_{\nu}Y^{\mu} - Y^{\nu}\partial_{\nu}X^{\mu} = X^{\nu}\nabla_{\nu}Y^{\mu} - (\nabla_{\nu}X^{\mu})Y^{\nu}. \qquad (5.144)$$

This is true because of the symmetry of the connection

$$
\begin{aligned}
X^{\nu}\nabla_{\nu}Y^{\mu} - (\nabla_{\nu}X^{\mu})Y^{\nu} &= X^{\nu}(\partial_{\nu}Y^{\mu} + \Gamma^{\mu}_{\ \nu\lambda}Y^{\lambda}) - (\partial_{\nu}X^{\mu} + \Gamma^{\mu}_{\ \nu\lambda}X^{\lambda})Y^{\nu} \\
&= X^{\nu}\partial_{\nu}Y^{\mu} - Y^{\nu}\partial_{\nu}X^{\mu} + \underbrace{(\Gamma^{\mu}_{\ \nu\lambda} - \Gamma^{\mu}_{\ \lambda\nu})}_{=0}X^{\nu}Y^{\lambda}. \qquad (5.145)
\end{aligned}
$$

Rewriting Eq. (5.144) in a covariant form, we obtain

$$\nabla_X Y - \nabla_Y X = \mathcal{L}_X Y = [X, Y]. \qquad (5.146)$$

The generalization of Eq. (5.144) for (p, q)-tensors is

$$
\begin{aligned}
\mathcal{L}_X T^{\mu_1 \cdots \mu_p}_{\nu_1 \cdots \nu_q} &= (X^{\lambda}\nabla_{\lambda})T^{\mu_1 \cdots \mu_p}_{\nu_1 \cdots \nu_q} - (\nabla_{\lambda}X^{\mu_1})T^{\lambda\mu_2 \cdots \mu_p}_{\nu_1 \cdots \nu_q} - \cdots - (\nabla_{\lambda}X^{\mu_p})T^{\mu_1 \cdots \mu_{p-1}\lambda}_{\nu_1 \cdots \nu_q} \\
&\quad + (\nabla_{\nu_1}X^{\lambda})T^{\mu_1 \cdots \mu_p}_{\lambda\nu_2 \cdots \nu_q} + \cdots + (\nabla_{\nu_q}X^{\lambda})T^{\mu_1 \cdots \mu_p}_{\nu_1 \cdots \nu_{q-1}\lambda}. \qquad (5.147)
\end{aligned}
$$

5.7 Metric Connection and Levi-Civita Connection

So far, we did not use the metric anywhere when we discussed the affine connection and its properties. Now we impose an additional condition for the connection, making reference to a metric g.

Let c be an arbitrary curve and let V be its tangent vector. Let X and Y be any pair of vector fields which are both parallel-transported along the curve, $\nabla_V X = \nabla_V Y = 0$. Now we require additionally that the parallel transport should be such that it preserves the relative angle between the vectors and their lengths. This leads to the condition

$$\nabla_V(g(X, Y)) = 0, \qquad \text{when} \quad \nabla_V X = 0 \text{ and } \nabla_V Y = 0. \qquad (5.148)$$

If the connection ∇ satisfies the condition (5.148), then it is said to be a *metric connection*. Note that, since

$$\nabla_V (g(X, Y)) = (\nabla_V g)(X, Y) + g(\underbrace{\nabla_V X}_{=0}, Y) + g(X, \underbrace{\nabla_V Y}_{=0}) = 0, \qquad (5.149)$$

the condition (5.148) is equivalent to

$$\nabla_V g = 0. \qquad (5.150)$$

Owing to the linearity of the connection, Eq. (5.150) implies $\nabla_{\mu}g_{\alpha\beta} = 0$, which, according to Eq. (5.136) is the equation $\nabla g = 0$ in component form. Thus we can also say that a metric connection is a connection which is *covariantly constant*, meaning that it must satisfy

$$\nabla g = 0. \qquad (5.151)$$

In fact, Eq. (5.151) can be used as an alternative and equivalent condition to define a metric connection.[1]

We can now derive the form of the components of the metric connection in a coordinate basis. When rewritten in components, Eq. (5.150) takes the form

$$(\nabla_\mu g)_{\alpha\beta} = \partial_\mu g_{\alpha\beta} - \Gamma^\lambda_{\mu\alpha} g_{\lambda\beta} - \Gamma^\lambda_{\mu\beta} g_{\alpha\lambda} = 0. \tag{5.152}$$

Cyclic permutation of the indices μ, α and β gives two additional equations,

$$(\nabla_\alpha g)_{\beta\mu} = \partial_\alpha g_{\beta\mu} - \Gamma^\lambda_{\alpha\beta} g_{\lambda\mu} - \Gamma^\lambda_{\alpha\mu} g_{\beta\lambda} = 0 \tag{5.153}$$

and

$$(\nabla_\beta g)_{\mu\alpha} = \partial_\beta g_{\mu\alpha} - \Gamma^\lambda_{\beta\mu} g_{\lambda\alpha} - \Gamma^\lambda_{\beta\alpha} g_{\mu\lambda} = 0. \tag{5.154}$$

Recalling the symmetrization of indices $\Gamma^\gamma_{(\alpha\beta)} \equiv \frac{1}{2}(\Gamma^\gamma_{\alpha\beta} + \Gamma^\gamma_{\beta\alpha})$, and subtracting Eq. (5.152) from the sum of Eq. (5.153) and Eq. (5.154), leads to

$$-\partial_\mu g_{\alpha\beta} + \partial_\alpha g_{\beta\mu} + \partial_\beta g_{\mu\alpha} + T^\lambda_{\mu\alpha} g_{\lambda\beta} + T^\lambda_{\mu\beta} g_{\lambda\alpha} - 2\Gamma^\lambda_{(\alpha\beta)} g_{\lambda\mu} = 0, \tag{5.155}$$

or, equivalently,

$$\Gamma^\lambda_{(\alpha\beta)} g_{\lambda\mu} = \frac{1}{2}\left\{ (\partial_\alpha g_{\beta\mu} + \partial_\beta g_{\mu\alpha} - \partial_\mu g_{\alpha\beta}) + T^\lambda_{\mu\alpha} g_{\lambda\beta} + T^\lambda_{\mu\beta} g_{\lambda\alpha} \right\}. \tag{5.156}$$

Thus the symmetric part of the connection is

$$\Gamma^\kappa_{(\alpha\beta)} = \left\{ \begin{matrix} \kappa \\ \alpha\beta \end{matrix} \right\} + \frac{1}{2}(T_\alpha{}^\kappa{}_\beta + T_\beta{}^\kappa{}_\alpha). \tag{5.157}$$

In the above, we introduced the *Christoffel symbols*, named after Elwin Bruno Christoffel,

$$\left\{ \begin{matrix} \kappa \\ \alpha\beta \end{matrix} \right\} = \frac{1}{2} g^{\kappa\mu}(\partial_\alpha g_{\beta\mu} + \partial_\beta g_{\mu\alpha} - \partial_\mu g_{\alpha\beta}), \tag{5.158}$$

which are completely determined by the metric. Note also the notation in Eq. (5.157),

$$T_\alpha{}^\kappa{}_\beta = g_{\alpha\lambda} g^{\kappa\mu} T^\lambda_{\mu\beta}. \tag{5.159}$$

Summing the antisymmetric and symmetric parts, we obtain the coefficients of a metric connection,

$$\Gamma^\kappa_{\alpha\beta} = \Gamma^\kappa_{(\alpha\beta)} + \Gamma^\kappa_{[\alpha\beta]} = \left\{ \begin{matrix} \kappa \\ \alpha\beta \end{matrix} \right\} + \frac{1}{2}\left(T_\alpha{}^\kappa{}_\beta + T_\beta{}^\kappa{}_\alpha + T^\kappa{}_{\alpha\beta} \right), \tag{5.160}$$

and one can see that the result includes a part that contains the Christoffel symbols, determined by the metric, and a part that depends on the torsion tensor, called the *contorsion* $K^\kappa{}_{\alpha\beta}$ and defined as

$$K^\kappa{}_{\alpha\beta} = \frac{1}{2}\left(T_\alpha{}^\kappa{}_\beta + T_\beta{}^\kappa{}_\alpha + T^\kappa{}_{\alpha\beta} \right). \tag{5.161}$$

What is the effect of the torsion tensor? As we stated above, parallel transport with the metric connection must preserve the angle between any pair of vectors and their lengths, or, for a pseudo-Riemannian manifold, $g(X, Y)$, but it is possible that the

---

[1] The metric connection is also an affine connection, so must satisfy the earlier definitions.

vectors get rotated when they are parallel-transported, as was the case in a previous example; see Eq. (5.87).

Example   Let us consider again the manifold $M = \mathbb{R}^2$ with the Euclidean metric δ_{ij} and the connection defined by Eq. (5.45). Since the components of the metric are constants, the Christoffel symbols vanish. Nevertheless, the connection is nonvanishing, as the torsion is nonzero. As shown in Eq. (5.87), in that case parallel transport along a straight line parallel to the x^2-axis rotates all vectors in a similar way, preserving their lengths and relative angles.

If the torsion tensor vanishes, $T^\kappa{}_{\alpha\beta} = 0$, then the metric connection is called the *Levi-Civita connection*, which is named after Tullio Levi-Civita, and its components reduce to the Christoffel symbols:

$$\Gamma^\kappa{}_{\alpha\beta} = \left\{ {\kappa \atop \alpha\beta} \right\}. \tag{5.162}$$

The Levi-Civita connection can be directly determined on a (pseudo-)Riemannian manifold, since it is the unique torsion-free metric connection. This is the following *fundamental theorem of Riemannian geometry*. Note that, despite the name, it also applies to the pseudo-Riemannian case. Although we have already shown the existence and uniqueness of the Levi-Civita connection using coordinates, the theorem can be proven in a more elegant, coordinate-independent way.

Theorem 5.1 (Fundamental theorem of Riemannian geometry)   *Given a Riemannian or pseudo-Riemannian manifold M with a metric g, there exists a unique torsion-free connection ∇ such that the metric is covariantly constant, $\nabla g = 0$.*

Proof   We start with three arbitrary vector fields X, Y, and Z on the manifold M. Since $g(Y, Z)$ is a function, one has $\nabla_X g(Y, Z) = X g(Y, Z)$. Then the Leibniz rule implies

$$X g(Y, Z) = \nabla_X g(Y, Z) = \underbrace{(\nabla_X g)(Y, Z)}_{=0} + g(\nabla_X Y, Z) + g(Y, \nabla_X Z), \tag{5.163}$$

where $\nabla_X g(Y, Z) = 0$, because we assumed $\nabla g = 0$, so $\nabla_X g = 0$ for any X. Cyclic permutations of X, Y and Z then give two additional equations

$$Y g(Z, X) = g(\nabla_Y Z, X) + g(Z, \nabla_Y X), \tag{5.164}$$

$$Z g(X, Y) = g(\nabla_Z X, Y) + g(X, \nabla_Z Y). \tag{5.165}$$

Subtracting Eq. (5.165) from the sum of Eqs. (5.163) and (5.164), and using the symmetry and linearity of the metric to rewrite, e.g.,

$$g(\nabla_X Y, Z) + g(Z, \nabla_Y X) = g(\nabla_X Y + \nabla_Y X, Z), \tag{5.166}$$

one obtains

$$X g(Y, Z) + Y g(Z, X) - Z g(X, Y) = g(\nabla_X Y + \nabla_Y X, Z)$$
$$- g(\nabla_Z X - \nabla_X Z, Y) + g(\nabla_Y Z - \nabla_Z Y, X). \tag{5.167}$$

Since the metric was assumed to be torsion-free, one can apply Eq. (5.146),

$$\nabla_Z X - \nabla_X Z = [Z, X], \qquad \nabla_X Y + \nabla_Y X = 2\nabla_X Y - [X, Y], \qquad (5.168)$$

to write

$$Xg(Y, Z) + Yg(Z, X) - Zg(X, Y) = 2g(\nabla_X Y, Z)$$
$$- g([X, Y], Z) - g([Z, X, Y]) + g([Y, Z], X), \qquad (5.169)$$

and finally

$$g(\nabla_X Y, Z) = \frac{1}{2} [Xg(Y, Z) + Yg(Z, X) - Zg(X, Y)$$
$$+ g([X, Y], Z) + g([Z, X], Y) - g([Y, Z], X)]. \qquad (5.170)$$

Equation (5.170) is the *Koszul formula*, which is named after Jean-Louis Koszul. It implies that the connection $\nabla_X Y$ is uniquely determined (if it exists). This statement can be proven by a *reductio ad absurdum*, by showing that assuming that it is not true leads to a contradiction. Thus, assume that the connection is not unique, so that there is a different connection $\tilde{\nabla}$ that also satisfies Eq. (5.170). Then,

$$g(\nabla_X Y, Z) - g(\tilde{\nabla}_X Y, Z) = g(\nabla_X Y - \tilde{\nabla}_X Y, Z) = 0. \qquad (5.171)$$

At this point, one distinguishes two cases.

- If g is a Riemannian metric, it is positive definite, so if $Z \neq 0$ then $\nabla_X Y - \tilde{\nabla}_X Y = 0$ for every X and every Y, which means that $\nabla = \tilde{\nabla}$, in contradiction with the assumption $\nabla \neq \tilde{\nabla}$.
- If g is pseudo-Riemannian, then it is semi-definite, but Eq. (5.171) holds for all Z so $\nabla_X Y - \tilde{\nabla}_X Y = 0$ and again this means $\nabla = \tilde{\nabla}$, in contradiction with the assumption $\nabla \neq \tilde{\nabla}$.

For the existence of the connection to be guaranteed, we need to check that ∇ defined by the Koszul formula satisfies the properties of a connection. We check only the property $\nabla_{fX} Y = f\nabla_X Y$. The remaining properties can be verified by similar, and somewhat tedious, calculations, and are left as an exercise for the reader. Starting from $g(\nabla_{fX} Y, Z)$, a sequence of algebraic manipulations will yield

$$g(\nabla_{fX} Y, Z) = \frac{1}{2} [fXg(Y, Z) + Yg(Z, fX) - Zg(fX, Y)$$
$$+ g([fX, Y], Z) + g([Z, fX], Y) - g([Y, Z], fX)]$$
$$= \frac{1}{2} [fXg(Y, Z) + Y(fg(Z, X)) - Z(fg(X, Y))$$
$$+ g([fX, Y], Z) + g([Z, fX], Y) - fg([Y, Z], X)]$$
$$= \frac{1}{2} [fXg(Y, Z) + Y[f]g(Z, X) + fY(g(Z, X)) - Z[f]g(X, Y) - fZ(g(X, Y))$$
$$+ g(f[X, Y] - Y[f]X, Z) + g(f[Z, X] + Z[f]X, Y) - fg([Y, Z], X)]$$
$$= \frac{1}{2} [fXg(Y, Z) + fY(g(Z, X))) - fZ(g(X, Y)) + Y[f]g(Z, X) - Z[f]g(X, Y$$
$$+ fg([X,Y], Z) - Y[f]g(X, Z) + fg(f[Z,X], T) + Z[f]g(X,Y) - fg([Y, Z], X)]$$

$$= f \cdot \frac{1}{2} \left[Xg(Y,Z) + Yg(Z,X) - Zg(X,Y) + g([X,Y],Z) \right.$$

$$\left. + g([Z,X],Y) - g([Y,Z],X) \right]$$

$$= fg(\nabla_X Y, Z) = g(f\nabla_X Y, Z). \tag{5.172}$$

Thus $g(\nabla_{fX} Y - f\nabla_X Y, Z) = 0$ for all X, Y, Z, which means $\nabla_{fX} Y = f\nabla_X Y$. $\qquad\square$

5.8 Covariant Derivative with the Levi-Civita Connection

One benefit of having a covariantly constant (i.e., metric-compatible connection) is that it commutes with raising and lowering of indices, for example

$$\nabla_\alpha V_\beta = \nabla_\alpha (g_{\beta\mu} V^\mu) = \underbrace{(\nabla_\alpha g_{\beta\mu})}_{=0} V^\mu + g_{\beta\mu} \nabla_\alpha V^\mu = g_{\beta\mu} \nabla_\alpha V^\mu. \tag{5.173}$$

Another benefit is that we can derive some useful formulas for the connection and the covariant derivative. We begin with the matrix determinant identity

$$\det M = e^{\operatorname{tr}\ln M}, \tag{5.174}$$

which one can show by a straightforward calculation, assuming a basis where the matrix is diagonal (we assume for simplicity that M is symmetric and $\ln M$ is well defined). We will be interested in a determinant of a Riemannian metric g,

$$\det g = \det(g_{\alpha\beta}), \tag{5.175}$$

viewing the components $g_{\alpha\beta}$ as those of a symmetric matrix. Taking a partial derivative, we obtain

$$\partial_\mu(\det g) = \partial_\mu e^{\operatorname{tr}\ln g} = e^{\operatorname{tr}\ln g} \partial_\mu(\operatorname{tr}\ln g) = (\det g)\operatorname{tr}(g^{-1}\partial_\mu g) = (\det g)g^{\alpha\beta}\partial_\mu g_{\beta\alpha}. \tag{5.176}$$

The equation

$$\partial_\mu(\det g) = (\det g)\operatorname{tr}(g^{-1}\partial_\mu g) \tag{5.177}$$

is also known as (a particular version of) *Jacobi's formula*. We will apply it to derive a useful equation for partially contracted components of the Levi-Civita connection. Let us recall the equation for its components

$$\Gamma^\kappa_{\alpha\beta} = \frac{1}{2}g^{\kappa\mu}(\partial_\alpha g_{\beta\mu} + \partial_\beta g_{\alpha\mu} - \partial_\mu g_{\alpha\beta}). \tag{5.178}$$

We shall soon need the result for a contracted upper and lower index (note the symmetry in the lower indices, so the choice does not matter),

$$\Gamma^\kappa_{\kappa\beta} = \frac{1}{2}g^{\kappa\mu}(\partial_\kappa g_{\beta\mu} + \partial_\beta g_{\kappa\mu} - \partial_\mu g_{\kappa\beta})$$

$$= \frac{1}{2}g^{\kappa\mu}\partial_\beta g_{\kappa\mu}, \tag{5.179}$$

where in the first line, the first and third terms cancel each other due to the symmetry of the metric components. If we compare the result with the right-hand side of (5.176), we notice that we can also write

$$\Gamma_{\kappa\beta}^{\kappa} = \frac{1}{2}(\det g)^{-1}\partial_{\beta}(\det g) = \frac{1}{\sqrt{\det g}}\partial_{\beta}\sqrt{\det g}. \qquad (5.180)$$

We now use the result to write a useful expression for the covariant divergence of a vector:

$$\nabla_{\mu}V^{\mu} = \partial_{\mu}V^{\mu} + \Gamma_{\mu\lambda}^{\mu}V^{\lambda} = \partial_{\mu}V^{\mu} + V^{\mu}\frac{1}{\sqrt{\det g}}\partial_{\mu}\sqrt{\det g}$$

$$= \frac{1}{\sqrt{\det g}}\sqrt{\det g}\,\partial_{\mu}V^{\mu} + V^{\mu}\frac{1}{\sqrt{\det g}}\partial_{\mu}\sqrt{\det g}$$

$$= \frac{1}{\sqrt{\det g}}\partial_{\mu}\left(\sqrt{\det g}V^{\mu}\right). \qquad (5.181)$$

The benefit of the end result is that instead of working out different components of the connection coefficients, we need to only compute the determinant of the metric and then an ordinary partial derivative, which is often simpler.

We continue by deriving a formula for the Laplacian, which you have probably seen in earlier courses on calculus, in the context of curvilinear coordinates such as the spherical coordinates. Let us choose as the vector V^{μ} the covariant derivative of a function f (which reduces to an ordinary partial derivative):

$$V^{\mu} = \nabla^{\mu}f = \partial^{\mu}f. \qquad (5.182)$$

Substituting into Eq. (5.181) gives

$$\nabla_{\mu}\nabla^{\mu}f = \frac{1}{\sqrt{\det g}}\partial_{\mu}(\sqrt{\det g}\partial^{\mu}f) = \frac{1}{\sqrt{\det g}}\partial_{\mu}(\sqrt{\det g}g^{\mu\nu}\partial_{\nu})f. \qquad (5.183)$$

In mathematics the Laplacian is usually denoted by Δ, in physics by ∇^2:

$$\Delta f \equiv \nabla^2 f = \nabla_{\mu}\nabla^{\mu}f. \qquad (5.184)$$

Example Let us apply the above formulas for the Euclidean metric in $\mathbb{R}^3$ in spherical coordinates (r, θ, φ),

$$ds^2 = dr^2 + r^2(d\theta^2 + \sin^2\theta d\varphi^2). \qquad (5.185)$$

The metric and its inverse are diagonal matrices

$$g = \text{diag}(1, r^2, r^2\sin^2\theta)$$
$$g^{-1} = \text{diag}(1, r^{-2}, r^{-2}\sin^{-2}\theta), \qquad (5.186)$$

with the determinant and its square root

$$\det g = r^4\sin^2\theta, \quad \sqrt{\det g} = r^2\sin\theta. \qquad (5.187)$$

For the covariant divergence of a vector we obtain

$$\nabla_\mu V^\mu = \frac{1}{r^2 \sin \theta} \partial_r (r^2 \sin \theta V^r) + \frac{1}{r^2 \sin \theta} \partial_\theta (r^2 \sin \theta V^\theta)$$
$$+ \frac{1}{r^2 \sin \theta} \partial_\varphi (r^2 \sin \theta V^\varphi)$$
$$= \frac{1}{r^2} \partial_r (r^2 V^r) + \frac{1}{\sin \theta} \partial_\theta (\sin \theta V^\theta) + \partial_\varphi (V^\varphi), \qquad (5.188)$$

the usual formula for the divergence of a vector $\vec{V} = (V^r, V^\theta, V^\varphi)$.

For the Laplacian, we similarly obtain

$$\nabla^2 f = \frac{1}{r^2 \sin \theta} \partial_r (r^2 \sin \theta g^{rr} \partial_r) f + \frac{1}{r^2 \sin \theta} \partial_\theta (r^2 \sin \theta g^{\theta\theta} \partial_\theta) f$$
$$+ \frac{1}{r^2 \sin \theta} \partial_\varphi (r^2 \sin \theta g^{\varphi\varphi} \partial_\varphi) f$$
$$= \frac{1}{r^2} \partial_r (r^2 \partial_r) f + \frac{1}{r^2 \sin \theta} \partial_\theta \left(r^2 \sin \theta \frac{1}{r^2} \partial_\theta \right) f$$
$$+ \frac{1}{r^2 \sin \theta} \partial_\varphi \left(\frac{1}{\sin \theta} \partial_\varphi \right) f$$
$$= \frac{1}{r^2} \partial_r (r^2 \partial_r) f + \frac{1}{r^2 \sin \theta} \partial_\theta (\sin \theta \partial_\theta) f + \frac{1}{r^2 \sin^2 \theta} \partial_\varphi^2 f, \qquad (5.189)$$

which is the familiar formula from calculus.

5.9 Geodesics of Levi-Civita Connections

Given a curve $c(s) = (x^\mu(s))$, its length $I(c)$ is defined by

$$I(c) = \int_c ds = \int_c \sqrt{g_{\mu\nu} \frac{dx^\mu}{ds'} \frac{dx^\nu}{ds'}} ds' \equiv \int_c L ds'. \qquad (5.190)$$

One can normalize s' such that $L = 1$, so $s' = s$. Curves with extremal (minimum or maximum) length satisfy a variational principle, often represented with the notation $\delta I = 0$, regarding the length of the curve: They satisfy the Euler–Lagrange equation

$$\frac{d}{ds} \left(\frac{\partial L}{\partial x'^\mu} \right) - \frac{\partial L}{\partial x^\mu} = 0, \qquad \text{where } x'^\mu = \frac{dx^\mu}{ds}. \qquad (5.191)$$

L is called the *Lagrange function*, or, more simply, the Lagrangian. Instead of L, which contains a square root, one can equivalently use a simpler Lagrange function

$$F = \frac{1}{2} g_{\mu\nu} \frac{dx^\mu}{ds} \frac{dx^\nu}{ds} = \frac{1}{2} L^2, \qquad (5.192)$$

because

$$\frac{d}{ds} \left(\frac{\partial F}{\partial x'^\mu} \right) - \frac{\partial F}{\partial x^\mu} = L \underbrace{\left(\frac{d}{ds} \left(\frac{\partial L}{\partial x'^\mu} \right) - \frac{\partial L}{\partial x^\mu} \right)}_{=0} + \frac{\partial L}{\partial x'^\mu} \underbrace{\frac{dL}{ds}}_{=0} = 0, \qquad (5.193)$$

where we used the fact that L is constant, which implies $\frac{dL}{ds} = 0$, when $x^\mu(s)$ satisfies the Euler–Lagrange equation (5.191). Then $\delta(\int F ds) = 0$ gives

$$\frac{d}{ds}\left(g_{\lambda\mu}\frac{dx^\mu}{ds}\right) - \frac{1}{2}\frac{\partial g_{\mu\nu}}{\partial x^\lambda}\frac{dx^\mu}{ds}\frac{dx^\nu}{ds} = 0, \tag{5.194}$$

from which follows

$$\frac{\partial g_{\lambda\mu}}{\partial x^\nu}\frac{dx^\mu}{ds}\frac{dx^\nu}{ds} + g_{\lambda\mu}\frac{d^2x^\mu}{ds^2} - \frac{1}{2}\frac{\partial g_{\mu\nu}}{\partial x^\lambda}\frac{dx^\mu}{ds}\frac{dx^\nu}{ds} = 0, \tag{5.195}$$

which, in turn, implies

$$g_{\lambda\mu}\frac{d^2x^\mu}{ds^2} + \frac{1}{2}\left(\frac{\partial g_{\lambda\mu}}{\partial x^\nu} + \frac{\partial g_{\lambda\nu}}{\partial x^\mu} - \frac{\partial g_{\mu\nu}}{\partial x^\lambda}\right)\frac{dx^\mu}{ds}\frac{dx^\nu}{ds} = 0. \tag{5.196}$$

Multiplying Eq. (5.196) by $g^{\kappa\lambda}$ and summing over λ, one obtains

$$\frac{d^2x^\kappa}{ds^2} + \left\{\begin{matrix}\kappa\\\mu\nu\end{matrix}\right\}\frac{dx^\mu}{ds}\frac{dx^\nu}{ds} = 0, \tag{5.197}$$

which is the geodesic equation with a Levi-Civita connection. The quantity $I = \int F ds$ sometimes provides a convenient starting point for computing the Christoffel symbols $\left\{\begin{matrix}\kappa\\\mu\nu\end{matrix}\right\}$: By plugging the metric into I, it is possible to derive the Euler–Lagrange equations and read off the Christoffel symbols simply by comparing the Euler–Lagrange equations with Eq. (5.197).

Note that, previously, when we discussed the geodesic equation for a general connection, we said that geodesics are the "straightest" possible curves. Now, with the Levi-Civita connection, which is only based on the metric, we have found that the geodesics also extremize the length. By checking that a solution of the geodesic equation is a (global) minimum of the length, we can also identify it as the shortest possible curve.

Example As a simple example of using equations of motion to compute the components of the Levi-Civita connection, we consider $\mathbb{R}^3$ with cylindrical coordinates $x = (r, \phi, z)$. Now the Euclidean metric takes the form

$$g_{\mu\nu}dx^\mu dx^\nu = dr^2 + r^2 d\phi^2 + dz^2. \tag{5.198}$$

The metric is of course flat, but the use of curvilinear coordinates makes the Levi-Civita connection nonvanishing. (It is good to remember that the connection does not transform like a tensor – if it did, in this case the connection would vanish in all coordinates, because one would obtain a product of Jacobian factors multiplying zero.) We compute the nonvanishing components by identifying the free particle equations of motion as the geodesic equation. From the Lagrangian

$$L = \frac{1}{2}\left(r'^2 + r^2\phi'^2 + z'^2\right) \tag{5.199}$$

we obtain the Euler–Lagrange equations of motion

$$r'' - r\phi'\phi' = 0,$$

$$\phi'' + \frac{1}{r}\left(\phi'r' + r'\phi'\right) = 0,$$

$$z'' = 0, \tag{5.200}$$

from which one gets the nonvanishing connection components

$$\Gamma^r_{\phi\phi} = -r, \qquad \Gamma^\phi_{\phi r} = \Gamma^\phi_{r\phi} = \frac{1}{r}. \tag{5.201}$$

The reason why this is a quick way is that we automatically compute only the nontrivial components. Had we used Eq. (5.178) instead, we would have done much additional work only to find additional vanishing components.

Note also that one can explicitly redefine the integral that gives the length of the curve, by multiplying it by a constant parameter m:

$$I = m \int \sqrt{\pm g_{\mu\nu} \frac{dx^\mu}{ds'} \frac{dx^\nu}{ds'}} ds'. \tag{5.202}$$

In physics, Eq. (5.202), with the choice of a minus sign inside the square root, is the relativistic action associated with a free massive pointlike particle with mass m, moving in a curved spacetime, i.e., free point particles are assumed to move along timelike geodesics. If $m^2 > 0$, we say that the corresponding geodesics on a pseudo-Riemannian manifold are *timelike*. On the other hand, the geodesics of massless particles, such as the photon, are called *null*, or *lightlike*, geodesics, with $ds^2 = 0$. Finally, note that the $m^2 < 0$ case, which would correspond to *spacelike* geodesics, is not realized by any type of physical particles.

Example   Consider a $(1 + 2)$-dimensional Minkowski spacetime $\mathbb{R}^{1,2}$. We write the spatial part of the metric in polar coordinates,

$$ds^2 = -dt^2 + dr^2 + r^2 d\varphi^2. \tag{5.203}$$

Let us find the solutions for radial null geodesics with constant angle φ. For the radial null geodesic,

$$0 = ds^2 = -dt^2 + dr^2 + 0 \tag{5.204}$$

so

$$\frac{dr}{dt} = \pm 1, \tag{5.205}$$

so we find two possible solutions:

$$r = t + \text{const} \tag{5.206}$$

for null geodesics directed radially outward from the origin, and

$$r = -t + \text{const} \tag{5.207}$$

for null geodesics directed radially inward towards the origin. Varying the angle $\varphi_0 \in S^1$ and setting the constants to zero, the locus of the two types of geodesics gives the light cone at the origin.

5.10 Curvature and Torsion

In this section we derive some tensors that are particularly important in the context of general relativity: the Riemann tensor, the Ricci tensor, the scalar curvature, and the Einstein tensor. These tensors are used in the famous Einstein equation connecting the geometry of a spacetime to the distribution of matter, describing how matter curves the spacetime.

We shall proceed in two different ways. First, we just introduce some definitions *by fiat*, and check that they do satisfy the properties of a tensor. Then, we will proceed in an alternative, more geometrical, fashion, to show the effect of curvature of a manifold when vectors are parallel-transported. Finally, we will verify that this procedure yields the same tensors that we introduced at the beginning.

To start with, we define the *Riemann curvature tensor R* as the map

$$R : \chi(M) \times \chi(M) \times \chi(M) \to \chi(M),$$
$$R(X, Y, Z) \equiv R(X, Y)Z \equiv \nabla_X \nabla_Y Z - \nabla_Y \nabla_X Z - \nabla_{[X,Y]} Z. \tag{5.208}$$

Then we define a *torsion tensor T* as the map

$$T : \chi(M) \times \chi(M) \to \chi(M),$$
$$T(X, Y) \equiv \nabla_X Y - \nabla_Y X - [X, Y]. \tag{5.209}$$

As we will show, this will be the same torsion tensor previously defined in Eq. (5.65), so that the name is justified. First, however, we check that the definitions (5.208) and (5.209) really define tensors, i.e., multilinear maps. For the Riemann curvature tensor, from its definition one can easily verify that, for example, $R(X + X', Y, Z) = R(X, Y, Z) + R(X', Y, Z)$, and analogous relations for the other arguments, holds, but it is less obvious that $R(fX, gY, hZ) = fgh R(X, Y, Z)$ for arbitrary functions $f, g, h \in \mathcal{F}(M)$. To verify this property, we first note that

$$[fX, gY] = fX[g]Y - gY[f]X + fg[X, Y]. \tag{5.210}$$

Using Eq. (5.210), one obtains

$$R(fX, gY)(hZ) = f\nabla_X(g\nabla_Y(hZ)) - g\nabla_Y(f\nabla_X(hZ))$$
$$- fX[g]\nabla_Y(hZ) + gY[f]\nabla_X(hZ) - fg\nabla_{[X,Y]}(hZ). \tag{5.211}$$

The first term on the right-hand side of Eq. (5.211) is

$$f\nabla_X(g\nabla_Y(hZ)) = f\nabla_X(gY[h]Z + gh\nabla_Y Z)$$
$$= fX[g]Y[h]Z + fg(X[Y[h]])Z + fgY[h]\nabla_X Z$$
$$+ fgX[h]\nabla_Y Z + fhX[g]\nabla_Y Z + fgh\nabla_X \nabla_Y Z. \tag{5.212}$$

The second term on the right-hand side of Eq. (5.211) can be obtained from Eq. (5.212) by interchanging $X \leftrightarrow Y$ and $f \leftrightarrow g$, and changing the sign. As a consequence, Eq. (5.211) can be rewritten as

$$
\begin{aligned}
R(fX, gY)(hZ) &= f\nabla_X(gY[h]Z + gh\nabla_Y Z)\\
&= fX[g]Y[h]Z + fg(X[Y[h]])Z + fgY[h]\nabla_X Z\\
&\quad + fgX[h]\nabla_Y Z + fhX[g]\nabla_Y Z + fgh\nabla_X\nabla_Y Z\\
&\quad - gY[f]X[h]Z - fg(Y[X[h]])Z - fgX[h]\nabla_Y Z\\
&\quad - fgY[h]\nabla_X Z - ghY[f]\nabla_X Z - fgh\nabla_Y\nabla_X Z\\
&\quad - fX[g]\nabla_Y(hZ) + gY[f]\nabla_X(hZ) - fg\nabla_{[X,Y]}(hZ)\\
&= fgh(\nabla_X\nabla_Y Z - \nabla_Y\nabla_X Z - \nabla_{[X,Y]}Z)\\
&= fghR(X, Y)Z,
\end{aligned}
\tag{5.213}
$$

where we used the fact that the covariant derivative of a function is the same as the directional derivative. Thus R is a linear map. Expanding in a basis, $X = X^\mu e_\mu$, $Y = Y^\nu e_\nu$, and $Z = Z^\lambda e_\lambda$, we have

$$
R(X, Y)Z = X^\mu Y^\nu Z^\lambda R(e_\mu, e_\nu)e_\lambda.
\tag{5.214}
$$

Since R maps three vector fields to a vector field, it is a $(1,3)$-tensor. We then expand the vector field $R(e_\mu, e_\nu)e_\lambda$ in the same basis to extract the components of the Riemann tensor,

$$
R(e_\mu, e_\nu)e_\lambda = R^\kappa{}_{\lambda\mu\nu}e_\kappa.
\tag{5.215}
$$

Note the placement of indices. Finally, we derive the components $R^\kappa{}_{\lambda\mu\nu}$. To this purpose, we use the fact that $[e_\mu, e_\nu] = 0$ and $dx^\kappa(e_\sigma) = \delta^\kappa{}_\sigma$, and get

$$
\begin{aligned}
R^\kappa{}_{\lambda\mu\nu} &= dx^\kappa(R(e_\mu, e_\nu)e_\lambda)\\
&= dx^\kappa(\nabla_\mu\nabla_\nu e_\lambda - \nabla_\nu\nabla_\mu e_\lambda)\\
&= dx^\kappa\left(\nabla_\mu(\Gamma^\eta{}_{\nu\lambda}e_\eta) - \nabla_\nu(\Gamma^\eta{}_{\mu\lambda}e_\eta)\right)\\
&= dx^\kappa\left((\partial_\mu\Gamma^\eta{}_{\nu\lambda})e_\eta + \Gamma^\eta{}_{\nu\lambda}\Gamma^\rho{}_{\mu\eta}e_\rho - (\partial_\nu\Gamma^\eta{}_{\mu\lambda})e_\eta - \Gamma^\eta{}_{\mu\lambda}\Gamma^\rho{}_{\nu\eta}e_\rho\right).
\end{aligned}
\tag{5.216}
$$

The final equation for the components of the Riemann tensor is

$$
R^\kappa{}_{\lambda\mu\nu} = \partial_\mu\Gamma^\kappa{}_{\nu\lambda} - \partial_\nu\Gamma^\kappa{}_{\mu\lambda} + \Gamma^\eta{}_{\nu\lambda}\Gamma^\kappa{}_{\mu\eta} - \Gamma^\eta{}_{\mu\lambda}\Gamma^\kappa{}_{\nu\eta}.
\tag{5.217}
$$

The number of independent components of the Riemann tensor in an n-dimensional manifold is $\frac{n^2(n^2-1)}{12}$.

For the $(1,2)$-tensor T defined in Eq. (5.209), a similar (but shorter) calculation shows that $T(fX, gY) = fgT(X, Y)$, so $T(X, Y) = X^\mu Y^\nu T(e_\mu, e_\nu)$. Expanding the vector field

$$
T(e_\mu, e_\nu) = T^\lambda{}_{\mu\nu}e_\lambda,
\tag{5.218}
$$

we derive the equation for the components $T^\lambda{}_{\mu\nu}$:

$$
T^\lambda{}_{\mu\nu} = dx^\lambda(T(e_\mu, e_\nu)) = dx^\lambda(\nabla_\mu e_\nu - \nabla_\nu e_\mu) = dx^\lambda(\Gamma^\eta{}_{\mu\nu}e_\eta - \Gamma^\eta{}_{\nu\mu}e_\eta),
\tag{5.219}
$$

therefore

$$
T^\lambda{}_{\mu\nu} = \Gamma^\lambda{}_{\mu\nu} - \Gamma^\lambda{}_{\nu\mu},
\tag{5.220}
$$

which agrees with Eq. (5.65) that we had previously used to define the components of the torsion tensor, thus Eq. (5.209) indeed defines the same torsion tensor.

We complete the list of definitions by defining the *Ricci tensor*, named after Gregorio Ricci-Curbastro,

$$\text{Ric} : \chi(M) \times \chi(M) \to \Omega(M)$$
$$(X, Y) \to \text{Ric}(X, Y) = dx^\lambda(R(e_\lambda, Y)X). \tag{5.221}$$

The components $(\text{Ric})_{\mu\nu}$ are often denoted as $R_{\mu\nu}$, and are obtained from the components of the Riemann tensor by contracting a pair of indices (note their placement):

$$R_{\mu\nu} = \text{Ric}(e_\mu, e_\nu) = R^\lambda{}_{\mu\lambda\nu}. \tag{5.222}$$

The *scalar curvature R* is a function defined as

$$R = g^{\mu\nu}R_{\mu\nu} = R^{\lambda\nu}{}_{\lambda\nu}. \tag{5.223}$$

Finally, the *Einstein tensor* is defined in component form as the combination

$$G_{\mu\nu} = R_{\mu\nu} - \frac{1}{2}Rg_{\mu\nu}. \tag{5.224}$$

Geometric Interpretation of the Riemann Curvature Tensor

After the formal definitions, we now proceed to explain how the Riemann tensor is related with the concept of "curvature" of a manifold. In particular, one of the questions that we are going to address is what happens to a vector when it is parallel-transported along a closed curve, and how it has been changed when it arrives back to the starting point.

Figure 5.5 depicts how even torsion-free parallel-transport along a closed triangle-shaped path on a curved manifold (a two-sphere S^2) has an effect on the orientation of the vector, resulting from curvature. Note that the length of the vector and its relative angle with respect to the path is preserved during parallel-transport. You may also try this out experimentally by drawing such a triangular curve, e.g., on a basketball, and parallel-transport a pencil along it. We will now study this effect quantitatively and in more detail.

To begin with, it is helpful to introduce some new concepts. We have already discussed differential forms, whose components were those of an antisymmetric tensor, each component being a smooth function. We can now consider a more general situation, in which, rather than being functions, the components themselves can be arbitrary tensors. As a simple example, consider a two-form F in an explicit basis

$$F = F_{\mu\nu}dx^\mu \wedge dx^\nu, \tag{5.225}$$

where the components $F_{\mu\nu} = F_{\mu\nu}(x)$ are smooth functions. If these smooth functions are replaced by vector fields $F_{\mu\nu}$, which are expanded in the coordinate basis

$$F_{\mu\nu} = F^\lambda{}_{\mu\nu}(x)\frac{\partial}{\partial x^\lambda}, \tag{5.226}$$

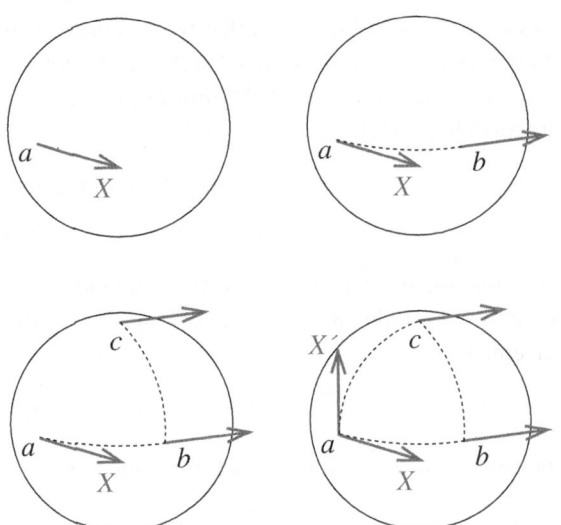

Fig. 5.5 Parallel-transporting (without torsion) a vector along a loop on a two-sphere results into a vector whose direction deviates from that of the original one. In the figure, the vector X, initially at the point a (top-left panel), is first parallel-transported to the point b (top-right panel), then to the point c (bottom-left panel), and finally back to the point a (bottom-right panel): The resulting vector X' has a direction different from the direction that X originally had.

then

$$F = F_{\mu\nu}dx^{\mu} \wedge dx^{\nu} = F^{\lambda}_{\mu\nu}(x)dx^{\mu} \wedge dx^{\nu} \otimes \frac{\partial}{\partial x^{\lambda}}. \qquad (5.227)$$

is a *vector-valued two-form*. Conversely, one could interpret F as a vector field, with components that are two-forms, rather than functions,

$$F = F^{\lambda}(x)\frac{\partial}{\partial x^{\lambda}}, \quad F^{\lambda}(x) = F^{\lambda}_{\mu\nu}(x)dx^{\mu} \wedge dx^{\nu}, \qquad (5.228)$$

so one could equally well call F a "two-form-valued vector field." In analogy with this example, one defines a *tensor-valued k-form field* (or a k-form-valued tensor field) as a (p, q)-tensor

$$T = T^{\mu_1 \cdots \mu_p}_{\nu_1 \cdots \nu_q} e^{*\nu_1} \otimes \cdots \otimes e^{*\nu_q} \otimes e_{\mu_1} \otimes \cdots \otimes e_{\mu_p}, \qquad (5.229)$$

where we used $e^{*\nu} = dx^{\nu}, e_{\mu} = \partial/\partial x^{\mu}$, the components of which are k-form fields. Using shortened notation

$$T^{\vec{\mu}}_{\vec{\nu}} \equiv T^{\mu_1 \cdots \mu_p}_{\nu_1 \cdots \nu_q}, \qquad (5.230)$$

the k-form field components are

$$T^{\vec{\mu}}_{\vec{\nu}} = \left(T^{\vec{\mu}}_{\vec{\nu}}\right)_{\lambda_1 \cdots \lambda_k} dx^{\lambda_1} \wedge \cdots \wedge dx^{\lambda_k}. \qquad (5.231)$$

Having promoted the components of k-forms to be vectors, one also needs to generalize the exterior-derivative operator d. This is required, because for ordinary k-forms the exterior derivative acts on the components by partial differentiation: For example, for a one form $\eta = \eta_{\nu}dx^{\mu}$, the exterior derivative is obtained as $d\eta = \partial_{\mu}\eta_{\nu}dx^{\mu} \wedge dx^{\nu}$. For tensor-valued forms, the partial derivatives acting on the components are replaced by covariant derivatives: For example, replacing the components η_ν with vector-valued components η_ν^λ it is natural to consider an expression like

$$\partial_\mu \eta_\nu \rightarrow \nabla_\mu \eta_\nu^\lambda = \partial_\mu \eta_\nu^\lambda + \Gamma_{\mu\kappa}^\lambda \eta_\nu^\kappa. \tag{5.232}$$

The crucial reason to adopt this generalization of the operator d is that it allows one to map a tensor-valued form to another tensor-valued form. Indeed, while the partial derivative $\partial_\mu \eta^\lambda$ is not a tensor, the covariant derivative $\nabla_\mu \eta^\lambda$ *is a tensor*. This leads us to define a covariant exterior derivative. Instead of considering its action on the components, we define its action directly for forms. We present the definition first for a vector-valued form, then we generalize it to more arbitrary forms. For the definition, we need to recall the connection one-forms

$$\omega_\nu^\mu = \Gamma_{\lambda\nu}^\mu dx^\lambda. \tag{5.233}$$

The *covariant exterior derivative D* is an operator acting on tensor-valued k-forms. For a vector-valued k-form v^μ (presenting the vector index only) the covariant exterior derivative is defined as

$$Dv^\mu = dv^\mu + \omega_\nu^\mu \wedge v^\nu. \tag{5.234}$$

Example   Let v^μ be a vector-valued one-form, $v^\mu = v_\lambda^\mu dx^\lambda$. Then,

$$Dv^\mu = d(v_\lambda^\mu dx^\lambda) + \Gamma_{\kappa\nu}^\mu v_\lambda^\nu dx^\kappa \wedge dx^\lambda = (\partial_\kappa v_\lambda^\mu + \Gamma_{\kappa\nu}^\mu v_\lambda^\nu)dx^\kappa \wedge dx^\lambda. \tag{5.235}$$

While the exterior derivative squares to zero (i.e., is a nilpotent operator), $d^2 = 0$, the same is not true for the covariant exterior derivative. For example, operating on a vector-valued k-form,

$$D^2 v^\mu = D(dv^\mu + \omega_\nu^\mu \wedge v^\nu) = d(dv^\mu + \omega_\nu^\mu \wedge v^\nu) + \omega_\sigma^\mu \wedge (dv^\sigma + \omega_\nu^\sigma \wedge v^\nu)$$

$$= (d\omega_\nu^\mu) \wedge v^\nu - \cancel{\omega_\nu^\mu \wedge dv^\nu} + \cancel{\omega_\sigma^\mu \wedge dv^\sigma} + \omega_\sigma^\mu \wedge \omega_\nu^\sigma \wedge v^\nu$$

$$= (d\omega_\nu^\mu + \omega_\sigma^\mu \wedge \omega_\nu^\sigma) \wedge v^\nu \equiv R_\nu^\mu v^\nu, \tag{5.236}$$

where we introduced *curvature two-forms R_ν^μ*,

$$R_\nu^\mu = d\omega_\nu^\mu + \omega_\sigma^\mu \wedge \omega_\nu^\sigma. \tag{5.237}$$

We leave it as an exercise to show that using the transformation rule of the connection one-forms, given by Eq. (5.71), in Eq. (5.237), R_ν^μ indeed transforms as a $(1,1)$-tensor:

$$R_\nu^\mu = (J^{-1})_\sigma^\mu R_\rho^\sigma J_\nu^\rho, \tag{5.238}$$

where J is the Jacobian of the coordinate transformation, and J^{-1} is its inverse. Expanding in the coordinate basis, we extract the components of the curvature two-form,

$$R_\nu^\mu \equiv \frac{1}{2} R_{\nu\lambda\kappa}^\mu dx^\lambda \wedge dx^\kappa, \tag{5.239}$$

so the components are antisymmetric[2] under interchange of the last two indices:

$$R^{\mu}{}_{\nu\kappa\lambda} = -R^{\mu}{}_{\nu\lambda\kappa}. \tag{5.240}$$

We can identify the components as the components of the Riemann tensor by evaluating the curvature two-form on a pair of vectors U, V,

$$
\begin{aligned}
R^{\mu}{}_{\nu}(U, V) &= \left(\partial_{\lambda}\Gamma^{\mu}{}_{\kappa\lambda} + \Gamma^{\mu}{}_{\lambda\sigma}\Gamma^{\sigma}{}_{\kappa\nu}\right) dx^{\lambda}(U) \wedge dx^{\kappa}(V) \\
&= \left(\partial_{\lambda}\Gamma^{\mu}{}_{\kappa\nu} + \Gamma^{\mu}{}_{\lambda\sigma}\Gamma^{\sigma}{}_{\kappa\nu}\right) (dx^{\lambda}(U) \otimes dx^{\kappa}(V) - dx^{\kappa}(V) \otimes dx^{\lambda}(U)) \\
&= U^{\lambda}V^{\kappa}\left(\partial_{\lambda}\Gamma^{\mu}{}_{\kappa\nu} - \partial_{\kappa}\Gamma^{\mu}{}_{\lambda\nu} + \Gamma^{\mu}{}_{\lambda\sigma}\Gamma^{\sigma}{}_{\kappa\nu} - \Gamma^{\mu}{}_{\kappa\sigma}\Gamma^{\sigma}{}_{\lambda\nu}\right) \\
&= R^{\mu}{}_{\nu\lambda\kappa}U^{\lambda}V^{\kappa},
\end{aligned}
\tag{5.241}
$$

making contact with Eq. (5.217) derived previously.

We showed in Fig. 5.5 that parallel transport of a vector along a closed curve on a curved manifold like the two-dimensional sphere causes the vector to rotate with respect to its original orientation. Then the holonomy operator for such transport is nontrivial. Now we can study this in more detail, and show how the nontrivial transport properties are related to the curvature.

The difference ΔX between a vector parallel-transported along the closed curve and the original vector is given by

$$\Delta X^{\mu} = \left[P\exp\left(-\oint dt M(t)\right)\right]^{\mu}{}_{\nu} X^{\nu} - X^{\mu} \tag{5.242}$$

$$= \left[-\oint dt M^{\mu}{}_{\nu}(t) + \oint dt M^{\mu}{}_{\rho}(t)\int_{0}^{t} ds M^{\rho}{}_{\nu}(s) + \cdots\right] X^{\nu}. \tag{5.243}$$

As a closed curve, we choose an infinitesimal parallelogram, with sides a and c having antiparallel constant tangent vectors U and $-U$, and sides b and d having antiparallel constant tangent vectors V and $-V$, as sketched in Fig. 5.6.

For simplicity, we let the curve parameter t run by the same infinitesimal amount ϵ for each segment. We claim that in the limit when the infinitesimal parallelogram shrinks to zero size, the infinitesimal change in ΔX depends on the Riemann tensor, as follows.

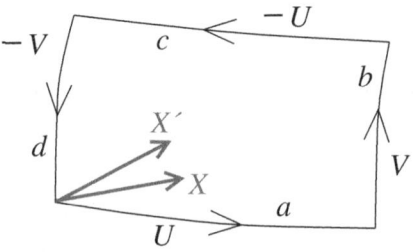

Fig. 5.6   A vector X parallel transported along a parallelogram to the vector $X' = X + \Delta X$.

---

[2] Recall that $dx^{\lambda} \wedge dx^{\kappa} = dx^{\lambda} \otimes dx^{\kappa} - dx^{\kappa} \otimes dx^{\lambda}$.

Proposition 5.1 *In the limit $\epsilon \to 0$, the variation of the vector X after its parallel transport along the curve is given by*

$$\lim_{\epsilon \to 0} \frac{(\Delta X)^\mu}{\epsilon^2} = -R^\mu_{\ \nu}(U, V)X^\nu = -R^\mu_{\ \nu\kappa\sigma}X^\nu U^\kappa V^\sigma, \qquad (5.244)$$

where $R^\mu_{\ \nu\kappa\sigma}$ are the components of the Riemann tensor.

Proof Denoting the whole closed contour by $x(t)$, let us assume that in the segment $a: t \in [0, \epsilon]$, in the segment $b: t \in [\epsilon, 2\epsilon]$, in the segment $c: t \in [2\epsilon, 3\epsilon]$, and finally along the segment $d: t \in [3\epsilon, 4\epsilon]$. The curve is thus

$$x(t) \approx \begin{cases} x(0) + tU & t \in [0, \epsilon] \\ x(\epsilon) + V(t - \epsilon) \approx x(0) + \epsilon U + V(t - \epsilon) & t \in [\epsilon, 2\epsilon] \\ x(0) + \epsilon(U + V) - U(t - 2\epsilon) & t \in [2\epsilon, 3\epsilon] \\ x(0) + \epsilon V - V(t - 3\epsilon) & t \in [3\epsilon, 4\epsilon] \end{cases}. \qquad (5.245)$$

We compute first the holonomy operator for the transport along the segment a. We need first

$$M^\mu_{\ \nu}(t) = \omega^\mu_{\ \nu}(U) = \Gamma^\mu_{\kappa\nu}(t)U^\kappa. \qquad (5.246)$$

Note that $\Gamma^\mu_{\kappa\nu}$ is evaluated at points along the segment a, so it depends on the curve parameter t. Through a Taylor expansion in t, one can write

$$\Gamma^\mu_{\kappa\nu}(t) = \Gamma^\mu_{\kappa\nu}(0) + tU^\lambda\partial_\lambda\Gamma^\mu_{\kappa\nu}(0) + \cdots. \qquad (5.247)$$

The exponent of the holonomy operator along a is a matrix, with components

$$-\int_0^\epsilon dt' M^\mu_{\ \nu}(t') = -\epsilon\Gamma^\mu_{\kappa\nu}(0)U^\kappa - \frac{\epsilon^2}{2}U^\lambda\partial_\lambda\Gamma^\mu_{\kappa\nu}(0)U^\kappa + \mathcal{O}(\epsilon^3), \qquad (5.248)$$

where $(\Gamma^\mu_{\kappa\nu}U^\kappa)$ is a matrix with entries labeled by the μ and ν indices. Similarly, for the segment b one obtains

$$M^\mu_{\ \nu}(t) = \omega^\mu_{\ \nu}(V) = \Gamma^\mu_{\kappa\nu}(t)V^\kappa. \qquad (5.249)$$

In expanding $\Gamma^\mu_{\kappa\nu}(t)$ along the segment b one has to take into account that this segment starts from the point where the previous segment ends, thus

$$\begin{aligned} \Gamma^\mu_{\kappa\nu}(t) &= \Gamma^\mu_{\kappa\nu}(\epsilon) + (t - \epsilon)V^\lambda\partial_\lambda\Gamma^\mu_{\kappa\nu}(0) + \cdots \\ &= \Gamma^\mu_{\kappa\nu}(0) + \epsilon U^\lambda\partial_\lambda\Gamma^\mu_{\kappa\nu}(0) + (t - \epsilon)V^\lambda\partial_\lambda\Gamma^\mu_{\kappa\nu}(0) + \cdots, \end{aligned} \qquad (5.250)$$

and we obtain the exponent

$$-\int_\epsilon^{2\epsilon} dt' M^\mu_{\ \nu}(t') = -\epsilon\left(\Gamma^\mu_{\kappa\nu}(0) + \epsilon U^\lambda\partial_\lambda\Gamma^\mu_{\kappa\nu}(0)\right)V^\kappa - \frac{\epsilon^2}{2}V^\lambda\partial_\lambda\Gamma^\mu_{\kappa\nu}(0)V^\kappa + \mathcal{O}(\epsilon^3). \qquad (5.251)$$

For the remaining segments c and d, the calculations are similar. Along c, one obtains

$$-\int_0^\epsilon dt' M^\mu_{\ \nu}(t') = \epsilon\left(\Gamma^\mu_{\kappa\nu}(0) + \epsilon(U + V)^\lambda\partial_\lambda\Gamma^\mu_{\kappa\nu}(0)\right)U^\kappa$$
$$- \frac{\epsilon^2}{2}U^\lambda\partial_\lambda\Gamma^\mu_{\kappa\nu}(0)U^\kappa + \mathcal{O}(\epsilon^3), \qquad (5.252)$$

whereas along d one finds

$$-\int_0^\epsilon dt' M^\mu_{\ \nu}(t') = \epsilon \left(\Gamma^\mu_{\kappa\nu}(0) + \epsilon V^\lambda \partial_\lambda \Gamma^\mu_{\kappa\nu}(0) \right) V^\kappa - \frac{\epsilon^2}{2} V^\lambda \partial_\lambda \Gamma^\mu_{\kappa\nu}(0) V^\kappa + \mathcal{O}(\epsilon^3).$$
$$(5.253)$$

Summing up the contributions gives the integral around the contour $abcd$. Terms that are linear in ϵ cancel out, leaving

$$-\oint dt' M^\mu_{\ \nu}(t') = -\int_0^{4\epsilon} dt' M^\mu_{\ \nu}(t')$$
$$= -\frac{\epsilon^2}{2} U^\lambda \partial_\lambda \Gamma^\mu_{\kappa\nu}(0) U^\kappa - \epsilon^2 U^\lambda \partial_\lambda \Gamma^\mu_{\kappa\nu}(0) V^\kappa - \frac{\epsilon^2}{2} V^\lambda \partial_\lambda \Gamma^\mu_{\kappa\nu}(0) V^\kappa$$
$$+ \epsilon^2 (U + V)^\lambda \partial_\lambda \Gamma^\mu_{\kappa\nu}(0) U^\kappa - \frac{\epsilon^2}{2} U^\lambda \partial_\lambda \Gamma^\mu_{\kappa\nu}(0) U^\kappa$$
$$+ \epsilon^2 V^\lambda \partial_\lambda \Gamma^\mu_{\kappa\nu}(0) V^\kappa - \frac{\epsilon^2}{2} V^\lambda \partial_\lambda \Gamma^\mu_{\kappa\nu}(0) V^\kappa + \mathcal{O}(\epsilon^3)$$
$$= \epsilon^2 \left(-U^\lambda V^\kappa + V^\lambda U^\kappa \right) \partial_\lambda \Gamma^\mu_{\kappa\nu}(0) + \mathcal{O}(\epsilon^3).$$
$$(5.254)$$

At the second order, in (5.242) we also need to split the integration to segments:

$$\oint dt M^\mu_{\ \rho}(t) \int_0^t ds M^\rho_{\ \nu}(s) = \left(\int_0^\epsilon dt + \int_\epsilon^{2\epsilon} dt + \cdots \int_{3\epsilon}^{4\epsilon} \right) M^\mu_{\ \rho}(t) \int_0^t ds M^\rho_{\ \nu}(s),$$
$$(5.255)$$

with the splitting in the s-integral depending on the value of t. We show explicitly some of the initial steps in the derivation, from which one can notice a pattern. Denoting the first integral as

$$\int_0^\epsilon dt M^\mu_{\ \rho}(t) \int_0^t ds M^\rho_{\ \nu}(s) \equiv I_1,$$
$$(5.256)$$

one has

$$I_1 = \int_0^\epsilon dt\, U^\kappa (\Gamma^\mu_{\kappa\rho}(0) + t U^\lambda \partial_\lambda \Gamma^\mu_{\kappa\rho}(0) + \cdots) \int_0^t ds\, U^\sigma (\Gamma^\rho_{\sigma\nu}(0) + s U^\alpha \partial_\alpha \Gamma^\rho_{\sigma\nu}(0) + \cdots)$$
$$= \int_0^\epsilon dt\, U^\kappa \left[\Gamma^\mu_{\kappa\rho}(0) + t U^\lambda \partial_\lambda \Gamma^\mu_{\kappa\rho}(0) + \cdots \right] U^\sigma \left[t\Gamma^\rho_{\sigma\nu}(0) + \frac{1}{2} t^2 U^\alpha \partial_\alpha \Gamma^\rho_{\sigma\nu}(0) + \cdots \right]$$
$$= \frac{1}{2} \epsilon^2 U^\kappa U^\sigma \Gamma^\mu_{\kappa\rho}(0) \Gamma^\rho_{\sigma\nu}(0) + \mathcal{O}(\epsilon^3)$$
$$(5.257)$$

and one notices that terms involving partial derivatives of connection coefficients $\partial_\alpha \Gamma$ contribute at higher orders in ϵ. This also holds for the contributions from the other terms: Defining

$$\int_\epsilon^{2\epsilon} dt M^\mu_{\ \rho}(t) \int_0^t ds M^\rho_{\ \nu}(s) \equiv I_2,$$
$$(5.258)$$

one obtains

$$I_2 = \int_{\epsilon}^{2\epsilon} dt \, V^\kappa (\Gamma^\mu_{\kappa\rho}(0) + t U^\lambda \partial_\lambda \Gamma^\mu_{\kappa\rho}(0) + \cdots)$$

$$\left[\int_0^\epsilon ds \, U^\sigma (\Gamma^\rho_{\sigma\nu}(0) + s U^\alpha \partial_\alpha \Gamma^\rho_{\sigma\nu}(0) + \cdots) \right.$$

$$\left. + \int_\epsilon^t ds \, V^\sigma (\Gamma^\rho_{\sigma\nu}(0) + (s - \epsilon) U^\alpha \partial_\alpha \Gamma^\rho_{\sigma\nu}(0) + \cdots) \right]$$

$$= \epsilon^2 U^\kappa U^\sigma \Gamma^\mu_{\kappa\rho}(0) \Gamma^\rho_{\sigma\nu}(0) + \cdots + V^\kappa U^\sigma \Gamma^\mu_{\kappa\rho}(0) \Gamma^\rho_{\sigma\nu}(0) \int_\epsilon^{2\epsilon} dt \, t + \mathcal{O}(\epsilon^3)$$

$$= \epsilon^2 V^\kappa \left(U^\sigma + \frac{1}{2} V^\sigma \right) \Gamma^\mu_{\kappa\rho}(0) \Gamma^\rho_{\sigma\nu}(0) + \mathcal{O}(\epsilon^3). \tag{5.259}$$

Proceeding in a similar manner, when the first integral is taken along the segment c, one obtains

$$\int_{2\epsilon}^{3\epsilon} dt M^\mu{}_\rho(t) \int_0^t ds M^\rho{}_\nu(s) = \epsilon^2 (-U^\kappa) \left(U^\sigma + V^\sigma - \frac{1}{2} U^\sigma \right) \Gamma^\mu_{\kappa\rho}(0) \Gamma^\rho_{\sigma\nu}(0) + \mathcal{O}(\epsilon^3), \tag{5.260}$$

and, when the first integral is taken along the segment d,

$$\int_{3\epsilon}^{4\epsilon} dt M^\mu{}_\rho(t) \int_0^t ds M^\rho{}_\nu(s) = \epsilon^2 (-V^\kappa) \left(U^\sigma + V^\sigma - U^\sigma - \frac{1}{2} V^\sigma \right) \Gamma^\mu_{\kappa\rho}(0) \Gamma^\rho_{\sigma\nu}(0) + \mathcal{O}(\epsilon^3). \tag{5.261}$$

Combining all the contributions together yields

$$\oint dt M^\mu{}_\rho(t) \int_0^t ds M^\rho{}_\nu(s) = -\epsilon^2 (U^\kappa V^\sigma - V^\kappa U^\sigma) \Gamma^\mu_{\kappa\rho}(0) \Gamma^\rho_{\sigma\nu}(0) + \mathcal{O}(\epsilon^3), \tag{5.262}$$

from which follows

$$\frac{\Delta X^\mu}{\epsilon^2} = (U^\kappa V^\sigma - V^\kappa U^\sigma)(\partial_\sigma \Gamma^\mu_{\kappa\nu} - \Gamma^\mu_{\kappa\rho} \Gamma^\rho_{\lambda\nu}) X^\nu + \mathcal{O}(\epsilon)$$

$$= U^\kappa V^\sigma (\partial_\sigma \Gamma^\mu_{\kappa\nu} - \partial_\kappa \Gamma^\mu_{\sigma\nu} + \Gamma^\mu_{\kappa\rho} \Gamma^\rho_{\lambda\nu} - \Gamma^\mu_{\sigma\rho} \Gamma^\rho_{\kappa\nu}) X^\nu + \mathcal{O}(\epsilon)$$

$$= -R^\mu{}_{\nu\kappa\sigma} U^\kappa V^\sigma + \mathcal{O}(\epsilon). \tag{5.263}$$

$\square$

One can also define a vector-valued *torsion two-form*. In the coordinate basis, the torsion two-form can be defined as

$$T^\mu \equiv D dx^\mu = d(dx^\mu) + \omega^\mu{}_\nu \wedge dx^\nu = \omega^\mu{}_\nu \wedge dx^\nu. \tag{5.264}$$

Extracting the components

$$T^\mu \equiv T^\mu{}_{\kappa\lambda} dx^\kappa \wedge dx^\lambda \tag{5.265}$$

and substituting $\omega^\mu{}_\nu = \Gamma^\mu_{\kappa\nu} dx^\kappa$ into Eq. (5.264), one recovers the familiar formula:

$$T^\mu{}_{\kappa\nu} = \frac{1}{2} \left(\Gamma^\mu_{\kappa\nu} - \Gamma^\mu_{\nu\kappa} \right). \tag{5.266}$$

The torsion two-form is often defined in the so-called non-coordinate basis, which we discuss in Section 5.11.

5.11 Non-coordinate Basis

So far we have been using coordinate bases $e_\mu = \frac{\partial}{\partial x^\mu}$ for the vector fields and dual coordinate bases dx^μ for the one-forms. There is another important choice, called the *non-coordinate basis*. Given a point $p \in M$, let $\{e_\mu\}$ be a coordinate basis of T_pM. With a metric g, the coordinate basis is in general not orthonormal, since

$$g(e_\mu, e_\nu) = g_{\mu\nu}. \tag{5.267}$$

We form a new set of linearly independent vectors, then orthonormalize this set to define a basis $\{\hat{e}_a\}$. Orthonormality means that

$$g(\hat{e}_a, \hat{e}_b) = \bar{\delta}_{ab}, \tag{5.268}$$

where for a Riemannian manifold $\bar{\delta}$ is the Cartesian metric $\delta = \mathrm{diag}(1, \ldots, 1)$, while for a pseudo-Riemannian manifold of signature (n_-, n_+) $\bar{\delta}$ is the metric

$$\eta = \mathrm{diag}(\underbrace{-1, \ldots, -1}_{n_- \text{ terms}}, \underbrace{1, \ldots, 1}_{n_+ \text{ terms}}). \tag{5.269}$$

The new orthonormal vectors can also be expanded in the original coordinate basis,

$$\hat{e}_a = e_a^\mu e_\mu = e_a^\mu \frac{\partial}{\partial x^\mu} \tag{5.270}$$

with coefficients e_a^μ. Substituting into Eq. (5.268) gives the relation

$$g_{\mu\nu} e_a^\mu e_b^\nu = \bar{\delta}_{ab}. \tag{5.271}$$

The orthonormal basis $\{\hat{e}_a\}$ is called the *non-coordinate basis*. Other names are *vielbein* (from the German word for "multi-legged," in the case of a manifold of arbitrary dimension), *zweibein* (from the German for "biped," when $\dim M = 2$), or *vierbein* (from the German for "quadruped," when $\dim M = 4$). In four dimensions, in the theory of general relativity, the name *tetrad* is also common.

The dual basis of the non-coordinate basis is denoted as $\{\hat{e}^{*a}\}$ and satisfies

$$\hat{e}^{*a}(\hat{e}_b) = \bar{\delta}_b^a = \delta_b^a. \tag{5.272}$$

We will use a shorter notation $\{\hat{e}^a\}$ for the dual non-coordinate basis dropping the asterisk, as the Latin upper index is enough to identify this as a dual basis for one-forms. Similarly expanding in the dual coordinate basis,

$$\hat{e}^a = e_\mu^a e^{*\mu} = e_\mu^a dx^\mu. \tag{5.273}$$

Then, Eq. (5.272) implies that the coefficients e_μ^a and e_a^μ satisfy the relation

$$e_\mu^a e_b^\mu = \delta_b^a. \tag{5.274}$$

This equation can be written in matrix form: If we define a matrix $E = (e_a^\mu)$, then its inverse is $E^{-1} = (e_\mu^a)$. A convenient feature of the non-coordinate basis is that the metric g takes a simple form

$$g = g_{\mu\nu} dx^\mu \otimes dx^\nu = \bar{\delta}_{ab}\, \hat{e}^a \otimes \hat{e}^b \tag{5.275}$$

and it is possible to verify that, due to Eq. (5.272), this reproduces Eq. (5.268).

Working with the non-coordinate basis and its dual, the operation of raising and lowering the indices of tensor components involves the components $\bar{\delta}_{ab}$ and its inverse $\bar{\delta}^{ab}$, for example

$$T^{ab}_{cd} = \bar{\delta}^{ae}\bar{\delta}_{cf}T_e^{bf}_{d}.$$ (5.276)

In particular, on a Riemannian manifold this is particularly simple, since $\bar{\delta}_{ab}$ are just the usual Kronecker delta δ_{ab}; therefore indices can be raised and lowered trivially:

$$T^{ab}_{cd} = T_a^{bc}_{d} \qquad \text{on a Riemannian manifold.}$$ (5.277)

A non-coordinate basis is not unique. It is always possible to define a new non-coordinate basis $\hat{e}'_a = O_a^{b}\hat{e}_b$, provided that O preserves Eq. (5.268). For a Riemann manifold of dimension dim $M = n$, this means

$$O_a^{c}O_b^{d}\delta_{cd} = O_a^{c}O_b^{c} = \delta_{ab},$$ (5.278)

which can be written in matrix form as

$$OO^T = \mathbb{1},$$ (5.279)

meaning that O is a orthogonal matrix: $O \in O(n)$. Likewise, for a pseudo-Riemannian manifold of type $(1, n - 1)$, we would obtain the requirement

$$O\eta O^T = \eta,$$ (5.280)

which means that O is a Lorentz transformation.

Although we defined a non-coordinate frame (and its dual) at a point p, often they can be extended into larger coordinate neighborhoods of M. We shall shortly see an example. Non-coordinate frames have two additional important characteristic features to remember.

(i) In general, the basis vectors are not necessarily commuting with each other, i.e., their commutators are non-vanishing and have expansions

$$[\hat{e}_a, \hat{e}_b] = C^c_{ab}\hat{e}_c,$$ (5.281)

where at least some of the coefficients C^c_{ab} are not identically vanishing.

(ii) The dual basis vectors are in general not exact, so in general one cannot find functions y^a on M such that one could write $\hat{e}^a = dy^a$ for all $\hat{e}^a$. This explains why this type of basis is called a "non-coordinate" basis: There are no coordinates y^a in which one could express all of the basis vectors and their duals just like in a coordinate basis.

Let us now discuss an example, to see how all this works in practice.

Example   Consider the usual metric of a two-sphere S^2 of radius r,

$$g = r^2(d\theta \otimes d\theta + \sin^2 \theta d\phi \otimes d\phi).$$ (5.282)

Defining

$$\hat{e}^1 = rd\theta, \quad \hat{e}^2 = r\sin\theta d\phi,$$ (5.283)

one can rewrite the metric as

$$g = \hat{e}^1 \otimes \hat{e}^2 + \hat{e}^2 \otimes \hat{e}^2.$$ (5.284)

In this case the nonzero coefficients e_μ^a appearing in Eq. (5.273) are

$$e_\theta^1 = r, \quad e_\phi^2 = r\sin\theta. \qquad (5.285)$$

Note that $\hat{e}^2$ is neither a closed nor an exact one-form. The corresponding non-coordinate basis vector fields are

$$\hat{e}_1 = \frac{1}{r}\frac{\partial}{\partial\theta}, \quad \hat{e}_2 = \frac{1}{r\sin\theta}\frac{\partial}{\partial\phi} \qquad (5.286)$$

with nonzero coefficients e_a^μ

$$e_1^\theta = \frac{1}{r}, \quad e_2^\phi = \frac{1}{r\sin\theta}. \qquad (5.287)$$

Finally, there is one nonvanishing commutator,

$$[\hat{e}_1, \hat{e}_2] = -\frac{\cos\theta}{r\sin^2\theta}\frac{\partial}{\partial\phi} = -\cot\theta\,\hat{e}_2. \qquad (5.288)$$

It is often customary to define the connection one-forms, the torsion two-form, and the curvature two-form in a non-coordinate basis. First, the connection coefficients are given as before,

$$\nabla_{\hat{e}_a}\hat{e}_b = \Gamma^c_{\ ab}\hat{e}_c, \qquad (5.289)$$

so the connection one-forms are

$$\omega^a_{\ b} = \Gamma^a_{\ cb}\hat{e}^c. \qquad (5.290)$$

For a metric connection, it is possible to prove the following antisymmetry relation

$$\omega_{ab} = -\omega_{ba}, \qquad (5.291)$$

where indices are lowered as in Eq. (5.276):

$$\omega_{ab} = \bar{\delta}_{ac}\omega^c_{\ b}. \qquad (5.292)$$

Thus, on a Riemannian manifold one can simply write

$$\omega^a_{\ b} = -\omega^b_{\ a}. \qquad (5.293)$$

To show Eq. (5.291), we first note that

$$\nabla\hat{e}_b = \omega^c_{\ b}\hat{e}_c, \qquad \nabla\hat{e}^c = -\omega^c_{\ b}\hat{e}^b. \qquad (5.294)$$

Then, if the connection is metric-compatible,

$$\begin{aligned}
0 = \nabla g &= \nabla(\bar{\delta}_{ab}\hat{e}^a \otimes \hat{e}^b) = \bar{\delta}_{ab}((\nabla\hat{e}^a) \otimes \hat{e}^b + \hat{e}^a \otimes (\nabla\hat{e}^b)) \\
&= \bar{\delta}_{ab}(-\omega^a_{\ c}\hat{e}^c \otimes \hat{e}^b - \omega^b_{\ c}\hat{e}^a \otimes \hat{e}^c) \\
&= -(\omega_{ab} + \omega_{ba})\hat{e}^a \otimes \hat{e}^b,
\end{aligned} \qquad (5.295)$$

which implies Eq. (5.291).

We can now define the torsion two-forms and the curvature two-forms in the non-coordinate basis as:

$$T^a \equiv D\hat{e}^a = d\hat{e}^a + \omega^a_{\ b} \wedge \hat{e}^b \qquad (5.296)$$

$$R^a_{\ b} = d\omega^a_{\ b} + \omega^a_{\ c} \wedge \omega^c_{\ b}. \qquad (5.297)$$

Note that, since $\hat{e}^a$ is not in general closed, the term $d\hat{e}^a$ in Eq. (5.296) can be nonvanishing. Equation (5.296) is known as *Cartan's first structure equation*, while Eq. (5.297) is *Cartan's second structure equation*.

The non-coordinate basis and the structure equations give an alternative way to compute the components of the Riemann tensor for a Levi-Civita connection, the metric connection with no torsion, using the antisymmetry property encoded in Eq. (5.291) and the Cartan structure equations.

Example In the previous example, for a two-sphere of radius r_0 with the usual metric, we obtained

$$\hat{e}^1 = r_0 d\theta, \quad \hat{e}^2 = r_0 \sin\theta d\phi. \tag{5.298}$$

Requiring the torsion to vanish, Cartan's first structure equation (5.296) gives

$$0 = T^1 = d(r_0 d\theta) + \omega^1{}_1 \wedge (r_0 d\theta) + \omega^1{}_2 \wedge (r_0 \sin\theta d\phi), \tag{5.299}$$
$$0 = T^2 = d(r_0 \sin\theta d\phi) + \omega^2{}_1 \wedge (r_0 d\theta) + \omega^2{}_2 \wedge (r_0 \sin\theta d\phi). \tag{5.300}$$

On the other hand, the antisymmetry property (5.291) implies

$$\omega^1{}_1 = -\omega^1{}_1 = 0, \tag{5.301}$$

and, similarly, $\omega^2{}_2 = 0$. Then, as r_0 is a nonzero constant and $d(r_0 d\theta) = 0$, Eq. (5.299) and Eq. (5.300) respectively reduce to

$$\omega^1{}_2 \wedge \sin\theta d\phi = 0, \tag{5.302}$$
$$d(\sin\theta d\phi) + \omega^2{}_1 \wedge d\theta = \cos\theta d\theta \wedge d\phi + \omega^2{}_1 \wedge d\theta = 0, \tag{5.303}$$

where we divided out the constant factors r_0. From Eq. (5.303), using the antisymmetry $\omega^1{}_2 = -\omega^2{}_1$, one obtains

$$\omega^2{}_1 = \cos\theta d\phi = -\omega^1{}_2. \tag{5.304}$$

Next, we derive the curvature two-forms using the second structure equation (5.297):

$$\begin{aligned}
R^1{}_1 &= d\omega^1{}_1 + \omega^1{}_1 \wedge \omega^1{}_1 + \omega^1{}_2 \wedge \omega^2{}_1 = 0, \\
R^1{}_2 &= d\omega^1{}_2 + \omega^1{}_1 \wedge \omega^1{}_2 + \omega^1{}_2 \wedge \omega^2{}_2 = d(-\cos\theta d\phi) = \sin\theta d\theta \wedge d\phi, \\
R^2{}_1 &= d\omega^2{}_1 + \omega^2{}_1 \wedge \omega^1{}_1 + \omega^2{}_2 \wedge \omega^2{}_1 = d(\cos\theta d\phi) = -\sin\theta d\theta \wedge d\phi, \\
R^2{}_2 &= d\omega^2{}_2 + \omega^2{}_1 \wedge \omega^1{}_2 + \omega^2{}_2 \wedge \omega^2{}_2 = 0.
\end{aligned} \tag{5.305}$$

We then read off the coefficients of the Riemann tensor expanding the curvature two-forms in the non-coordinate basis:

$$R^a{}_b = \frac{1}{2} R^a{}_{bcd} \hat{e}^c \wedge \hat{e}^d. \tag{5.306}$$

One can see that the following components vanish:

$$R^1{}_{1cd} = R^2{}_{2cd} = 0. \tag{5.307}$$

Expanding $R^1{}_2$,

$$R^1{}_2 = \sin\theta \wedge d\phi = \frac{1}{r_0^2}(r_0 d\theta) \wedge (r_0 \sin\theta d\phi) = \frac{1}{2}\frac{1}{r_0^2}\left[\hat{e}^1 \wedge \hat{e}^2 - \hat{e}^2 \wedge \hat{e}^1\right], \tag{5.308}$$

we read off

$$R^1{}_{212} = \frac{1}{r_0^2} = -R^1{}_{221}.$$ (5.309)

Similarly, expanding $R^2{}_1$ we obtain

$$R^2{}_{112} = -\frac{1}{r_0^2} = -R^2{}_{121}.$$ (5.310)

For the Ricci tensor $R_{ab} = R^c{}_{acb}$ we get

$$R_{11} = R^1{}_{111} + R^2{}_{121} = \frac{1}{r_0^2},$$

$$R_{12} = R^1{}_{112} + R^2{}_{122} = 0,$$

$$R_{21} = R^1{}_{211} + R^2{}_{221} = 0,$$

$$R_{22} = R^1{}_{212} + R^2{}_{222} = \frac{1}{r_0^2}.$$ (5.311)

Note that $R_{ab} = R_{ba}$.

Finally, the scalar curvature is

$$R = R^1{}_1 + R^2{}_2 = R_{11} + R_{22} = \frac{2}{r_0^2}.$$ (5.312)

5.12 Hypersurfaces and Extrinsic Curvature

The curvature that we discussed in Section 5.10 is an *intrinsic property* of the (pseudo-)Riemannian manifold (M, g). The intrinsic curvature differs from the *extrinsic curvature* that can arise from embedding a manifold as a submanifold into a higher-dimensional manifold. For example, the Euclidean plane $(\mathbb{R}^2, \delta)$ has no intrinsic curvature, but we can map it to a curved surface in the three-dimensional space $(\mathbb{R}^3, \delta)$. Consider the following example which shows that second derivatives give information about the curvature.

Example Equation (5.313)

$$f(x, y, z)) \equiv z - \frac{1}{2}(x^2 + y^2) = 0$$

$$\Leftrightarrow z = z(x, y) = \frac{1}{2}(x^2 + y^2)$$ (5.313)

defines a two-dimensional paraboloid surface in $\mathbb{R}^3$, which we can think of as the image of $\mathbb{R}^2$. At the origin, the first derivatives $\partial_i z$ vanish, as this is a local minimum, but the curvature is revealed by the second derivatives, the Hessian matrix

$$H = \begin{pmatrix} H_{xx} & H_{xy} \\ H_{yx} & H_{yy} \end{pmatrix} = \begin{pmatrix} \partial_x^2 z & \partial_x \partial_y z \\ \partial_y \partial_x z & \partial_y^2 z \end{pmatrix}$$

$$= \begin{pmatrix} \partial_x^2 z & \partial_x \partial_y z \\ \partial_y \partial_x z & \partial_y^2 z \end{pmatrix} = \begin{pmatrix} 1 & 0 \\ 0 & 1 \end{pmatrix},$$ (5.314)

which shows that the surface is convex at every point.

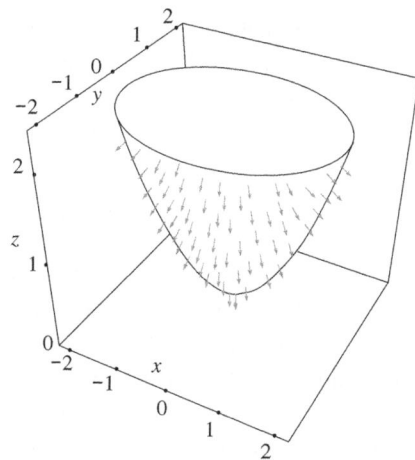

Fig. 5.7   A visualization the surface $\frac{1}{2}(x^2 + y^2) - z = 0$ and its (outward directed) unit normal vector field.

Recall also that the normal vector $\vec{n}$ of a surface defined by $f = f(x, y, z) =$ constant (an equipotential surface) is obtained by the gradient $\vec{n} = \nabla f$ or, in component form, $n_i = \partial_i f$. To obtain a unit normal vector field, we rescale $\vec{n}$ by $[(\nabla f)^2]^{-1/2}$. Finally, we can reverse the orientation multiplying by -1, to make the unit normal vector field point outwards. For the above example, we obtain $\vec{n} = (1 + x^2 + y^2)^{-1/2}(x, y, -1)$; the surface and the normal vector field are depicted in Fig. 5.7.

5.12.1 Hypersurfaces and the First Fundamental Form

After these preliminary observations, we now move on to consider surfaces in a (pseudo-)Riemannian manifold (M, g). Let Σ be a submanifold of M, $\dim M = m$, $\dim \Sigma = n$ with $0 < n < m$, $k \equiv m - n$. We say that the submanifold is of *codimension* k. If Σ is of codimension one ($\dim \Sigma = \dim M - 1$) we say that Σ is a *hypersurface* in M. As in the previous example, one way to construct a hypersurface is to choose a non-injective smooth function f in M and set it to be a constant f_0, then

$$\Sigma = \{x \in M | f(x) = f_0\} \tag{5.315}$$

is a hypersurface in M. We restrict our attention to hypersurfaces defined in this way. A vector field $V_\perp$ normal to the surface is obtained as in the above, but now we must remember to use the inverse metric to raise indices; the components of the normal vector field are given by

$$V_\perp^\mu(x) = g^{\mu\nu}(x)\frac{\partial f}{\partial x^\nu}. \tag{5.316}$$

On a Lorentzian manifold (a spacetime) the squared length of the normal vector field at point $x \in \Sigma$ has three alternatives

$$1.\ (V_\perp)^2(x) = g_{\mu\nu}(x)V_\perp^\mu(x)V_\perp^\nu(x) < 0, \quad V_\perp(x) \text{ is timelike} \tag{5.317}$$

$$2.\ (V_\perp)^2(x) = g_{\mu\nu}(x)V_\perp^\mu(x)V_\perp^\nu(x) = 0, \quad V_\perp(x) \text{ is lightlike} \tag{5.318}$$

$$3.\ (V_\perp)^2(x) = g_{\mu\nu}(x)V_\perp^\mu(x)V_\perp^\nu > 0, \quad V_\perp(x) \text{ is spacelike;} \tag{5.319}$$

consequently we say that Σ is respectively timelike, lightlike, or spacelike at x. If this property is unchanged $\forall x \in \Sigma$, we say that the hypersurface Σ is timelike, lightlike, or spacelike. A lightlike hypersurface is alternatively called *null*.

We can rescale the length of the normal vector field at any point x by multiplying the components by a nonzero function $h(x)$:

$$V_\perp^\mu \mapsto h(x) V_\perp^\mu(x). \tag{5.320}$$

We assume here that $h(x)$ is smooth and nonvanishing. Since it is also continuous, $h(x) > 0$ or $h(x) < 0$ everywhere. Thus the rescaling has no effect on the classification into timelike, etc., surfaces. However, the choice of the sign of $h(x)$ gives two opposite directions to the normal vector. In the context of a hypersurface in a Riemannian manifold, the normal vector field is either *inward directed* or *outward directed*, depending on the orientation and shape of the surface. For a spacelike surface in a Lorentzian manifold, such as the constant time slice $t = t_0$ in Minkowski space, the normal vector field is either *future-directed* or *past-directed* (for example, $\partial/\partial t$ or $-\partial/\partial t$). For other than null hypersurfaces, we can rescale and define a unit normal vector field:

$$n = \pm \frac{V_\perp}{\sqrt{|V_\perp^2|}}. \tag{5.321}$$

Given a hypersurface Σ defined by $f = f_0$, we find a normal vector field by Eq. (5.316). What about the converse statement: Suppose that we are given a vector field $v = v^\mu \partial_\mu$, under what conditions does it define a hypersurface for which it is normal? We quote a simplified version of a theorem by Ferdinand Georg Frobenius that provides an answer. Suppose that we find a candidate orthonormal hypersurface Σ by finding that its tangent vectors X are orthogonal to v, $v_\mu X^\mu = 0$ at every point on the surface. Note that v_μ can be viewed as components of a one-form field corresponding to the vector field (lowering the index).

Theorem 5.2 (Simplified version of Frobenius's theorem)   *The vector field $v^\mu \partial_\mu$ will define a hypersurface Σ orthogonal to it if and only if the corresponding one-form satisfies*

$$\nabla_{[\mu} v_{\nu]} X^\mu Y^\nu = 0 \tag{5.322}$$

for every pair of vectors X, Y normal to v ($v_\mu X^\mu = v_\mu Y^\mu = 0$) at every point on the surface.

Proof   Omitted.   □

There is one criterion that ensures that the vector field v satisfies the above condition (5.322). If the corresponding one-form v_μ satisfies

$$v_{[\mu} \nabla_\sigma v_{\nu]} = 0, \tag{5.323}$$

then we find

$$0 = v_{[\mu} \nabla_\sigma v_{\nu]} X^\mu Y^\sigma = \frac{1}{3} (\overbrace{X^\mu v_\mu}^{=0} \nabla_{[\sigma} v_{\nu]} Y^\sigma + \overbrace{Y^\sigma v_\sigma}^{=0} \nabla_{[\nu} v_{\mu]} X^\mu) + \frac{1}{3} v_\nu \nabla_{[\mu} v_{\sigma]} X^\mu Y^\sigma$$

$$= \frac{1}{3} v_\nu \nabla_{[\mu} v_{\sigma]} X^\mu Y^\sigma \quad \Rightarrow \quad \nabla_{[\mu} v_{\sigma]} X^\mu Y^\sigma = 0, \tag{5.324}$$

so that (5.322) is satisfied. We also note that if the vector field v^μ is derived from smooth functions h and f by $v^\mu = hg^{\mu\nu}\partial_\nu f$, then (see Problem 5.4) it is possible to prove that the corresponding one-form satisfies the condition (5.323), so by Eq. (5.324) it also satisfies the Frobenius condition (5.322) $\nabla_{[\mu}v_{\nu]}X^\mu Y^\nu = 0$ for all pairs of vectors X and Y orthogonal to v. Not only is this a nice consistency check, we also need this result in the following Section 5.12.2. Before moving on, we note that the condition expressed by Eq. (5.322), or, alternatively, Eq. (5.323), is referred to as the *hypersurface orthogonality condition*. One can also show that Eq. (5.323) implies that locally in a neighborhood of Σ there exist functions h, f such that $v_\mu = h\partial_\mu f$, so the hypersurface can be defined by the condition $f = $ constant.

Since the $(d-1)$-dimensional surface Σ is a submanifold, it can be parameterized by an intrinsic local coordinate system. Let us denote the local coordinates on the surface Σ by $(y^i) = (y^1, \ldots, y^{d-1})$, and the local coordinates on the embedding manifold M by $(x^\mu) = x^1, \ldots, x^d$. Thus the embedding can specified by the function

$$F : \Sigma \to M, \quad F(y) = (F^\mu(y)) \equiv (x^\mu(y)). \tag{5.325}$$

Example The paraboloid of the earlier example can be parameterized by y^1, y^2 and we write the embedding as

$$F : \mathbb{R}^2 \equiv \Sigma \mapsto \mathbb{R}^3,$$

$$y \mapsto F(y) = (x^1(y), x^2(y), x^3(y)) = \left(y^1, y^2, \frac{1}{2}[(y^1)^2 + (y^2)^2]\right). \tag{5.326}$$

We can then pull back tensors T from M to tensors $\hat{T}$ on Σ by $\hat{T} = F^*T$. Likewise, we can push vectors on Σ to vectors on M by $\hat{v} \mapsto v = F_*\hat{v}$. As an example of a pullback map, we can obtain the *induced metric* on Σ as a pullback of the metric g on M. We prefer to follow a frequent convention and denote the induced metric as h rather than $\hat{g}$:

$$h = F^*g, \quad \text{with components } h_{ij}(y) = \frac{\partial x^\mu}{\partial y^i}\frac{\partial x^\nu}{\partial y^j}g_{\mu\nu}(x(y)). \tag{5.327}$$

Example We can derive the induced metric on the paraboloid surface more simply by moving to cylindrical coordinates where the Euclidean metric on $\mathbb{R}^3$ reads

$$\delta = dr^2 + r^2 d\phi^2 + dz^2, \tag{5.328}$$

then use $(y^1, y^2) \equiv (r, \phi)$ as the local coordinates on Σ so the embedding is $f(r, \phi) = (r, \phi, z(r, \phi)) = (r, \phi, z = \frac{1}{2}r^2)$, and derive the induced metric by

$$h = F^*\delta = dr^2 + r^2 d\phi^2 + \left[d(\frac{1}{2}r^2)\right]^2 = dr^2 + r^2 d\phi^2 + r^2 dr^2$$

$$= (1 + r^2)dr^2 + r^2 d\phi^2, \tag{5.329}$$

so the components of the induced metric are $h_{rr} = 1 + r^2$ and $h_{\phi\phi} = r^2$.

Vector fields on Σ are naturally tangent to Σ (they are elements of the tangent bundle $T\Sigma$). We can expand a tangent vector field X in local coordinates y as

$$X = X(y)^i \frac{\partial}{\partial y^i}. \tag{5.330}$$

Note that the push vector field F_*X is still tangential to and supported on Σ, the only difference is that we are expanding the vector field in the coordinates x of M by

$$F_*X = X^i(y(x))\frac{\partial x^\mu}{\partial y^i}\frac{\partial}{\partial x^\mu} = (F_*X)^\mu\frac{\partial}{\partial x^\mu}. \qquad (5.331)$$

In particular, F_*X must be orthogonal to a normal vector field n:

$$0 = (F_*X) \cdot n = g_{\mu\nu}(F_*X)^\mu n^\nu = g_{\mu\nu}\frac{\partial x^\mu}{\partial y^i}X^i n^\nu. \qquad (5.332)$$

For later reference, consider specifically the basis vector fields $e_i \equiv \partial/\partial y^i$ on Σ. For their push

$$F_*e_i = \frac{\partial x^\mu}{\partial y^i}\frac{\partial}{\partial x^\mu} \qquad (5.333)$$

the orthogonality condition (5.332) becomes

$$0 = g_{\mu\nu}\frac{\partial x^\mu}{\partial y^i}n^\nu. \qquad (5.334)$$

Finally, note the two ways of expressing a scalar product of two vector fields X, Y on Σ. We can either interpret it as a scalar product based on h:

$$X \cdot Y = h_{ij}X^i Y^i \qquad (5.335)$$

or one based on g, for the push vector fields:

$$(F_*X) \cdot (F_*Y) = g_{\mu\nu}(F_*X)^\mu(F_*Y)^\nu = g_{\mu\nu}\frac{\partial x^\mu}{\partial y^i}\frac{\partial x^\nu}{\partial y^j}X^i Y^j$$

$$= h_{ij}X^i Y^j. \qquad (5.336)$$

The First Fundamental Form (the Induced Metric)

We know that one way to construct a vector field V on Σ is by starting with a curve $c(t) \in \Sigma$ and take its tangent, then join such vectors smoothly to a vector field X. We could then use the embedding F of Σ to M and think of F_*V as a vector field in M tangential to, and supported by, the surface Σ. Another possibility that we consider next is to consider a generic vector field X in the neighborhood of Σ and project it to a tangent vector field. We construct a *projection tensor* P with components $P_{\mu\nu}$ by

$$P_{\mu\nu} = g_{\mu\nu} - \sigma n_\mu n_\nu \qquad (5.337)$$

associated with a (non-null) hypersurface Σ in (M, g), with a unit normal vector field n, with the length squared $\sigma = g_{\mu\nu}n^\mu n^\nu = \pm 1$. The projection tensor P is also called the *first fundamental form* of the hypersurface. Sometimes one sees the notation $\perp$ instead of P. Note that the projection tensor is symmetric,

$$P_{\mu\nu} = P_{\nu\mu}. \qquad (5.338)$$

We first establish that Eq. (5.337) defines a projection by checking its idempotence (squaring to itself):

$$(P^2)_{\mu\nu} = g^{\alpha\beta}P_{\mu\alpha}P_{\beta\nu} = g^{\alpha\beta}(g_{\mu\alpha}g_{\beta\nu} - \sigma g_{\mu\alpha}n_\beta n_\nu - \sigma n_\mu n_\alpha g_{\beta\nu} + n_\mu n_\alpha n_\beta n_\nu)$$

$$= g_{\mu\alpha}\delta^\alpha_\nu - \sigma g_{\mu\alpha}n^\alpha n_\nu - \sigma n_\mu n_\alpha\delta^\alpha_\nu + \sigma^3 n_\mu n_\nu$$

$$= g_{\mu\nu} - 2\sigma n_\mu n_\nu + \sigma n_\mu n_\nu = P_{\mu\nu}. \qquad (5.339)$$

Then, we check the second property, that the projection PX of a generic vector field X is a tangent vector field, by showing that the projection is orthogonal to the (unit) normal field n:

$$
\begin{aligned}
n \cdot (PX) &= g_{\alpha\beta} n^\alpha (PX)^\beta = g_{\alpha\beta} n^\alpha P_\nu^\beta X^\nu \\
&= g_{\alpha\beta} n^\alpha (\delta_\nu^\beta - \sigma n^\beta n_\nu) X^\nu \\
&= \left(n_\nu - \sigma g_{\alpha\beta} n^\alpha n^\beta n_\nu \right) X^\nu = \left(n_\nu - \sigma^2 n_\nu \right) X^\nu = 0.
\end{aligned}
\tag{5.340}
$$

Having constructed a projection, we can apply it to generic tensors T and project by acting on its components:

$$
T^{\mu_1 \cdots \mu_q}_{\nu_1 \cdots \nu_r} \mapsto P^{\mu_1}_{\alpha_1} \cdots P^{\mu_q}_{\alpha_q} P^{\beta_1}_{\nu_1} \cdots P^{\beta_r}_{\nu_r} T^{\alpha_1 \cdots \alpha_q}_{\beta_1 \cdots \beta_r}.
\tag{5.341}
$$

In particular, we can apply the projection to the metric tensor, and find

$$
\begin{aligned}
P^\alpha_\mu P^\beta_\nu g_{\alpha\beta} &= (\delta^\alpha_\mu - \sigma n_\mu n^\alpha)(\delta^\beta_\nu - \sigma n^\beta n_\nu) g_{\alpha\beta} \\
&= g_{\mu\nu} - \sigma n_\mu n^\alpha g_{\alpha\nu} - \sigma g_{\mu\beta} n^\beta n_\nu + \sigma^3 n_\mu n_\nu = g_{\mu\nu} - \sigma n_\mu n_\nu \\
&= P_{\mu\nu}.
\end{aligned}
\tag{5.342}
$$

Thus we can think of $P_{\mu\nu}$ as a (projected) metric on the hypersurface Σ. In fact, it is an easy exercise (see Problem 5.7) to show that for two tangent vectors V, W, the projection tensor acts like the metric,

$$
P_{\mu\nu} V^\mu W^\nu = g_{\mu\nu} V^\mu W^\nu.
\tag{5.343}
$$

There remains a potential confusion that has to be clarified. We know that the embedding F also induces a metric h_{ij} on the hypersurface; thus one may suspect $P_{\mu\nu}$ and h_{ij} to be somehow related, but on the face of it the definition of P (5.337) looks quite different from the definition of the induced metric h according to Eq. (5.327). The resolution is that $P_{\mu\nu}$ is expressed in the coordinates x of M: Since it is also a projection of the metric onto the hypersurface, it is natural to express it in the local coordinates y^i of Σ: we thus need to calculate the pullback $\hat{P} = F^*P$, with components

$$
\begin{aligned}
\hat{P}_{ij} &= \frac{\partial x^\mu}{\partial y_i} \frac{\partial x^\nu}{\partial y_j} P_{\mu\nu} = \frac{\partial x^\mu}{\partial y^i} \frac{\partial x^\nu}{\partial y^j} (g_{\mu\nu} - \sigma n_\mu n_\nu) \\
&= h_{ij} - \sigma \underbrace{\frac{\partial x^\mu}{\partial y^i} n_\mu}_{=0} \underbrace{\frac{\partial x^\nu}{\partial y^j} n_\nu}_{=0} \\
&= h_{ij},
\end{aligned}
\tag{5.344}
$$

where in the last step we used Eq. (5.334). In this way, we can identify P with the induced metric h. For this reason in the literature one often sees the notation $h_{\mu\nu}$ instead of $P_{\mu\nu}$. In the context of spacelike hypersurfaces in spacetimes, the projection tensor is sometimes called the *spatial metric*.

Example Let us check the identification (5.344) for the paraboloid. In cylindrical coordinates it is the surface $f = z - \frac{1}{2}r^2 = 0$. A normal vector field $\nabla f = (-r, 0, 1)$ is converted to an outward directed unit normal field n with components

$$
n^r = \frac{r}{\sqrt{1 + r^2}} = n_r, \quad n^\phi = 0, \quad n^z = \frac{-1}{\sqrt{1 + r^2}} = n_z.
\tag{5.345}
$$

We previously derived the induced metric to be

$$h = (h_{ij}) = \mathrm{diag}(h_{rr}, h_{\phi\phi}) = (1 + r^2, r^2). \tag{5.346}$$

The projection tensor $P_{\mu\nu}$ has the nonvanishing components

$$P_{rr} = g_{rr} - n_r n_r = 1 - \frac{r^2}{1 + r^2} = \frac{1}{1 + r^2},$$

$$P_{rz} = P_{zr} = -n_r n_z = \frac{r}{1 + r^2},$$

$$P_{\phi\phi} = g_{\phi\phi} = r^2,$$

$$P_{zz} = g_{zz} - n_z n_z = 1 - \frac{1}{1 + r^2} = \frac{r^2}{1 + r^2}. \tag{5.347}$$

Note that the components of P form a non-diagonal 3×3 matrix, while h is a diagonal 2×2 matrix. However, when we pull back P to the paraboloid by

$$\hat{P}_{rr} = P_{rr} + 2(\partial_r z)P_{rz} + (\partial_r z)^2 P_{zz} = \frac{1 + 2r^2 + r^4}{1 + r^2} = 1 + r^2$$

$$\hat{P}_{r\phi} = P_{r\phi} + 0 = 0$$

$$\hat{P}_{\phi\phi} = P_{\phi\phi} = r^2, \tag{5.348}$$

we find $\hat{P}_{ij} = h_{ij}$, as expected.

5.12.2 Extrinsic Curvature and the Second Fundamental Form

We now introduce the *second fundamental form* or, as it is usually called, the *extrinsic curvature* of a hypersurface Σ. In distinction to the intrinsic curvature that, as the name suggests, is an intrinsic property of the geometry of the manifold, the extrinsic curvature reflects the embedding to a larger-dimensional manifold.

As our heuristic introduction in the beginning of the section suggests, it involves second derivatives. However, for the definition we proceed slightly differently. First, we imagine a small neighborhood of the hypersurface Σ, and extend its unit normal field into this neighborhood. We could imagine moving the hypersurface Σ into the neighborhood following the unit normal field, in the two opposite directions, and in this way generate a family of hypersurfaces, all sharing the same unit normal vector field. We can now ask how the projection tensor P varies in the direction of the unit normal vector field, and compute the rate of change by the Lie derivative[3],

$$K_{\mu\nu} \equiv \frac{1}{2}\mathcal{L}_n P_{\mu\nu}. \tag{5.349}$$

Equation (5.349) defines the *extrinsic curvature* $K_{\mu\nu}$ of the hypersurface Σ. Let us study some of its properties. Firstly, we note that it is symmetric,

$$K_{\mu\nu} = K_{\nu\mu}. \tag{5.350}$$

---

[3] One can show that the result does not depend on how the unit normal vector field was extended to a small neighborhood of Σ.

What is less obvious is that we can also derive the extrinsic curvature by starting with the Lie derivative of the metric, and then project the outcome to the hypersurface: We can rewrite,

$$K_{\mu\nu} = \frac{1}{2} P_\mu^\alpha P_\nu^\beta \mathcal{L}_n g_{\alpha\beta}. \tag{5.351}$$

To see this, we first observe that

$$n^\alpha \nabla_\mu n_\alpha = \frac{1}{2} \nabla_\mu (n^\alpha n_\alpha) = \pm \frac{1}{2} \nabla_\mu 1 = 0, \tag{5.352}$$

which will simplify calculations. On the other hand, the Lie derivative of the metric is

$$\mathcal{L}_n g_{\alpha\beta} = n^\lambda \overbrace{\nabla_\lambda g_{\alpha\beta}}^{=0} + (\nabla_\alpha n^\lambda) g_{\lambda\beta} + (\nabla_\beta n^\lambda) g_{\alpha\lambda}$$

$$= \nabla_\alpha n_\beta + \nabla_\beta n_\alpha. \tag{5.353}$$

Starting from Eq. (5.351), one obtains

$$K_{\mu\nu} = \frac{1}{2} (\delta_\mu^\alpha - \sigma n_\mu n^\alpha)(\delta_\nu^\beta - \sigma n_\nu n^\beta)(\nabla_\alpha n_\beta + \nabla_\beta n_\alpha)$$

$$= \frac{1}{2}(\nabla_\mu n_\nu + \nabla_\nu n_\mu) + -\frac{1}{2}\sigma n_\mu n^\alpha \nabla_\alpha - \frac{1}{2}\sigma n_\nu n^\beta \nabla_\beta n_\mu$$

$$+ \text{ terms that vanish by virtue of Eq. (5.352)}$$

$$= \frac{1}{2}\mathcal{L}_n g_{\mu\nu} - \frac{1}{2}\sigma n_\mu \mathcal{L}_n n_\nu - \frac{1}{2}\sigma n_\nu \mathcal{L}_n n_\nu = \frac{1}{2}\mathcal{L}_n(g_{\mu\nu} - \sigma n_\mu n_\mu)$$

$$= \frac{1}{2}\mathcal{L}_n P_{\mu\nu}, \tag{5.354}$$

so that we recover our earlier definition. Note that the above lines involved a combination for which we give a new notation

$$a^\mu \equiv n^\alpha \nabla_\alpha n^\mu \tag{5.355}$$

called the *acceleration* of the hypersurface. The name can be motivated by considering an integral curve $x^\mu(s)$ so that $n^\mu(s) = dx^\mu(s)/ds$ is the tangent vector (the unit length can be obtained by rescaling the affine parameter s). Then (5.355) reads

$$a^\mu = \frac{dx^\alpha}{ds}\frac{\partial}{\partial x^\alpha}\frac{dx^\mu}{ds} + \frac{dx^\alpha}{ds}\Gamma^\mu_{\alpha\lambda}\frac{dx^\lambda}{ds}$$

$$= \frac{d^2 x^\mu}{ds^2} + \Gamma^\mu_{\alpha\lambda}\frac{dx^\alpha}{ds}\frac{dx^\lambda}{ds}. \tag{5.356}$$

Vanishing acceleration $a^\mu = 0$ thus leads to a geodesic equation. Note that, for a spatial hypersurface, the normal vector and its integral curve $x(\tau)$ are timelike, and the acceleration is a "true" acceleration. If the condition $a^\mu = 0$ is satisfied everywhere in a neighborhood of Σ, the normal vector field is said to be *geodesic everywhere*. We continue more generally without assuming this choice and keep a^μ in the following formulas. Now we claim that there is yet another way to rewrite the extrinsic curvature. The virtue of the formula is that it is simple to compute, but has the drawback of obscuring the symmetry with respect to indices. We will find

$$K_{\mu\nu} = \nabla_\mu n_\nu - \sigma n_\mu a_\nu. \tag{5.357}$$

To see how this formula arises, we will follow a sequence of observations, and we will also need a result that we derived when we discussed the implications of Frobenius's theorem. Let us rewrite the above $K_{\mu\nu}$ by separating it into its symmetric and antisymmetric parts,

$$K_{\mu\nu} = K_{(\mu\nu)} + K_{[\mu\nu]} = \frac{1}{2}\left[\nabla_\mu n_\nu - \sigma n_\mu a_\nu + (\mu \leftrightarrow \nu)\right] + K_{[\mu\nu]}. \qquad (5.358)$$

We then anticipate that the antisymmetric part must vanish, since we know from our previous formulas that $K_{\mu\nu}$ is symmetric. Moreover, it is straightforward to check, by comparing with Eq. (5.354), that the symmetric part is equal to the previous definition of $K_{\mu\nu}$. To show that $K_{[\mu\nu]} = 0$, we consider acting on an arbitrary pair of vector fields: If

$$K_{[\mu\nu]} V^\mu W^\nu = 0 \qquad (5.359)$$

then we are done. We split the two vectors into a component that is tangent to the hypersurface, and a component that is normal to it. The normal part of V will then be proportional to the unit normal field by some function f:

$$V = PV + (1 - P)V = V_{||} + V_\perp = V_{||} + fn, \qquad (5.360)$$

and similarly $W = W_{||} + hn$ by some function h. Now we see that

$$\begin{aligned}
V^\mu(\nabla_\mu - \sigma n_\mu a_\nu) &= fn^\mu \nabla_\mu n_\nu + V^\mu_{||}\nabla_\mu n_\nu - \sigma fn^\mu n_\mu a_\nu - \sigma \overbrace{V^\mu_{||} n_\mu}^{=0} a_\nu \\
&= fa_\nu + V^\mu_{||}\nabla_\mu a_\nu - fa_\nu = V^\mu_{||}\nabla_\mu n_\nu.
\end{aligned} \qquad (5.361)$$

Then we find

$$\begin{aligned}
V^\mu(\nabla_\mu - \sigma n_\mu a_\nu)W^\nu &= V^\mu_{||}\nabla_\mu n_\nu(W^\nu_{||} + hn^\nu) \\
&= V^\mu_{||}(\nabla_\mu n_\nu)W^\nu_{||} + hW^\mu_{||}\underbrace{n^\nu \nabla_\mu n_\nu}_{=0} \\
&= V^\mu_{||}(\nabla_\mu n_\nu)W^\nu_{||}.
\end{aligned} \qquad (5.362)$$

These steps imply that

$$K_{[\mu\nu]}V^\mu W^\nu = \nabla_{[\mu}n_{\nu]}V^\mu_{||}W^\nu_{||}. \qquad (5.363)$$

The last step is to recall that the unit normal vector field has the form $n_\mu = h\partial_\mu f$ where h is the normalizing factor function and f is the function that defines the hypersurface. From Problem 5.10 we know that it then satisfies Eq. (5.323) $n_{[\mu}\nabla_\nu n_{\sigma]} = 0$, which in turns guarantees by Eq. (5.324) that

$$\nabla_{[\mu}n_{\nu]}V^\mu_{||}W^\nu_{||} = 0 \qquad (5.364)$$

since the tangent components are orthogonal to n, $n_\mu V^\mu_{||} = n_\mu W^\mu_{||} = 0$. Hence we have shown that Eq. (5.357) is an equivalent definition of the extrinsic curvature. We can now also easily check that it is orthogonal to the normal vector field n of Σ,

$$n^\mu K_{\mu\nu} = n^\mu(\nabla_\mu n_\nu - \sigma n_\mu a_\nu) = a_\nu - \sigma^2 a_\nu = 0. \qquad (5.365)$$

Finally, we can compute a scalar K from the extrinsic curvature:

$$K = g^{\mu\nu}K_{\mu\nu}. \qquad (5.366)$$

The extrinsic curvature and the scalar K are important in general relativity, for example in its classical action in the Lagrangian formulation, appearing in the Gibbons–Hawking–York boundary term. We leave it as an exercise (see Problem 5.10) to show that the scalar K is equal to the covariant divergence of the normal vector field,

$$K = \nabla_\mu n^\mu \tag{5.367}$$

on the hypersurface. Suppose that we have a single hypersurface which is not part of a continuous family. In this case we can often extend the unit normal field to a small neighborhood of the surface by solving the geodesic equation, and thus arrange that the unit normal field satisfies $a^\mu = 0$ in the neighborhood, i.e., choose the unit normal field so that it is geodesic everywhere. This is a convenient choice, often used in the literature, since it simplifies the calculation of the extrinsic curvature: Eq. (5.357) reduces to

$$K_{\mu\nu} = \nabla_\mu n_\nu = \nabla_{(\mu} n_{\nu)}, \tag{5.368}$$

and the last form makes the symmetry again explicit. When the unit normal field is geodesic everywhere, the scalar extrinsic curvature can also be interpreted as the *expansion* of the geodesics, their relative deviation as they are followed off the hypersurface. In this context one often denotes $K = \theta$, the latter being the symbol for expansion.

Another convention that is often found in the literature is to define the extrinsic curvature to be an object on the hypersurface Σ. More precisely, this means that we define the pullback of the extrinsic curvature from M to Σ

$$\hat{K} = F^* K, \quad \hat{K}_{ij}(y) = \frac{\partial x^\mu}{\partial y^i} \frac{\partial x^\nu}{\partial y^j} K_{\mu\nu}(x(y)) \tag{5.369}$$

just like we defined an induced metric $h = F^* g$ on the hypersurface. In jest, we could therefore call the pullback $\hat{K}_{ij}$ the extrinsic curvature, "intrinsic" to the hypersurface (remembering that it is associated with the embedding of the surface).

Let us now practice the above definitions and formulas in an example. We compute the extrinsic curvature of the paraboloid surface. To simplify the calculations, we again use rotational symmetry about the z-axis and cylindrical coordinates.

Example Let us study the extrinsic curvature of the paraboloid. We use cylindrical coordinates $x = (r, \phi, z)$ in $\mathbb{R}^3$. The normal vector vector field has components

$$n_r = n^r = \frac{r}{\sqrt{1 + r^2}}, \quad n^\phi = 0, \quad n_z = n^z = \frac{-1}{\sqrt{1 + r^2}}. \tag{5.370}$$

We have computed the nonvanishing components of the Levi-Civita connection in a previous example. From Eq. (5.201) we obtain

$$\Gamma^r_{\phi\phi} = -r, \quad \Gamma^\phi_{\phi r} = \Gamma^\phi_{r\phi} = \frac{1}{r}. \tag{5.371}$$

For the acceleration, we first need to compute the covariant derivatives $\nabla_\alpha n^\mu = \partial_\alpha n^\mu + \Gamma^\mu_{\alpha\lambda} n^\lambda$. One obtains

$$\nabla_r n^r = \partial_r n^r + \overbrace{\Gamma^r_{r\lambda}}^{=0} n^\lambda = \frac{1}{(1 + r^2)^{3/2}}$$

$$\nabla_\phi n^r = \partial_\phi n^r + \Gamma^r_{\phi\phi} n^\phi = 0$$

$$\vdots$$

$$\nabla_\phi n^\phi = \partial_\phi n^\phi + \Gamma^\phi_{\phi r} n^r = \frac{1}{r} \frac{r}{\sqrt{1 + r^2}} = \frac{1}{\sqrt{1 + r^2}}$$

$$\nabla_r n^z = \frac{r}{(1 + r^2)^{3/2}} \tag{5.372}$$

omitting the rest of the vanishing components. Next we compute the acceleration $a^\mu = n^\alpha \nabla_\alpha n^\mu$:

$$a^r = n^\alpha \nabla_\alpha n^r = n^r \nabla_r n^r = \frac{r}{(1 + r^2)^2}$$

$$a^\phi = 0 + n^\phi \nabla_\phi n^\phi + n^z \nabla_z n^\phi = 0$$

$$a^z = n^r \nabla_r n^z + 0 + 0 = \frac{r^2}{(1 + r^2)^2}. \tag{5.373}$$

Lowering the indices (which is nontrivial only for a_ϕ, but it is zero) we can move to compute the extrinsic curvature:

$$K_{rr} = \nabla_r n_r - n_r a_r = \frac{1}{(1 + r^2)^{5/2}}$$

$$K_{r\phi} = \nabla_r n_\phi - n_r a_\phi = 0$$

$$K_{rz} = \nabla_r n_z - n_r a_z = \frac{r}{(1 + r^2)^{5/2}}$$

$$K_{\phi\phi} = \nabla_\phi n_\phi - 0 = \frac{r^2}{(1 + r^2)^{1/2}}$$

$$K_{\phi z} = 0$$

$$K_{zz} = \nabla_z n_z - n_z a_z = \frac{r^2}{(1 + r^2)^{5/2}}. \tag{5.374}$$

The scalar curvature is then

$$K = g^{rr} K_{rr} + 2 \underbrace{g^{rz}}_{=0} K_{rz} + g^{\phi\phi} K_{\phi\phi} + g^{zz} K_{zz}$$

$$= K_{rr} + \frac{1}{r^2} K_{\phi\phi} + K_{zz}$$

$$= \frac{1}{(1 + r^2)^{5/2}} + \frac{1}{(1 + r^2)^{1/2}} + \frac{r^2}{(1 + r^2)^{5/2}}$$

$$= \frac{2 + r^2}{(1 + r^2)^{3/2}} \tag{5.375}$$

Note that K is strictly positive, and that $K \to 2$ when $r \to 0$, whereas $K \to 0$ for $r \to \infty$. Finally, the pullback to the paraboloid gives

$$\hat{K}_{rr} = K_{rr} + 2(\partial_r z) K_{rz} + (\partial_r z)^2 K_{zz}$$

$$= \frac{1}{(1 + r^2)^{5/2}} + 2r \frac{r}{(1 + r^2)^{5/2}} + r^2 \frac{r^2}{(1 + r^2)^{5/2}}$$

$$= \frac{(1 + r^2)^2}{(1 + r^2)^{5/2}} = \frac{1}{\sqrt{1 + r^2}}$$

$$\hat{K}_{r\phi} = K_{r\phi} + (\partial_r z)K_{z\phi} = 0$$

$$\hat{K}_{\phi\phi} = K_{\phi\phi} = \frac{r^2}{\sqrt{1 + r^2}}. \tag{5.376}$$

The extrinsic curvature is positive in both r and ϕ directions. In the r direction, constant ϕ sections are parabolas; in the ϕ direction, constant r sections are circles with increasing radius.

5.13 Isometries

Isometries are a very important concept. They are symmetries of a Riemannian manifold. In particular, a physical theory defined on a spacetime which is a Riemannian manifold is usually required to be invariant under the isometries of the spacetime.

Let (M, g) be a (pseudo-)Riemannian manifold. A diffeomorphism $f: M \to M$ is an *isometry* if it preserves the metric,

$$f^* g_{f(p)} = g_p, \tag{5.377}$$

for all points $p \in M$.

By interpreting the metric as a map on vector fields, the requirement encoded in Eq. (5.377) means

$$g_{f(p)}(f_* X, f_* Y) = g_p(X, Y) \tag{5.378}$$

for all tangent vectors $X, Y \in T_p M$. In component form, Eq. (5.378) reads

$$\frac{\partial y^\alpha}{\partial x^\mu} \frac{\partial y^\beta}{\partial x^\nu} g_{\alpha\beta}(f(p)) = g_{\mu\nu}(p), \tag{5.379}$$

where x are the coordinates of the point p, while y are the coordinates of $f(p)$. The meaning of Eq. (5.378) is that an isometry preserves the angles between all tangent vectors and their lengths.

Example A simple example of an isometry is a rotation in Euclidean space. Consider $\mathbb{R}^n$ with the Cartesian metric δ,

$$\delta = dx^1 \otimes dx^1 + \cdots + dx^n \otimes dx^n = \delta_{ij} dx^i \otimes dx^j. \tag{5.380}$$

Then a rotation is a diffeomorphism $f: \mathbb{R}^n \to \mathbb{R}^n$

$$y^i = f^i(x) = O^i_j x^j, \tag{5.381}$$

where O^i_j are the components of an orthogonal matrix $O \in O(n)$. Then

$$f^* \delta = \delta_{ij} dy^i \otimes dy^j = \delta_{ij} O^i_k O^j_l dx^k \otimes dx^l = (O^T)^k_i O^i_l dx^k \otimes dx^l$$
$$= (O^T O)^k_l dx^k \otimes dx^l = \delta_{kl} dx^k \otimes dx^l = \delta. \tag{5.382}$$

Hence the metric is preserved. The relation

$$O_k^i O_l^j \delta_{ij} = \delta_{kl} \tag{5.383}$$

is an example of Eq. (5.379).

Note that the identity map is an isometry (in fact, a trivial one). Moreover, given two isometries f and g, the composite map $f \circ g$ is also an isometry. Finally, if f is an isometry, so is its inverse f^{-1}. These three properties imply that the isometries form a group, with composition of maps as the product; this group is called the *isometry group*. The isometry group is a group of symmetries of a (pseudo)-Riemannian manifold.

Example   Let (M, g) be the Euclidean space $(\mathbb{R}^n, \delta)$ with the Euclidean metric. All translations $x^\mu \mapsto x^\mu + a^\mu$ along some direction $a = (a^\mu)$ are isometries, and so are rotations. The isometry group of this space is made of all translations, rotations, and their combinations, and is called the *Euclidean group*, or *Galilean group*, and denoted as E^n.

Example   Let (M, g) be the $(d + 1)$-dimensional Minkowski spacetime $(\mathbb{R}^{1,d}, \eta)$ with the Minkowski metric η. As in the previous example, all spacetime translations $x^\mu \mapsto x^\mu + a^\mu$ are isometries, and additional isometries of this spacetime are space rotations, boosts, and combinations thereof. The isometry group is called the *Poincaré group*.

At scales typical of laboratory experiments, our spacetime is approximately flat (i.e., is a Minkowski spacetime), so its approximate isometry group is the Poincaré group. Because of this, the theories describing physical phenomena in the laboratory are required to be invariant under the isometries of special relativity, i.e., under the Poincaré group. More precisely, that requirement is a strictly necessary one for theories describing experiments which involve velocities where relativistic effects become important, i.e., velocities comparable with the speed of light in vacuum c, whose value is approximately $c \simeq 3 \cdot 10^8$ m/s. For phenomena which only involve velocities $v \ll c$, time "decouples" and one can make a further approximation, in which only the Euclidean isometries of the spacelike directions are relevant. Recall also that symmetries such as time translations and space translations lead to conservation laws, like the conservation of energy and momentum. As you can see, important physical principles can be viewed as a reflection of the isometries of the spacetime.

5.14 Killing Vector Fields

Let us now consider the limit of "small" isometries, i.e., infinitesimal displacements of the form

$$x = p \mapsto f(p) = y \approx x + \epsilon X, \tag{5.384}$$

where ϵ is an infinitesimal parameter and X is a vector field along the direction of the infinitesimal displacement. If the map defined in Eq. (5.384) is an isometry, the vector field X is called a *Killing vector field*, after the German mathematician Wilhelm Karl Joseph Killing. Since the infinitesimal displacement is an isometry, Eq. (5.379) must be satisfied, which means

$$\frac{\partial(x^\alpha + \epsilon X^\alpha)}{\partial x^\mu} \frac{\partial(x^\beta + \epsilon X^\beta)}{\partial x^\nu} g_{\alpha\beta}(x + \epsilon X) = g_{\mu\nu}(x). \tag{5.385}$$

Expanding the left-hand side of this equation in powers of ϵ, and requiring the term of order ϵ to vanish (for consistency with the right-hand side, in which there is no dependence on ϵ), one obtains the equation

$$X^\xi \partial_\xi g_{\mu\nu} + \partial_\mu X^\alpha g_{\alpha\nu} + \partial_\nu X^\beta g_{\mu\beta} = 0. \tag{5.386}$$

On the left-hand side of Eq. (5.386), one can recognize a Lie derivative, so that the equation can be rewritten as

$$\mathcal{L}_X g_{\mu\nu} = 0. \tag{5.387}$$

Expressing $\mathcal{L}_X g_{\mu\nu}$ in terms of the covariant derivative, one obtains

$$\mathcal{L}_X g_{\mu\nu} = X^\lambda \overbrace{\nabla_\lambda g_{\mu\nu}}^{=0} + (\nabla_\mu X^\lambda) g_{\lambda\nu} + (\nabla_\nu X^\lambda) g_{\mu\lambda} = 0, \tag{5.388}$$

where we used the fact that $\nabla_\lambda g_{\mu\nu} = 0$ for a metric connection. Thus, the Killing vector field satisfies

$$\nabla_\mu X_\nu + \nabla_\nu X_\mu = 0, \tag{5.389}$$

which is called the *Killing equation*.

Given two Killing vector fields X and Y, one can verify that the following properties hold.

(i) All linear combinations $aX + bY$ with a, $b \in \mathbb{R}$ are also Killing vector fields.
(ii) The Lie bracket $[X, Y]$ is also a Killing vector field.

It then follows that the Killing vector fields form an algebra, the *Lie algebra of the isometry group*. Note that this algebra is a Lie algebra provided the isometry group is a Lie group, which is usually the case.

Now let $x^\mu(t)$ be a geodesic, let its tangent vector be $U^\mu = \frac{dx^\mu}{dt}$, and let V^μ be a Killing vector. Then,

$$(U^\nu \nabla_\nu)(U^\mu V_\mu) = \underbrace{U^\mu U^\nu \nabla_\nu V_\mu}_{=\frac{1}{2} U^\mu U^\nu (\nabla_\mu V_\nu + \nabla_\nu V_\mu)} + V_\mu \underbrace{U^\nu \nabla_\nu U^\mu}_{=0 \ (\text{geodesic})} = 0. \tag{5.390}$$

Thus $U^\mu V_\mu = U \cdot V$ is a *constant on a geodesic*.

An m-dimensional manifold M can have at most $\frac{m(m+1)}{2}$ linearly independent Killing vector fields. A manifold with the maximum number of Killing vector fields is said to be a *maximally symmetric manifold*.

Example $\mathbb{R}^m$ with the Cartesian metric δ is a maximally symmetric manifold. As the connection coefficients vanish, the Killing equation is $\partial_\mu V_\nu + \partial_\nu V_\mu = 0$. Its solutions are

$$\begin{cases} V^{\mu}_{(i)} = \delta^{\mu}_i & (m \text{ of them}) \\ V_{\mu} = a_{\mu\nu}x^{\nu}, \text{ with } a_{\mu\nu} = -a_{\nu\mu} = \text{ constant} \neq 0 & \left(\frac{m(m-1)}{2} \text{ of them}\right) \end{cases} \tag{5.391}$$

so that in total there are $m + \frac{m(m-1)}{2} = \frac{m(m+1)}{2}$ linearly independent Killing vector fields.

Besides $\mathbb{R}^m$, other examples of maximally symmetric m-dimensional manifolds are the Minkowski spacetime, the de Sitter space dS_m, and the anti-de Sitter space AdS_m.

Problems

5.1   Given the two-dimensional torus T^2, parameterized in terms of the two real variables $x^1 \in [0, 2\pi)$ and $x^2 \in [0, 2\pi)$, consider its embedding into the three-dimensional Euclidean space $\mathbb{R}^3$ with the metric $\delta = \mathrm{diag}(1, 1, 1)$ given by

$$f: T^2 \mapsto \mathbb{R}^3,$$
$$(x^1, x^2) \mapsto \begin{pmatrix} A\cos x^1 + B\cos x^1 \cdot \cos x^2 \\ A\sin x^1 + B\sin x^1 \cdot \cos x^2 \\ B\sin x^2 \end{pmatrix}, \tag{5.392}$$

with $0 < B < A$, and determine the induced metric $g = f^*\delta$ on T^2.

5.2   Let θ and ϕ be the polar coordinates. Introduce the complex numbers z and $\bar{z}$, where

$$z = e^{i\phi}\tan(\theta/2) \equiv \xi + i\eta, \tag{5.393}$$

and ξ and η are real numbers. Show that the metric of the two-sphere transforms as

$$\begin{aligned} ds^2 &= d\theta \otimes d\theta + \sin^2\theta d\phi \otimes d\phi \\ &= \frac{2}{(1+|z|^2)^2}(d\bar{z} \otimes dz + dz \otimes d\bar{z}) \\ &= \frac{2}{(1+\xi^2+\eta^2)^2}(d\xi \otimes d\xi + d\eta \otimes d\eta) \end{aligned} \tag{5.394}$$

and the area two-form ω transforms as

$$\begin{aligned} \omega &= \sin\theta d\theta \wedge d\phi \\ &= \frac{2i}{(1+|z|^2)^2}(dz \wedge d\bar{z}) \\ &= \frac{4}{(1+\xi^2+\eta^2)^2}(d\xi \wedge d\eta). \end{aligned} \tag{5.395}$$

5.3   Let X and Y be vector fields and f a function on M. Compute the "double covariant derivatives"

(i) $\nabla_X\nabla_Y f$,

(ii) $\nabla_{\mu}\nabla_{\nu}f$,

writing them as sums of terms that involve partial derivatives and (if needed) connection coefficients.

5.4 Let $\Gamma^\alpha_{\mu\nu}$ be the Levi-Civita connection (implying the symmetry $\Gamma^\alpha_{\mu\nu} = \Gamma^\alpha_{\nu\mu}$).
(i) Show that $\nabla_\mu \nabla_\nu f = \nabla_\nu \nabla_\mu f$.
(ii) Consider a one-form field v with coefficients of the form

$$v_\mu = h\nabla_\mu f = h\partial_\mu f, \qquad (5.396)$$

where $f = f(x), h = h(x)$ are some smooth functions. Show that

$$v_{[\mu}\nabla_\sigma v_{\nu]} = 0. \qquad (5.397)$$

5.5 **Geodesics on a torus.** Let the metric on a two-dimensional torus T^2 with radii r and $R > r$ be

$$g = r^2 d\theta \otimes d\theta + (R + r\cos\theta)^2 d\phi \otimes d\phi, \qquad (5.398)$$

where θ and ϕ take values in the interval $[0, 2\pi)$. Find the geodesic equation(s) on a torus with the metric (5.398), by doing the following.
(i) Using the variational principle and the action of a free massive point particle. As a first step, substitute the torus metric into the appropriate Lagrangian.
(ii) Calculating the Christoffel symbols and substituting them into the general geodesic equation.

Which method do you expect to be easier? Next, do the following.
(iii) Solve the equations. Are there conserved quantities?
(iv) Sketch some examples of geodesics on a torus.

5.6 **Spatially homogeneous and isotropic universe.** The Robertson–Walker metric

$$g = -dt \otimes dt + a^2(t)\left(\frac{dr \otimes dr}{1 - kr^2} + r^2(d\theta \otimes d\theta + \sin^2\theta \, d\phi \otimes d\phi)\right), \qquad (5.399)$$

where k can be -1, 0, or 1, describes a spatially homogenous and isotropic spacetime. Calculate the Riemann tensor, the Ricci tensor, and the scalar curvature associated with the metric (5.399).

5.7 Show that, given two tangent vectors V and W of a hypersurface Σ, contraction with the projection tensor $P_{\mu\nu}$ reduces to a scalar product:

$$P_{\mu\nu}V^\mu W^\nu = g_{\mu\nu}V^\mu V^\nu = V \cdot W. \qquad (5.400)$$

5.8 **Symmetry of a sphere.** The metric of the S^2 sphere of unit radius can be written as

$$g = d\theta \otimes d\theta + \sin^2\theta \, d\phi \otimes d\phi. \qquad (5.401)$$

Show that

$$L_1 = -\cos\phi\frac{\partial}{\partial\theta} + \cot\theta\sin\phi\frac{\partial}{\partial\phi} \tag{5.402}$$

$$L_2 = \sin\phi\frac{\partial}{\partial\theta} + \cot\theta\cos\phi\frac{\partial}{\partial\phi} \tag{5.403}$$

$$L_3 = \frac{\partial}{\partial\phi} \tag{5.404}$$

are its Killing vectors. Calculate the commutators $[L_a, L_b]$ and identify the associated symmetry.

5.9   **Three-dimensional sphere.** Consider the following embedding of the three-dimensional sphere of unit radius into the four-dimensional Euclidean space $(\mathbb{R}^4, \delta)$ with the metric $\delta = \mathrm{diag}(1, 1, 1, 1)$:

$$f: S^3 \mapsto \mathbb{R}^4$$
$$(\theta, \phi_1, \phi_2) \mapsto (\sin\theta\cos\phi_1, \sin\theta\sin\phi_1, \cos\theta\cos\phi_2, \cos\theta\sin\phi_2),$$
$$\tag{5.405}$$

where $0 \le \theta \le \frac{\pi}{2}$, $\phi_a \in S^1$.
 (i) Find the induced metric $ds_3^2 \equiv g = f^*\delta$ on S^3.
 (ii) Consider the spacetime $\mathbb{R} \times S^3$, with the metric

$$ds^2 = -dt^2 + ds_3^2 \tag{5.406}$$

and calculate the corresponding Christoffel symbols

$$\begin{Bmatrix} \mu \\ \alpha\beta \end{Bmatrix} \tag{5.407}$$

in the coordinates y^α, with $\alpha \in \{0, 1, 2, 3\}$, and $(y^0, y^1, y^2, y^3) = (t, \theta, \phi_1, \phi_2)$.
 (iii) Find three Killing vectors X_i, with $i \in \{1, 2, 3\}$ for the metric (5.406); verify that one of them is timelike and the two others are generators of the Cartan subalgebra (see Section 6.5) of the Lie algebra of the isometry group $SO(4)$ of S^3.
 (iv) Write the Killing vectors in the form

$$X_i = X_i^\alpha \frac{\partial}{\partial y^\alpha}. \tag{5.408}$$

Write the spacetime metric in the form $ds^2 = h_{\alpha\beta}dy^\alpha dy^\beta$, where the components $h_{\alpha\beta}$ form a diagonal 4×4 matrix. Compute $\sqrt{-h} \equiv \sqrt{-\det(h_{\alpha\beta})}$ and substitute these quantities into the formula

$$R = \int d^3y\sqrt{-h}\, T^{0\alpha}h_{\alpha\beta}X^\beta. \tag{5.409}$$

Show that one obtains the following three quantities, one for each of the three Killing vectors X_i:

$$E \equiv R_1 = \int d\theta d\phi_1 d\phi_2 \cos\theta \sin\theta \, T^{00}, \tag{5.410}$$

$$L_1 \equiv R_2 = \int d\theta d\phi_1 d\phi_2 \cos\theta \sin^3\theta \, T^{02}, \tag{5.411}$$

$$L_2 \equiv R_3 = \int d\theta d\phi_1 d\phi_2 \cos^3\theta \sin\theta \, T^{03}. \tag{5.412}$$

These are the conserved quantities associated with the symmetries of the spacetime (total energy, and two angular momenta) when $T^{\alpha\beta}$ is the energy-momentum tensor.

5.10 Show that the scalar K, which, according to Eq. (5.366), is constructed from the extrinsic curvature as $K = g^{\mu\nu}K_{\mu\nu}$, is equal to the covariant divergence of the normal vector field, $K = \nabla_\mu n^\mu$, on the hypersurface.

5.11 **The two-dimensional anti-de Sitter spacetime** AdS$_2$ **and its isometries**. The two-dimensional anti-de Sitter spacetime can be constructed as a hyperboloid in a three-dimensional flat spacetime. Consider $\mathbb{R}^{2,1}$, the three-dimensional manifold with a "two-time" metric

$$\eta \equiv ds^2 = -dX_1^2 - dX_2^2 + dX_3^2, \tag{5.413}$$

where we embed a hyperboloid H_2 of equation

$$-X_1^2 - X_2^2 + X_3^2 = -R^2 \tag{5.414}$$

as follows. Let H_2 be equipped with coordinates t and r, and consider the embedding $f: H_2 \to \mathbb{R}^{2,1}$:

$$f: H_2 \mapsto \mathbb{R}^{2,1}, \quad (t,r) \mapsto \begin{pmatrix} X_1 = \frac{R^2}{2r} + \frac{r}{2R^2}(R^2 - t^2) \\ X_2 = \frac{r}{R}t \\ X_3 = \frac{R^2}{2r} - \frac{r}{2R^2}(R^2 + t^2) \end{pmatrix}. \tag{5.415}$$

With the induced metric ds_P^2 on H_2, the result is the two-dimensional anti-de Sitter spacetime AdS$_2$, an example of a maximally symmetric pseudo-Riemannian manifold, with the metric expressed in Poincaré coordinates t and r.

(i) Show that the induced metric $ds_P^2 \equiv g = f^*\eta$ on H_2 is

$$ds_P^2 = -\frac{r^2}{R^2}dt^2 + \frac{R^2}{r^2}dr^2. \tag{5.416}$$

(ii) Show that by the coordinate transformation

$$z = \frac{R^2}{r} \tag{5.417}$$

and setting $R = \frac{1}{2}$ the metric can be written in the form

$$ds_P^2 = \frac{1}{4z^2}(-dt^2 + dz^2). \tag{5.418}$$

(iii) Show that, by denoting $\partial_t = \frac{\partial}{\partial t}$, etc., the vector fields

$$H = i\partial_t, \tag{5.419}$$

$$D = i(t\partial_t + z\partial_z), \tag{5.420}$$

$$K = i\left[(t^2 + z^2)\partial_t + 2tz\partial_z\right] \tag{5.421}$$

are Killing vectors in AdS$_2$, recalling that a generic Killing vector field X satisfies

$$\mathcal{L}_X g_{\mu\nu} = X^\lambda \partial_\lambda g_{\mu\nu} + (\partial_\mu X^\lambda)g_{\lambda\nu} + (\partial_\nu X^\lambda)g_{\lambda\mu} = 0. \tag{5.422}$$

(iv) Derive the commutation relations

$$i[D, H] = H, \tag{5.423}$$

$$i[K, H] = 2D. \tag{5.424}$$

(v) There is an additional commutation relation $i[D, K] = -K$ which you do not need to show. Instead, consider the linear combinations

$$R \equiv \frac{1}{2}\left(\frac{K}{a} + aH\right), \qquad L_\pm \equiv \frac{1}{2}\left(\frac{K}{a} - aH\right) \pm iD, \tag{5.425}$$

where $a > 0$ is a free parameter. Using the commutation relations for H, D, and K, show that R, L_+, and L_- satisfy the commutation relations

$$[R, L_\pm] = \pm L_\pm, \tag{5.426}$$

$$[L_-, L_+] = 2R. \tag{5.427}$$

Equations (5.426) and (5.427) are the commutation relations of the sl(2, $\mathbb{R}$) $\equiv$ so(2, 1) Lie algebra in the Cartan–Weyl basis, named after Élie Cartan and Hermann Weyl, (see Section 6.5), corresponding to the three isometries of the AdS$_2$ spacetime.

Semisimple Lie Algebras and their Unitary Representations

This chapter introduces the semisimple Lie algebras. We have a two-fold focus: On one hand we are introducing some general concepts about properties of Lie algebras and groups, and their unitary representations, while on the other hand we choose to focus with more detail on the irreducible representations of the $su(N)$ Lie algebras and of the corresponding $SU(N)$ Lie groups. This choice is motivated in part by elementary particle physics, where the Standard Model is based on the $U(1) \times SU(2) \times SU(3)$ gauge symmetry, with various proposed extensions involving higher-dimensional $SU(N)$ groups. In addition, $SU(N)$ Lie groups also have important applications in condensed matter theory and other areas of physics. We conclude the chapter by a discussion of the spacetime symmetry of the theory of special relativity, namely the Lorentz–Poincaré group and its representations.

6.1 Lie Groups and Algebras

A *Lie group* G is a differentiable manifold with a group structure, which, as discussed in Chapter 2, can be summarized as follows:

(i) there exists a product $G \times G \rightarrow G$, $(g_1, g_2) \mapsto g_1 g_2$, that is associative $g_1(g_2 g_3) = (g_1 g_2) g_3$,

(ii) there exists a unit element $e \in G$ such that $eg = ge = g$ for every $g \in G$, and

(iii) every element of $g \in G$ has an inverse element $g^{-1} \in G$, such that $g g^{-1} = g^{-1}g = e$.

In addition to these properties that define a generic group, for a Lie group it is also required that the multiplication and inverse operations are smooth. The multiplication being smooth means that the map $G \times G \rightarrow G$, $(g_1, g_2) \mapsto g_1 g_2$ must be smooth. The smoothness of both the inverse and the multiplication is guaranteed by requiring the map $(g_1, g_2) \mapsto g_1^{-1} g_2$ to be smooth for all $g_1, g_2 \in G$. A subgroup H of a Lie group G is a *Lie subgroup*, if the inclusion map $i : H \rightarrow G, i(h) = h$ is an injective immersion (i.e., an embedding of H into G as a submanifold) and a group homomorphism.

We already know some examples of Lie groups, namely the matrix Lie groups $GL(n, \mathbb{R})$ and $GL(n, \mathbb{C})$, and their subgroups $SL(n, \mathbb{R})$, $SL(n, \mathbb{C})$, $O(n)$, $U(n)$, $SO(n, \mathbb{R})$, and $SU(n)$. We will narrow our focus to *compact* Lie groups, i.e., those with a topologically compact group manifold. This set of compact Lie groups contains the $U(1)$, $O(n)$, $SO(n)$, $U(n)$, $SU(n)$, and $Sp(n)$ groups, the compact forms of the so-called exceptional Lie groups $\mathbb{G}_2$, $\mathbb{F}_4$, $\mathbb{E}_6$, $\mathbb{E}_7$, and $\mathbb{E}_8$, and all of their finite direct product groups.

Next, we define a Lie algebra. First, recall that an algebra is a vector space V over a field K (usually $K = \mathbb{R}$ or $K = \mathbb{C}$) which also has a bilinear product: $(aX + bY)Z = aXZ + bYZ$ and $X(aY + bZ) = aXY + bXZ$ for all elements X, Y, and Z in V, and for every a and b in K. The product does not need to be commutative. A *Lie algebra* satisfies some additional requirements. The product of X and Y, which in a Lie algebra is denoted by $[X, Y]$ and called *Lie multiplication*, *commutator*, or *Lie bracket*, must satisfy the following requirements:

(i) it must be linear in the first argument, i.e., it must satisfy $[aX + bY, Z] = a[X, Z] + b[Y, Z]$ for all X, Y, and Z in V, and for all a and b in K,

(ii) it must be antisymmetric: $[X, Y] = -[Y, X]$ for every X and Y in V, and

(iii) it must satisfy the *Jacobi identity*: $[X, [Y, Z]] + [Y, [Z, X]] + [Z, [X, Y]] = 0$ for all X, Y, and Z in V.

Note that the first two requirements above imply linearity also in the second argument, as required for an algebra. If the product is commutative, i.e., if $[X, Y] = 0$ for all X and for all Y in V, the Lie algebra is said to be *Abelian*.

If a subspace $W \subset V$ is closed under Lie multiplication, i.e., if $[X, Y] \in W$ for all X and Y in W, it is called a *Lie subalgebra* of V. If the commutator of every element X of W and every element Y of V is always an element of the subalgebra W, namely $[X, Y] \in W$, then W is called an *ideal*, or an *invariant Lie subalgebra*. Note that V itself is always its own ideal, but this is called the *trivial ideal*. Conversely, a *nontrivial ideal* is an ideal that is a proper subset to V. The possible existence of nontrivial ideals leads us to introduce the following two definitions.

(i) When an algebra has no nontrivial ideals, it is called *simple*.

(ii) If an algebra has no nontrivial Abelian ideals, it is called *semisimple*.

Note that all simple Lie algebras are also semisimple, while the converse statement is not true.

In general, semisimple algebras are direct sums of simple algebras.

Associated with the simple and semisimple Lie algebras are simple and semisimple Lie groups. We will later give a more technical definition, and say that a simple Lie algebra *generates* a simple Lie group. Similarly, a semisimple Lie algebra generates a semisimple Lie group. First, we will proceed in the opposite direction, and see how to obtain a Lie algebra from a Lie group.

6.2 Lie Algebra of a Lie Group

To define a Lie algebra, we first need a vector space. Since a Lie group G is a (smooth) manifold, we can use the space of vector fields on G. However, we will add one powerful additional requirement and restrict to a specific set of vector fields.

Let $a \in G$ be a given element. We first consider the left translation

$$L_a : G \to G, \quad g : L_a(g) = ag, \tag{6.1}$$

which is a diffeomorphism (since the product must be a smooth map, and we fix one of the elements in the product). Then we are going to use the associated push map

$(L_a)_*$, which maps a tangent vector at g to a tangent vector at ag. In general, given a vector field X on G, we could look at its values at points g and ag to obtain two tangent vectors. In general, these two have no simple relation. Now we are going to require that $X|_g$ is related to $X|_{ag}$ by the pushforward map: Then we say that a vector field X on G is *left-invariant*, if the pushforward satisfies

$$(L_a)_* X|_g = X|_{ag}. \tag{6.2}$$

The condition in Eq. (6.2) turns out to be a powerful requirement. In local coordinates, it means

$$(L_a)_* X|_g = X^\mu(g) \frac{\partial x^\alpha(ag)}{\partial x^\mu(g)} \frac{\partial}{\partial x^\alpha}\bigg|_{ag} = X|_{ag} = X^\alpha(ag) \frac{\partial}{\partial x^\alpha}\bigg|_{ag}, \tag{6.3}$$

and thus

$$X^\alpha(ag) = X^\mu(g) \frac{\partial x^\alpha(ag)}{\partial x^\mu(g)}. \tag{6.4}$$

In other words, if we know the components of $X|_g$, we can calculate the components of $X|_{ag}$. It then follows that a left-invariant vector field is *uniquely defined* by its value at a *single point*, for example, at the unit element $e \in G$, because

$$X|_g = (L_g)_* X_e \equiv L_{g*} V, \tag{6.5}$$

where $V = X|_e \in T_e G$. This means that we can construct the full left-invariant vector field on G starting from a single tangent vector at the "origin," the unit element e. This is not true for generic vector fields, by knowing its value at one point we cannot determine its value in some other arbitrary point.[1]

Let us denote the set of left-invariant vector fields by $\mathfrak{g}$. This set is a vector space (since the pushforward L_{g*} is a linear map) and is isomorphic to $T_e G$, hence its dimension is $\dim \mathfrak{g} = \dim G$.

Example Consider a left-invariant vector field on $GL(n, \mathbb{R})$. We start with a tangent vector V at the unit element e

$$V = V^{ij} \frac{\partial}{\partial x^{ij}}\bigg|_e \in T_e GL(n, \mathbb{R}), \tag{6.6}$$

where V^{ij} is an arbitrary $n \times n$ real matrix. Next we construct a left-invariant vector field X. We obtain its value at an arbitrary point g from V,

$$X|_g = L_{g*} V = V^{ij} \overbrace{\frac{\partial(x^{kl}(g)x^{lm}(e))}{\partial x^{ij}(e)}}^{=x^{km}(g)} \frac{\partial}{\partial x^{km}(g)} = V^{ij} x^{kl}(g) \delta_i^l \delta_j^m \frac{\partial}{\partial x^{km}(g)}$$

$$= V^{ij} x^{ki}(g) \frac{\partial}{\partial x^{kj}(g)} = \underbrace{x^{ki}(g) V^{ij}}_{(gV)^{kj}} \frac{\partial}{\partial x^{kj}(g)} = (gV)^{kj} \frac{\partial}{\partial x^{kj}(g)}, \tag{6.7}$$

and we are using as coordinates x on $GL(n, \mathbb{R})$ the entries of the matrix, $x^{ij}(g) = g^{ij}$.

---

[1] For example, by knowing the wind velocity at the North pole one cannot predict the wind velocity in, say, Beijing.

Since $\mathfrak{g}$ is a collection of vector fields, we can compute their commutators. The result is again left-invariant:

$$L_{a*} \, [X, Y]|_g = [L_{a*}X|_g, L_{a*}Y|_g]. \tag{6.8}$$

In turn, using left-invariance, we have $[L_{a*}X|_g, L_{a*}Y|_g] = [X|_{ag}, Y|_{ag}]$, hence we obtain

$$L_{a*} \, [X, Y]|_g = [X|_{ag}, Y|_{ag}] \equiv [X, Y]|_{ag}. \tag{6.9}$$

Thus, if X and Y are two arbitrary elements of $\mathfrak{g}$, then also $[X, Y]$ is an element of $\mathfrak{g}$.

We are now ready to define the *Lie algebra of a Lie group* G as the set of left-invariant vector fields $\mathfrak{g}$ with the commutator $[\, , \,] : \mathfrak{g} \times \mathfrak{g} \to \mathfrak{g}, \;\; (X, Y) \mapsto [X, Y]$.

Note that, since $\mathfrak{g}$ is isomorphic to the tangent space of the Lie group at the identity $T_e G$, it is justified and common to consider the latter as the Lie algebra associated with the Lie group. A common notation is to denote a Lie group with capital letters and the associated Lie algebra with small letters, e.g., the Lie group SU(2) and the Lie algebra su(2).

Recall also that in Eq. (4.16) we defined a tangent vector V as an operator associated with a curve $c(t)$, or an equivalence class of curves giving the same tangent vector, at a point $c(0) = p$, and there was a one-to-one correspondence between equivalence classes of curves and tangent vectors. The following example uses this to find the Lie algebras associated with some common matrix Lie groups. We will use the following helpful matrix identities:

$$\det M = \exp(\operatorname{tr} \ln M), \tag{6.10}$$

for a positive definite matrix M, so

$$\det(\mathbb{1}_n + A) = \exp[\operatorname{tr} \ln(\mathbb{1}_n + A)] \simeq \exp\left[\operatorname{tr}\left(A - \frac{1}{2}A^2 + \cdots\right)\right] \simeq 1 + \operatorname{tr} A + \cdots. \tag{6.11}$$

Example If we identify the Lie algebra with the tangent space at unity, we can construct the Lie algebras of common matrix groups.

1. For the Lie algebra gl$(n, \mathbb{R})$, we simply have:

$$\mathrm{gl}(n, \mathbb{R}) = \{A | \; A \text{ is an } n \times n \text{ real matrix}\}. \tag{6.12}$$

2. For sl$(n, \mathbb{R})$, we take a curve $c(t)$ that passes through the unit element (identity matrix) $e = \mathbb{1}_n \in \mathrm{SL}(n, \mathbb{R})$ with $c(0) = e$ and derive the tangent vector at $t = 0$. For small t,

$$c(t) \simeq \mathbb{1}_n + tA + \cdots, \tag{6.13}$$

so we find the tangent vector as

$$\left.\frac{dc}{dt}\right|_{t=0} = A \in T_e \, \mathrm{SL}(n, \mathbb{R}). \tag{6.14}$$

Now we use the unit determinant condition and the identity above to write:

$$1 = \det c(t) = \det(\mathbb{1}_n + tA) \simeq 1 + t \operatorname{tr} A + \cdots. \tag{6.15}$$

Thus

$$\operatorname{tr} A = 0 \tag{6.16}$$

and we can identify

$$\mathrm{sl}(n, \mathbb{R}) = \{A \mid A \text{ is an } n \times n \text{ real matrix, with } \operatorname{tr} A = 0\}. \tag{6.17}$$

3. For so(n), we start with a curve $c(t) = \mathbb{1}_n + tA$ and require $c(t)$ to be orthogonal:

$$\mathbb{1} = c(t)c(t)^T = (\mathbb{1} + tA)(\mathbb{1} + tA^T) = \mathbb{1} + t(A + A^T) + O(t^2). \tag{6.18}$$

Thus we need to have $A = -A^T$, so

$$\mathrm{so}(n) = \{A \mid A \text{ is an antisymmetric, real } n \times n \text{ matrix}\}. \tag{6.19}$$

For complex matrices, the coordinates are taken to be the real and imaginary parts of the matrix elements.

4. For u(n), we start from $c(t) = \mathbb{1}_n + tA$ and impose the unitarity condition

$$\mathbb{1} = c(t)c(t)^\dagger = (\mathbb{1} + tA)(\mathbb{1} + tA^\dagger) = \mathbb{1} + t(A + A^\dagger) + O(t^2) \tag{6.20}$$

to find the constraint $A = -A^\dagger$. Hence u(n) can be identified with the set of anti-Hermitian $n \times n$ complex matrices. Note, however, that in physics (motivated by quantum mechanics, where observables correspond to Hermitian operators) it is more frequent to use conventions in which t is multiplied by the imaginary unit i, i.e., to write $c(t) = \mathbb{1}_n + itA$. This leads to the constraint $A^\dagger = A$. With this convention:

$$\mathrm{u}(n) = \{A \mid A \text{ is a Hermitian } n \times n \text{ complex matrix}\}. \tag{6.21}$$

5. For su(n), imposing also the unit-determinant constraint, we find, with the physics conventions:

$$\mathrm{su}(n) = \{A \mid A \text{ is a Hermitian, traceless } n \times n \text{ complex matrix}\}. \tag{6.22}$$

6.2.1 The Exponential Map

The previous example showed how to find the Lie algebra associated with a (matrix) Lie group. Conversely, one can construct elements of the group from elements of the algebra using the *exponential map*. Recall that a vector field X on a manifold gives rise to a flow, and a one-parameter group of transformations satisfying the group properties (4.92):

$$\sigma_t = e^{tX}, \tag{6.23}$$

with $\sigma_t \circ \sigma_s = \sigma_{t+s}$, which defines how the points are mapped along the flow: $x \mapsto \sigma_t(x) = e^{tX}x$. This one-parameter group (for a fixed X) is an Abelian group. Conversely, given a one-parameter group σ_t one can find the associated vector field X by

$$X = \left.\frac{d\sigma_t}{dt}\right|_{t=0}. \tag{6.24}$$

Now in particular let X be an element of the Lie algebra $\mathfrak{g}$ associated with a Lie group G. Then $\exp(tX)$ defines a curve in G with the tangent vector X at unity, for $t = 0$.

Note that if we rescale X by some nonzero number λ, $X \mapsto X' \equiv \lambda X$, then X' gives rise to the same one-parameter subgroup. This motivates us to define the *exponential map* (denoted by exp) as the mapping from a Lie algebra $\mathfrak{g}$ to the associated Lie group G through

$$\exp : \mathfrak{g} \to G, \qquad X \mapsto \exp(X) = e^X. \tag{6.25}$$

Let $\dim \mathfrak{g} = \dim G = n$. Since the Lie algebra is also a vector space, we can choose a basis $\{X_1, X_2, \ldots, X_n\}$ and expand any element X in this basis,

$$X = \alpha_a X_a = \alpha_1 X_1 + \cdots + \alpha_n X_n, \tag{6.26}$$

where α_a are real numbers (recalling that the physics convention includes a factor i). Using the exponential map

$$\exp(X) = e^{\alpha_a X_a} \tag{6.27}$$

and varying the parameters α_a we can clearly obtain a subset U of elements of the Lie group around the unit element $e \in U$. Moreover, within U, we can use the *Baker–Campbell–Hausdorff formula*, named after Henry Frederick Baker, John Edward Campbell, and Felix Hausdorff,

$$\exp(X)\exp(Y) = \exp\left(X + Y + \frac{1}{2}[X, Y] + \frac{1}{12}([X, [X, Y]] + [Y, [Y, X]]) + \cdots\right) \tag{6.28}$$

to construct the product of two group elements.

In general, the image $\mathrm{Im}(\mathfrak{g})$ of the Lie algebra under exponentiation may be a proper subset of G, i.e., the exponential map may not be a surjection. However, it is a surjection at least in the following two important cases:

(i) G is compact and connected (as a manifold),
(ii) G is $\mathrm{GL}(n, \mathbb{R})$.

In these cases every element of G can be represented as e^X with some element of the Lie algebra.

Note that in physics conventions, one often includes a factor i in the exponent of the exponential map, writing the group elements as $e^{iX} = e^{i\alpha_a X_a}$.

Example   The group $\mathrm{SU}(2)$ is the manifold S^3, which is compact and connected. Every element $h \in \mathrm{SU}(2)$ can be written as e^{-iX}. The matrix group $\mathrm{O}(n)$ is not connected, it is a disjoint union of two connected components $\mathrm{SO}(n) = \mathrm{O}^+(n) = \{A \in \mathrm{O}(n) | \det A = +1\}$ and $\mathrm{O}^-(n) = \{A \in \mathrm{O}(n) | \det A = -1\}$, the former containing the unit element. Every element of $\mathrm{SO}(n)$ can be written in the form e^{-iX}, but the exponential map is not an injection.

The exponential map is also compatible with Lie group homomorphisms in the following way. Let $f : G \to H$ be a Lie group homomorphism (a smooth homomorphism). The associated pushforward f_* is a map between the Lie algebras, $f_* : \mathfrak{g} \to \mathfrak{h}$. One can show that

$$f(\exp(X)) = \exp(f_*(X)), \quad \forall X \in \mathfrak{g}. \tag{6.29}$$

Let us return to the basis $\{X_1, \ldots, X_n\}$. From now on, we call X_a the *generators of the Lie algebra*. If we think of the Lie algebra as the tangent space T_eG, and its elements as vectors, we can expand the commutator of a pair of basis vectors in the same basis,

$$[X_a, X_b] = f_{ab}{}^c X_c, \tag{6.30}$$

with coefficients $f_{ab}{}^c$. If instead we think of the Lie algebra as the space of left-invariant vector fields, there also exists a basis $\{X_a\}$ and we can expand a commutator in the same basis as in Eq. (6.30).

However, now a potential ambiguity could arise. For vector fields, the $f_{ab}{}^c$ need not be constant numbers: In general, one would expect them to be some smooth functions on the group manifold. If that was the case, there would exist two conflicting interpretations of the Lie algebra. In reality, however, the $f_{ab}{}^c$ turn out to be constants, known as *structure constants* of the Lie algebra. We discuss this in more detail next.

Note that the structure constants really define the Lie algebra. Consider the commutator $[Y, Z]$ of any pair of elements of the algebra. Expanding in the basis $\{X_a\}$,

$$Y = Y^a X_a, \qquad Z = Z^a X_a, \tag{6.31}$$

where the Y^a and Z^a are the components, we get

$$[Y, Z] = Y^a Z^b [X_a, X_b] = f_{ab}{}^c Y^a Z^b X_c \tag{6.32}$$

which determines the commutator uniquely, by determining its coefficients.

Structure Constants of the Lie Algebra

Let $\{V_1, \ldots, V_n\}$ be a basis of T_eG (assuming $\dim G = n < \infty$). Then $\{X_a|_g = L_{g*}V_i\}$, for $a = 1, \ldots, n$, is a basis of T_gG (usually not a coordinate basis). Since the vectors $\{V_1, \ldots, V_n\}$ are linearly independent, $\{X_1|_g, \ldots, X_n|_g\}$ are also linearly independent. The pushforward L_{g*} is an isomorphism between T_eG and T_gG, and $(L_{g*})^{-1} = L_{g^{-1}*}$. Since V_a are basis vectors of T_eG, one can expand the commutator

$$[V_a, V_b] = f_{ab}{}^c V_c. \tag{6.33}$$

Let us then push this to T_gG; on the one hand

$$L_{g*}[V_a, V_b] = [L_{g*}V_a, L_{g*}V_b] = [X_a|_g, X_b|_g]; \tag{6.34}$$

on the other hand

$$L_{g*}(f_{ab}{}^c V_c) = f_{ab}{}^c X_c|_g. \tag{6.35}$$

Thus, one obtains

$$[X_a|_g, X_b|_g] = f_{ab}{}^c X_c|_g. \tag{6.36}$$

Letting g vary over all G, we get the same equation everywhere on G with the same numbers $f_{ab}{}^c$. Thus we can write

$$[X_a, X_b] = f_{ab}{}^c X_c. \tag{6.37}$$

The constant coefficients $f_{ab}{}^c$ are called the *structure constants* of the Lie algebra.

Note that the antisymmetry of the commutator $[X_a, X_b] = -[X_b, X_a]$ implies that the structure constants have the antisymmetry property

$$f_{ab}{}^c = -f_{ba}{}^c. \tag{6.38}$$

Moreover, from the Jacobi identity for the commutator one can also derive the following identity for the structure constants,

$$f_{ab}{}^c f_{cd}{}^e + f_{bd}{}^c f_{ca}{}^e + f_{da}{}^c f_{cb}{}^e = 0. \tag{6.39}$$

Note that we have been using both upper and lower indices, and we ask if indices can be raised and lowered. As we will later learn, there is a metric to use for this purpose. However, in the case of semisimple Lie algebras that are our present focus, the metric turns out to be the flat Euclidean metric δ_{ab}. Consequently, the placement of the indices does not matter, i.e. they can be raised and lowered freely. Hence we will rewrite

$$f_{abc} \equiv f_{ab}{}^c, \tag{6.40}$$

so Eq. (6.39) can be rewritten as

$$f_{abc} f_{cde} + f_{bdc} f_{cae} + f_{dac} f_{cbe} = 0. \tag{6.41}$$

6.3 Irreducible Representations of Lie Algebras

We begin by studying some general features of representations of Lie algebras associated with compact simple Lie groups. For finite groups, Theorem 3.5, that we discussed in Chapter 3, showed that every finite-dimensional representation is unitary, and unitary representations are completely reducible. The corresponding statements for compact Lie groups are found in the following two theorems.

Theorem 6.1 *All finite-dimensional representations of a compact Lie group are equivalent to unitary representations.*

Proof Omitted.   □

Then, the complete reducibility of unitary representations is stated by the second part of the Peter–Weyl theorem. This theorem, formulated by Herman Weyl and Franz Peter in 1924, consists of three parts or separate statements; here, we consider only the second part.

Theorem 6.2 (Peter–Weyl theorem, part II) *Every unitary representation of a compact group G on a complex Hilbert space is completely reducible, i.e., can be fully decomposed into a direct sum of irreducible finite-dimensional unitary representations of G.*

Proof Omitted.   □

From the above theorems, it is well motivated to focus on m-dimensional unitary irreducible representations of a compact Lie group G. Let N denote the dimension of G, meaning that the elements of G are parameterized by N real parameters α_a, with $a = 1, \ldots, N$. In an m-dimensional representation D, the group elements $g(\alpha)$ are represented by $m \times m$ matrices $D(\alpha)$, labeled by the parameters $\alpha = (\alpha_a)$. We assume that the origin of the parameters has been chosen so that the unit element e corresponds to $D(\alpha)|_{\alpha_a=0} = \mathbb{1}_m$. The associated Lie algebra $\mathfrak{g}$ is given by the tangent space at the unit element. The elements of a basis $\{X_a\}$ with $a = 1, \ldots, N$ are the *generators* of the corresponding representation of the Lie algebra.

Since the group is compact, we can use the exponential map to express the group elements (albeit not necessarily uniquely). In the physics convention, which includes a factor i in the exponent, we write

$$g(\alpha) = \exp(i\alpha_a X_a), \qquad (6.42)$$

where we identified the parameters α_a with the components of the Lie algebra element $X = -i \ln g(\alpha)$ expanded in the generator basis. Moving to the representation D, the matrices $D(\alpha)$ corresponding to the group elements are unitary, and the parameters α_a are real, so the generators X_a are represented by $m \times m$ matrices, denoted as $T(X_a)$, which must be Hermitian: $T(X_a)^\dagger = T(X_a)$. These matrices give a representation T of the Lie algebra $\mathfrak{g}$.

Let us discuss the relation between the representations D of the group and T of the algebra in more detail. A representation T of a Lie algebra $\mathfrak{g}$ is a linear map

$$T : \mathfrak{g} \to \mathrm{End}(V), \quad X \mapsto T(X), \qquad (6.43)$$

where $T(X)$ is an endomorphism of the vector space V, that is, a linear map $V \to V$. As before, the linear map $T(X)$ is identified with a matrix (of coefficients), once a basis in V has been chosen. An endomorphism need not be an automorphism, i.e., it is not necessarily invertible. For example, the matrices $T(X)$ may have zero eigenvalues, so that the determinant vanishes. T must also be a Lie algebra homomorphism, mapping commutators to commutators:

$$T([X, Y]) = T(X)T(Y) - T(Y)T(X). \qquad (6.44)$$

One can realize that Eq. (6.42) implies a relation between the unitary representation D of the Lie group and the Hermitian representation T of the Lie algebra. On one hand, we can construct the representation T of the Lie algebra $\mathfrak{g}$ from the representation D, using curves through the unit element. Let X be the tangent vector of a curve $g(\alpha(t))$ through the unit element at $t = 0$, namely $g(\alpha(0)) = e$. Using the representation D we then construct $T(X)$ as

$$T(X) = \left.\frac{dD\left(g(\alpha(t))\right)}{dt}\right|_{t=0}. \qquad (6.45)$$

Repeating this for every X yields a representation T. The dimension of T is the same as that of D. Conversely, using first the exponential map to write a Lie group element $g(\alpha)$ as in Eq. (6.42), we can construct $D(g(\alpha))$ using the representation T for $T(X_a)$, rewriting Eq. (6.42) as

$$D\left(g(\alpha)\right) = D\left(\exp(i\alpha_a X_a)\right) = \exp\left(i\alpha_a T(X_a)\right). \tag{6.46}$$

Furthermore, one can show that if D_1 and D_2 are equivalent representations of a Lie group, the corresponding representations T_1 and T_2, obtained through Eq. (6.45), are also equivalent. The converse statement holds for a connected Lie group. Also, if T is irreducible, the corresponding D through Eq. (6.46) is also irreducible, and, again, the converse is true for a connected Lie group. We conclude that, at least as long as we are interested in simple, compact and connected Lie groups, we can use the same symbol D for the representation of the Lie algebra and the Lie group.

In what follows, we mostly adopt a short notation and denote the matrix representing a generator with the same symbol X_a, instead of $T(X_a)$ or $D(X_a)$. In addition to the convention of including an i factor into the exponent of the exponential map, we modify the structure constants by factoring out an i term, i.e.,

$$[X_a, X_b] = if_{abc}X_c. \tag{6.47}$$

With this convention, we note that the structure constants must be real numbers: This can be proven by noting that, since the generators are Hermitian, one has $X_a^\dagger = X_a$ and $X_b^\dagger = X_b$. On the one hand,

$$[X_a, X_b]^\dagger = (X_a X_b - X_b X_a)^\dagger = X_b^\dagger X_a^\dagger - X_a^\dagger X_b^\dagger = X_b X_a - X_a X_b$$
$$= [X_b, X_a] = -[X_a, X_b] = -if_{abc}X_c. \tag{6.48}$$

On the other hand, Eq. (6.47) implies

$$[X_a, X_b]^\dagger = \left(if_{abc}X_c\right)^\dagger = -if_{abc}^\star X_c^\dagger = -if_{abc}^\star X_c. \tag{6.49}$$

Comparing Eq. (6.48) and Eq. (6.49), we see that the structure constants must be real:

$$f_{abc}^\star = f_{abc}. \tag{6.50}$$

6.4 The Defining Representation and the Adjoint Representation

6.4.1 The Defining Representation

There are two important irreducible representations to be discussed first. The first one we have already seen: The matrix Lie groups $GL(n, \mathbb{C})$ and their various subgroups were defined using matrices, which are of course a representation of the underlying abstract group structure. This is the *defining representation*. For example, in the case of the SU(2) Lie group and su(2) Lie algebra, the three generators (usually called J_a) are represented by 2×2 matrices, proportional to the Pauli matrices:

$$J_1 = \frac{\sigma_1}{2} = \frac{1}{2}\begin{pmatrix} 0 & 1 \\ 1 & 0 \end{pmatrix}, \quad J_2 = \frac{\sigma_2}{2} = \frac{1}{2}\begin{pmatrix} 0 & -i \\ i & 0 \end{pmatrix}, \quad J_3 = \frac{\sigma_3}{2} = \frac{1}{2}\begin{pmatrix} 1 & 0 \\ 0 & -1 \end{pmatrix}. \tag{6.51}$$

It should be remarked that there is an arbitrariness involved in the choice of the normalization of the generators (which is reflected in the values of the structure constants), so the defining representation is not unique. The choice (6.51) is the choice of a basis of a defining representation of su(2) for which the structure constants defined as in Eq. (6.47) are $f_{abc} = \epsilon_{abc}$.

The dimension m of the defining representation is usually not the same as the dimension N of the Lie group and algebra. However, that is the case for another important representation, the *adjoint representation*. We define it first through a long and detailed route, using what we have learned about manifolds, vector fields, and group action. In the end we arrive at a simple recipe for constructing the matrices representing the generators of the Lie algebra, which offers another, more accessible, way to introduce the adjoint representation. The impatient reader may skip the longer discussion and jump to Eq. (6.62).

6.4.2 The Adjoint Representation

We begin by considering the action of the Lie group G on itself by conjugation. Let b be some element of G, $b \in G$. We denote the conjugation by b as the map ad_b,

$$\mathrm{ad}_b : G \to G, \quad \mathrm{ad}_b(g) \equiv \mathrm{ad}_b\, g = bgb^{-1}. \tag{6.52}$$

This is a homomorphism: $\mathrm{ad}_b\, g_1 \cdot \mathrm{ad}_b\, g_2 = \mathrm{ad}_b(g_1 g_2)$, and $\mathrm{ad}_b \cdot \mathrm{ad}_c = \mathrm{ad}_{bc}$, $(\mathrm{ad}_b)^{-1} = \mathrm{ad}_{b^{-1}}$, $\mathrm{ad}_e = \mathrm{id}_G$. (Note that this is really a composite map: $\mathrm{ad}_b \cdot \mathrm{ad}_c \equiv \mathrm{ad}_b \circ \mathrm{ad}_c$.) The differential map ad_{b*} pushes vectors from $T_g G$ to $T_{\mathrm{ad}_b\, g} G$. If $g = e$, $\mathrm{ad}_b\, e = beb^{-1} = e$, so ad_{b*} maps $T_e G$ to itself. It is also linear and invertible, with $(\mathrm{ad}_{b*})^{-1} = (\mathrm{ad}_{b^{-1}})_*$, thus an automorphism. Let us denote this map by Ad_b:

$$\mathrm{Ad}_b : T_e G \to T_e G, \quad \mathrm{Ad}_b = \mathrm{ad}_{b*}|_{T_e G}. \tag{6.53}$$

One can easily show that $(f \circ g)_* = f_* \circ g_*$, thus $\mathrm{ad}_{b*}\, \mathrm{ad}_{c*} = \mathrm{ad}_{bc*}$. It then follows that with Ad_b we can construct a representation of G in the Lie algebra $\mathfrak{g} \cong T_e G$, the *adjoint representation*:

$$\mathrm{Ad} : G \to \mathrm{Aut}(\mathfrak{g}), \quad b \mapsto \mathrm{Ad}_b. \tag{6.54}$$

Let us assume that G is compact, and $Y \in T_e G \cong \mathfrak{g}$. We use the exponential map to generate a curve through the unit element, $h(t) = \exp(tY)$. Then, taking a derivative of

$$\mathrm{ad}_g\, h(t) = g e^{tY} g^{-1} \approx g(\mathbb{1} + tY + \cdots)g^{-1} \tag{6.55}$$

at $t = 0$ gives

$$\mathrm{Ad}_g\, Y = gYg^{-1}. \tag{6.56}$$

Given a basis $\{X_a\}$ of $\mathfrak{g}$, the group elements g are represented in the adjoint representation D^{adj} by the matrices $D^{\mathrm{adj}}(g)$. We follow a convention and define their action from the right by

$$gX_a g^{-1} = X_b \left(D^{\mathrm{adj}}(g)\right)^b{}_a. \tag{6.57}$$

Continuing as in the above, we replace the conjugating element g by an element $g(t) = e^{tX}$ belonging to a curve through the unit element generated by $X \in \mathfrak{g}$, and then take the derivative at $t = 0$:

$$g(t)Yg(t)^{-1} = e^{tX}Ye^{-tX} \approx Y + t[X, Y] + \cdots, \tag{6.58}$$

where the derivative defines the map

$$\text{ad}_X : \mathfrak{g} \to \mathfrak{g}, \qquad Y \mapsto [X, Y], \tag{6.59}$$

which in turn defines the *adjoint representation of the Lie algebra* $\mathfrak{g}$ in $\mathfrak{g}$,

$$\text{ad} : \mathfrak{g} \to End(\mathfrak{g}), \qquad X \mapsto \text{ad}_X. \tag{6.60}$$

This is a rather abstract way to define the adjoint representation. Let us derive something more user-friendly. Consider a basis of generators X_a. Using the structure constants f_{abc} we get

$$\text{ad}_{X_a} X_b = [X_a, X_b] = if_{abc}X_c. \tag{6.61}$$

Comparing Eq. (6.61) with the right-hand side of Eq. (6.56), substituting a curve $g(t)$ and taking a derivative with respect to t at the origin, we introduce a matrix $D^{\text{adj}}(X_a)$ with the components

$$[D^{\text{adj}}(X_a)]_{bc} = -if_{abc}. \tag{6.62}$$

We see that the adjoint representation of the Lie algebra means simply representing the generators X_a with the $N \times N$ matrices in Eq. (6.62); the action in $\mathfrak{g}$ in Eq. (6.61) becomes simple matrix multiplication. Note that since the matrix elements are defined by the structure constants which depend on the choice of basis, there is some ambiguity. There exists a preferred choice of basis, and we will introduce it shortly.

Example   A preferred basis for su(2) is one in which the commutation relations are

$$[J_a, J_b] = i\epsilon_{abc}J_c. \tag{6.63}$$

In the adjoint representation we represent the generators by the 3×3 matrices with components $(J_a) = -i\epsilon_{abc}$. Explicitly:

$$J_1 = \begin{pmatrix} 0 & 0 & 0 \\ 0 & 0 & -i \\ 0 & i & 0 \end{pmatrix}, \quad J_2 = \begin{pmatrix} 0 & 0 & i \\ 0 & 0 & 0 \\ -i & 0 & 0 \end{pmatrix}, \quad J_3 = \begin{pmatrix} 0 & -i & 0 \\ i & 0 & 0 \\ 0 & 0 & 0 \end{pmatrix}. \tag{6.64}$$

With the adjoint representation, we can define a scalar product in the Lie algebra. We define a bilinear form

$$B : \mathfrak{g} \times \mathfrak{g} \to \mathbb{C}, \quad (X, Y) \mapsto B(X, Y) \tag{6.65}$$

by setting

$$B(X, Y) = \text{tr}(\text{ad}_X \circ \text{ad}_Y). \tag{6.66}$$

In terms of the matrices $D^{\text{adj}}(X)$ that represent the elements X, the composite map is just the matrix product, and we get a simpler definition

$$B(X, Y) = \text{tr}(D^{\text{adj}}(X)D^{\text{adj}}(Y)). \tag{6.67}$$

The bilinear form B is called the *Killing form*. Specifying a basis of generators X_a, the Killing form is defined by the matrix with coefficients

$$B_{ab} \equiv B(X_a, X_b) = \text{tr}\left(D^{\text{adj}}(X_a)D^{\text{adj}}(X_b)\right). \tag{6.68}$$

In the preferred basis where f_{abc} is completely antisymmetric, we get

$$B_{ab} = \text{tr}\left(D^{\text{adj}}(X_a)D^{\text{adj}}(X_b)\right) = (-if_{acd})(-if_{bdc}) = f_{acd}f_{bcd}. \tag{6.69}$$

Example Using the adjoint representation of $su(2)$ above, explicit calculation of the matrix products and traces yields

$$B_{ab} = 2\delta_{ab}. \tag{6.70}$$

As we will discuss in detail below, in the preferred basis we get

$$B_{ab} = \lambda_{\text{adj}}g_{ab}, \tag{6.71}$$

where λ_{adj} is the *Dynkin index* of the adjoint representation, named after Eugene Dynkin, and g_{ab} can be thought of as a metric. For compact semisimple Lie algebras, it is just the Euclidean metric $g_{ab} = \delta_{ab}$. The metric and its inverse can be used to raise and lower indices, as we learned in Chapter 5,

$$g^{ab} = (g^{-1})_{ab}, \qquad g^{ab}g_{bc} = \delta_c^a, \qquad X^a = g^{ab}X_c. \tag{6.72}$$

When the metric is the simple Euclidean metric, indices can be raised and lowered trivially. This justifies for us the notations $f_{abc} = f_{ab}{}^c$ and $X_a = X^a$ for the structure constants and generators.

6.4.3 Dynkin Index, Casimir Operators, and Anomalies

Let us assume that we are working in a finite-dimensional representation D where the $N = \dim \mathfrak{g}$ generators X_a are matrices.[2] From now on, we will simply write X_a instead of $D(X_a)$. We define a $N \times N$ matrix M with elements

$$M_{ab} = \text{tr}(X_a X_b). \tag{6.73}$$

M is symmetric and real, so we can diagonalize it by an orthogonal matrix O:

$$M_{\text{diag}} = OMO^T, \tag{6.74}$$

and move to a new basis of generators $X'_a = (OX)_a$. The eigenvalues in M_{diag} are real. For the nonzero eigenvalues λ_a we can do a further rescaling of the corresponding generator, $X'_a \to rX'_a$ with $r \in \mathbb{R} \setminus \{0\}$; this will rescale $\lambda_a \to r^2\lambda_a$ without changing its sign. In the case of compact semisimple Lie groups and algebras, all λ_a will be positive, so by rescaling the generators we can adjust them to the same positive value λ_D. In other words, we can choose a basis of generators where

$$\text{tr}(X_a X_b) = \lambda_D \delta_{ab}, \qquad \lambda_D > 0, \tag{6.75}$$

where we dropped the primes from our notation for the generators. Importantly, the numerical value of λ_D is not independent of the representation D, as different representations involve matrices of different size, which has an effect on the trace,

---

[2] Note that in Section 6.3 we used the notation $T(X_a)$ for a representation of the algebra, and D for a representation of the group, while here we use D for a representation of the algebra.

and we have already fixed the basis of the generators and the possibility to rescale them.[3] The number λ_D is called the *Dynkin index of the representation D.*

Example For the su(2) Lie algebra, it is common to use a basis where the generators J_a, with $a \in \{1, 2, 3\}$, are related in the defining representation D to the Pauli matrices σ_a by $J_a = \frac{1}{2}\sigma_a$, according to Eq. (6.51). Then, the Dynkin index for the defining representation D is

$$\text{tr}(J_a J_b) = \frac{1}{4}\,\text{tr}(\sigma_a \sigma_b) = \frac{1}{4}[\delta_{ab}\,\text{tr}(\mathbb{1}_2) + i\epsilon_{abc}\,\text{tr}(\sigma_c)] = \frac{1}{2}\delta_{ab} \tag{6.76}$$

so the Dynkin index for the defining representation is $\lambda_D = \frac{1}{2}$.

On the other hand, in the adjoint representation in Eq. (6.64) we get (e.g., by an explicit calculation) $\lambda_{D^{\text{adj}}} = 2$. This can be checked either by explicit calculation, or by using the property

$$\sum_{l=1}^{3} \epsilon_{aml}\epsilon_{bol} = \delta_{ab}\delta_{mo} - \delta_{ao}\delta_{mb} \tag{6.77}$$

and the fact that in the adjoint su(2) representation one has $(J_a)_{bc} = -i\epsilon_{abc}$, so that

$$
\begin{aligned}
\text{tr}(J_a J_b) &= \sum_{l=1}^{3} (J_a J_b)_{ll} \\
&= \sum_{l=1}^{3}\sum_{m=1}^{3} (J_a)_{lm}(J_b)_{ml} \\
&= (-i)^2 \sum_{l=1}^{3}\sum_{m=1}^{3} \epsilon_{alm}\epsilon_{bml} \\
&= -\sum_{l=1}^{3}\sum_{m=1}^{3} \epsilon_{alm}\epsilon_{bml} \\
&= \sum_{l=1}^{3}\sum_{m=1}^{3} \epsilon_{aml}\epsilon_{bml} \\
&= \sum_{m=1}^{3} (\delta_{ab}\delta_{mm} - \delta_{am}\delta_{mb}) \\
&= 3\delta_{ab} - \delta_{ab} \\
&= 2\delta_{ab}.
\end{aligned} \tag{6.78}
$$

For a direct-sum representation $D_1 \oplus D_2$ the Dynkin index satisfies

$$\lambda_{D_1 \oplus D_2} = \lambda_{D_1} + \lambda_{D_2} \tag{6.79}$$

and for a tensor-product representation $D_1 \otimes D_2$,

$$\lambda_{D_1 \otimes D_2} = (\dim D_1)\lambda_{D_2} + (\dim D_2)\lambda_{D_1}. \tag{6.80}$$

We leave it as an exercise to prove these properties (see Problem 6.5).

---

[3] Note that there is an arbitrariness in λ_D, corresponding to the arbitrariness in the choice of the X_as that we mentioned at the beginning of this section.

In this choice of basis, the structure constants f_{abc} are completely antisymmetric. This can be seen explicitly by writing

$$f_{abc} = \frac{-i}{\lambda_D} \, \text{tr}([X_a, X_b] X_c). \tag{6.81}$$

We define next the *second-order Casimir operator* (or *quadratic Casimir element*) $C^{(2)}$, named after Hendrik Casimir, as

$$C^{(2)} = \sum_{a=1}^{N} X_a X_a = X_1^2 + \cdots + X_N^2. \tag{6.82}$$

One can check (see Problem 6.6) that it commutes with all the generators,

$$\left[C^{(2)}, X_a \right] = 0, \tag{6.83}$$

and hence with all the elements of the Lie algebra. By the corollary of Schur's lemma, it must then be proportional to the identity operator. The constant of proportionality depends on the representation. Let D be an irreducible representation with $\dim D = n$. In D then

$$C^{(2)} = C_2(D) \, \mathbb{1}_n, \tag{6.84}$$

the constant $C_2(D)$ is called the *quadratic Casimir* (or the *eigenvalue* of the second-order Casimir operator) of the representation D. There exists a relation between the Dynkin index λ_D, the quadratic Casimir, the dimension of the representation, and the dimension of the algebra.[4] We derive it as follows:

$$\lambda_D \dim \mathfrak{g} = \lambda_D \sum_{a=1}^{\dim \mathfrak{g}} \delta_{aa} = \sum_{a=1}^{\dim \mathfrak{g}} \text{tr}(X_a X_a)$$

$$= \text{tr}(C^{(2)}) = C_2(D) \, \text{tr}(\mathbb{1}_n) = n C_2(D). \tag{6.85}$$

We thus find that

$$\lambda_D = \frac{\dim D}{\dim \mathfrak{g}} C_2(D). \tag{6.86}$$

Example We will later find that the irreducible representations of su(2) are the spin-j representations with $j = 0, \frac{1}{2}, 1, \frac{3}{2}, \ldots$, having dimension $2j + 1$ and the quadratic Casimir $C_2(j) = j(j + 1)$. Let us check the validity of Eq. (6.86) for the defining and for the adjoint representation. The $j = \frac{1}{2}$ representation is the defining representation with dimension 2. Applying Eq. (6.86) we should get

$$\lambda_{\frac{1}{2}} = \frac{2}{3} \cdot \frac{1}{2} \left(\frac{1}{2} + 1 \right) = \frac{1}{2}, \tag{6.87}$$

which matches our earlier result. The $j = 1$ representation has dimension 3, it is the adjoint representation, with quadratic Casimir $C_2(j = 1) = 2$. Now we should get the Dynkin index

---

[4] Note that the arbitrariness in the normalization of the generators that we mentioned above, induces an arbitrariness in the value of the quadratic Casimir.

$$\lambda_1 = \frac{3}{3} \cdot 2 = 2, \tag{6.88}$$

again matching our earlier result.

As the name suggests, the second-order Casimir operator generalizes to nth-order Casimir operators. They are defined as follows. Let $g^{a_1 \cdots a_n}$ be a totally symmetric tensor with respect to its indices, $g^{a_{p(1)} \cdots a_{p(n)}} = g^{a_1 \cdots a_n}$ for any permutation $p \in S_n$. If the operator

$$C^{(n)} = g^{a_1 \cdots a_n} X_{a_1} \cdots X_{a_n} \tag{6.89}$$

commutes with all generators X_a, and hence all elements of the Lie algebra $\mathfrak{g}$, it is called a *nth-order Casimir operator*. For example, for the su(N) Lie algebra, there exist Casimir operators for orders $n = 2, 3, \ldots, N$. Note that, in general, higher-order Casimir operators are not unique. For example, one can modify the fourth-order Casimir operator by adding a second-order polynomial in the second-order Casimir operator. While higher-order Casimir operators are beyond the scope of this book, it is worthwhile making some comments about third-order Casimir operators.

Let us begin with the generators X_a represented in a choice of defining representation. We define a $(3, 0)$-tensor d_{abc} by

$$d_{abc} = \text{tr}(\{X_a, X_b\} X_c), \tag{6.90}$$

where

$$\{X_a, X_b\} = X_a X_b + X_b X_a \tag{6.91}$$

is the *anticommutator* of X_a and X_b. One can check that d_{abc} is completely symmetric. Then, moving to generators in a generic representation D, we find that

$$\text{tr}\left(\{D(X_a), D(X_b)\} D(X_c)\right) = A(D) d_{abc}, \tag{6.92}$$

and we get a relation modified by a representation-dependent coefficient $A(D)$. It is called the *anomaly* of the representation. The name derives from gauge field theories. Those are theories that are constructed starting from a classical symmetry, described by a Lie group, which is "gauged" by making the symmetry transformation parameters spacetime-dependent (i.e., promoted to functions). Elementary particles are then described as excitations of quantum fields, which are chosen from the irreducible representations of the symmetry group. The invariance of the theory under the symmetry is valid at the classical level, but may be destroyed in the quantization of the theory, by the regularization and renormalization process that is involved. If that happens, then the symmetry is said to be *anomalous*. For the quantum theory to be well-behaved and consistent, it is usually crucial that the gauge symmetry holds at the quantum level and does not become anomalous. Typically, checking for the absence of an anomaly can be done by calculations involving Eq. (6.92), and it becomes a simple algebraic task to check if an anomaly can be avoided by computing the sum of anomalies of all of the involved representations, and verifying if the total anomaly is zero.

Given two representations D_1 (of dimension $\dim D_1$) and D_2 (of dimension $\dim D_2$), the anomaly of their direct sum and the anomaly of their tensor product are respectively given by

$$A(D_1 \oplus D_2) = A(D_1) + A(D_2)$$
$$A(D_1 \otimes D_2) = (\dim D_1) \cdot A(D_2) + (\dim D_2) \cdot A(D_1), \tag{6.93}$$

which are analogous to those of the Dynkin index, Eq. (6.79) and Eq. (6.80).

Moreover, anticipating the definition of the *complex-conjugate representation* $\overline{D}$ of a generic representation D (in which the generators are represented by the matrices X_a) as the one in which the generators are represented by the matrices $-X_a^*$, the anomaly also satisfies the following property:

$$A\left(\overline{D}\right) = -A(D). \tag{6.94}$$

6.4.4 The Algebra of the SU(2) Group

The generators of the special unitary group SU(2) make up the smallest simple non-Abelian Lie algebra, which we can denote as su(2). As we shall see, it provides us with a simple example where we learn properties that can be generalized later to other Lie algebras.

Conventionally, however, it is more common to define the su(2) algebra in terms of three generators denoted as J_a, with $a \in \{1, 2, 3\}$. The J_a generators satisfy the commutation relations

$$[J_a, J_b] = i\epsilon_{abc}J_c, \quad \text{i.e.} f_{abc} = \epsilon_{abc}, \tag{6.95}$$

which means $[J_1, J_2] = iJ_3$ and cyclic permutations thereof: $[J_2, J_3] = iJ_1$ and $[J_3, J_1] = iJ_2$.

To find the finite-dimensional irreducible representations, let us suppose that J_3 is represented by an $N \times N$ Hermitian matrix. Let us diagonalize J_3, and let us denote the eigenvalues and eigenvectors as j and $\vec{v}_j$, respectively:

$$J_3\vec{v}_j = j\vec{v}_j. \tag{6.96}$$

Note that the eigenvectors are generally complex (while the eigenvalues are real, because J_3 is Hermitian). As some eigenvectors could be degenerate, let us include an additional label η:

$$J_3\vec{v}_{j,\eta} = j\vec{v}_{j,\eta}, \quad \eta \in \{1, 2, \ldots\}. \tag{6.97}$$

We define the scalar product between complex N-component vectors:

$$\langle \vec{v} | \vec{w} \rangle = \sum_{i=1}^{N} v_i^{\star} w_i, \tag{6.98}$$

and take the eigenvectors $|j, \eta\rangle \equiv \vec{v}_{j,\eta}$ to be orthonormal under the scalar product:

$$\langle j, \eta | j, \beta \rangle \equiv \langle \vec{v}_{j,\eta} | \vec{v}_{j,\beta} \rangle = \delta_{\eta\beta}. \tag{6.99}$$

Now, we define the two linear combinations of J_1 and J_2:

$$J^{\pm} = \frac{J_1 \pm iJ_2}{\sqrt{2}}, \tag{6.100}$$

such that $(J^{\pm})^{\dagger} = J^{\mp}$. They satisfy the commutation relations

$$[J_3, J^{\pm}] = \pm J^{\pm} \tag{6.101}$$

and

$$[J^+, J^-] = J_3. \tag{6.102}$$

From Eq. (6.101) it follows that J^+ maps an eigenvector of J_3 with eigenvalue j to an eigenvector with eigenvalue $j + 1$:

$$\begin{aligned} J_3 J^+ |j, \eta\rangle &= ([J_3, J^+] + J^+ J_3)|j, \eta\rangle \\ &= J^+ |j, \eta\rangle + J^+ J_3 |j, \eta\rangle \\ &= (j + 1)J^+ |j, \eta\rangle, \end{aligned} \tag{6.103}$$

so that J^+ is said to be a *raising operator*.

Similarly, Eq. (6.101) also implies that J^- maps an eigenvector of J_3 with eigenvalue j to an eigenvector with eigenvalue $j - 1$:

$$\begin{aligned} J_3 J^- |j, \eta\rangle &= ([J_3, J^-] + J^- J_3)|j, \eta\rangle \\ &= -J^- |j, \eta\rangle + J^- J_3 |j, \eta\rangle \\ &= (j - 1)J^- |j, \eta\rangle, \end{aligned} \tag{6.104}$$

and J^- is said to be a *lowering operator*. Raising and lowering operators are sometimes called *ladder operators*.

Let us now consider the *highest vector*, i.e., the eigenvector of J_3 with the highest eigenvalue j, that we denote as $|j, \eta\rangle$. If there is more than one such highest vector, we assume them to be orthonormalized: $\langle j, \eta | j, \rho \rangle = \delta_{\eta, \rho}$.

As j is the highest eigenvalue of J_3, and since J^+ increases the eigenvalue of J_3 by a unit, one must have

$$J^+ |j, \eta\rangle = 0. \tag{6.105}$$

On the other hand, acting on $|j, \eta\rangle$ with J^-, one obtains an eigenvector of J_3 with eigenvalue $J_3 = j - 1$:

$$J^- |j, \eta\rangle \equiv N_{j,\eta} |j - 1, \eta\rangle. \tag{6.106}$$

States with different η are again orthogonal to each other:

$$\begin{aligned} N_{j,\beta}^\star N_{j,\eta} \langle j - 1, \beta | j - 1, \eta \rangle &= \langle j, \beta | J^+ J^- | j, \eta \rangle \\ &= \langle j, \beta | [J^+, J^-] | j, \eta \rangle \\ &= \langle j, \beta | J_3 | j, \eta \rangle \\ &= j \delta_{\eta, \beta}. \end{aligned} \tag{6.107}$$

Thus, choosing $N_{j,\eta} = \sqrt{j} \equiv N_j$, the $|j - 1, \eta\rangle$ states are orthonormalized. Then

$$J^+ |j - 1, \eta\rangle = \frac{1}{N_j} J^+ J^- |j, \eta\rangle = \frac{1}{N_j} [J^+, J^-]|j, \eta\rangle = N_j |j, \eta\rangle. \tag{6.108}$$

Note that J^+ and J^- increase or decrease j, but leave η unchanged.

Similarly, one can then construct states $|j - 2, \eta\rangle$:

$$J^- |j - 1, \eta\rangle = N_{j-1} |j - 2, \eta\rangle, \quad J^+ |j - 2, \eta\rangle = N_{j-1} |j - 1, \eta\rangle, \tag{6.109}$$

$$J^- |j - 2, \eta\rangle = N_{j-2} |j - 3, \eta\rangle, \quad J^+ |j - 3, \eta\rangle = N_{j-2} |j - 2, \eta\rangle, \tag{6.110}$$

$\ldots,$

etc., to

$$\mathcal{J}^-|j-k,\eta\rangle = N_{j-k}|j-k-1,\eta\rangle, \qquad \mathcal{J}^+|j-k-1,\eta\rangle = N_{j-k}|j-k,\eta\rangle. \quad (6.111)$$

The Ns satisfy the recursion relation $N_{j-k}^2 = j - k + N_{j-k+1}^2$; this can be proven by noting that

$$
\begin{aligned}
N_{j-k}^2 &= \langle j-k, \eta | \mathcal{J}^+ \mathcal{J}^- | j-k, \eta \rangle \\
&= \langle j-k, \eta | [\mathcal{J}^+, \mathcal{J}^-] | j-k, \eta \rangle + \langle j-k, \eta | \mathcal{J}^- \mathcal{J}^+ | j-k, \eta \rangle \\
&= j-k + N_{j-k+1}^2.
\end{aligned}
\quad (6.112)
$$

The recursion relation can be rewritten as $N_{j-k}^2 - N_{j-k+1}^2 = j - k$. Summing over the values of k from 0 to an arbitrary i, one obtains

$$\sum_{k=0}^{i} \left(N_{j-k}^2 - N_{j-k+1}^2 \right) = \sum_{k=0}^{i} (j-k) = j(i+1) - \frac{i(i+1)}{2} = \frac{(2j-i)(i+1)}{2}. \quad (6.113)$$

On the other hand, in the sum in the first term of Eq. (6.113) the first term in the parenthesis for each value of k cancels against the second term in the parenthesis for the following value of k, leaving only the second term in the parenthesis for the first value of k (i.e., for $k = 0$) and the first term in the parenthesis for the last value of k (i.e., for $k = i$):

$$\sum_{k=0}^{i} \left(N_{j-k}^2 - N_{j-k+1}^2 \right) = -N_{j+1}^2 + N_{j-i}^2. \quad (6.114)$$

Using the fact that $N_{j+1}^2 = 0$, the $\sum_{k=0}^{i} \left(N_{j-k}^2 - N_{j-k+1}^2 \right)$ sum reduces to N_{j-i}^2, so we end up with

$$N_{j-i}^2 = \frac{(2j-i)(i+1)}{2}. \quad (6.115)$$

The finiteness of the dimension of the representation implies that there must exist an integer, nonnegative value of i, that we can call q, such that $\mathcal{J}^-|j-q,\eta\rangle = 0$. This requires $q = 2j$, hence the possible values for the highest eigenvalue of J_3 in a finite-dimensional representation of the su(2) algebra are

$$j = \frac{q}{2}, \qquad q \in \mathbb{N}. \quad (6.116)$$

As a consequence, the representation breaks up into subspaces, labeled by η, of dimension $2j + 1$ (the eigenvalues of J_3 being $j, j - 1, \ldots, -j$) and decomposes as the direct sum $\oplus_\eta V^\eta$, with dim $V^\eta = 2j + 1$. This shows that the representation, in general, is reducible. However, for every irreducible representation, there can be only one η. In conclusion: Unitary irreducible representations of SU(2) are $2j + 1$ dimensional, characterized by their highest J_3 eigenvalue j (called the highest weight of the representation), with j being integer or half-integer, $j = q/2$, with $q \in \mathbb{N}$.

In quantum mechanics, a representations with highest weight j is called a *spin-j representation*, since j is associated with the spin angular momentum of a particle measured in units of the reduced Planck constant $\hbar$:

$$J^2 = j(j+1), \qquad (J^2 = J_1^2 + J_2^2 + J_3^2). \quad (6.117)$$

All reducible representations reduce to a direct sum of spin-j representations.

The minimal value that j can take is $j = 0$. This corresponds to the *trivial representation*; this is a representation of dimension one, in which all elements of the Lie algebra are represented by zero and, correspondingly, all group elements are represented by the unit. The simplest nontrivial representation has spin $j = 1/2$, and is generated by

$$J_a = \frac{1}{2}\sigma_a, \tag{6.118}$$

where σ_a are the Hermitian, traceless Pauli matrices:

$$\sigma_1 = \begin{pmatrix} 0 & 1 \\ 1 & 0 \end{pmatrix}, \quad \sigma_2 = \begin{pmatrix} 0 & -i \\ i & 0 \end{pmatrix}, \quad \sigma_3 = \begin{pmatrix} 1 & 0 \\ 0 & -1 \end{pmatrix}. \tag{6.119}$$

6.5 Roots and Weights

We now address the classification of unitary irreducible representations for a simple or semisimple Lie algebra $\mathfrak{g}$, of dimension dim $\mathfrak{g} = N$. In the previous su(2) example, we learned that we can classify the irreducible representations by the highest eigenvalue of the generator J_3, which is diagonal, and in the representation we can move between the eigenvectors associated with the highest and with the lowest eigenvalue with the raising and lowering operators $J^{\pm}$. We will find that similar structures can be found more generally. We can find a set of mutually commuting and hence simultaneously diagonalizable generators $H_1, \ldots, H_m$, analogous to J_3, and classify the irreducible representations by a "highest" m-component eigenvalue vector $(\mu_1, \mu_2, \ldots, \mu_m)$. Then, among all eigenvalue vectors of the representation, called "weights," we can move by ladder operators $E_{\pm\vec{\alpha}}$, which come in pairs and form the remaining set of generators, in analogy with the $J^{\pm}$ generators of su(2). The vectors $\pm\vec{\alpha}$ are called *roots* and the $E_{\pm\vec{\alpha}}$ are *root generators*.

Let us consider a unitary irreducible representation D of $\mathfrak{g}$. Let $\{X_a\}_{a=1}^{n}$ denote the generators of $\mathfrak{g}$ in the representation D. In the vector space spanned by these generators, we search for the largest set of linear combinations $H_i = \sum_a r_{ia}X_a$ (with real coefficients r_{ia}), such that

$$[H_i, H_j] = 0 \qquad \forall i,j. \tag{6.120}$$

Note that we can always find at least one H_i, since it commutes with itself. The reality of the r_{ia} coefficients and the Hermiticity of the X_a generators implies that the H_i are Hermitian, too. By a suitable choice of linear combinations and rescalings, one can diagonalize the matrix tr(H_iH_j) and adjust all the eigenvalues to be the same, to arrive at

$$\text{tr}(H_iH_j) = k_D\delta_{ij}, \tag{6.121}$$

where the coefficient k_D depends on the representation D.

The H_is satisfying Eq. (6.120) are called the *Cartan generators* of the Lie algebra, and their maximal number, often denoted as m, is called the *rank* of the Lie algebra (or of the Lie group). Note that Eq. (6.120) implies that the Cartan generators form an Abelian subalgebra of $\mathfrak{g}$.

Example In su(2), from the three mutually noncommuting generators we can construct at most one Cartan generator (the usual choice is J_3). Thus, the rank of the su(2) Lie algebra is one.

As a consequence of Eq. (6.120), the m independent Cartan generators in the given representation D can be diagonalized simultaneously: Since the H_is are Hermitian, their eigenvalues are real numbers that we denote as μ_i. The corresponding eigenvectors are denoted as $|\vec{\mu}, D\rangle$, and they are specific of the representation D. The m-component vector $\vec{\mu}$ is called a *weight vector* for that representation. For the adjoint representation, the weight vectors are also called *root vectors*.

Let us now choose a basis for $\mathfrak{g}$, in which the first m basis vectors are precisely the H_is, that is, $X_i = H_i$ for $i = 1, \ldots, m$. Let us consider the adjoint representation: As its dimension is equal to dim $\mathfrak{g}$, it is possible to build a one-to-one mapping between the generators X_a and the basis vectors of the adjoint representation. Then we can use the generators as labels of the basis vectors and denote the latter by $|X_b\rangle$. We define a scalar product in this space via the Hilbert–Schmidt inner product:

$$\langle X_a | X_b \rangle \equiv \mathrm{tr}(X_a^\dagger X_b) = \mathrm{tr}(X_a X_b). \tag{6.122}$$

Moreover, we choose the basis such that the generators satisfy

$$X_a |X_b\rangle = i f_{abc} |X_c\rangle, \tag{6.123}$$

where the coefficients appearing on the right-hand side of Eq. (6.123) are the Lie algebra structure constants. We can then adopt a concise notation

$$|[X_a, X_b]\rangle \equiv i f_{abc} |X_c\rangle \tag{6.124}$$

so that we can use the commutator as the label of the vector that appears on the right-hand side of Eq. (6.123). We will then repeatedly use the notation

$$X_a |X_b\rangle = |[X_a, X_b]\rangle. \tag{6.125}$$

Thus, as a consequence of Eq. (6.120) one has

$$H_i |H_j\rangle = 0 \tag{6.126}$$

for all i and j in $\{1, \ldots, m\}$. The remaining $n - m$ basis vectors for $\mathfrak{g}$ can then be chosen as the other eigenvectors of the H_is: We denote them as $|E_{\vec{\alpha}}\rangle$, where their label $\vec{\alpha}$ is the list of eigenvalues

$$H_i |E_{\vec{\alpha}}\rangle = \alpha_i |E_{\vec{\alpha}}\rangle. \tag{6.127}$$

Note that no $\vec{\alpha}$ can be a null vector, otherwise the corresponding $E_{\vec{\alpha}}$ would be a Cartan generator itself, in contradiction with our assumption that it is not. Also note that, since the Cartan generators are Hermitian, all $\vec{\alpha}$ are real-component vectors.

Example Let us illustrate the above notations with the adjoint representation of su(2), where the generators J_a, for $a \in \{1, 2, 3\}$, are represented by Eq. (6.64), but we show them again here for convenience:

$$J_1 = \begin{pmatrix} 0 & 0 & 0 \\ 0 & 0 & -i \\ 0 & i & 0 \end{pmatrix}, \quad J_2 = \begin{pmatrix} 0 & 0 & i \\ 0 & 0 & 0 \\ -i & 0 & 0 \end{pmatrix}, \quad J_3 = \begin{pmatrix} 0 & -i & 0 \\ i & 0 & 0 \\ 0 & 0 & 0 \end{pmatrix}. \tag{6.128}$$

We choose J_3 as the Cartan generator. Note that it is not diagonal yet. We first choose a basis vector $|J_3\rangle$ which satisfies $J_3|J_3\rangle = 0$ as required by Eq. (6.126). It is simple to see that

$$|J_3\rangle = \begin{pmatrix} 0 \\ 0 \\ 1 \end{pmatrix} \tag{6.129}$$

works. If choose the remaining two basis states as

$$|J_1\rangle = \begin{pmatrix} 1 \\ 0 \\ 0 \end{pmatrix}, \qquad |J_2\rangle = \begin{pmatrix} 0 \\ 1 \\ 0 \end{pmatrix}, \tag{6.130}$$

it is straightforward to verify that with the matrices above, this choice of basis satisfies

$$J_a|J_b\rangle = i\epsilon_{abc}|J_c\rangle = |[J_a, J_b]\rangle, \tag{6.131}$$

as in Eq. (6.125). The adjoint representation is the $j = 1$ irreducible representation. In the above basis, the missing two eigenvectors of J_3 are

$$|J^+\rangle = \frac{1}{\sqrt{2}} \begin{pmatrix} 1 \\ i \\ 0 \end{pmatrix}, \qquad |J^-\rangle = \frac{1}{\sqrt{2}} \begin{pmatrix} 1 \\ -i \\ 0 \end{pmatrix}, \tag{6.132}$$

with $J_3|J^\pm\rangle = \pm|J^\pm\rangle$. Thus the vectors in Eq. (6.132) are the vectors $|E_{\pm a}\rangle$ labeled by the two root generators of su(2) with roots $\pm\alpha = \pm 1$. The eigenvalues of J_3 are $m = 1$, $m = 0$, and $m = -1$. It is possible to diagonalize J_3 via a change of basis:

$$\tilde{J}_a = U^\dagger J_a U, \tag{6.133}$$

where U is a unitary matrix having the orthonormalized eigenvectors of J_3 as its columns

$$U = \begin{pmatrix} \frac{1}{\sqrt{2}} & 0 & \frac{1}{\sqrt{2}} \\ \frac{i}{\sqrt{2}} & 0 & -\frac{i}{\sqrt{2}} \\ 0 & 1 & 0 \end{pmatrix}, \tag{6.134}$$

such that

$$\tilde{J}_1 = \frac{1}{\sqrt{2}} \begin{pmatrix} 0 & -1 & 0 \\ -1 & 0 & 1 \\ 0 & 1 & 0 \end{pmatrix},$$

$$\tilde{J}_2 = \frac{1}{\sqrt{2}} \begin{pmatrix} 0 & i & 0 \\ -i & 0 & -i \\ 0 & i & 0 \end{pmatrix},$$

$$\tilde{J}_3 = \begin{pmatrix} 1 & 0 & 0 \\ 0 & 0 & 0 \\ 0 & 0 & -1 \end{pmatrix}.$$

Now the basis vectors are

$$|j = 1, m = +1\rangle = U^\dagger |J^+\rangle,$$
$$|j = 1, m = 0\rangle = U^\dagger |J_3\rangle,$$
$$|j = 1, m = -1\rangle = U^\dagger |J^-\rangle.$$

We leave it as an exercise (see Problem 6.7) to verify that

$$(\tilde{J}_a)_{mm'} = \langle j = 1, m|\tilde{J}_a|j = 1, m'\rangle. \tag{6.135}$$

Applying Eq. (6.125), the left-hand side of Eq. (6.127) can be rewritten as $H_i|E_{\vec{\alpha}}\rangle = |[H_i, E_{\vec{\alpha}}]\rangle$. In turn, by linearity one can rewrite the right-hand side as $|\alpha_i E_{\vec{\alpha}}\rangle$, thus one obtains $|[H_i, E_{\vec{\alpha}}]\rangle = |\alpha_i E_{\vec{\alpha}}\rangle$. Equivalently, the latter equation can be rewritten as an equality

$$[H_i, E_{\vec{\alpha}}] = \alpha_i E_{\vec{\alpha}}, \tag{6.136}$$

which is an algebraic relation, and thus must hold true in all representations. Note that Eq. (6.136) can be thought of as a generalization of Eq. (6.101), which is valid for the su(2) algebra, to an arbitrary Lie algebra. As we already remarked, the Hermiticity of the Cartan generators implies that the α_is are real; in addition, it implies that the $E_{\vec{\alpha}}$s are non-Hermitian, and that they appear in pairs with opposite eigenvalues $E_{\pm\vec{\alpha}}$, with $E_{\vec{\alpha}}^\dagger = E_{-\vec{\alpha}}$: This can be proven by taking the adjoint of Eq. (6.136), see Problem 6.8.

Let us study the effect of acting on one of the eigenvectors of the Cartan generators $|\vec{\mu}, D\rangle$ with one of the $E_{\vec{\alpha}}$. We can prove that this yields an eigenvector of the Cartan generators with a different eigenvalue

$$\begin{aligned} H_i E_{\vec{\alpha}}|\vec{\mu}, D\rangle &= [H_i, E_{\vec{\alpha}}]|\vec{\mu}, D\rangle + E_{\vec{\alpha}} H_i|\vec{\mu}, D\rangle \\ &= \alpha_i E_{\vec{\alpha}}|\vec{\mu}, D\rangle + \mu_i E_{\vec{\alpha}}|\vec{\mu}, D\rangle \\ &= (\mu_i + \alpha_i) E_{\vec{\alpha}}|\vec{\mu}, D\rangle, \end{aligned} \tag{6.137}$$

which means that $E_{\vec{\alpha}}|\vec{\mu}, D\rangle$ is proportional to the eigenvector of H_i with eigenvalue $\mu_i + \alpha_i$:

$$E_{\vec{\alpha}}|\vec{\mu}, D\rangle = N_{\vec{\alpha},\vec{\mu}}|\vec{\mu} + \vec{\alpha}, D\rangle. \tag{6.138}$$

Equation (6.138) is true for every representation D, but $N_{\vec{\alpha},\vec{\mu}}$ depends on D.

Before determining $N_{\vec{\alpha},\vec{\mu}}$, we focus on the adjoint representation, and use the mapping between elements of the algebra and vector states, which implies that $E_{\vec{\alpha}}|E_{-\vec{\alpha}}\rangle$ has weight zero, because

$$\begin{aligned} H_i E_{\vec{\alpha}}|E_{-\vec{\alpha}}\rangle &= [H_i, E_{\vec{\alpha}}]|E_{-\vec{\alpha}}\rangle + E_{\vec{\alpha}} H_i|E_{-\vec{\alpha}}\rangle \\ &= \alpha_i E_{\vec{\alpha}}|E_{-\vec{\alpha}}\rangle - \alpha_i E_{\vec{\alpha}}|E_{-\vec{\alpha}}\rangle = 0. \end{aligned} \tag{6.139}$$

This implies that $E_{\vec{\alpha}}|E_{-\vec{\alpha}}\rangle$ is a linear combination of the zero-eigenvalue vectors associated with the Cartan subalgebra:

$$E_{\vec{\alpha}}|E_{-\vec{\alpha}}\rangle = \sum_{i=1}^m c_i|H_i\rangle. \tag{6.140}$$

Since in the adjoint representation the generators are normalized as $\text{tr}(X_a X_b) = \lambda \delta_{ab}$, by multiplying $\langle H_i|$ by Eq. (6.140) one can obtain

$$\begin{aligned}
c_i &= \frac{\langle H_i|E_{\vec\alpha}|E_{-\vec\alpha}\rangle}{\langle H_i|H_i\rangle} = \frac{\langle H_i|[E_{\vec\alpha}, E_{-\vec\alpha}]\rangle}{\langle H_i|H_i\rangle} = \frac{\text{tr}(H_i[E_{\vec\alpha}, E_{-\vec\alpha}])}{\text{tr}(H_i H_i)} = \frac{\text{tr}(H_i[E_{\vec\alpha}, E_{-\vec\alpha}])}{\lambda} \\
&= \frac{\text{tr}(H_i E_{\vec\alpha} E_{-\vec\alpha}) - \text{tr}(H_i E_{-\vec\alpha} E_{\vec\alpha})}{\lambda} = \frac{\text{tr}(E_{-\vec\alpha} H_i E_{\vec\alpha}) - \text{tr}(E_{-\vec\alpha} E_{\vec\alpha} H_i)}{\lambda} \\
&= \frac{\text{tr}(E_{-\vec\alpha}[H_i, E_{\vec\alpha}])}{\lambda} = \frac{\alpha_i \, \text{tr}(E_{-\vec\alpha} E_{\vec\alpha})}{\lambda} = \alpha_i,
\end{aligned} \tag{6.141}$$

having used the invariance of the trace of a product of matrices under cyclic permutations of the factors. Thus $E_{\vec\alpha}|E_{-\vec\alpha}\rangle = \sum_{i=1}^{m} \alpha_i |H_i\rangle$ or, equivalently,

$$[E_{\vec\alpha}, E_{-\vec\alpha}] = \sum_{i=1}^{m} \alpha_i H_i. \tag{6.142}$$

From this latter result, it is then straightforward to derive the value of the coefficient $N_{\vec\alpha,\vec\mu}$ appearing on the right-hand side of Eq. (6.138), for a generic representation D:

$$\langle \vec\mu, D|[E_{\vec\alpha}, E_{-\vec\alpha}]|\vec\mu, D\rangle = \alpha_i \langle \vec\mu, D|H_i|\vec\mu, D\rangle = \vec\alpha \cdot \vec\mu. \tag{6.143}$$

On the other hand, this quantity is also equal to

$$\begin{aligned}
\langle \vec\mu, D|[E_{\vec\alpha}, E_{-\vec\alpha}]|\vec\mu, D\rangle &= \langle \vec\mu, D|E_{\vec\alpha} E_{-\vec\alpha}|\vec\mu, D\rangle - \langle \vec\mu, D|E_{-\vec\alpha} E_{\vec\alpha}|\vec\mu, D\rangle \\
&= |N_{-\vec\alpha,\vec\mu}|^2 - |N_{\vec\alpha,\vec\mu}|^2,
\end{aligned} \tag{6.144}$$

where

$$\begin{aligned}
N_{-\vec\alpha,\vec\mu} &= \langle \vec\mu - \vec\alpha, D|E_{-\vec\alpha}|\vec\mu, D\rangle = \langle \vec\mu - \vec\alpha, D|E_{\vec\alpha}^\dagger|\vec\mu, D\rangle \\
&= \langle \vec\mu, D|E_{\vec\alpha}|\vec\mu - \vec\alpha, D\rangle^\star = N_{\vec\alpha,\vec\mu-\vec\alpha}^\star,
\end{aligned} \tag{6.145}$$

so that

$$|N_{\vec\alpha,\vec\mu-\vec\alpha}|^2 - |N_{\vec\alpha,\vec\mu}|^2 = \vec\alpha \cdot \vec\mu. \tag{6.146}$$

As we remarked above, applying $E_{\vec\alpha}$ to a generic vector $|\vec\mu, D\rangle$ turns it into a new vector, which is linearly independent from $|\vec\mu, D\rangle$. If the representation D is finite-dimensional, iterating this procedure cannot generate arbitrarily many linearly independent vectors: This means that, starting from an arbitrary vector $|\vec\mu, D\rangle$, there must exist a finite maximum natural number p such that $E_{\vec\alpha}^p|\vec\mu, D\rangle \propto |\vec\mu + p\vec\alpha, D\rangle$ is finite, but $E_{\vec\alpha}|\vec\mu + p\vec\alpha, D\rangle = 0$. Thus, p represents the maximum number of times that the raising operator $E_{\vec\alpha}$ (which increases $\vec\mu \to \vec\mu + \vec\alpha$) can be applied to the vector $|\vec\mu, D\rangle$, obtaining a linearly independent vector. Note that, in view of Eq. (6.138), this means that $N_{\vec\alpha,\vec\mu+p\vec\alpha} = 0$.

Similarly, for the lowering operator $E_{-\vec\alpha}$, which decreases $\vec\mu \to \vec\mu + \vec\alpha$, there must exist a finite maximum natural number q, such that the $E_{-\vec\alpha}^q|\vec\mu, D\rangle \propto |\vec\mu - q\vec\alpha, D\rangle$ is nonzero, but $E_{-\vec\alpha}|\vec\mu - q\vec\alpha, D\rangle = 0$, i.e., $N_{-\vec\alpha,\vec\mu-q\vec\alpha} = 0$.

Now, let us consider Eq. (6.146) for $|\vec\mu, D\rangle$, as well as for all vectors obtained by applying the $E_{\vec\alpha}$ operator on it once, twice, ..., p times, and for those obtained by applying the $E_{\vec{\alpha}}$ operator once, twice, ..., q times. Summing all of these equations together, we obtain

$$\sum_{k=-q}^{p} \left(|N_{\vec{\alpha},\vec{\mu}+(k-1)\vec{\alpha}}|^2 - |N_{\vec{\alpha},\vec{\mu}+k\vec{\alpha}}|^2 \right) = \sum_{k=-q}^{p} \vec{\alpha} \cdot \left(\vec{\mu} + k\vec{\alpha} \right). \tag{6.147}$$

On the left-hand side of Eq. (6.147), the second term of each summand cancels against the first term of the next summand. Thus, the sum reduces to the difference of the first term in the first summand, minus the second term in the last summand, i.e., to $|N_{\vec{\alpha},\vec{\mu}-(q+1)\vec{\alpha}}|^2 - |N_{\vec{\alpha},\vec{\mu}+p\vec{\alpha}}|^2$. As remarked above, however, both of these quantities vanish, so the left-hand side of Eq. (6.147) is zero.

The right-hand side of Eq. (6.147) can be rewritten as follows:

$$\sum_{k=-q}^{p} \vec{\alpha} \cdot \left(\vec{\mu} + k\vec{\alpha} \right) = \left(\sum_{k=-q}^{p} \vec{\alpha} \cdot \vec{\mu} \right) + \vec{\alpha} \cdot \vec{\alpha} \sum_{k=-q}^{p} k$$

$$= \vec{\alpha} \cdot \vec{\mu}(q+p+1) + \vec{\alpha} \cdot \vec{\alpha} \frac{(p+q+1)(p-q)}{2}. \tag{6.148}$$

Thus, Eq. (6.147) implies

$$(q+p+1)\vec{\alpha} \cdot \vec{\mu} + \frac{(p+q+1)(p-q)}{2} \alpha^2 = 0, \tag{6.149}$$

or, using the fact that $q + p + 1 > 0$, since both q and p are natural numbers,

$$\vec{\alpha} \cdot \vec{\mu} + \frac{\alpha^2}{2} (p-q) = 0, \tag{6.150}$$

i.e., defining $M = q - p$, which, being the difference of two natural numbers, must be an integer,

$$\frac{\vec{\alpha} \cdot \vec{\mu}}{\alpha^2} = \frac{M}{2}. \tag{6.151}$$

When considering the adjoint representation, Eq. (6.151) must hold for any pair of different roots $E_{\vec{\alpha}}$ and $E_{\vec{\beta}}$, i.e., one must have

$$\frac{\vec{\alpha} \cdot \vec{\beta}}{\alpha^2} = \frac{M}{2} \tag{6.152}$$

as well as

$$\frac{\vec{\beta} \cdot \vec{\alpha}}{\beta^2} = \frac{M'}{2} \tag{6.153}$$

for another integer M'. Multiplying Eq. (6.152) and Eq. (6.153), one obtains

$$\frac{\left(\vec{\alpha} \cdot \vec{\beta} \right)^2}{\alpha^2 \beta^2} = \frac{MM'}{4}, \tag{6.154}$$

namely, denoting the angle between $\vec{\alpha}$ and $\vec{\beta}$ as $\theta_{\alpha\beta}$,

$$4\cos^2 \theta_{\alpha\beta} = MM'. \tag{6.155}$$

Equation (6.155) is a polynomial equation for which only integer solutions are considered: An equation of this type is called a *Diophantine equation*. We note that, being the product of two integers, the right-hand side of Eq. (6.155) is necessarily an integer; moreover, since $0 \leq \cos^2 \theta_{\alpha\beta} \leq 1$, the left-hand side is nonnegative and never larger than 4. These conditions imply that the only admissible solutions for Eq. (6.155) are:

- $MM' = 0$, corresponding to $\theta_{\alpha\beta} = \pi/2$,
- $MM' = 1$, corresponding to $\theta_{\alpha\beta} = \pi/3$ or $\theta_{\alpha\beta} = 2\pi/3$,
- $MM' = 2$, corresponding to $\theta_{\alpha\beta} = \pi/4$ or $\theta_{\alpha\beta} = 3\pi/4$,
- $MM' = 3$, corresponding to $\theta_{\alpha\beta} = \pi/6$ or $\theta_{\alpha\beta} = 5\pi/6$,
- $MM' = 4$, corresponding to $\theta_{\alpha\beta} = 0$ or $\theta_{\alpha\beta} = \pi$.

The last case, however, can be discarded, as it would imply that $\vec{\alpha}$ and $\vec{\beta}$ are proportional to each other, in which case they could not be linearly independent.

6.5.1 An su(2) Subalgebra

In the following, we will simplify the notation by dropping the vector signs. We note that we can choose a root α and a Cartan generator H_i and construct a triplet of generators which form an su(2) subalgebra. Let $|\alpha|$ be the positive square root of $\alpha \cdot \alpha$, and define the three rescaled generators

$$E^\pm \equiv \frac{1}{|\alpha|} E_{\pm\alpha}, \qquad E_3 \equiv \frac{\alpha_i}{\alpha \cdot \alpha} H_i. \tag{6.156}$$

They satisfy the commutation relations

$$[E_3, E^\pm] = \pm E^\pm, \qquad [E^+, E^-] = E_3, \tag{6.157}$$

which are the same as the commutation relations for the J^+, J^-, and J_3 generators of the su(2) algebra expressed by Eq. (6.101) and Eq. (6.102).

If $|\mu, D\rangle$ is an eigenvector of $E_{-\alpha}E_\alpha$, then the vectors

$$|\mu \pm k\alpha, D\rangle \propto (E_{\pm\alpha})^k |\mu, D\rangle \tag{6.158}$$

form an irreducible representation of the su(2) subalgebra. Note that the condition $E_{-\alpha}E_\alpha |\mu, D\rangle \propto |\mu, D\rangle$ ensures that $|\mu, D\rangle$ is not a linear combination of vectors from different representations but with the same μ.

The vectors $|\mu + k\alpha, D\rangle$ satisfy

$$E_3 |\mu + k\alpha, D\rangle = \frac{(\alpha \cdot \mu + k\alpha^2)}{|\alpha|^2} |\mu + k\alpha, D\rangle. \tag{6.159}$$

For the spin-j representation, the highest E_3 eigenvalue, i.e., the one that corresponds to $|\mu + p\alpha, D\rangle$, must be

$$j = \frac{\alpha \cdot \mu}{\alpha^2} + p. \tag{6.160}$$

Conversely, the lowest E_3 eigenvalue, corresponding to $|\mu - q\alpha, D\rangle$ is

$$-j = \frac{\alpha \cdot \mu}{\alpha^2} - q. \tag{6.161}$$

Thus

$$0 = j - j = 2\frac{\alpha \cdot \mu}{\alpha^2} + (p - q), \tag{6.162}$$

from which follows

$$\frac{\alpha \cdot \mu}{\alpha^2} = -\frac{p-q}{2}. \tag{6.163}$$

On the other hand, we also see that

$$-q \le p - q = -2\frac{\alpha \cdot \mu}{\alpha^2} \le p, \tag{6.164}$$

so the irreducible representation must contain an eigenvector $|\mu + k\alpha\rangle$ of weight $\mu + k\alpha$ with

$$k = -2\frac{\alpha \cdot \mu}{\alpha^2}. \tag{6.165}$$

Weyl Reflections and Root Systems

Since we started with an arbitrary weight μ and an arbitrary root α, we can define a map s_α

$$s_\alpha : \mu \mapsto s_\alpha(\mu) \equiv \mu - 2\frac{\alpha \cdot \mu}{\alpha^2}\alpha \tag{6.166}$$

that associates a weight μ with a new weight $s_\alpha(\mu)$. The map defined in Eq. (6.166) is called a *Weyl reflection*. The name is not accidental, since, in general, the reflection of a vector v in $\mathbb{R}^m$ with respect to the hyperplane through the origin perpendicular to the vector a is defined through the function

$$s_a(v) = v - 2\frac{a \cdot v}{a^2}a. \tag{6.167}$$

Thus Eq. (6.166) defines the reflection of the weight μ with respect to a hyperplane orthogonal to the root α.

In particular, since the roots are the nonzero weights in the adjoint representation, one can apply Weyl reflections to map a root to another root. As will be shown shortly, this gives rise to a symmetry associated with the Lie algebra. First, we define a *root system* Φ to be the set of all roots α with the following properties.

(i) The roots span the Euclidean space $\mathbb{R}^m$.
(ii) If $\alpha \in \Phi$, then also $-\alpha \in \Phi$.
(iii) The set Φ is closed under Weyl reflections s_α for all $\alpha \in \Phi$, i.e., if $\alpha, \beta \in \Phi$, then $s_\alpha(\beta) \in \Phi$.
(iv) For all pairs of roots $\alpha, \beta \in \Phi$, the number $2\frac{\alpha \cdot \beta}{\alpha^2}$ is an integer.

The property (iv) results from applying Eq. (6.165), with a root β in place of μ. Weyl reflections generate a group, which acts faithfully in the root system Φ. The group is called the *Weyl group*, and is a finite group.

For every root α, we can define a *coroot* $\alpha^\vee$ by

$$\alpha^\vee \equiv 2\frac{\alpha}{\alpha^2}. \tag{6.168}$$

Thus

$$(\alpha^\vee, \beta) \equiv 2\frac{\alpha \cdot \beta}{\alpha^2} \tag{6.169}$$

is an integer for every coroot–root pair. The coroots form a root system $\Phi^\vee$ called the *dual root system*.

6.5.2 The Algebra of the SU(3) Group

The SU(3) group is defined as the group of 3×3 complex unitary matrices U having determinant equal to 1, with the matrix multiplication as the group multiplication. The dimension of the algebra (or of the group) is $\dim SU(3) = 8$. The generators of su(3) are eight Hermitian traceless matrices. A common choice for a defining representation is in terms of a basis given by the *Gell-Mann matrices* λ_a, named after Murray Gell-Mann, which are the analogue of the Pauli matrices for the algebra of generators of the SU(2) group:

$$
\lambda_1 = \begin{pmatrix} 0 & 1 & 0 \\ 1 & 0 & 0 \\ 0 & 0 & 0 \end{pmatrix}, \quad
\lambda_2 = \begin{pmatrix} 0 & -i & 0 \\ i & 0 & 0 \\ 0 & 0 & 0 \end{pmatrix}, \quad
\lambda_3 = \begin{pmatrix} 1 & 0 & 0 \\ 0 & -1 & 0 \\ 0 & 0 & 0 \end{pmatrix},
$$

$$
\lambda_4 = \begin{pmatrix} 0 & 0 & 1 \\ 0 & 0 & 0 \\ 1 & 0 & 0 \end{pmatrix}, \quad
\lambda_5 = \begin{pmatrix} 0 & 0 & -i \\ 0 & 0 & 0 \\ i & 0 & 0 \end{pmatrix}, \quad
\lambda_6 = \begin{pmatrix} 0 & 0 & 0 \\ 0 & 0 & 1 \\ 0 & 1 & 0 \end{pmatrix},
$$

$$
\lambda_7 = \begin{pmatrix} 0 & 0 & 0 \\ 0 & 0 & -i \\ 0 & i & 0 \end{pmatrix}, \quad
\lambda_8 = \frac{1}{\sqrt{3}} \begin{pmatrix} 1 & 0 & 0 \\ 0 & 1 & 0 \\ 0 & 0 & -2 \end{pmatrix}. \tag{6.170}
$$

The generators X_a of the su(3) Lie algebra are then

$$
X_a = \frac{1}{2} \lambda_a. \tag{6.171}
$$

The generators are normalized as

$$
\mathrm{tr}(X_a X_b) = \frac{1}{2} \delta_{ab}, \tag{6.172}
$$

so the Dynkin index of the defining representation is

$$
\lambda_{\mathrm{def}} = \frac{1}{2}. \tag{6.173}
$$

In analogy with the Pauli matrices, a product of two Gell-Mann matrices can be written as

$$
\lambda_a \lambda_b = \frac{1}{2}[\lambda_a, \lambda_b] + \frac{1}{2}\{\lambda_a, \lambda_b\} = \frac{1}{2}\left(2i \sum_{c=1}^{8} f_{abc} \lambda_c \right) + \frac{1}{2}\left(\frac{4}{3}\delta_{ab} \mathbb{1} + 2 \sum_{c=1}^{8} d_{abc} \lambda_c \right)
$$

$$
= \frac{2}{3}\delta_{ab} \mathbb{1} + \sum_{c=1}^{8} (d_{abc} + i f_{abc}) \lambda_c. \tag{6.174}
$$

In addition, the Gell-Mann matrices satisfy a *Fierz completeness relation*, named after Markus Fierz:

$$
\sum_{a=1}^{8} (\lambda_a)_{ij}(\lambda_a)_{kl} = 2\delta_{il}\delta_{kj} - \frac{2}{3}\delta_{ij}\delta_{kl}. \tag{6.175}
$$

Note that the generators X_1, X_2, and X_3 form an su(2) subalgebra of the su(3) Lie algebra. Accordingly, by exponentiation one obtains an SU(2) subgroup of the

SU(3) group.[5] There are also two other ways to obtain an su(2) subalgebra, using either $\frac{1}{2}\lambda_4$, $\frac{1}{2}\lambda_5$, and $\left(\frac{1}{4}\lambda_3 + \frac{\sqrt{3}}{4}\lambda_8\right)$ as generators, or $\frac{1}{2}\lambda_6$, $\frac{1}{2}\lambda_7$, and $\left(-\frac{1}{4}\lambda_3 + \frac{\sqrt{3}}{4}\lambda_8\right)$.

The rank of the su(3) algebra is 2. In the basis of generators according to Eq. (6.171), X_3 and X_8 are diagonal. Thus they are the Cartan generators. We follow the convention to label $H_1 = X_3$ and $H_2 = X_8$. Since they are diagonal, their common eigenvectors are

$$e_1 = \begin{pmatrix} 1 \\ 0 \\ 0 \end{pmatrix}, \quad e_2 = \begin{pmatrix} 0 \\ 1 \\ 0 \end{pmatrix}, \quad e_3 = \begin{pmatrix} 0 \\ 0 \\ 1 \end{pmatrix}. \qquad (6.176)$$

The eigenvalues of H_1 and H_2 are the weights μ_1 and μ_2, forming the weight vector $\mu = (\mu_1, \mu_2)$. Moving to the Dirac notation, we use the weights to label the eigenvectors. For e_1, we have the eigenvalues $\mu_1 = \frac{1}{2}$, $\mu_2 = \frac{1}{2\sqrt{3}}$, so $e_1 = \left|\frac{1}{2}, \frac{1}{2\sqrt{3}}\right\rangle$. Proceeding similarly, we find

$$e_1 = \left|\frac{1}{2}, \frac{1}{2\sqrt{3}}\right\rangle, \quad e_2 = \left|-\frac{1}{2}, \frac{1}{2\sqrt{3}}\right\rangle, \quad e_3 = \left|0, -\frac{1}{\sqrt{3}}\right\rangle. \qquad (6.177)$$

If we plot the weight vectors on the (μ_1, μ_2) plane, they form an equilateral triangle.

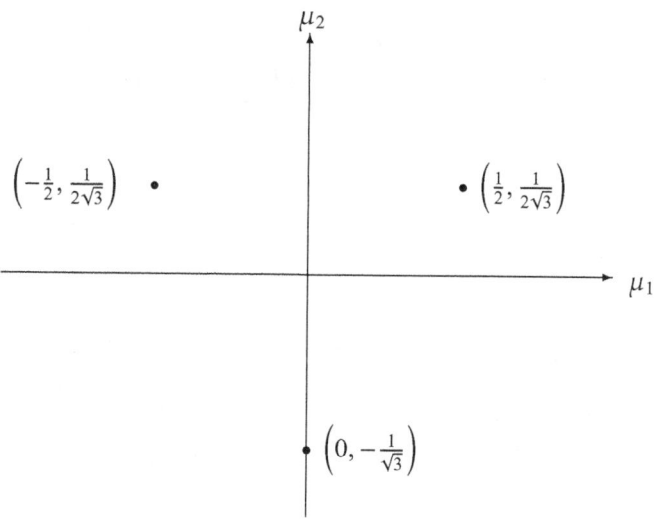

There will be $8 - 2 = 6$ root vectors, i.e., three pairs $\pm\alpha$. To find them, we use the fact that the corresponding generators $E_{\pm\alpha}$ map one weight to another. Thus the α vectors are found as differences of weights. In the plot of the weight vectors on the (μ_1, μ_2) plane, they correspond to the vectors from one weight to another. Thus it is simple to read off that the possible α vectors are

$$\alpha \in \left\{ \pm(1, 0), \quad \pm\left(\frac{1}{2}, \frac{\sqrt{3}}{2}\right), \quad \pm\left(-\frac{1}{2}, \frac{\sqrt{3}}{2}\right) \right\}. \qquad (6.178)$$

---

[5] In elementary-particle physics applications, this subgroup is sometimes called the *isospin group*.

The corresponding generators must be matrices mapping the eigenvectors e_i to each other. Such matrices will have only a single nonvanishing off-diagonal element. The first pair of such matrices can be found in the su(2) subalgebra, where they correspond to the pair $J^\pm$ except that now they are labeled with the root pair $\pm\alpha = (\pm 1, 0)$:

$$E_{(1,0)} = X_1 + iX_2 = \begin{pmatrix} 0 & 1 & 0 \\ 0 & 0 & 0 \\ 0 & 0 & 0 \end{pmatrix},$$

$$E_{(-1,0)} = X_1 - iX_2 = \begin{pmatrix} 0 & 0 & 0 \\ 1 & 0 & 0 \\ 0 & 0 & 0 \end{pmatrix}. \tag{6.179}$$

The other ones are found similarly,

$$E_{\left(\pm\frac{1}{\sqrt{2}},\pm\frac{\sqrt{3}}{2}\right)} = X_4 \pm iX_5, \qquad E_{\left(\mp\frac{1}{\sqrt{2}},\pm\frac{\sqrt{3}}{2}\right)} = X_6 \pm iX_7, \tag{6.180}$$

each matrix having a single nonzero off-diagonal entry, equal to 1. The roots form a regular hexagon, centered at the origin.

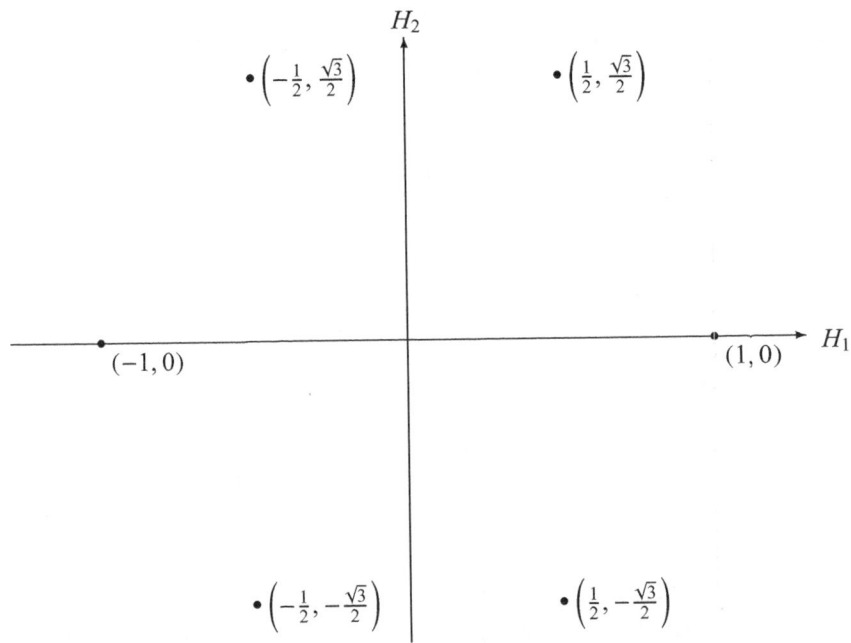

Note that all angles between root vectors of the su(3) algebra are multiples of $\frac{\pi}{3}$.

6.5.3 Ordering the Weight Vectors

Consider a basis $\{H_1, H_2, \ldots, H_m\}$ for the Cartan subalgebra, and let $\mu_1, \mu_2, \ldots, \mu_m$ be the corresponding weights. For the weight vector $\mu = (\mu_1, \mu_2, \ldots, \mu_m)$ there are two common alternative definitions. We may define that μ is a *positive weight vector* when its first nonzero entry is positive. Conversely, μ is said to be a *negative weight vector* when its first nonzero entry is negative. Finally, μ is said to be a *zero weight vector* if all of its entries are zero.

Note that in the literature, sometimes an alternative definition of a positive weight vector is one for which the last (instead of the first) nonzero entry is positive and, similarly, a negative weight vector is one for which the last (instead of the first) nonzero entry is negative: This alternative definition is more convenient in certain contexts. Throughout this book, we will mostly use the definition of a positive weight vector that is one for which the first nonzero entry is positive, except in the discussion in Section 6.9.

We then define a strict order relation among vectors. First, let us recall that a *strict order relation* $>$ is a binary relation among elements of a set S, such that the following conditions hold:

(i) irreflexivity: $\nexists a \in S$, such that $a > a$,
(ii) antisymmetry: for every pair of elements $a, b \in S$, if $a > b$ is true, then $b > a$ is false,
(iii) transitivity: for every triple of elements $a, b, c \in S$, if $a > b$ and $b > c$, then also $a > c$.

We now introduce an order relation, which we denote by $>$, among weight vectors, that is defined as follows: Given two weight vectors μ and μ', we write

$$\mu > \mu' \quad \text{when } \mu - \mu' \text{ is a positive weight vector.} \tag{6.181}$$

This order relation allows one to order all states $|\mu, D\rangle$. For a finite representation, it is then possible to identify the *highest weight vector*, which is the weight vector that is greater than all others. The highest weight vector is unique, and does not depend on the chosen basis $\{H_1, H_2, \ldots, H_m\}$ for the Cartan subalgebra.

Example The weights of the su(3) defining representation of are ordered as

$$\left(\frac{1}{2}, \frac{1}{2\sqrt{3}}\right) > \left(0, -\frac{1}{\sqrt{3}}\right) > \left(-\frac{1}{2}, \frac{1}{2\sqrt{3}}\right). \tag{6.182}$$

Thus, the highest weight vector is $\left(\frac{1}{2}, \frac{1}{2\sqrt{3}}\right)$.

6.5.4 Complex Conjugate Representations

Let X_a be the Hermitian generators, represented by Hermitian $n \times n$ matrices in an irreducible n-dimensional representation D with the basis vectors e_i, for $i = 1, \ldots, n$. We will use upper and lower indices to label the entries of the matrices X_a, such that X_a acting on a basis vector maps it to a new vector as follows:

$$e_i \mapsto (X_a)_i^j e_j \tag{6.183}$$

with a sum over the upper/lower index pair j, according to Einstein's summation convention. Consider then a generic vector v in the representation D; using Eq. (6.183), the action of X_a on v can be expressed as

$$v = v^i e_i \mapsto v^i (X_a)_i^j e_j, \tag{6.184}$$

which we can interpret as a transformation rule for the components:

$$v^j \mapsto (X_a)^j_i v^i. \tag{6.185}$$

Consider, then, the corresponding group elements in the same representation; by the exponential map, they are matrices $U = \exp(i\alpha_a X_a)$ with components

$$U^j_i = (\mathbb{1} + i\alpha_a X_a + \cdots)^j_i. \tag{6.186}$$

Representing the basis vectors e_i as column vectors as before,

$$e_i = \begin{pmatrix} 0 \\ \vdots \\ 1 \\ \vdots \\ 0 \end{pmatrix}, \tag{6.187}$$

where the ith entry is 1, consistently with Eq. (6.185) we then have

$$v = v^j e_j = \begin{pmatrix} v^1 \\ \vdots \\ v^n \end{pmatrix} \mapsto Uv = \begin{pmatrix} U^1_i v^i \\ \vdots \\ U^n_i v^i \end{pmatrix}. \tag{6.188}$$

Products of the form $v^\dagger w$ remain invariant under Eq. (6.188):

$$v^\dagger w \mapsto v^\dagger U^\dagger U w = v^\dagger w. \tag{6.189}$$

An adjoint vector is a row vector, so we introduce a dual basis

$$e^{\star i} = (0, \ldots, 1, \ldots, 0) \tag{6.190}$$

and expand

$$v^\dagger = (v^*_1, \ldots, v^*_n) = v^*_i e^{\star i}, \tag{6.191}$$

denoting the components of $v^\dagger$ with a lower index to be consistent with the upper/lower index pair contraction convention. For the components of $v^\dagger$ we find the transformation relation

$$v^*_i \mapsto (v^\dagger U^\dagger)_i = v^*_j (\mathbb{1} - i\alpha_a X_a + \cdots)^j_i, \tag{6.192}$$

having used the fact that $X^\dagger_a = X_a$ and that the Lie group parameters α_a are real.

We then introduce a new concept. Let us consider the complex conjugate of the Lie algebra commutation relations:

$$[X_a, X_b]^* = (if_{abc} X_c)^*. \tag{6.193}$$

The left-hand side of Eq. (6.193) can be rewritten as

$$[X_a, X_b]^* = (X_a X_b - X_b X_a)^* = X^*_a X^*_b - X^*_b X^*_a. \tag{6.194}$$

Multiplying each factor in the two products appearing in the last term by (-1), we obtain

$$[X_a, X_b]^* = (-X^*_a)(-X^*_b) - (-X^*_b)(-X^*_a) = [-X^*_a, -X^*_b]. \tag{6.195}$$

On the other hand, as a consequence of the reality of the structure constants f_{abc}, the right-hand side of Eq. (6.193) can be rewritten as

$$(if_{abc}X_c)^* = -if_{abc}X_c^* = if_{abc}(-X_c^*).$$ (6.196)

Comparing Eq. (6.193) with Eq. (6.195) and Eq. (6.196), we find that

$$[-X_a^*, -X_b^*] = if_{abc}(-X_c^*),$$ (6.197)

i.e., we can define a new set of generators $\overline{X}_a$,

$$\overline{X}_a \equiv -X_a^*,$$ (6.198)

that satisfy the same Lie algebra commutation relations as the X_a do. Note that

$$(\overline{X}_a)^i_j = (-X_a^*)^i_j = -(X_a^T)^i_j = -(X_a)^j_i.$$ (6.199)

Motivated by Eq. (6.192), for every dual vector $u = u_i e^{*i}$ we define a vector $\bar{u} = \bar{u}^j e_j$ with components

$$\bar{u}^j = u_j^*.$$ (6.200)

Under $\overline{X}_a$, the vector $\bar{u}$ transforms as

$$\bar{u}^i \mapsto (\overline{X}_a)^i_j \bar{u}^j = -(X_a)^j_i u_j^*.$$ (6.201)

Under the action of the group element obtained by the exponential map $\exp(i\alpha_a \overline{X}_a)$, we find

$$\bar{v}^i \mapsto \exp(i\alpha_a \overline{X}_a)^i_j \bar{v}^j = \left(\mathbb{1} + i\alpha_a \overline{X}_a + \cdots\right)^i_j \bar{v}^j.$$ (6.202)

The matrices $\overline{X}_a = -X_a^*$ in general are different from X_a; they define the *complex conjugate representation* $\overline{D}$ which may or may not be unitarily equivalent to the original representation D. By construction, D and $\overline{D}$ have the same dimension. If they are equivalent, we say that D is a *real representation* $\overline{D} = D$. If not, D is said to be a *complex representation*. To investigate the equivalence, we recall that irreducible representations are uniquely labeled by their highest weights. In an irreducible representation D, the Cartan matrices H_i are all diagonal. The diagonal entries are the weights, which are all real numbers. Thus they are related to the Cartan matrices $\overline{H}_i$ in the complex conjugate representation by

$$\overline{H}_i = -H_i.$$ (6.203)

Therefore, the weight vectors $\bar{\mu}$ of the complex conjugate representation $\overline{D}$ are related to weight vectors μ of D by

$$\bar{\mu} = -\mu.$$ (6.204)

In particular, the highest weight of $\overline{D}$ is minus the lowest weight of D, and vice versa. Thus, if the lowest weight of D is minus the highest weight of D, D is real. If not, D is complex.

We summarize for convenience the important transformation rules that we discussed here: The components v^i of vectors transform under the group and the algebra by

$$v^i \mapsto U^i_j v^j, \qquad v^i \mapsto (X_a)^i_j v^j,$$ (6.205)

whereas the components $w_i = v_i^*$ of dual vectors transform as

$$w_i \mapsto w_j(U^\dagger)_i^j, \qquad w_i \mapsto w_j(-X_a)_i^j. \tag{6.206}$$

6.5.5 The Antifundamental Representation of su(3)

In the defining representation of the su(3) Lie algebra, we can easily see that the highest weight $\left(\frac{1}{2}, \frac{1}{2\sqrt{3}}\right)$ is not minus the lowest weight $\left(-\frac{1}{2}, \frac{1}{2\sqrt{3}}\right)$. Thus the su(3) defining representation is a complex representation. In physics, it is customary to denote the defining representation of the su(3) Lie algebra by **3** and call it the *fundamental representation*. Its complex conjugate representation is denoted $\overline{\mathbf{3}}$ and called the *antifundamental representation*, with highest weight $\left(\frac{1}{2}, -\frac{1}{2\sqrt{3}}\right)$. Its weights are equal and opposite to those of the fundamental representation, and are shown in the plot below.

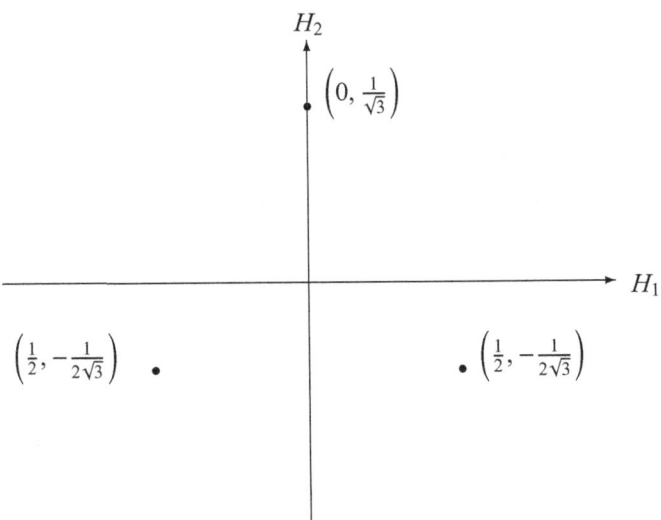

6.6 Simple Roots

As we have mentioned, the nonzero weight vectors of the adjoint representation are called root vectors. The m weight vectors corresponding to the Cartan generators are zero vectors. Thus, there are dim $\mathfrak{g} - m$ root vectors, coming in pairs[6] of opposite elements $\pm\alpha$. We define a *simple root* as a positive root vector that cannot be written as the sum of two positive roots.

Example   The positive roots of su(3) are $(1,0)$, $\left(\frac{1}{2}, \frac{\sqrt{3}}{2}\right)$, $\left(\frac{1}{2}, -\frac{\sqrt{3}}{2}\right)$. Since $(1,0) = \left(\frac{1}{2}, \frac{\sqrt{3}}{2}\right) + \left(\frac{1}{2}, -\frac{\sqrt{3}}{2}\right)$, a sum of positive roots, $(1,0)$ is not simple.

---

[6] Note that the fact that the roots come in pairs implies that the difference between the dimension and the rank of a Lie algebra must necessarily be an even number.

Lemma 6.3   *The difference of any two distinct simple roots is not a root.*

Proof   Let α and β be two distinct simple roots. $\alpha - \beta$ cannot be zero, otherwise one would have $\alpha = \beta$, in contradiction with the assumption that α and β are distinct. If $\alpha - \beta$ is positive, then α can be written as the sum of two positive roots, $\alpha = (\alpha - \beta) + \beta$, in contradiction with the assumption that α be a simple root. Conversely, if $\beta - \alpha$ is positive, then β can be written as the sum of two positive roots, $\beta = (\beta - \alpha) + \alpha$, in contradiction with the assumption that β be a simple root.   □

Consider two simple roots α and β in the adjoint representation; the previous lemma implies that $\beta - \alpha$ is not a root, therefore $E_{-\alpha}|\beta\rangle = 0$. This means that, for this representation and for this choice of roots, one has $q = 0$. Similarly, interchanging α and β, one obtains that also the q corresponding to this case vanishes. Accordingly, from Eq. (6.152), recalling that $M = q - p$, one obtains

$$\frac{2\alpha \cdot \beta}{\alpha^2} = \frac{2|\alpha||\beta|\cos\theta_{\alpha\beta}}{\alpha^2} = \frac{2|\beta|\cos\theta_{\alpha\beta}}{|\alpha|} = -p \leq 0. \tag{6.207}$$

As both $|\alpha|$ and $|\beta|$ are positive, this implies that the angle between any pair of simple roots is in the range

$$\frac{\pi}{2} \leq \theta_{\alpha\beta} < \pi. \tag{6.208}$$

Lemma 6.4   *Every set of simple root vectors, such that the relative angle between each pair of them is in the $[\pi/2, \pi)$ range, is a set of linearly independent vectors.*

Proof   This lemma can be demonstrated through a proof by contradiction: If the simple roots α were not linearly independent, then there would exist a set of non-vanishing coefficients x_α such that

$$\sum_\alpha x_\alpha \alpha = 0. \tag{6.209}$$

Then Eq. (6.209) would imply

$$\gamma \equiv \sum_{x_\alpha > 0} x_\alpha \alpha = -\sum_{x_\beta < 0} x_\beta \beta = \sum_{x_\beta < 0} \left(-x_\beta\right) \beta \equiv \delta. \tag{6.210}$$

Note that γ is a positive vector, since it is defined as a linear combination of positive root vectors with positive coefficients. As a consequence, $\gamma^2 > 0$. Next, note that, from the equality $\gamma = \delta$ it follows that

$$\gamma^2 = \gamma \cdot \delta = \sum_{x_\alpha > 0} \sum_{x_\beta < 0} x_\alpha \left(-x_\beta\right) \alpha \cdot \beta < 0. \tag{6.211}$$

The last inequality follows from the fact that in each term of the sum both x_α and $(-x_\beta)$ are positive, while $\alpha \cdot \beta$ is negative by hypothesis. Then γ^2 should be simultaneously strictly positive and strictly negative, which is absurd.   □

We can now prove the following lemma.

Lemma 6.5 *Every positive root ϕ can be written as a linear combination of simple roots α with nonnegative integer coefficients k_α:*

$$\phi = \sum_\alpha k_\alpha \alpha. \tag{6.212}$$

Proof If ϕ is a simple root, then Eq. (6.212) is trivially true. Conversely, if ϕ is not simple, then it is a sum of two positive roots, $\phi = \phi_1 + \phi_2$. If ϕ_1 and ϕ_2 are both simple, Eq. (6.212) is true. If not, then the same argument can be repeated for $\phi_1 = \phi_{11} + \phi_{12}$, or for $\phi_2 = \phi_{21} + \phi_{22}$, or for both. The argument can then be iterated, until every summand is written as a sum of two positive roots. The final result is that ϕ can be written as a sum of a certain (integer and nonnegative) number of simple roots. □

It is then easy to show that the following lemma holds.

Lemma 6.6 *The number of simple roots of a simple Lie algebra is equal to its rank, m.*

Proof As the simple roots α are weights, i.e., linearly independent m-component vectors, their number cannot be larger than m. Were their number $t < m$, then it would be possible to introduce a basis where all of these m-component vectors have their first $(m - t)$ components equal to zero. As a consequence of Eq. (6.212), then, the first $(m - t)$ components of every root ϕ would also be equal to zero and

$$[H_i, E_\phi] = \overset{=0}{\phi_i}\, E_\phi = 0 \qquad \forall i \in \{1, 2, \ldots, m - t\} \tag{6.213}$$

for all roots ϕ. Then, an arbitrary nonvanishing linear combination of the form $p = \sum_{i=1}^{m-t} p_i H_i$ would commute with E_ϕ for every root ϕ. In addition, p being a linear combination of Cartan generators, it would also commute with all Cartan generators. As a consequence, p would commute with all generators of the algebra, and the algebra would not be simple, in contrast with the assumption that the algebra be simple. □

Example The rank of the su(3) algebra is 2. Of the three positive roots, $(1, 0)$ is not simple. The remaining two positive roots, $\left(\frac{1}{2}, \frac{\sqrt{3}}{2}\right)$ and $\left(\frac{1}{2}, -\frac{\sqrt{3}}{2}\right)$, must then be the simple roots of su(3).

At this point, let us then consider all possible nonvanishing linear combinations of simple roots with natural coefficients,

$$S_\alpha = \sum_\alpha k_\alpha \alpha, \quad k_\alpha \in \mathbb{N}, \tag{6.214}$$

and ask which of them are roots. Let us define $K = \sum k_\alpha$, and work by induction, noting that $K \geq 1$ ($K = 0$ being ruled out, because otherwise the linear combination would be vanishing). For $K = 1$, all coefficients of the sum on the right-hand side of Eq. (6.214) must vanish, except for one, which is equal to 1; then, S_α is just one of the simple roots themselves. Let us now assume that $K = 2$: then either all coefficients on the right-hand side of Eq. (6.214) must vanish, except for one, which is equal to 2, in which case S_α is just twice one of the simple roots, or all coefficients vanish, except for two, which are equal to 1. In this latter case, S_α reduces to the sum of two simple roots $\alpha + \beta$, which is a root if $\alpha \cdot \beta < 0$; then

$$\frac{2\alpha \cdot \beta}{\alpha^2} = -(p - q) = -p < 0, \qquad (6.215)$$

having used the fact that $q = 0$ because α and β are simple. Thus, $\alpha + p\beta$ will still be a root. Consider now the more general case, in which $\gamma = \sum_\alpha k_\alpha \alpha$, with $\sum k_\alpha = K_0$, was found to be a root. By acting on that γ with E_α, where α is a simple root, one obtains $\gamma + \alpha$. From the knowledge of how γ was constructed, one knows the value of q such that $\gamma - q\alpha$ is still a root. Knowing q, one can evaluate $\gamma \cdot \alpha$ and use the formula

$$\frac{2\alpha \cdot \gamma}{\alpha^2} = -(p - q) \qquad (6.216)$$

to determine if $p > 0$. If p is positive, then $\gamma + \alpha$ is a root since $\gamma + p\alpha$ with $p \geq 1$ is still a root. This gives all the roots with $K = K_0 + 1$; if there were some root ρ that does not have the form $\gamma + \alpha$ for any α, then, since $\rho - \alpha$ is not a root, one would have $q = 0$ and

$$\frac{2\alpha \cdot \rho}{\alpha^2} = -p \leq 0 \quad \forall \alpha, \qquad (6.217)$$

but this would imply that ρ would be linearly independent from all the simple roots, which is a contradiction.

To summarize, knowing the simple roots α, one can construct *all* roots.

6.6.1 Cartan Matrix and Dynkin Diagrams

Let us recapitulate some of the main findings of the previous subsections. As discussed above, the angles between root vectors must be solutions of the Diophantine Eq. (6.155): This constrains them to the very limited number of possibilities that we listed in Section 6.5. Furthermore, earlier in Section 6.6 we found that the simple roots of the algebra, whose number is m, the rank of the algebra, i.e., the number of Cartan generators, are at relative angles between $\pi/2$ and π, and can be used to find all the roots. In turn, we can use the roots to write down the algebra by finding the $N_{\alpha,\beta}$s.

The properties of the simple roots can be encoded into a matrix, called the *Cartan matrix*. Given a Lie algebra g of rank m, the elements of the Cartan matrix are defined as

$$A^{ij} = 2\frac{\alpha^i \cdot \alpha^j}{|\alpha^j|^2}. \qquad (6.218)$$

Introducing the coroots of the simple roots,

$$\alpha^{j\vee} \equiv \frac{2\alpha^j}{|\alpha^j|^2}, \qquad (6.219)$$

Eq. (6.218) can be written more concisely as

$$A^{ij} = (\alpha^i, \alpha^{j\vee}), \qquad (6.220)$$

where (,) denotes the scalar product. Without giving a detailed proof, we simply state that the Cartan matrix has the following properties:

(i) $A^{ij} \in \mathbb{Z}$, with $A^{ij} < 0$ for $i \neq j$,

(ii) $\det A \neq 0$, so A is invertible,

(iii) all the diagonal entries $A^{ii} = 2$ (where no summation over i is assumed),

(iv) A is generally not symmetric, but if $A^{ij} = 0$, then $A^{ji} = 0$,

(v) the product of two off-diagonal entries that are symmetric under reflection with respect to the main diagonal is a natural number less than 4, namely: $A^{ij} A^{ji} \in \{0, 1, 2, 3\}$.

The last property (v) follows from the previous analysis for the angles between simple roots.

Example We found that the simple roots of the su(3) Lie algebra are $\alpha^1 = \left(\frac{1}{2}, \frac{\sqrt{3}}{2}\right)$ and $\alpha^2 = \left(\frac{1}{2}, -\frac{\sqrt{3}}{2}\right)$. The associated coroots are

$$\alpha^{1\vee} = \left(1, \sqrt{3}\right), \qquad \alpha^{2\vee} = \left(1, -\sqrt{3}\right). \tag{6.221}$$

Thus, the Cartan matrix of su(3) is

$$A = \begin{pmatrix} 2 & -1 \\ -1 & 2 \end{pmatrix}. \tag{6.222}$$

There is also a concise diagrammatic notation to encode the same information, provided by the *Dynkin diagram*: a diagram that can be constructed from the Cartan matrix A through the following rules.

(i) Draw m open circles (one for each simple root α^i).

(ii) Connect the pair of circles representing the simple roots α^i, α^j with the number of lines given by the corresponding product $A^{ij} A^{ji}$. Do this for all pairs.

(iii) In the cases when $A^{ij} A^{ji} = 2$ or 3, if $A^{ij} > A^{ji}$ then fill the circle corresponding to α^i if $A^{ij} > A^{ji}$ (or, alternatively, draw an arrow pointing from the circle corresponding to α^i to α^j, or vice versa), or fill the circle corresponding to α^j if $A^{ji} > A^{ij}$ (or draw an arrow pointing from the circle corresponding to α^j to α^i).

In a Dynkin diagram, the number of lines between two circles is related to the angle between the roots to which the circles correspond; in particular, the number of lines is increasing with the angle. More precisely, one has:

◯ ◯	if the angle is $\pi/2$,
◯—◯	if the angle is $2\pi/3$,
◯═◯	if the angle is $3\pi/4$,
◯≡◯	if the angle is $5\pi/6$.

The filling rule (or, equivalently, the arrow rule) (iii) for circles reflects the fact that for some Lie algebras (other than su(N)) the simple roots may have different lengths: the filled circle (or the one the arrow starts from) is the one associated with the longer of the two simple roots.

6.6.2 Fundamental Weights and Fundamental Representations

Let us label the simple roots by an index i, running from 1 to the rank of the algebra m, as α^i. Given an arbitrary irreducible representation D, we define a weight vector μ in D as the *highest weight* in that representation if and only if, for every positive root ϕ, $\mu + \phi$ is not a weight in the representation. This is equivalent to requiring that $\mu + \alpha^i$ is not a weight for all the simple roots α^i. As a consequence, when E_{α^i} acts on μ, the corresponding p^i must be zero, i.e., Eq. (6.151) must reduce to

$$\frac{2\alpha^i \cdot \mu}{(\alpha^i)^2} = q^i, \tag{6.223}$$

where the q^is are natural numbers.

Equation (6.223) can be interpreted as a relation giving the components of μ on the basis of the simple roots α^is. Introducing the coroots

$$\alpha^{i\vee} \equiv \frac{2\alpha^i}{(\alpha^i)^2} \tag{6.224}$$

of the simple roots, Eq. (6.223) takes an elegant form

$$\left(\alpha^{i\vee}, \mu\right) = q^i. \tag{6.225}$$

Knowing the highest vector μ, one can construct also the other vectors in that representation by acting on μ with the lowering operators $E_{-\alpha^i}$. This means that every irreducible representation of a simple Lie group of rank m can be uniquely identified by the m natural numbers q^i.

It is then particularly interesting to consider a set of m irreducible representations D^j with highest vectors μ^j, such that

$$\frac{2\alpha^i \cdot \mu^j}{(\alpha^i)^2} = \delta^{ij}. \tag{6.226}$$

Each highest vector satisfying Eq. (6.226) is called a *fundamental weight*. Using the definition of the coroots of the simple roots given in Eq. (6.224), Eq. (6.226) can be interpreted as an orthonormality relation

$$(\alpha^{i\vee}, \mu^j) = \delta^{ij}. \tag{6.227}$$

Each irreducible representation D^j having a fundamental weight μ^j as highest vector is called a *fundamental representation*.

The fundamental weights provide a basis on which to expand the highest vector of every irreducible representation in a unique way

$$\mu = \sum_{j=1}^{m} q^j \mu^j, \tag{6.228}$$

with coefficients q^j which are natural numbers. Alternatively, we can label the irreducible representations by just listing the integers $(q^1, q^2, \ldots, q^m)$; this is called the *Dynkin label* of the representation. With this labeling, the fundamental representations have the labels $(1, 0, \ldots, 0)$, $(0, 1, 0, \ldots, 0)$, $\ldots$, $(0, 0, \ldots, 1)$. Later, we will see that it is possible to build the representation with highest weight μ by taking the tensor product of q^1 fundamental representations of highest weight μ^1 with q^2

fundamental representations of highest weight μ^2, and so on. Finally, we note that for the su(3) algebra the Dynkin label is sometimes denoted as (n, m) rather than (q^1, q^2), to avoid too many superscripts in further equations.

6.6.3 Irreducible Representations of su(3)

As we discussed earlier in this section, the simple roots of the su(3) Lie algebra are $\alpha^1 = \left(\frac{1}{2}, \frac{\sqrt{3}}{2}\right)$ and $\alpha^2 = \left(\frac{1}{2}, -\frac{\sqrt{3}}{2}\right)$; the associated coroots are

$$\alpha^{1\vee} = \left(1, \sqrt{3}\right), \qquad \alpha^{2\vee} = \left(1, -\sqrt{3}\right). \tag{6.229}$$

Let us find the fundamental weight $\mu^1 = (a, b)$ that satisfies the orthonormality relations expressed in Eq. (6.227):

$$\begin{cases} (\alpha^{1,\vee}, \mu^1) = a + b\sqrt{3} = 1 \\ (\alpha^{2,\vee}, \mu^1) = a - b\sqrt{3} = 0 \end{cases}. \tag{6.230}$$

The solution is

$$\mu^1 = \left(\frac{1}{2}, \frac{1}{2\sqrt{3}}\right), \tag{6.231}$$

which we recognize as the highest weight of the fundamental representation $\mathbf{3} = (1, 0)$. Similarly, the other fundamental weight is found to be

$$\mu^2 = \left(\frac{1}{2}, -\frac{1}{2\sqrt{3}}\right), \tag{6.232}$$

which is the highest weight of the antifundamental representation $\overline{\mathbf{3}} = (0, 1)$. Other irreducible representations are then classified by their highest weights $\mu = n\mu^1 + m\mu^2$, labeled by the Dynkin labels (n, m), where n and m are natural numbers.[7] The label $(0, 0)$ denotes the trivial, also called *singlet*, representation. In elementary particle physics it is customary to label the irreducible representations by their dimension, as we did for the fundamental and antifundamental representation. Note, however, that in general the dimension does not uniquely identify a representation. We will later show that the dimension $D(n, m)$ of the (n, m) representation is given by the formula

$$D(n, m) = \frac{(n + 1)(m + 1)(n + m + 2)}{2}. \tag{6.233}$$

Thus, for example, the $(1, 1)$ representation has dimension $D = 8$. It is the adjoint representation, whose nonzero weights are the roots in (6.178). The highest weight (root) was $(1, 0) = \left(\frac{1}{2}, \frac{\sqrt{3}}{2}\right) + \left(\frac{1}{2}, -\frac{\sqrt{3}}{2}\right) = \mu^1 + \mu^2$ so the Dynkin label is $(1, 1)$. In elementary particle physics the adjoint representation is denoted $\mathbf{8}$.

There exists a simple way to determine if a representation (n, m) of the su(3) algebra is real or complex. Recall that the highest weight μ^2 of the antifundamental representation is minus the lowest weight of the fundamental representation. Likewise, $-\mu^1$ is the lowest weight of the antifundamental representation. Thus, the

---

[7] Note that here and in the rest of the discussion for the su(3) algebra, the natural number m in the Dynkin label (n, m) is *not* the rank of the algebra.

lowest weight of the (n, m) representation is $-n\mu^2 - m\mu^1$. The representation is real if its lowest weight equals minus its highest weight, i.e.,

$$-n\mu^2 - m\mu^1 = -(n\mu^1 + m\mu^2) \Leftrightarrow n = m, \qquad (6.234)$$

implying that the (n, m) representation is real if and only if $n = m$. Thus, all representations (n, n) are real. Likewise, we also observe that minus the lowest weight of the (n, m) representation, i.e., $m\mu^1 + n\mu^2$, is the highest weight of the complex conjugate representation. Thus $\overline{(n, m)} = (m, n)$.

Let us look at the representation $(2, 0)$, which has $q^1 = 2$ and $q^2 = 0$. It has highest weight

$$2\mu^1 = \left(1, \frac{1}{\sqrt{3}}\right). \qquad (6.235)$$

We know that $2\mu^1 - \alpha^1$ and $2\mu^1 - 2\alpha^1$ are weights corresponding to unique states, but that $2\mu^1 - \alpha^2$ is not a weight. We can then complete the weight diagram by applying Weyl reflections, with the result shown below.

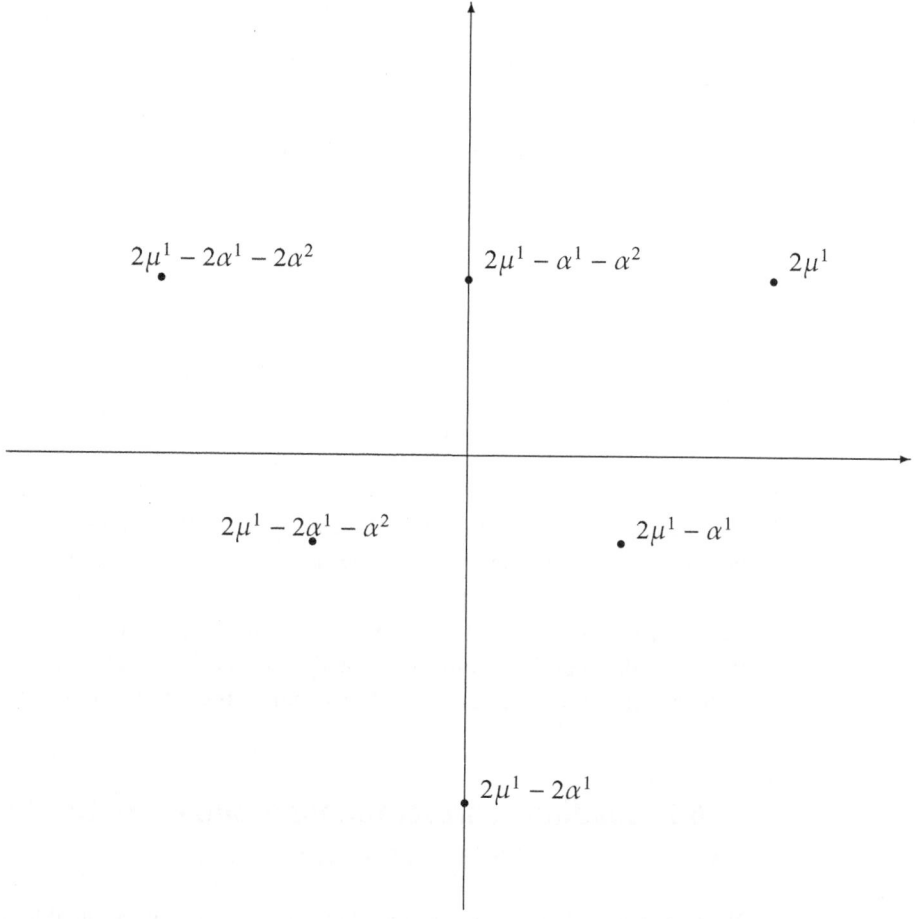

This is a six-dimensional representation, so $(2, 0) = \mathbf{6}$. Its complex conjugate is $(0, 2) = \bar{\mathbf{6}}$, its weight diagram is the upside-down inverted image of the diagram above.

It can be shown that the weights of a general SU(3) representation form either triangles or hexagons. For example, consider the representation $(2, 1)$, also denoted as **15**, shown below.

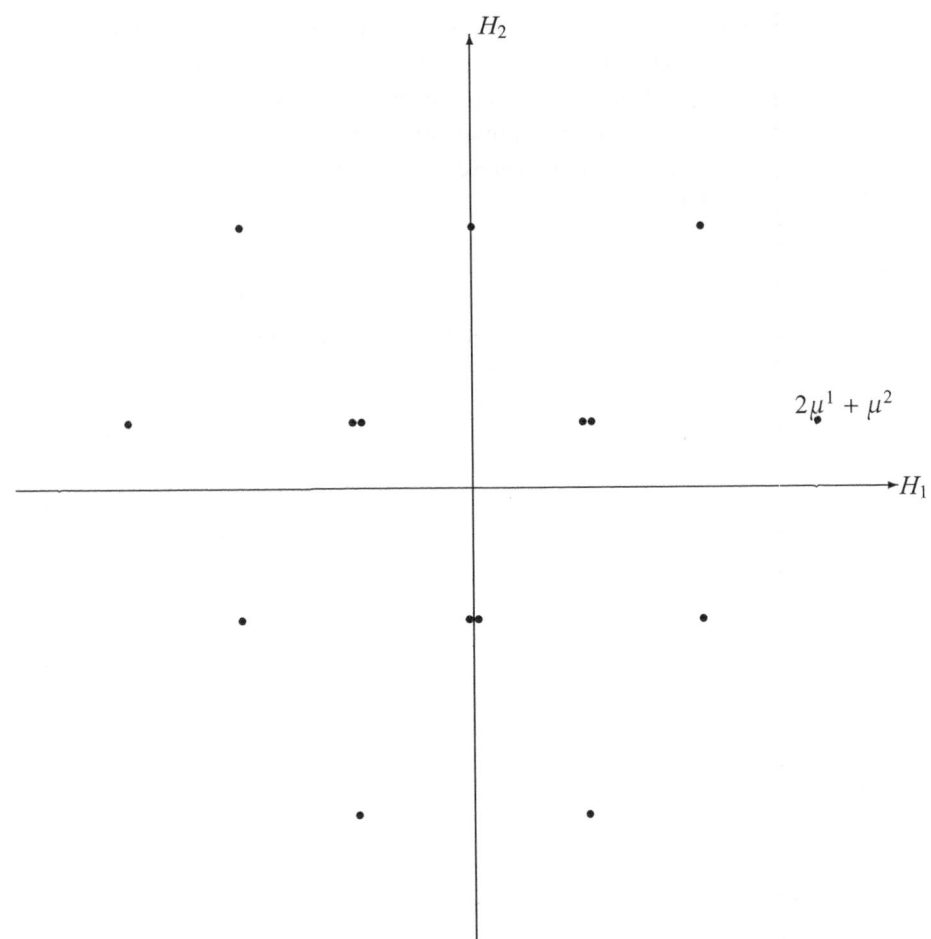

The general pattern is that the outermost layer of weights is unique. Then, the next layer is doubly degenerate. The degeneracy increases by one unit each time one moves in from an outer hexagonal layer. Reaching each triangular layer, the degeneracy remains constant as one moves further inwards. Thus, for instance, in the triangular representations, $(n, 0)$ and $(0, n)$, each weight corresponds to a unique state, as for the representations **3**, **6**, **10**, and their complex conjugates.

6.7 Building Representations of su(2) with Tensor Products

We have learnt that the highest weight vector μ of every irreducible representation D of a Lie algebra of rank r has the unique expansion given in Eq. (6.228)

$$\mu = q_1\mu_1 + \cdots + q_r\mu_r \qquad (6.236)$$

in the basis of the fundamental weights μ_i, the highest weights of the fundamental representations, where q_i are natural numbers. Note that, for later convenience, here we are denoting the algebra rank as r (instead of m). For the su(2) Lie algebra, the fundamental representation is the spin-$\frac{1}{2}$ representation, so the above equation reduces to the simple form

$$j = q \cdot \frac{1}{2}, \tag{6.237}$$

where j is the highest weight of a spin-j representation. This is consistent with the fact that the allowed values of j are $l/2$, where l is a natural number.

Next we study how one can use tensor products to build larger representations of the su(2) algebra. Consider two irreducible representations with spins j_i and $i = 1$ and 2, with the $2j_i + 1$ eigenvectors of J_3

$$|j_1 m_i\rangle, \qquad m_i \in \{j_i, j_i - 1, \ldots, -j_i\} \tag{6.238}$$

as the basis vectors of the vector space V_i of the representation. Now consider their tensor products

$$|j_1 m_1\rangle \, |j_2 m_2\rangle \equiv |j_1 m_1\rangle \otimes |j_2 m_2\rangle. \tag{6.239}$$

The three su(2) generators are linear operators, represented by matrices J_i^a (with $a \in \{1, 2, 3\}$), acting in V_i. We can construct linear combinations of tensor-product operators

$$J^a = J_1^a \otimes \mathbb{1}_2 + \mathbb{1}_1 \otimes J_2^a \tag{6.240}$$

acting on the tensor-product vectors

$$J^a \, |j_1 m_1\rangle \, |j_2 m_2\rangle = \left(J_1^a \, |j_1 m_1\rangle\right) |j_2 m_2\rangle + |j_1 m_1\rangle \left(J_2^a \, |j_2 m_2\rangle\right). \tag{6.241}$$

The J^a satisfy the su(2) algebra commutation relations:

$$[J^a, J^b] = i\epsilon_{abc} J^c. \tag{6.242}$$

We then form the raising and lowering operators $J^\pm = J^1 \pm iJ^2$. Next, we show that the $(2j_1 + 1)(2j_2 + 1)$ vectors $|j_1 m_1\rangle \, |j_2 m_2\rangle$ form a representation of the su(2) Lie algebra (6.242) that can be decomposed into irreducible spin-j representations. Consider first the case $j_1 = j_2 = \frac{1}{2}$. We start with the product of the highest-weight vectors $m_1 = m_2 = \frac{1}{2}$ and act on it with J^3:

$$J^3 \left|\frac{1}{2}\frac{1}{2}\right\rangle \left|\frac{1}{2}\frac{1}{2}\right\rangle = \left(J_1^3 \left|\frac{1}{2}\frac{1}{2}\right\rangle\right) \left|\frac{1}{2}\frac{1}{2}\right\rangle + \left|\frac{1}{2}\frac{1}{2}\right\rangle \left(J_2^3 \left|\frac{1}{2}\frac{1}{2}\right\rangle\right)$$

$$= \left(\frac{1}{2} + \frac{1}{2}\right) \left|\frac{1}{2}\frac{1}{2}\right\rangle \left|\frac{1}{2}\frac{1}{2}\right\rangle = 1 \cdot \left|\frac{1}{2}\frac{1}{2}\right\rangle, \tag{6.243}$$

and with J^+:

$$J^+ \left|\frac{1}{2}\frac{1}{2}\right\rangle \left|\frac{1}{2}\frac{1}{2}\right\rangle = \left(J_1^+ \left|\frac{1}{2}\frac{1}{2}\right\rangle\right) \left|\frac{1}{2}\frac{1}{2}\right\rangle + \left|\frac{1}{2}\frac{1}{2}\right\rangle \left(J_2^+ \left|\frac{1}{2}\frac{1}{2}\right\rangle\right) = 0. \tag{6.244}$$

We conclude that $\left|\frac{1}{2}\frac{1}{2}\right\rangle \left|\frac{1}{2}\frac{1}{2}\right\rangle$ is an eigenvector with eigenvalue 1, the highest weight eigenvector of the spin-1 irreducible representation of the su(2) algebra:

$$|j = 1, m = 1\rangle = \left|\frac{1}{2}\frac{1}{2}\right\rangle \left|\frac{1}{2}\frac{1}{2}\right\rangle. \tag{6.245}$$

What about the other states? Recalling that $J^{\pm}|j,m\rangle = \sqrt{\frac{(j\mp m)(j\pm m+1)}{2}}|j,m\pm 1\rangle$, so that, for example, $J^-\left|\frac{1}{2},\frac{1}{2}\right\rangle = \sqrt{\frac{(1)(0+1)}{2}}\left|\frac{1}{2},-\frac{1}{2}\right\rangle = \frac{1}{\sqrt{2}}\left|\frac{1}{2},-\frac{1}{2}\right\rangle$, we can lower the weight acting with J^-:

$$J^-\left|\frac{1}{2}\frac{1}{2}\right\rangle\left|\frac{1}{2}\frac{1}{2}\right\rangle = \frac{1}{\sqrt{2}}\left|\frac{1}{2}-\frac{1}{2}\right\rangle\left|\frac{1}{2}\frac{1}{2}\right\rangle + \frac{1}{\sqrt{2}}\left|\frac{1}{2}\frac{1}{2}\right\rangle\left|\frac{1}{2}-\frac{1}{2}\right\rangle, \qquad (6.246)$$

we check that this is an eigenvector of J^3 with eigenvalue 0:

$$J^3\left(\frac{1}{\sqrt{2}}\left|\frac{1}{2}-\frac{1}{2}\right\rangle\left|\frac{1}{2}\frac{1}{2}\right\rangle + \frac{1}{\sqrt{2}}\left|\frac{1}{2}\frac{1}{2}\right\rangle\left|\frac{1}{2}-\frac{1}{2}\right\rangle\right)$$
$$= \left(-\frac{1}{2}+\frac{1}{2}+\frac{1}{2}-\frac{1}{2}\right)(\ldots) = 0, \qquad (6.247)$$

so it is the eigenvector $|j=1,m=0\rangle$. Proceeding with J^-:

$$J^-\frac{1}{\sqrt{2}}\left(\left|\frac{1}{2}-\frac{1}{2}\right\rangle\left|\frac{1}{2}\frac{1}{2}\right\rangle + \left|\frac{1}{2}\frac{1}{2}\right\rangle\left|\frac{1}{2}-\frac{1}{2}\right\rangle\right)$$
$$= \frac{1}{\sqrt{2}}\left(0+(\frac{1}{\sqrt{2}}\left|\frac{1}{2}-\frac{1}{2}\right\rangle\left|\frac{1}{2}-\frac{1}{2}\right\rangle + \frac{1}{\sqrt{2}}\left|\frac{1}{2}-\frac{1}{2}\right\rangle\left|\frac{1}{2}-\frac{1}{2}\right\rangle + 0\right)$$
$$= \left|\frac{1}{2}-\frac{1}{2}\right\rangle\left|\frac{1}{2}-\frac{1}{2}\right\rangle, \qquad (6.248)$$

which has J^3 eigenvalue $-\frac{1}{2}-\frac{1}{2} = -1$. Thus we obtained the eigenvectors

$$|j=1,m=1\rangle = \left|\frac{1}{2}\frac{1}{2}\right\rangle\left|\frac{1}{2}\frac{1}{2}\right\rangle$$
$$|j=1,m=0\rangle = \frac{1}{\sqrt{2}}\left(\left|\frac{1}{2}-\frac{1}{2}\right\rangle\left|\frac{1}{2}\frac{1}{2}\right\rangle + \left|\frac{1}{2}\frac{1}{2}\right\rangle\left|\frac{1}{2}-\frac{1}{2}\right\rangle\right) \qquad (6.249)$$
$$|j=1,m=-1\rangle = \left|\frac{1}{2}-\frac{1}{2}\right\rangle\left|\frac{1}{2}-\frac{1}{2}\right\rangle$$

of the spin $j=1$ irreducible representation. The missing fourth vector that can be built from the tensor products of the two states is the orthogonal combination

$$\frac{1}{\sqrt{2}}\left(\left|\frac{1}{2}-\frac{1}{2}\right\rangle\left|\frac{1}{2}\frac{1}{2}\right\rangle - \left|\frac{1}{2}\frac{1}{2}\right\rangle\left|\frac{1}{2}-\frac{1}{2}\right\rangle\right)$$

for which we find $J^3 = J^+ = 0$. It is therefore the highest weight vector with $j=m=0$: the vector $|j=0,m=0\rangle$ of the spin-0 (scalar) irreducible representation. We have thus established the following decomposition of the tensor product representation:

$$D^{(j=\frac{1}{2})} \otimes D^{(j=\frac{1}{2})} = D^{(j=1)} \oplus D^{(j=0)}. \qquad (6.250)$$

This procedure can be generalized to $j_1,j_2 \geq \frac{1}{2}$. Consider first the tensor product of two vectors

$$|j_1 j_1\rangle |j_2 j_2\rangle;$$

it can be seen to be the highest weight eigenvector of the spin-(j_1+j_2) representation. Then, as above, act on it with J^- repeatedly to build the rest of the eigenvectors of the

$j_1 + j_2$ representation. Then construct the orthogonal complements of these eigenvectors, act on them repeatedly with J^-, again construct orthogonal complements, and so on. The process gives the reduction

$$D^{(j_1)} \otimes D^{(j_2)} = D^{(j_1+j_2)} \oplus D^{(j_1+j_2-1)} \oplus \cdots \oplus D^{(|j_1-j_2|)}. \tag{6.251}$$

One can check that the dimensions of the representations match between both sides:

$$(2j_1 + 1)(2j_2 + 1) = \sum_{k=|j_1-j_2|}^{j_1+j_2} (2k + 1). \tag{6.252}$$

Another way to do the decomposition is the one that is usually presented in quantum mechanics textbooks. Let us denote the (orthonormal) tensor product vectors as

$$|j_1 m_1 j_2 m_2\rangle \equiv |j_1 m_1\rangle \otimes |j_2 m_2\rangle, \tag{6.253}$$

which we want to compare with the orthonormal eigenvectors $|jm\rangle$. The basis (6.253) satisfies the completeness relation

$$\mathbb{1} = \bigoplus_{j_1,j_2} \left(\mathbb{1}_{2j_1+1} \otimes \mathbb{1}_{2j_2+1} \right) = \sum_{j_1,j_2} \sum_{m_1 m_2} |j_1 m_1 j_2 m_2\rangle \langle j_1 m_1 j_2 m_2| \tag{6.254}$$

so then we can obtain

$$|jm\rangle = \mathbb{1} |jm\rangle = \sum_{j_1,j_2} \sum_{m_1,m_2} |j_1 m_1 j_2 m_2\rangle \langle j_1 m_1 j_2 m_2 | jm\rangle$$

$$= \sum_{j_1,j_2} \sum_{m_1,m_2} \langle j_1 m_1 j_2 m_2 | jm\rangle \, |j_1 m_1 j_2 m_2\rangle \tag{6.255}$$

with only a finite number of nonzero coefficients $\langle j_1 m_1 j_2 m_2 | jm\rangle$. We can then interpret them as elements of a unitary matrix that rotates the orthonormal basis $|j_1 m_1 j_2 m_2\rangle$ to the orthonormal basis $|jm\rangle$, decomposing the tensor-product representation to a direct sum of irreducible representations. There remains an undetermined phase factor, since the eigenvectors $|jm\rangle$ can always be redefined by multiplying them by $e^{i\varphi}$. A common convention is to fix the phase by demanding that the matrix elements with $m = j$, $m_1 = j_1$, and $m_2 = (j - j_1)$ are positive:

$$\langle jj | j_1 j_1 j_2 (j - j_1) \rangle > 0. \tag{6.256}$$

This is called the *Condon–Shortley phase convention*. In physics, the matrix elements $\langle j_1 m_1 j_2 m_2 | jm \rangle$ are called *Clebsch–Gordan coefficients*.

6.8 Tensor Methods to Build su(3) Representations

We now discuss the construction of higher-dimensional representations of the su(3) Lie algebra using tensor products of irreducible representations, generalizing the previous discussion of the su(2) case. Consider first the tensor product $\mathbf{3} \otimes \mathbf{3}$. We start with the simple tensors

$$v \otimes w = v^i w^j e_i \otimes e_j, \tag{6.257}$$

where $v^i w^j$ are the components. As we did for the su(2) algebra, we can form new su(3) generators

$$X^a = X_1^a \otimes \mathbb{1}_2 + \mathbb{1}_1 \otimes X_2^a \qquad (6.258)$$

that satisfy the su(3) commutation relations and act on the $v \otimes w$ tensor as

$$X^a(v \otimes w) = X_1^a v \otimes \mathbb{1}_2 w + \mathbb{1}_1 v \otimes X_2^a w, \qquad (6.259)$$

or, in components,

$$v^i w^j \mapsto (X^a)_k^i v^k w^j + v^i (X^a)_k^j w^k. \qquad (6.260)$$

We then want to understand how to decompose the tensor product into irreducible representations of the X^a operators defined in Eq. (6.258) that generate the composite su(3) algebra. Following the strategy used in the su(2) case, one could first identify the tensor corresponding to the highest weight, and then act on it by (composite) lowering operators E_α, and so on. However, we will proceed in a different way.

Before discussing the actual construction, consider the tensor product $\mathbf{3} \otimes \bar{\mathbf{3}}$; now the simple tensors involve a vector v and a dual vector ω,

$$v \otimes \omega = v^i \omega_j e_i \otimes e^{*j}. \qquad (6.261)$$

In this case, the transformation rule expressed by Eq. (6.260) has to be modified, in order to account for the vector and dual-vector transformation rules that are respectively given in Eq. (6.205) and in Eq. (6.206):

$$v^i \omega_j \mapsto (X^a)_k^i v^k \omega_j - v^i \omega_k (X^a)_j^k. \qquad (6.262)$$

Note that the tensor-product vector spaces do not consist only of simple tensors like those in Eq. (6.257) and in Eq. (6.261); for example, $\mathbf{3} \otimes \bar{\mathbf{3}}$ also contains $(1, 1)$-tensors of the general form

$$T = T_j^i \, e_i \otimes e^{*j}. \qquad (6.263)$$

Likewise, in general the tensor product of n copies of the fundamental representation and m copies of the antifundamental representation $\mathbf{3}^{\otimes n} \otimes \bar{\mathbf{3}}^{\otimes m}$ contains (n, m)-tensors, with components having n upper and m lower indices,

$$T = T_{j_1 j_2 \dots j_m}^{i_1 i_2 \dots i_n} e_{i_1} \otimes \cdots \otimes e_{i_n} \otimes e^{*j_1} \otimes \cdots \otimes e^{*j_m}. \qquad (6.264)$$

The transformation rules for the components of these tensors under transformations in the group are

$$T_{j_1 j_2 \dots j_m}^{i_1 i_2 \dots i_n} \mapsto U_{k_1}^{i_1} \cdots U_{k_n}^{i_n} T_{l_1 \dots l_m}^{k_1 \dots k_n} (U^\dagger)_{j_1}^{l_1} \cdots (U^\dagger)_{j_m}^{l_m} \qquad (6.265)$$

while under transformations in the Lie algebra:

$$\begin{aligned}
T_{j_1 j_2 \dots j_m}^{i_1 i_2 \dots i_n} \mapsto & (X^a)_{k_1}^{i_1} T_{j_1 \dots j_m}^{k_1 i_2 \dots i_n} + \cdots + (X^a)_{k_n}^{i_n} T_{j_1 \dots j_m}^{i_1 i_2 \dots k_n} \\
& - T_{l_1 j_2 \dots j_m}^{i_1 i_2 \dots i_n} (X^a)_{j_1}^{l_1} - \cdots - T_{j_1 j_2 \dots l_m}^{i_1 i_2 \dots i_n} (X^a)_{j_m}^{l_m}.
\end{aligned} \qquad (6.266)$$

For example, the components of a $(1, 1)$-tensor like the one in Eq. (6.263) transform under the group

$$T_j^i \mapsto U_k^i T_l^k (U^\dagger)_j^l, \qquad \text{i.e.,} \quad T \mapsto U T U^\dagger. \qquad (6.267)$$

States in an irreducible representation are expected to correspond to tensors with particular properties. Then, the decomposition of a tensor product into a direct sum of irreducible representations corresponds to manipulating the initial tensor product into a sum of tensors with well-defined symmetry properties.

Example In quantum chromodynamics (the fundamental theory of the strong nuclear interaction) the gluons transform under the adjoint **8** representation. They are represented by a gauge field with components A_μ, for $\mu \in \{0, 1, 2, 3\}$, each being a Hermitian traceless 3×3 matrix which can be expanded in the basis of Gell-Mann matrices λ^a as $A_\mu = A_\mu^a X^a$, where $X^a = \lambda^a/2$. Conversely, the quarks are represented by a field ψ in the fundamental representation of the gauge group. The interaction of quarks with the gluon field can be considered as a generalization of the electromagnetic interaction of electrons with the photon field.

For the identification of the states in an irreducible representation as tensors, we must determine their symmetries and other relevant properties. We begin by noting that the tensors δ_j^i, i.e., the Kronecker delta, as well as ϵ^{ijk} and ϵ_{ijk}, namely the completely antisymmetric ϵ-symbols, are invariant under the action of the SU(3) group:

$$\delta_j^i \mapsto U_k^i \delta_l^k (U^\dagger)_j^l = (UU^\dagger)_j^i = \delta_j^i, \tag{6.268}$$

$$\epsilon^{ijk} \mapsto U_l^i U_m^j U_n^k \epsilon^{lmn} = (\det U)\epsilon^{ijk} = \epsilon^{ijk}, \tag{6.269}$$

$$\epsilon_{ijk} \mapsto \epsilon_{lmn}(U^\dagger)_i^l (U^\dagger)_j^m (U^\dagger)_k^n = (\det U^\dagger)\epsilon_{ijk} = \epsilon_{ijk}. \tag{6.270}$$

This means that we can multiply tensors by δ_j^i, by ϵ^{ijk}, or by ϵ_{ijk} without changing their transformation properties. Consider for example a scalar, that is a $(0, 0)$-tensor v, that is associated with the trivial representation:

$$v \text{ transforms equivalently to } \delta_j^i v. \tag{6.271}$$

This means, in particular, that the number of indices is not sufficient to identify a tensor with an irreducible representation. In the following, we denote the equivalence relation between tensors that transform equivalently to each other by the symbol $\leftrightarrow$. In addition to multiplying tensors with each other, we can also contract their indices with ϵ-symbols. Consider the three equivalent ways to associate tensors with the antifundamental $\bar{\mathbf{3}} = (0, 1)$ representation:

$$\omega_k \leftrightarrow \delta_j^i \omega_k \leftrightarrow \epsilon^{ijk}\omega_k, \tag{6.272}$$

where the last expression has two free upper indices. Let us check that it transforms in the same way as ω_k:

$$\epsilon^{ijk}\omega_k \mapsto U_l^i U_m^j \epsilon^{lmk}\omega_k = U_l^i U_m^j \epsilon^{lmn} U_n^p (U^\dagger)_p^k \omega_k$$
$$= (\det U)\epsilon^{ijp}\omega_k (U^\dagger)_p^k = \epsilon^{ijk}[\omega_l (U^\dagger)_k^l], \tag{6.273}$$

hence the essential part is the correct transformation rule for ω_k. These examples show that there is some freedom in the index arrangement. The example in the expression (6.272) shows that we can lift a lower index to an antisymmetric pair of upper indices with ϵ^{ijk} without altering the transformation properties. Likewise, we can trade an upper index for an antisymmetric pair of lower indices using ϵ_{ijk}. Contracting an upper–lower index pair as in the expression (6.271) just leaves a quantity with no indices, i.e., a $(0,0)$-tensor (namely a scalar).

Following these examples, we are now ready to state (without presenting its proof) the full rule for irreducible representations.

The irreducible representation (n, m) of su(3) corresponds to tensors $T^{i_1 \ldots i_n}_{j_1 \ldots j_m}$ which are:

(i) completely symmetric under every permutation of the upper indices $i_1, \ldots, i_n$,
(ii) completely symmetric under every permutation of the lower indices $j_1, \ldots, j_m$,
(iii) traceless in any contracted upper–lower index pair:

$$T^{k i_2 \ldots i_n}_{k j_2 \ldots j_m} = T^{k i_2 \ldots i_n}_{j_1 k j_3 \ldots j_m} = \cdots = 0.$$

"Completely symmetric" means that any reordering by a permutation is equal to the original ordering. With these properties, the number of independent components of the tensor $T^{i_1 \ldots i_n}_{j_1 \ldots j_m}$ is precisely $D(n, m)$, the dimension of the (n, m)-representation.

Now we are ready to return to the problem that we started with: the decomposition of the tensor product $\mathbf{3} \otimes \mathbf{3}$. We first need to extract the symmetric part of the product in Eq. (6.257):

$$v^i w^j = \frac{1}{2}(v^i w^j + v^j w^i) + \frac{1}{2}(v^i w^j - v^j w^i). \tag{6.274}$$

Then, using

$$\epsilon^{ijk} \epsilon_{klm} = \delta^i_l \delta^j_m - \delta^i_m \delta^j_l, \tag{6.275}$$

we reexpress the antisymmetric term appearing on the right-hand side of Eq. (6.274) as

$$\frac{1}{2}(v^i w^j - v^j w^i) = \frac{1}{2}\epsilon^{ijk} \epsilon_{klm} v^l w^m. \tag{6.276}$$

The first term $\frac{1}{2}(v^i w^j + v^j w^i)$ on the right-hand side of Eq. (6.274) is a completely symmetric tensor with two upper indices, hence it belongs to the $(2, 0)$ representation. Then, in Eq. (6.276) we can remove the first ϵ-symbol ϵ^{ijk}, as in (6.272), and we are left with $\frac{1}{2}\epsilon_{klm} v^l w^m$. This remaining tensor has one lower index (and thus it is trivially completely symmetric) and no free upper indices, so it is a tensor of type $(0, 1)$ transforming like ω_k, and thus belongs to the $(0, 1) = \bar{\mathbf{3}}$ representation. So we conclude

$$\mathbf{3} \otimes \mathbf{3} = (\mathbf{1}, \mathbf{0}) \otimes (\mathbf{1}, \mathbf{0}) = (\mathbf{2}, \mathbf{0}) \oplus (\mathbf{0}, \mathbf{1}) = \mathbf{6} \oplus \bar{\mathbf{3}}. \tag{6.277}$$

As a next example, we consider the $\mathbf{3} \otimes \bar{\mathbf{3}}$ product. In this case we split Eq. (6.261) as follows:

$$v^i \omega_j = \left(v^i \omega_j - \frac{1}{3}\delta^i_j v^k \omega_k\right) + \frac{1}{3}\delta^i_j v^k \omega_k. \tag{6.278}$$

The first term has one upper and one lower index, and is traceless, so it belongs to the $(1, 1)$ representation. The last term has a nonvanishing trace, but we can strip off the δ^i_j factor as in Eq. (6.271), and identify the remaining coefficient, which has no indices, as the scalar representation. Therefore we conclude

$$\mathbf{3} \otimes \bar{\mathbf{3}} = (\mathbf{1}, \mathbf{0}) \otimes (\mathbf{0}, \mathbf{1}) = (\mathbf{1}, \mathbf{1}) \oplus (\mathbf{0}, \mathbf{0}) = \mathbf{8} \oplus \mathbf{1}. \tag{6.279}$$

In a similar way, one can show that $\mathbf{3} \otimes \mathbf{6} = \mathbf{10} \oplus \mathbf{8}$. We leave the proof of this property as an exercise (see Problem 6.15). It then follows that

$$
\begin{aligned}
\mathbf{3} \otimes \mathbf{3} \otimes \mathbf{3} &= \mathbf{3} \otimes (\mathbf{3} \otimes \mathbf{3}) \\
&= \mathbf{3} \otimes \left(\mathbf{6} \oplus \overline{\mathbf{3}} \right) \\
&= (\mathbf{3} \otimes \mathbf{6}) \oplus \left(\mathbf{3} \otimes \overline{\mathbf{3}} \right) \\
&= \mathbf{10} \oplus \mathbf{8} \oplus \mathbf{8} \oplus \mathbf{1}.
\end{aligned}
\tag{6.280}
$$

Note that the final results of the decompositions in Eq. (6.279) and in Eq. (6.280) contain the singlet $\mathbf{1}$ representation. This has applications in quantum chromodynamics, where the physical states must transform as singlets of the SU(3) gauge symmetry: In particular, Eq. (6.279) implies that one way to construct a physical state is by combining a quark (which is described by a field in the fundamental representation $\mathbf{3}$) with an antiquark (associated with a field in the antifundamental representation $\overline{\mathbf{3}}$). Such states do exist in nature, and are called *mesons*:[8] One example of a meson is the particle called the *pion*. Similarly, from Eq. (6.280) follows that another possible way to obtain an SU(3)-singlet physical state is by combining three quarks. In this case, the experimentally observed states of this type are called *baryons*:[9] The constituents of atomic nuclei, i.e., the proton and the neutron, are two examples of baryons.

The drawback of the tensor method that we have presented here is that, in general, it is not immediately obvious how to split the original tensor into a sum of tensors with the right properties for irreducible representations. There are more straightforward and algorithmic techniques to carry out the decomposition. In Section 6.8.1 we discuss one of them: a diagrammatic method using Young tableaux.

6.8.1 Young Tableaux for su(3)

A *Young tableau* is a diagrammatic representation of an irreducible representation (n, m). We construct it as follows. Consider a tensor $T^{i_1 \ldots i_n}_{j_1 \ldots j_m}$. Recall that we can trade a lower index to an antisymmetric pair of upper indices by contracting it by means of the ϵ-symbol, and, similarly, an upper index can be replaced by an antisymmetrized pair of lower indices, by contracting it with the ϵ-symbol. Applying this to every lower index of $T^{i_1 \ldots i_n}_{j_1 \ldots j_m}$, one obtains

$$
T^{i_1 \ldots i_n}_{j_1 \ldots j_m} \leftrightarrow T^{i_1 \ldots i_n}_{j_1 \ldots j_m} \, \epsilon^{j_1 k_1 l_1} \cdots \epsilon^{j_m k_m l_m},
\tag{6.281}
$$

which can be used to obtain a new form for the tensors of the (n, m) representation, now with n "old" upper indices i and m antisymmetrized pairs of "new" indices kl. We then construct a diagram by drawing a single box for each of the indices i, and

---

[8] Strictly speaking, mesons are more complex than simple quark–antiquark states, owing to the highly nontrivial nature of quantum chromodynamics. Nevertheless, the interpretation of mesons as bound quark–antiquark pairs is the basis of the older and simpler *quark model* for the strong nuclear interaction.

[9] As for mesons, physical baryons actually have a richer structure than just three bound quarks, but this simple, approximate interpretation already captures several of their properties.

a column of two boxes for each of the pairs of indices kl, arranging the boxes as follows.

k_1	$\ldots$	k_m	i_1	$\ldots$	i_n
l_1	$\ldots$	l_m			

Thus, the first row has $m + n$ boxes and the second row has m boxes. The resulting diagram is the Young tableau of the (n, m) representation. We represent the trivial $(0, 0)$ representation by a bullet $\bullet$.

Example   For example, we have the following Young tableaux:

$$(1, 0) = \square, \qquad (2, 0) = \square\square,$$

$$(0, 1) = \begin{matrix}\square \\ \square\end{matrix}, \qquad (0, 2) = \begin{matrix}\square\square \\ \square\square\end{matrix},$$

$$(1, 1) = \begin{matrix}\square\square \\ \square\end{matrix}, \qquad (2, 1) = \begin{matrix}\square\square\square \\ \square\end{matrix}, \tag{6.282}$$

and so on.

For the su(3) Lie algebra, the maximum number of boxes in a column is two. A column with three boxes would correspond to three completely antisymmetrized upper indices, so

$$\begin{matrix}\square \\ \square \\ \square\end{matrix} = \epsilon^{ijk} \ \text{ or } \ \epsilon^{ijk}v \tag{6.283}$$

but this is invariant, i.e., it is the singlet $(0, 0)$ representation, and thus does not transform under the action of the group. Just as we could multiply a tensor with ϵ^{ijk} or drop that factor off, e.g.,

$$T^i_j \leftrightarrow T^i_j \epsilon^{klm}, \tag{6.284}$$

we can erase out all columns with three boxes from a Young tableaux. For example,

$$\begin{matrix}\square\square \\ \square\square \\ \square\end{matrix} \rightarrow \begin{matrix}\square\square \\ \square\end{matrix} = (1, 1) = \mathbf{8},$$

$$\begin{matrix}\square\square \\ \square\square \\ \square\square\end{matrix} \rightarrow \bullet = (0, 0) = \mathbf{1}. \tag{6.285}$$

There are no columns with four or more boxes for su(3), since they would correspond to ϵ-symbols with four or more indices, but those vanish because every index has only three possible values, so at least two of the indices would necessarily have the same value, and the ϵ-symbol would vanish, because of its antisymmetry under permutations of every pair of indices.

There is a simple graphic rule for relating the Young tableaux of complex conjugate representations (n, m) and $\overline{(n, m)} = (m, n)$; given the Young tableau of the representation (n, m), the one of the complex conjugate representation $\overline{(n, m)}$ is obtained by considering the boxes that have to be added below the columns of the

Young tableau of (n, m) to fill a rectangle with three rows and $n + m$ columns, and rotating the diagram they form by π around its center.

Example   The Young tableau of the representation $(n, m) = (2, 0)$ is

$$(2, 0) = \boxed{} \,. \qquad\qquad (6.286)$$

In order to fill a rectangle with three rows and $n + m = 2$ columns, like the following one

one needs to add two boxes in the first column and two boxes in the second column:

Rotating the latter diagram by π around its center, one obtains the Young tableau corresponding to the complex conjugate representation $\overline{(2, 0)} = (0, 2)$:

$$(0, 2) = \boxed{} \,. \qquad\qquad (6.287)$$

Example   The Young tableau of the representation $(n, m) = (2, 1)$ is

$$(2, 1) = \boxed{} \,. \qquad\qquad (6.288)$$

In order to fill a rectangle with three rows and $n + m = 3$ columns,

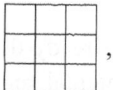

one needs to add one box in the first column, two boxes in the second column, and two boxes in the third column:

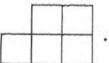

Rotating the latter diagram by π around its center, one obtains the Young tableau corresponding to the complex conjugate representation $\overline{(2, 1)} = (1, 2)$:

$$(1, 2) = \boxed{} \,. \qquad\qquad (6.289)$$

Using Young tableaux, it is possible to construct tensor products and to decompose them, with the following algorithm called the *Littlewood–Richardson rule*, that we present here for su(3) representations. A generalization of this rule for arbitrary su(N) algebras will be presented in Section 6.10.

Consider the tensor product of two irreducible representations A and B,

$$A \otimes B,$$

corresponding to Young tableaux that we will also denote as A and B.

Then, the decomposition of the tensor product can be carried out as follows.

1. Put letters a in the first row of B and letters b in the second row.

$$\begin{array}{|c|c|c|} \hline a & a & a \\ \hline b & b \\ \cline{1-2} \end{array}$$

2. Take boxes with a from B and attach them in all possible ways to the ends of the rows of A to get an expanded tableau A, such that there is no more than one box with a in each *column* of the expanded tableau. Note, however, that in the expanded tableau there can be more than one a per *row*.
3. Next, take the boxes with b from B and attach them to the previously obtained expanded tableau A, again with the same rule: There cannot be more than one b per column.
4. For every tableau, form a *word* as follows: Start reading the first row *from right to left*. At every step, add the letter that you encounter to the end of the word. After finishing the first row, move to the second row and repeat the process: Read from right to left adding letters to the word.
5. Now we need a definition: We say that the word is *admissible* if, by reading it *from left to right*, at every step at least as many as have occurred as bs up to that point; for example, *aabb* and *ababa* are admissible words, whereas *abba* and *baa* are not.[10]
6. Discard all tableaux with non-admissible words. The remaining ones form the correct decomposition of the tensor product.
7. Identify every tableau as a (n, m) irreducible representation of su(3).
8. Compute their dimensions $D(n, m)$.

This algorithm is best clarified by considering some examples of decompositions of tensor products of su(3) representations.

Example As we already discussed in the derivation of Eq. (6.277), the tensor product of the fundamental representation with itself can be decomposed into the sum of the **6** and the antifundamental ($\overline{\mathbf{3}}$) representations. Let us derive this result again, by using the Littlewood–Richardson rule:

$$\mathbf{3} \otimes \mathbf{3} = \square \otimes \boxed{a} = \boxed{\ \ a} \oplus \begin{array}{c}\square \\ \boxed{a}\end{array} = (2, 0) \oplus (0, 1) = \mathbf{6} \oplus \overline{\mathbf{3}}. \tag{6.290}$$

Example As a more involved example, let us study the decomposition of the tensor product of the adjoint representation with itself:

$$\mathbf{8} \otimes \mathbf{8} = (1, 1) \otimes (1, 1) = \ \ \otimes \begin{array}{c}\boxed{a\ \ a}\\ \boxed{b}\end{array}$$

---

[10] In the more general case of su(N) Lie algebras with arbitrary N, the definition of admissible word will be modified by additionally requiring that at every step at least as many bs have occurred as cs up to that point (if $N > 3$), and, moreover, if $N > 4$, that at least as many cs have occurred as ds up to that point, etc.

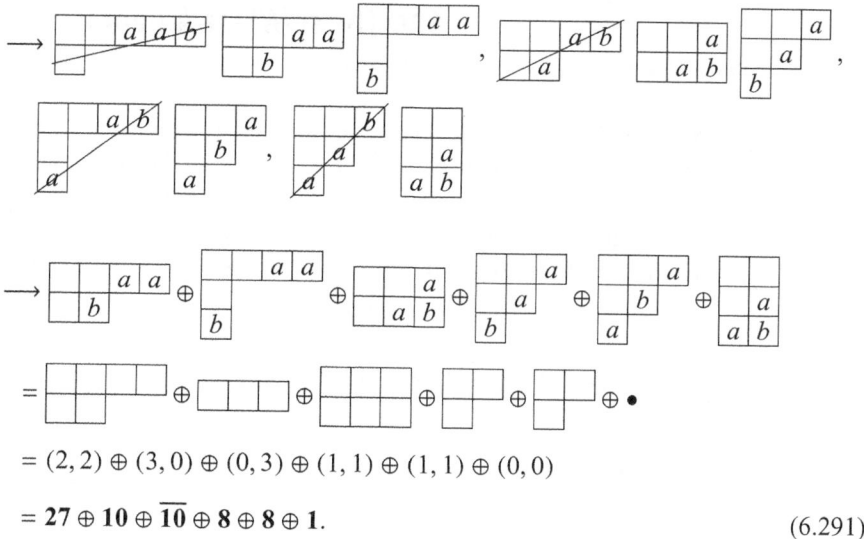

$$= (2,2) \oplus (3,0) \oplus (0,3) \oplus (1,1) \oplus (1,1) \oplus (0,0)$$

$$= \mathbf{27} \oplus \mathbf{10} \oplus \overline{\mathbf{10}} \oplus \mathbf{8} \oplus \mathbf{8} \oplus \mathbf{1}. \tag{6.291}$$

Note that we have discarded the four tableaux with the non-admissible word *baa* and kept the six tableaux with the admissible words *aab, aab, aba, aab, aba,* and *aba*.

Example Let us now consider the tensor product of the fundamental representation with the antifundamental one. As shown in Eq. (6.279), this tensor product decomposes into the sum of the adjoint (**8**) and the singlet (**1**) representations. Again, this result can also be derived following the Littlewood–Richardson rule:

$$\mathbf{3} \otimes \overline{\mathbf{3}} = \square \otimes \begin{array}{c}\boxed{a}\\\boxed{b}\end{array} \rightarrow \boxed{\ \ a\ }\ ,\quad \begin{array}{c}\square\\\boxed{a}\end{array}$$

"ba"   "ab" ,   "ba"   "ab"

$$= (1,1) \oplus (0,0) = \mathbf{8} \oplus \mathbf{1}. \tag{6.292}$$

As a rule of thumb, it turns out that, in the $A \otimes B$ product, it is generally convenient to take B to be the representation with the simpler of the two Young tableaux; recall that, although the tensor product is not a commutative operation, $A \otimes B$ and $B \otimes A$ are equivalent up to a redefinition of the order of rows and columns, hence the two representations $A \otimes B$ and $B \otimes A$ are unitarily equivalent.

Example Consider the derivation of the previous decomposition by writing the tensor product as $\overline{\mathbf{3}} \otimes \mathbf{3}$ (rather than $\mathbf{3} \otimes \overline{\mathbf{3}}$), i.e., having the **3** representation, with the simpler Young tableau, made of only one box, as the B factor:

$$\overline{\mathbf{3}} \otimes \mathbf{3} = \begin{array}{c}\square\\\square\end{array} \otimes \boxed{a} \rightarrow \begin{array}{c}\square\ \boxed{a}\\\square\end{array}\ ,\quad \begin{array}{c}\square\\\square\\\boxed{a}\end{array} = (1,1) \oplus (0,0) = \mathbf{8} \oplus \mathbf{1}, \tag{6.293}$$

which is a slightly faster way to derive the same result of Eq. (6.292).

6.9 The su(N) Lie Algebras

Now we move to the general case of su(N) Lie algebras with $N \geq 2$. The rank of su(N), which we denote here as r, is $r = N - 1$, so the algebras can be labeled by their rank as su($r + 1$). The generators are represented by traceless Hermitian matrices. In the defining representation by $N \times N$ matrices, one way to construct the r traceless diagonal matrices H_a, with $a \in \{1, \ldots, r\}$, representing the Cartan generators is

$$H_a = \frac{1}{\sqrt{2a(a+1)}} \operatorname{diag}\left(\underbrace{1, 1, \ldots, 1, 1}_{a \text{ entries}}, -a, \underbrace{0, 0, 0, \ldots, 0, 0}_{N - a - 1 \text{ entries}}\right), \tag{6.294}$$

or, explicitly,

$$H_1 = \frac{1}{2} \operatorname{diag}(1, -1, 0, 0, 0, \ldots, 0),$$

$$H_2 = \frac{1}{\sqrt{12}} \operatorname{diag}(1, 1, -2, 0, 0, \ldots, 0),$$

$$H_3 = \frac{1}{\sqrt{24}} \operatorname{diag}(1, 1, 1, -3, 0, \ldots, 0),$$

$$\vdots$$

$$H_{r-1} = \frac{1}{\sqrt{2(r-1)r}} \operatorname{diag}(1, \ldots, -(r-1), 0),$$

$$H_r = \frac{1}{\sqrt{2r(r+1)}} \operatorname{diag}(1, \ldots, 1, -r). \tag{6.295}$$

This construction generalizes the structure of the Cartan generators in the fundamental representation for the su(2) and for the su(3) algebras, which are given by half the diagonal Pauli and Gell-Mann matrices, respectively. Indeed, for rank $r = 1$, corresponding to su(2):

$$H_1 = \frac{1}{2} \operatorname{diag}(1, -1) \equiv \frac{1}{2}\sigma_3.$$

Conversely, for $r = 2$, corresponding to su(3), Eq. (6.294) leads to

$$H_1 = \frac{1}{2} \operatorname{diag}(1, -1, 0) \equiv \frac{1}{2}\lambda_3,$$

$$H_2 = \frac{1}{2}\left[\frac{1}{\sqrt{3}} \operatorname{diag}(1, 1, -2)\right] \equiv \frac{1}{2}\lambda_8, \tag{6.296}$$

where λ_3 and λ_8 are the diagonal su(3) Gell-Mann matrices, as expected. Let us now consider an additional example, of higher rank.

Example For $r = 4$, i.e., for the su(5) Lie algebra, the Cartan generators in the fundamental representation can be written according to Eq. (6.294) as:

$$H_1 = \frac{1}{2} \, \mathrm{diag}\, (1, -1, 0, 0, 0) \equiv \frac{1}{2}\Sigma_1,$$

$$H_2 = \frac{1}{\sqrt{12}} \, \mathrm{diag}\, (1, 1, -2, 0, 0) \equiv \frac{1}{2}\Sigma_2,$$

$$H_3 = \frac{1}{\sqrt{24}} \, \mathrm{diag}\, (1, 1, 1, -3, 0) \equiv \frac{1}{2}\Sigma_3,$$

$$H_4 = \frac{1}{\sqrt{40}} \, \mathrm{diag}\, (1, 1, 1, 1, -4) \equiv \frac{1}{2}\Sigma_4, \tag{6.297}$$

having defined the Σ_a matrices as

$$\Sigma_1 = \mathrm{diag}\, (1, -1, 0, 0, 0),$$

$$\Sigma_2 = \frac{1}{\sqrt{3}} \, \mathrm{diag}\, (1, 1, -2, 0, 0),$$

$$\Sigma_3 = \frac{1}{\sqrt{6}} \, \mathrm{diag}\, (1, 1, 1, -3, 0),$$

$$\Sigma_4 = \frac{1}{\sqrt{10}} \, \mathrm{diag}\, (1, 1, 1, 1, -4). \tag{6.298}$$

The common eigenvectors of the su(*N*) Cartan matrices are

$$e_1 = \begin{pmatrix} 1 \\ 0 \\ 0 \\ \vdots \\ 0 \\ 0 \end{pmatrix}, \quad e_2 = \begin{pmatrix} 0 \\ 1 \\ 0 \\ \vdots \\ 0 \\ 0 \end{pmatrix}, \quad \ldots, \, e_N = \begin{pmatrix} 0 \\ 0 \\ 0 \\ \vdots \\ 0 \\ 1 \end{pmatrix}. \tag{6.299}$$

Using the Dirac notation, we represent the eigenvectors e_i as kets labeled by the corresponding weight vectors, and in the following we denote them by ω^i, saving the notation μ^i for later use, to denote the fundamental weight vectors:

$$e_i = |\omega^i\rangle. \tag{6.300}$$

The weight vector ω^i has components (weights) given by the *i*th diagonal component of each Cartan matrix,

$$\omega^i = ((H_1)_{ii}, (H_2)_{ii}, \ldots, (H_{N-1})_{ii}), \tag{6.301}$$

or, explicitly,

$$\omega^1 = \left(\frac{1}{\sqrt{4}}, \frac{1}{\sqrt{12}}, \frac{1}{\sqrt{24}}, \ldots, \frac{1}{\sqrt{2(N-1)N}} \right),$$

$$\omega^2 = \left(-\frac{1}{\sqrt{4}}, \frac{1}{\sqrt{12}}, \frac{1}{\sqrt{24}}, \ldots, \frac{1}{\sqrt{2(N-1)N}} \right),$$

$$\omega^3 = \left(0, \frac{-2}{\sqrt{12}}, \frac{1}{\sqrt{24}}, \ldots, \frac{1}{\sqrt{2(N-1)N}} \right),$$

$$\vdots$$

$$\omega^{N-1} = \left(0,0,0,\ldots,\frac{-(N-2)}{\sqrt{2(N-2)(N-1)}},\frac{1}{\sqrt{2(N-1)N}}\right),$$

$$\omega^N = \left(0,0,0,\ldots,0,\frac{-(N-1)}{\sqrt{2(N-1)N}}\right). \tag{6.302}$$

Note that the weight vectors are not linearly independent; they satisfy

$$\omega^1 + \omega^2 + \cdots + \omega^N = 0. \tag{6.303}$$

At this point, it turns out to be more convenient to use the alternative definition of positivity for the weight vectors that we mentioned in Section 6.5.3, according to which a positive weight vector is one for which the last nonzero entry is positive. With this convention, we have a simple ordering $\omega^1 > \omega^2 > \cdots > \omega^N$, thus ω^1 is always the highest weight of the defining representation.

One can show that the weight vectors satisfy

$$\omega^i \cdot \omega^j = \frac{1}{2}\left(\delta_{ij} - \frac{1}{N}\right) = \begin{cases} \frac{N-1}{2N} & \text{if } i = j \\ -\frac{1}{2N} & \text{if } i \neq j \end{cases}. \tag{6.304}$$

The proof of Eq. (6.304) is left as an exercise, in Problem 6.19.

By taking a pair of subsequent weights in the above ordering, and subtracting the smaller weight from the larger one, one obtains the $\binom{N}{2}$ positive roots,

$$\alpha^{ij} \equiv \omega^i - \omega^j \quad \text{(with } i < j\text{)} \quad \text{is a positive root.} \tag{6.305}$$

Following the analogous construction that we carried out for the su(2) and su(3) algebras, the ladder generators $E_{\pm\alpha^{ij}}$ are then represented by $N \times N$ matrices with a single nonzero off-diagonal entry 1, completing the defining representation in the Cartan–Weyl basis.

One can prove (see Problem 6.20) that the only positive roots that cannot be written a sum of positive roots, i.e., the $N-1$ simple roots, are

$$\alpha^i \equiv \omega^i - \omega^{i+1}, \quad \text{for } i = 1, \ldots, N-1. \tag{6.306}$$

Their explicit form, in the representation that we introduced above, is

$$\alpha^1 = (1,0,0,0,\ldots,0,0),$$

$$\alpha^2 = \left(-\frac{1}{2},\frac{\sqrt{3}}{2},0,0,\ldots,0,0\right),$$

$$\alpha^3 = \left(0,-\frac{1}{\sqrt{3}},\sqrt{\frac{2}{3}},0,\ldots,0,0\right),$$

$$\vdots$$

$$\alpha^m = \left(0,0,0,0,\ldots,-\sqrt{\frac{m-1}{2m}},-\sqrt{\frac{m+1}{2m}},\ldots,0,0\right),$$

$$\vdots$$

$$\alpha^{N-1} = \left(0,0,0,0,\ldots,-\sqrt{\frac{N-2}{2(N-1)}},-\sqrt{\frac{N}{2(N-1)}}\right). \tag{6.307}$$

The scalar product of two simple roots can be evaluated using Eq. (6.304):

$$
\begin{aligned}
\alpha^j \cdot \alpha^i &= (\omega^j - \omega^{j+1}) \cdot (\omega^i - \omega^{i+1}) \\
&= \omega^j \cdot \omega^i - \omega^{j+1} \cdot \omega^i - \omega^j \cdot \omega^{i+1} + \omega^{j+1} \cdot \omega^{i+1} \\
&= \frac{1}{2}\left[\left(\delta_{j,i} - \frac{1}{N}\right) - \left(\delta_{j+1,i} - \frac{1}{N}\right) - \left(\delta_{j,i+1} - \frac{1}{N}\right) + \left(\delta_{j+1,i+1} - \frac{1}{N}\right)\right] \\
&= \frac{1}{2}\left(\delta_{j,i} + \delta_{j+1,i+1} - \delta_{j+1,i} - \delta_{j,i+1}\right) \\
&= \delta_{j,i} - \frac{1}{2}\left(\delta_{j,i-1} + \delta_{j,i+1}\right),
\end{aligned}
\tag{6.308}
$$

where in the last step we used the fact that, for every integer a, b, and n, one has $\delta_{a+n,b} = \delta_{a,b-n}$ (a property that follows trivially from the definition of the Kronecker delta).

From the knowledge of the scalar products of simple roots we can then construct the Cartan matrix according to Eq. (6.218):

$$
A = \begin{pmatrix}
2 & -1 & 0 & 0 & \dots & 0 & 0 \\
-1 & 2 & -1 & 0 & \dots & 0 & 0 \\
0 & -1 & 2 & -1 & \dots & 0 & 0 \\
\vdots & & \ddots & \ddots & \ddots & & \vdots \\
0 & \dots & & & -1 & 2 & -1 \\
0 & \dots & & & 0 & -1 & 2
\end{pmatrix},
\tag{6.309}
$$

from which, in turn, we can obtain the Dynkin diagram of the su(N) algebra,

which has $N - 1$ circles ($N - 1$ being the number of simple roots, i.e., the rank) connected by single lines.

There is another, more indirect way to construct the simple roots of the su(N) Lie algebra. We begin with the observation that the $N - 1$ linearly independent simple roots α^i form a basis of the root system Φ, which is a real vector space of dimension $r = N - 1$, namely $\Phi = \mathbb{R}^{N-1}$. We can consider it as a subspace of $\mathbb{R}^N$, as the following hyperplane. Given the vectors $\{e_i\}$ defined in Eq. (6.299), which are an orthonormal basis of $\mathbb{R}^N$, construct the vector $u = \sum_i^N e_i$. Then, Φ can be defined as the hyperplane orthogonal to u,

$$
\Phi = \{v \in \mathbb{R}^N | v \cdot u = 0\}.
\tag{6.310}
$$

The projections $v_i = e_i - \frac{1}{N}u$ of the e_is to Φ are not linearly independent, as they satisfy $v_1 + v_2 + \cdots + v_N = 0$. Their scalar products are

$$
v_i \cdot v_j = \delta_{ij} - \frac{1}{N}.
\tag{6.311}
$$

Note that the properties of v_i are similar to those in Eq. (6.303) and in Eq. (6.304), up to a different normalization, for the weight vectors ω_i, except that the v_is are N-component vectors, even though they are all elements of the hyperplane Φ, which has dimension $N - 1$, while the ω_i have $N - 1$ components. The two can be related by introducing a basis in Φ and thus eliminating the spurious additional component in v_i.

Again adopting the convention that a positive vector has a positive first nonzero component, we have the ordering $v_1 > v_2 > \cdots > v_N$. To continue as before, we construct the differences

$$\beta_{ij} = v_i - v_j \quad \text{with } i < j; \tag{6.312}$$

these $\binom{N}{2}$ vectors are now the positive roots in the root system Φ. By similar reasoning as before,

$$\beta_i = v_i - v_{i+1} \quad i = 1, \ldots, N-1 \tag{6.313}$$

are the simple roots. Their explicit form is

$$\beta_1 = (1, -1, 0, 0, \ldots, 0, 0, 0),$$
$$\beta_2 = (0, 1, -1, 0, \ldots, 0, 0, 0),$$
$$\beta_3 = (0, 0, 1, -1, \ldots, 0, 0, 0),$$

$$\vdots$$

$$\beta_{N-2} = (0, 0, 0, 0, \ldots, 1, -1, 0),$$
$$\beta_{N-1} = (0, 0, 0, 0, \ldots, 0, 1, -1), \tag{6.314}$$

so they satisfy

$$(\beta_i)^2 = 2, \tag{6.315}$$
$$\beta_i \cdot \beta_{i\pm 1} = -1, \tag{6.316}$$
$$\beta_i \cdot \beta_j = 0, \quad i \neq j \text{ and } j \neq i \pm 1, \tag{6.317}$$

leading, as expected, to the same Cartan matrix as in Eq. (6.309).

Recall that the two su(3) fundamental weights μ^1 and μ^2 correspond to the fundamental (defining) representation and the antifundamental representation, with three components for vectors v^i and dual vectors w_i. The lower index could be traded for an antisymmetrized pair of upper indices, $v^{ij} = \epsilon^{ijk} w_k$. One can show that this generalizes to the su(N) algebra, in such a way that each fundamental representation $D(\mu^k)$ with highest weight μ^k corresponds to type $(k, 0)$-tensors $T^{i_1 i_2 \cdots i_k}$ with the k upper indices being completely antisymmetric:

$$D(\mu^k) \leftrightarrow T^{i_1 i_2 \cdots i_i}, \tag{6.318}$$

with

$$T^{i_{p(1)} i_{p(2)} \cdots i_{p(k)}} = -T^{i_1 i_2 \cdots i_k}, \quad p \in S_k, \quad \text{sgn}(p) = -1. \tag{6.319}$$

The components of the antisymmetric tensor are nonvanishing only when the k values of the indices are all different, with $\binom{N}{k}$ possible choices. Thus

$$\dim D(\mu^k) = \binom{N}{k}. \tag{6.320}$$

6.10 Young Tableaux for su(N)

A generic irreducible representation D of su(N) has the highest weight vector μ, which can be uniquely expanded over the $N-1$ fundamental weights μ^k as

$$\mu = q^1 \mu^1 + \cdots + q^{N-1} \mu^{N-1}, \tag{6.321}$$

so that alternatively we can specify D with the Dynkin label $(q^1, \ldots, q^{N-1})$. Generalizing the previous discussion of su(3) representations, the irreducible representation D can be associated with a Young tableau with q^k columns of k boxes, for $k = 1, \ldots, N-1$:

Example The defining representation of su(5) is the fundamental representation $(1, 0, 0, 0) = \mathbf{5}$, with the simple Young tableau

$$(1, 0, 0, 0) = \mathbf{5} = \square . \tag{6.322}$$

The representations $(0, 0, 1, 1)$ and $(2, 1, 0, 1)$ have the tableaux

$$(0, 0, 1, 1) = \quad , \qquad (2, 1, 0, 1) = \quad . \tag{6.323}$$

Alternatively, we could label the tableau by the number of boxes in each *row*; in the kth row we have ℓ_k boxes, with

$$\ell_1 = q^{N-1} + q^{N-2} + \cdots + q^2 + q^1,$$
$$\ell_2 = q^{N-1} + q^{N-2} + \cdots + q^2,$$
$$\vdots$$
$$\ell_{N-1} = q^{N-1}. \tag{6.324}$$

The table is now labeled by the list $\{\ell_1, \ell_2, \ldots, \ell_{N-1}\}$, sometimes called the *partition*. Displaying the previous Young tableau again:

1	...	ℓ_{N-1}	...	...	ℓ_{N-2}	...	ℓ_3	...	ℓ_2	ℓ_2+1	...	ℓ_1
1	...	ℓ_{N-1}	...	...	ℓ_{N-2}	...	ℓ_3	...	ℓ_2			
$\vdots$	$\vdots$	$\vdots$	$\vdots$	$\vdots$	$\vdots$							
1	...	ℓ_{N-1}	...	...	ℓ_{N-2}							
1	...	ℓ_{N-1}										

Example   The Young tableaux of the previous example are labeled by partitions as follows:

$$\{1,0,0,0\} = \mathbf{5} = \square \, ,$$

$$\{2,2,2,1\} = \;\;\;,$$

$$\{4,2,1,1\} = \;\;\;. \tag{6.325}$$

The dimension of the irreducible representation can be calculated from its Young tableau using the *factors-over-hooks rule*. The rule works as follows. For an $su(N)$ tableau, put integers into boxes starting from the top left corner, where you put the number N. Then fill in the first row, increasing the integers by one at each step to the right. For the second row, start from the left with $N-1$, then fill in the row, increasing the integers by one at each step. The third row then starts with $N-2$, and is filled by increasing the numbers as before. This goes on until all boxes in the tableau are filled. Thus, the numbers change as follows:

$$\rightarrow +1$$
$$\downarrow$$
$$-1$$

and one then computes the product of all integers in the boxes; this product is denoted as F.

A *hook* is a set of boxes with a shape like the $\neg$ symbol, that starts from the last box in a row, goes left and then turns down at some point, and continues down through the end of the column below. There are many ways to do this. For each hook, you calculate the corresponding *hook factor* h_i, which is number of boxes that the hook line passes through. Then one calculates the product $H = \prod_i h_i$ over all hooks. The dimension of the representation is then the ratio

$$\dim D = \frac{F}{H}. \tag{6.326}$$

Example   Consider the Young tableau for an su(N) irreducible representation D with two boxes in the first row, and one box in the second row. To compute the factor F, we fill the numbers:

N	$N+1$
$N-1$	

giving $F = N(N+1)(N-1) = N(N^2 - 1)$. For the hooks, there are three alternatives: one crossing all three boxes, one crossing only the top right box, and one crossing the bottom box. Thus, $H = 3 \cdot 1 \cdot 1 = 3$. The dimension of the representation is thus

$$\dim D = \frac{F}{H} = \frac{N(N^2 - 1)}{3}. \tag{6.327}$$

Let us check the result for $N = 3$, in which case the $\{2, 1\} = (1, 1)$ tableau labels the **8** adjoint representation. Now

$$\frac{3(3^2 - 1)}{3} = 8, \tag{6.328}$$

so the dimension is computed correctly.

The decomposition of a tensor product of irreducible representations of the su(N) algebra can be worked out by a generalization of the previous su(3) rules. The main difference is that now there can be more than two rows, so we need more letters. Consider the tensor product of two irreducible representations $A \otimes B$, using the corresponding Young tableaux.

1. Put letters a in the boxes on the first row of the tableau for B, letters b in the second row, letters c in the third row, and so on, e.g.,

a	a	a	a
b	b	b	
c			
d			

2. Take boxes with a from B and attach them in all possible ways to the ends of the rows of A to get an expanded tableaux, such that there is no more than one box with a in each *column* of the expanded tableau A (there can be more than one a per *row*).
3. Next, take the boxes with b from B and attach them to the previously obtained expanded tableaux A, again enforcing that there is no more than one b per column.
4. Next, take the boxes with c from B and attach them to the previously obtained expanded tableaux A, again with the same rule, so that there is no more than one c per column. Repeat this procedure until all the letters have been used.
5. For every tableau, form a word as follows: Start reading the first row *from right to left*. At every step, add the letter that you encounter to the end of the word. After finishing the first row, move to the second row and repeat the process row by row: Read from right to left adding letters to the word.
6. We generalize our previous definition of an *admissible word* if, by reading it *from left to right*, at every step at least as many as have occurred as bs, cs, etc., up to

that point, at least as many bs have occurred as cs, ds, etc., up to that point, and so on. For example, $aabc$ and $ababc$ are admissible words, while $acba$ and bca are not.

7. Discard the tableaux with non-admissible words. The remaining ones form the decomposition of the tensor product.
8. Identify every tableau as a $(q^1, q^2, \ldots, q^{N-1})$ irreducible representation.
9. Compute their dimensions using the factors-over-hooks rule.

6.11 Representations of SU(N) × SU(M) × U(1)

In this section we discuss the representation of a product of unitary groups of the form $SU(N) \times SU(M) \times U(1)$. A motivation of interest in this particular type of group stems from the fact that the Standard Model of elementary-particle physics describes three of the four fundamental interactions of nature, i.e., the electromagnetic interaction, the weak nuclear interaction, and the strong nuclear interaction, in terms a particular type of invariance under transformations of an $SU(3) \times SU(2) \times U(1)$ group. This group is the *gauge group of the Standard Model*; it is a *gauge group* because the transformations of the group are *local* ones, i.e., they depend on spacetime.

Let us discuss the product of two special unitary groups first. Note that the symbol × denotes the direct product of groups, not their tensor product. Elements of $SU(N) \times SU(M)$ in the defining representation have a block-diagonal matrix structure of the form

$$U = \begin{pmatrix} U_N & 0 \\ 0 & U_M \end{pmatrix}, \qquad (6.329)$$

where $U_N \in SU(N)$ and $U_M \in SU(M)$. They act on vectors of their direct-sum vector space $\mathbb{R}^N \oplus \mathbb{R}^M$,

$$v = \begin{pmatrix} v_1 \\ v_2 \end{pmatrix} = \begin{pmatrix} v_1 \\ 0 \end{pmatrix} + \begin{pmatrix} 0 \\ v_2 \end{pmatrix}, \qquad (6.330)$$

where $v_1 \in \mathbb{R}^N$ while $v_2 \in \mathbb{R}^M$, and we have included additional zero entries to express v as the sum of two vectors of $\mathbb{R}^{N+M}$ that are isomorphic to v_1 and v_2. In physics, the frequent notation for the representations is $v \in (N, M)$, with $v_1 \in (N, 1)$ and $v_2 \in (1, M)$ where the 1 denotes that v_1 transforms trivially (namely, is a scalar) under transformations in the $SU(M)$ group, and similarly v_2 is a scalar under transformations of the $SU(N)$ group. Instead of the defining representations, we can then consider other unitary irreducible representations, e.g., a generic representation D_1 for $SU(N)$ and a generic representation D_2 for $SU(M)$. Let us label the representations by their dimensions as $\mathbf{d}_1$ and $\mathbf{d}_2$, and denote their complex conjugate representations as $\bar{\mathbf{d}}_1$ and $\bar{\mathbf{d}}_2$, respectively. Thus, an element of $SU(N) \times SU(M)$ in the $(\mathbf{d}_1, \mathbf{d}_2)$ representation would act as a $d_1 \times d_1$ special unitary matrix U_1 and a $d_2 \times d_2$ special unitary matrix U_2.

Now let us include the U(1) group as an additional factor in the direct product of groups. Since U(1) is an Abelian group, all of its unitary irreducible representations are one-dimensional, and are characterized by a U(1) *charge Y*, so that U(1) acts on a vector v as a phase factor:

$$v \mapsto \exp\left(iY\phi\right)v. \tag{6.331}$$

Let us consider $SU(N) \times SU(M) \times U(1)$ and its action on a vector (v_1, v_2), with $v_1 \in \mathbb{R}^{d_1}$ and $v_2 \in \mathbb{R}^{d_2}$. Under the action of a U(1) element, all components of v_1 must have the same charge Y_1, in order not to interfere with the action by an SU(N) matrix U_1; likewise, all components of v_2 must have the same charge Y_2, in order not to interfere with the SU(M) action. If we then set $Y_1 = Y_2 \equiv Y$, we can consider the joint action by $SU(N) \times SU(M) \times U(1)$, under which a vector v transforms as

$$v = \left(\begin{array}{c} v_1 \\ v_2 \end{array}\right) \mapsto e^{iY\phi}\left(\begin{array}{c} U_1 v_1 \\ U_2 v_2 \end{array}\right). \tag{6.332}$$

We denote this joint representation with the label $(\mathbf{d}_1, \mathbf{d}_2)_Y$; thus we write

$$v \in (\mathbf{d}_1, \mathbf{d}_2)_Y \tag{6.333}$$

to denote that the vector v transforms in the representation $\mathbf{d}_1$ under the action of an element of SU(N), in the representation $\mathbf{d}_1$ under the action of an element of SU(M), and as an object with charge Y under the action of U(1).

Note that if, instead, we choose two different U(1) charges $Y_1 \neq Y_2$, the joint group action becomes

$$v = \left(\begin{array}{c} v_1 \\ v_2 \end{array}\right) \mapsto e^{iY_1\phi}\left(\begin{array}{c} U_1 v_1 \\ 0 \end{array}\right) \oplus e^{iY_2\phi}\left(\begin{array}{c} 0 \\ U_2 v_2 \end{array}\right), \tag{6.334}$$

corresponding to a direct-sum representation, with $v \in (\mathbf{d}_1, \mathbf{1})_{Y_1} \oplus (\mathbf{1}, \mathbf{d}_2)_{Y_2}$.

The generators of the Lie algebra $su(N) \times su(M) \times u(1)$ consist of combinations of the traceless Hermitian $d_1 \times d_1$ matrices X_N^a and $d_2 \times d_2$ matrices X_M^b, representing the generators of $su(N) \times su(M)$,

$$\left(\begin{array}{cc} X_N^a & 0 \\ 0 & 0 \end{array}\right), \qquad \left(\begin{array}{cc} 0 & 0 \\ 0 & X_M^b \end{array}\right),$$

and the u(1) generator:

$$\left(\begin{array}{cc} Y_1 \mathbb{1}_{d_1} & 0 \\ 0 & Y_2 \mathbb{1}_{d_2} \end{array}\right).$$

Example The elementary particles of the Standard Model are divided into spin-$\frac{1}{2}$ fermionic matter particles, spin-1 gauge bosons which mediate the interactions, and the spin-0 Higgs boson. The matter particles are further divided into quarks (the constituents of protons and neutrons) and leptons (e.g., the electron). There are also the corresponding antiparticles, called antiquarks and antileptons. Furthermore, the matter particles appear in three *generations*, i.e., three sets of particles with identical representation properties. To discuss the representation properties, we focus on one generation of particles. The transformation properties under the gauge symmetry group $SU(3) \times SU(2) \times U(1)$ of the Standard Model are best discussed in terms of the quantum fields associated with the particles. We start with a single generation, where the quarks are the u (up) and d (down) quarks, and the corresponding antiquarks. These quarks are associated with three quantum fields: left-handed Weyl spinor fields[11] denoted by q, $\bar{u}$, and $\bar{d}$. The field q transforms under the fundamental representation **3**, whereas the fields $\bar{u}$ and $\bar{d}$ transform under the antifundamental representation $\bar{\mathbf{3}}$ of the SU(3) group that describes the strong nuclear interaction. As the local SU(3) group associated with this type of interaction is conventionally described in terms of *color* degrees of freedom,[12] this means that all the three fields have three color components. In the Standard Model, the weak nuclear interaction and the electromagnetic interaction are described jointly by the SU(2) × U(1) group:[13] Under the SU(2) factor of this symmetry group, the field q transforms as a "doublet," i.e., it transforms in the defining **2** representation, whereas the fields $\bar{u}$ and $\bar{d}$ transform trivially, in the "singlet," or **1** representation. Note that the transformation properties of these fields with respect to the SU(3) × SU(2) × U(1) gauge symmetry are independent of their transformation properties under transformations of the spacetime coordinates: Both u and d are left-handed Weyl spinor fields. To summarize, the transformation properties of the q, $\bar{u}$, and $\bar{d}$ fields under the SU(3) × SU(2) gauge symmetry groups of the Standard Model are:

$$q \in (\mathbf{3}, \mathbf{2}), \tag{6.335}$$

$$\bar{u} \in (\bar{\mathbf{3}}, \mathbf{1}), \tag{6.336}$$

$$\bar{d} \in (\bar{\mathbf{3}}, \mathbf{1}). \tag{6.337}$$

Particles (and antiparticles) correspond to the quantum excitations of these quantum fields. More precisely, to make contact with an interpretation in terms of familiar particles and antiparticles, we first give the names u and d to the two components of the doublet field q.

Thus we have four left-handed Weyl spinor fields u, d, $\bar{u}$, and $\bar{d}$. The latter two are then converted into right-handed Weyl spinor fields $\bar{u}^\dagger$ and $\bar{d}^\dagger$ by the adjoint operation, and combined with the two left-handed fields to form the following Dirac fields[14]

$$\mathcal{U} \equiv \begin{pmatrix} u \\ \bar{u}^\dagger \end{pmatrix}, \qquad \mathcal{D} \equiv \begin{pmatrix} d \\ \bar{d}^\dagger \end{pmatrix}.$$

Each of these Dirac fields then corresponds to a quark and its antiquark; for example, $\mathcal{U}$ describes the up quark and the corresponding antiquark.

The leptons and antileptons in a generation are treated similarly, but they belong to different representations of the gauge group of the Standard Model; in particular, as will be discussed shortly, this reflects the fact that, in contrast with quarks and antiquarks, leptons and antileptons are insensitive to the strong nuclear interaction. As before, we begin first with two left-handed Weyl spinor fields ℓ and $\bar{e}$ to discuss the representation properties, and make contact with particles later. Both fields transform

[11] Left-handed Weyl spinor fields belong in the representation $\left(\frac{1}{2}, 0\right)$ of the Lorentz algebra that will be discussed in Section 6.12.

[12] Note that the *color* alluded to here has nothing to do with the colors associated with electromagnetic radiation of frequencies in the visible part of the spectrum.

[13] For this reason, in the Standard Model the electromagnetic interaction and the weak interaction are said to be *unified* into the *electroweak interaction*.

[14] A Dirac field is a spin-$\frac{1}{2}$ field corresponding to the $\left(\frac{1}{2}, 0\right) \oplus \left(0, \frac{1}{2}\right)$ representation of the Lorentz group discussed in Section 6.12, and is a combination of left- and right-handed Weyl spinors.

trivially under the SU(3) group, i.e., are not affected by the strong nuclear interaction, but they differ in their transformation properties under SU(2); in particular, ℓ transforms as a doublet, while $\bar{e}$ transforms as a singlet. Thus

$$\ell \in (\mathbf{1}, \mathbf{2}), \tag{6.338}$$

$$\bar{e} \in (\mathbf{1}, \mathbf{1}). \tag{6.339}$$

We then rename the two components of the ℓ doublet as the left-handed Weyl spinor fields ν_e and e and construct a Dirac spinor field

$$\mathcal{E} \equiv \left(\begin{array}{c} e \\ \bar{e}^\dagger \end{array} \right) \tag{6.340}$$

corresponding to the electron and its antiparticle, the positron. The remaining field ν_e corresponds to the electron neutrino, which in the Standard Model is assumed to be massless.[15]

The last factor in the gauge group of the Standard Model is the U(1) group, whose representations are labeled by the corresponding charges Y. Note that the U(1) charge is called the *weak hypercharge* and, in general, is *not* the electric charge. The U(1) symmetry generator is also denoted as Y.

Including the values of the weak hypercharge, the representation content of a generation of left-handed Weyl matter fields can be summarized using the notation $(\mathbf{d}_1, \mathbf{d}_2)_Y$ as follows:

$$q \in (\mathbf{3}, \mathbf{2})_{1/6}, \tag{6.341}$$

$$\bar{u} \in (\bar{\mathbf{3}}, \mathbf{1})_{-2/3}, \tag{6.342}$$

$$\bar{d} \in (\bar{\mathbf{3}}, \mathbf{1})_{1/3}, \tag{6.343}$$

$$\ell \in (\mathbf{1}, \mathbf{2})_{-1/2}, \tag{6.344}$$

$$\bar{e} \in (\mathbf{1}, \mathbf{1})_1. \tag{6.345}$$

In addition to gauge boson fields and fermionic matter fields, the Standard Model also includes a quantum field whose excitations correspond to the Higgs boson. This filed is a complex scalar field transforming as $(\mathbf{1}, \mathbf{2})_{-1/2}$ under transformations of the gauge group of the Standard Model.

The *electric charge* Q of a particle is obtained as

$$Q = T_3 + Y, \tag{6.346}$$

where T_3 is the eigenvalue of the diagonal generator of the SU(2) group.[16] For example, for the electron neutrino $T^3 = \frac{1}{2}$, $Y = -\frac{1}{2}$, so that $Q = 0$, which means that the electron neutrino does not carry electric charge.

---

[15] By now, the hypothesis that neutrinos are massless particles has actually been disproven by experiments. While the Standard Model requires only minimal modifications to accommodate massive neutrinos, the exact values of the masses of these particles have not been determined yet.

[16] Note that, in some physics textbooks, the conventional normalization of the weak hypercharge differs from ours by a factor 2. In that convention, Eq. (6.346) has to be replaced by $Q = T_3 + \frac{1}{2}Y$.

6.11.1 Subalgebras, Embeddings, and Branching Rules

Consider a simple Lie algebra $\mathfrak{g}$ with rank r. Its generators are

- the Cartan generators H_i with $i = 1, \ldots, r$, and
- the pairs of root generators $E_{\pm\alpha}$, where α are the positive roots.

One simple way to find a subalgebra $\mathfrak{h}$ of $\mathfrak{g}$ is to leave out a pair $E_{\pm\alpha_s}$ of root generators, where α_s is one of the simple roots. This procedure actually also gives a u(1) subalgebra: Starting from the r simple roots that span a r-dimensional vector space, and keeping $r - 1$ of them, that generate a $(r - 1)$-dimensional subspace, there exists a vector a orthogonal to it, and then a linear combination $\sum_{i=1}^{r} a_i H_i$ is a generator that can be shown to commute with all generators of $\mathfrak{h}$. Hence this process breaks the Lie algebra $\mathfrak{g}$ as follows:

$$\mathfrak{g} \to \mathfrak{h} \times \mathrm{u}(1). \tag{6.347}$$

Recall that the Dynkin diagrams contained $N - 1$ circles representing the simple roots of su(N). Leaving out a simple root then either shortens the Dynkin diagram, for example starting from su(5):

su(5) ○—○—○—○ (↑ drop this) → ○—○—○ su(4)

or breaks it into two pieces, for example

su(5) ○—○—○—○ (↑ drop this) → ○—○ ○ su(3) su(2).

Owing to the reflection symmetry of the su(5) Dynkin diagram, these are the only two independent possibilities. Including the u(1) factor, we see that the first choice gives a breaking of su(5) to a su(4) × u(1) subalgebra, while the latter corresponds to a breaking to a su(3) × su(2) × u(1) subalgebra. For su(6) we have more choices:

su(6) ○—○—○—○—○ →
- ○—○—○—○   su(5) × u(1)
- ○—○—○ ○   su(4) × su(2) × u(1)
- ○—○ ○—○   su(3) × su(3) × u(1)

and so on.

It is interesting to study how the irreducible representations of su(N) break into irreducible representations of the subalgebras. Consider, for example, the $(1, 0, 0, 0) = \mathbf{5} = \square$ defining representation of su(5), whose generators are 5×5 traceless Hermitian matrices, while the group elements are 5×5 unitary matrices with unit determinant. The subalgebra su(3) × su(2) × u(1) consists of the following generators:

$$\begin{pmatrix} X_3^a & 0 \\ 0 & \mathbb{1}_2 \end{pmatrix}, \quad \begin{pmatrix} \mathbb{1}_3 & 0 \\ 0 & X_2^b \end{pmatrix}, \quad \begin{pmatrix} -\frac{1}{3}\mathbb{1}_3 & 0 \\ 0 & \frac{1}{2}\mathbb{1}_2 \end{pmatrix},$$

where $X_3^a = \frac{1}{2}\lambda^a$ (for $1 \le a \le 8$) are the generators of su(3) in the defining representation, with λ^a the Gell-Mann matrices, $X_2^b = \frac{1}{2}\sigma^b$ (for $1 \le b \le 3$) are the su(2) generators, with σ^b the Pauli matrices, and the last diagonal matrix is the traceless u(1) generator with charges $q_1 = -1/3$, $q_2 = 1/2$; note, however, that in the literature the overall normalization conventions may vary. Thus, we see that the **5** representation of the su(5) Lie algebra breaks into a direct sum of two irreducible representations of su(3) × su(2) × u(1):

$$\mathbf{5} \to (\mathbf{3}, \mathbf{1})_{-1/3} \oplus (\mathbf{1}, \mathbf{2})_{+1/2}. \tag{6.348}$$

Similarly, for the $\overline{\mathbf{5}}$ representation we have:

$$\overline{\mathbf{5}} \to (\overline{\mathbf{3}}, \mathbf{1})_{1/3} \oplus (\mathbf{1}, \mathbf{2})_{-1/2}. \tag{6.349}$$

These are examples of the so-called *branching rules*. They are important for the construction of a possible *grand-unified theory* of elementary-particle physics, with the goal of unifying the three fundamental interactions of the Standard Model, described by the SU(3)×SU(2)×U(1) gauge group, into a single type of interaction, corresponding to a larger gauge group, which could hypothetically be seen in future collider experiments, at sufficiently high energies. The example that we just discussed is based on a theoretical particle physics model that was presented in reference [3].

Let us now discuss how the branching rules can be combined with the decomposition of tensor products. As an example, let us consider the Lie algebra of the SU(5) group, and the tensor product

$$\mathbf{5} \otimes \overline{\mathbf{5}} = \mathbf{24} \oplus \mathbf{1}, \tag{6.350}$$

where **24** is the adjoint representation of the su(5) algebra. On the other hand, using Eq. (6.348) and Eq. (6.349), we can also write

$$
\begin{aligned}
\mathbf{5} \otimes \overline{\mathbf{5}} &\to [(\mathbf{3}, \mathbf{1})_{-1/3} \oplus (\mathbf{1}, \mathbf{2})_{+1/2}] \otimes [(\overline{\mathbf{3}}, \mathbf{1})_{1/3} \oplus (\mathbf{1}, \mathbf{2})_{-1/2}] \\
&= [(\mathbf{3}, \mathbf{1})_{-1/3} \otimes (\overline{\mathbf{3}}, \mathbf{1})_{1/3}] \oplus [(\mathbf{1}, \mathbf{2})_{+1/2} \otimes (\overline{\mathbf{3}}, \mathbf{1})_{1/3}] \\
&\oplus [(\mathbf{3}, \mathbf{1})_{-1/3} \otimes (\mathbf{1}, \mathbf{2})_{-1/2}] \oplus [(\mathbf{1}, \mathbf{2})_{+1/2} \otimes (\mathbf{1}, \mathbf{2})_{-1/2}].
\end{aligned} \tag{6.351}
$$

Using the su(3) decomposition $\mathbf{3} \otimes \overline{\mathbf{3}} = \mathbf{8} \oplus \mathbf{1}$, as well as the su(2) decomposition $\mathbf{2} \otimes \mathbf{2} = \mathbf{3} \oplus \mathbf{1}$ (note that, when expressed in terms of spin labels instead of the representation dimensions, the latter is $j_1 \otimes j_2 = \frac{1}{2} \otimes \frac{1}{2} = 1 \oplus 0$), and the addition of u(1) charges, Eq. (6.351) can be rewritten as

$$\mathbf{5} \otimes \overline{\mathbf{5}} \to (\mathbf{8}, \mathbf{1})_0 \oplus (\mathbf{1}, \mathbf{1})_0 \oplus (\overline{\mathbf{3}}, \mathbf{2})_{5/6} \oplus (\mathbf{3}, \mathbf{2})_{-5/6} \oplus (\mathbf{1}, \mathbf{3})_0 \oplus (\mathbf{1}, \mathbf{1})_0. \tag{6.352}$$

We note that the last term on the right-hand side of this expression is the branching of the su(5) singlet representation **1**. Thus, comparing Eq. (6.352) with Eq. (6.350) we obtain the branching rule for the adjoint representation:

$$\mathbf{24} \to (\mathbf{8}, \mathbf{1})_0 \oplus (\mathbf{1}, \mathbf{3})_0 \oplus (\mathbf{1}, \mathbf{1})_0 \oplus (\mathbf{3}, \mathbf{2})_{-5/6} \oplus (\overline{\mathbf{3}}, \mathbf{2})_{5/6}. \tag{6.353}$$

The first three terms on the right-hand side of Eq. (6.353) correspond to the adjoint representations of su(3), su(2), and u(1), and describe the gauge bosons of the strong and electroweak interactions. The fourth term and the fifth term, instead, correspond to additional representations, and, in the Georgi–Glashow grand-unification model, describe the so-called X and Y bosons, for which, so far, there is no experimental evidence. According to the model, all these bosons unify into the gauge boson (transforming in the adjoint **24** representation) of an interaction described by an SU(5) gauge theory. To accommodate the quarks and leptons of the Standard Model into this grand-unified model, we additionally need the branching rule of the **10** representation of the su(5) algebra (for which we skip the proof): The branching rules for fermionic matter fields are then

$$\overline{\mathbf{5}} \to (\overline{\mathbf{3}}, \mathbf{1})_{1/3} \oplus (\mathbf{1}, \mathbf{2})_{-1/2},$$
$$\mathbf{10} \to (\mathbf{3}, \mathbf{2})_{1/6} \oplus (\overline{\mathbf{3}}, \mathbf{1})_{-2/3} \oplus (\mathbf{1}, \mathbf{1})_1,$$
$$\mathbf{1} \to (\mathbf{1}, \mathbf{1})_0. \tag{6.354}$$

Comparing with Eq. (6.341), we see that the first line contains the representations of the quark $\overline{d}$ and lepton ℓ fields of the Standard Model, while the second line contains the quark q and $\overline{u}$ fields, as well as the $\overline{e}$ field. The last line is an additional $\overline{\nu}$ field, which could be interpreted as the additional field that is necessary to describe massive neutrinos. To summarize, each generation of quarks and leptons can be embedded into a generation corresponding to the $\overline{\mathbf{5}}$ and **10** (and **1**, if neutrinos are assumed to be massive) representations of the SU(5) gauge group of this grand-unified model.

6.12 The Lorentz–Poincaré Group

6.12.1 The Lorentz Group

The Lorentz group is the group of transformations of spacetime coordinates that is characteristic of a theory formulated by Albert Einstein in 1905, relating physical quantities measured by different observers in motion at constant speed with respect to each other. The reference frames of such observers are called *inertial frames*. This theory is called *special relativity theory*, and is based on the following two postulates.

1. All physical laws take the same form, in all inertial frames.
2. The speed of light in vacuum has the same value in all inertial frames.

Consider two inertial reference frames, with spacetime coordinates denoted as (t, x^1, x^2, x^3) and (t', x'^1, x'^2, x'^3), and assume, for simplicity, that at time $t = t' = 0$ they coincide (i.e., $x^i = x'^i$ for each $i \in \{1, 2, 3\}$). Then, according to the second postulate of special relativity, a ray of light emitted at the origin at that time and propagating to another point P at a later time should be such that the spacetime coordinates of P measured in the two inertial frames satisfy

$$(ct)^2 - x^i x^i = (ct')^2 - x'^i x'^i, \tag{6.355}$$

where c denotes the speed of light in vacuum, and summation over the repeated Latin indices i, ranging from 1 to 3, is understood.

In the following, we also introduce Greek indices (e.g., μ, ν, etc.), assuming that they range from 0 to 3, and define the *four-vector* $x^\mu = (x^0, x^1, x^2, x^3)$, with $x^0 = ct$, and, similarly, x'^μ, in terms of the spacetime coordinates in the other inertial reference frame, and the metric

$$g_{\mu\nu} = \begin{cases} 1 & \text{if } \mu = \nu = 0 \\ -1 & \text{if } \mu = \nu \neq 0. \\ 0 & \text{if } \mu \neq \nu \end{cases} \qquad (6.356)$$

Note that the metric defined in Eq. (6.356) corresponds to the mostly minus convention for the signature. While in Chapter 5 we used the mostly plus convention, which is most common in general relativity, here we switch to the mostly minus convention, which, as we pointed out in Section 5.1, is often used in elementary particle physics. It is useful for the reader to get used to both conventions. Equation (6.355) can be rewritten as

$$g_{\mu\nu}x^\mu x^\nu = g_{\mu\nu}x'^\mu x'^\nu. \qquad (6.357)$$

We now define an *event A* as a point in spacetime. Given two arbitrary events A and B, one can define the *squared relativistic interval*

$$(\Delta s_{AB})^2 = (x_A^0 - x_B^0)^2 - (x_A^i - x_B^i)^2 = g_{\mu\nu}(x_A - x_B)^\mu (x_A - x_B)^\nu, \qquad (6.358)$$

and this quantity is an example of a *relativistic invariant*, i.e., a quantity that has the same value in all inertial reference frames.

Assuming that the spacetime coordinates of an event expressed in two different inertial frames, x^α and x'^β, are related to each other by a set of linear transformations

$$x'^\beta = \Lambda^\beta{}_\alpha x^\alpha, \qquad (6.359)$$

Eq. (6.357) implies

$$g_{\mu\nu}x^\mu x^\nu = g_{\alpha\beta}\Lambda^\alpha{}_\mu x^\mu \Lambda^\beta{}_\nu x^\nu. \qquad (6.360)$$

As Eq. (6.360) must hold for an arbitrary x, it follows that

$$g_{\mu\nu} = g_{\alpha\beta}\Lambda^\alpha{}_\mu \Lambda^\beta{}_\nu. \qquad (6.361)$$

A spacetime coordinate transformation Λ satisfying Eq. (6.361) is called a *Lorentz transformation*.

We note that, for $\mu = \nu = 0$, Eq. (6.361) reduces to

$$1 = g_{\alpha\beta}\Lambda^\alpha{}_0 \Lambda^\beta{}_0 = \Lambda^0{}_0 \Lambda^0{}_0 - \Lambda^i{}_0 \Lambda^i{}_0, \qquad (6.362)$$

from which one deduces

$$\left| \Lambda^0{}_0 \right| = \sqrt{1 + \Lambda^i{}_0 \Lambda^i{}_0} \geq 1. \qquad (6.363)$$

Depending on the sign of $\Lambda^0{}_0$, the transformation represented by the matrix Λ is said to be

- an *orthochronous Lorentz transformation*, if $\Lambda^0{}_0 \geq 1$,
- a *non-orthochronous Lorentz transformation*, if $\Lambda^0{}_0 \leq -1$.

Moreover, we also note that, if one interprets x and x' as four-component, real-valued column vectors, and g and Λ as 4×4 real-valued matrices, then Eq. (6.361) can be thought of as the matrix relation

$$g = \Lambda^\mathsf{T} g \Lambda. \qquad (6.364)$$

Taking the determinant of both sides of Eq. (6.364), one deduces that $\det \Lambda = \pm 1$. Accordingly, we call Λ

- a *proper Lorentz transformation*, if $\det \Lambda = 1$,
- an *improper Lorentz transformation*, if $\det \Lambda = -1$.

One can note that Eq. (6.364) has an analogy with the one that characterizes the group of orthogonal transformations of size 4, denoted as O(4):

$$\mathbb{1} = O^{\mathrm{T}} \mathbb{1} O, \tag{6.365}$$

the difference being that $\mathbb{1}_{\mu\nu} = \delta_{\mu\nu}$, while the $g_{\mu\nu}$ components are defined by Eq. (6.356). Note that $g_{\mu\nu}$ is a tensor that can be used to "lower the indices," e.g.,

$$x_\mu = g_{\mu\nu} x^\nu, \tag{6.366}$$

which implies that $x_0 = x^0$, while $x_i = -x^i$ for $i = 1, 2,$ or 3. To stress the fact that three diagonal entries of $g_{\mu\nu}$ have the same sign, while one has the opposite sign, the set of matrices satisfying Eq. (6.364) is denoted as O(3, 1). These matrices, with the matrix row-by-column multiplication as the group product, form a group: the *Lorentz group*. The Lorentz group can be defined as the group of linear coordinate transformations leaving the quantity $(\Delta s)^2$, defined in Eq. (6.358), invariant. This can be contrasted with the matrices of O(4) that leave the squared modulus of a vector invariant. It is also interesting to note that, while the squared modulus of a four-component real vector is always nonnegative, and vanishes if and only if the vector is zero, this is not the case for the $(\Delta s)^2$ invariant of a relativistic four-vector, which, as a consequence of the different signs appearing on the right-hand side of Eq. (6.358), can be positive, zero, or negative. Accordingly, a four-vector x is said to be

- a *timelike four-vector*, if $(x^0)^2 - x^i x^i > 0$,
- a *spacelike four-vector*, if $(x^0)^2 - x^i x^i < 0$,
- a *lightlike four-vector* (or a *null four-vector*), if $(x^0)^2 - x^i x^i = 0$.

Note that, in particular, a null vector is not necessarily the zero vector (although the zero vector is a null vector).

Let us consider some examples of transformations Λ belonging to the Lorentz group.

- Matrices of the form

$$\Lambda = \begin{pmatrix} 1 & 0 & 0 & 0 \\ 0 & & & \\ 0 & & R & \\ 0 & & & \end{pmatrix}, \tag{6.367}$$

with R a 3×3 orthogonal matrix, describe transformations that leave the x^0 coordinate unchanged, and rotate the spatial axes of the reference frame; the transformation is an orthochronous Lorentz transformation, and is proper if $\det R = 1$, i.e., if $R \in$ SO(3), while it is improper for $\det R = -1$. Note that the inverse transformation, Λ^{-1}, has the same form as Λ, but with R replaced by R^{T}. As a particular case of Eq. (6.367), a rotation of the reference frame by an angle θ around the z-axis takes the form

$$\Lambda = \begin{pmatrix} 1 & 0 & 0 & 0 \\ 0 & \cos\theta & \sin\theta & 0 \\ 0 & -\sin\theta & \cos\theta & 0 \\ 0 & 0 & 0 & 1 \end{pmatrix}. \qquad (6.368)$$

When the θ angle is infinitesimally small, Eq. (6.368) takes the form

$$\Lambda = \begin{pmatrix} 1 & 0 & 0 & 0 \\ 0 & 1 & \theta & 0 \\ 0 & -\theta & 1 & 0 \\ 0 & 0 & 0 & 1 \end{pmatrix} + O(\theta^2)$$

$$= \mathbb{1} + i\theta \begin{pmatrix} 0 & 0 & 0 & 0 \\ 0 & 0 & -i & 0 \\ 0 & i & 0 & 0 \\ 0 & 0 & 0 & 0 \end{pmatrix} + O(\theta^2), \qquad (6.369)$$

from which one can read off the generator of rotations around the z-axis (to be denoted as J^3) as the matrix multiplied by $i\theta$ in the last term. With a similar argument, one can derive the generators of rotations around the other two spatial axes, finding

$$J^1 = \begin{pmatrix} 0 & 0 & 0 & 0 \\ 0 & 0 & 0 & 0 \\ 0 & 0 & 0 & -i \\ 0 & 0 & i & 0 \end{pmatrix},$$

$$J^2 = \begin{pmatrix} 0 & 0 & 0 & 0 \\ 0 & 0 & 0 & i \\ 0 & 0 & 0 & 0 \\ 0 & -i & 0 & 0 \end{pmatrix},$$

$$J^3 = \begin{pmatrix} 0 & 0 & 0 & 0 \\ 0 & 0 & -i & 0 \\ 0 & i & 0 & 0 \\ 0 & 0 & 0 & 0 \end{pmatrix}. \qquad (6.370)$$

The generators in Eq. (6.370) satisfy

$$[J^a, J^b] = i\epsilon_{abc} J^c, \qquad (6.371)$$

where ϵ_{abc} is antisymmetric under the interchange of any pair of its indices, and such that $\epsilon_{123} = 1$, i.e., the generators of rotations form an su(2) algebra.

Another particular case of Eq. (6.367) is the parity transformation, i.e., inversion of all spatial coordinates, corresponding to $R = -\mathbb{1}$ in Eq. (6.367), which is an improper, orthochronous Lorentz transformation. Note that the parity transformation does not depend on a continuous parameter, i.e., it is a *discrete* transformation. Furthermore, it is an *involution*, i.e., a transformation that squares to the identity (and that, as a consequence, is equal to its own inverse).

- Matrices describing Lorentz boosts, i.e., relativistic transformations of the space-time coordinates between two inertial reference frames moving at constant relative velocity of modulus v. If the reference frame of the primed coordinates is moving with respect to one of unprimed coordinates at velocity v along the x-direction, then the boost can be represented by the matrix

$$\Lambda = \begin{pmatrix} \gamma & -\beta\gamma & 0 & 0 \\ -\beta\gamma & \gamma & 0 & 0 \\ 0 & 0 & 1 & 0 \\ 0 & 0 & 0 & 1 \end{pmatrix}, \tag{6.372}$$

with $\beta = v/c$ and $\gamma = 1/\sqrt{1-\beta^2}$. Defining the *rapidity* $\eta = \operatorname{arcsinh}(\beta\gamma)$, Eq. (6.372) can be rewritten as

$$\Lambda = \begin{pmatrix} \cosh\eta & -\sinh\eta & 0 & 0 \\ -\sinh\eta & \cosh\eta & 0 & 0 \\ 0 & 0 & 1 & 0 \\ 0 & 0 & 0 & 1 \end{pmatrix}. \tag{6.373}$$

We also note that the inverse of the transformation described by Eq. (6.372) has the same form, but with β replaced by $-\beta$ (or, equivalently, η replaced by $-\eta$).

In order to give a proper definition of the generators of infinitesimal boosts, it is instructive to first recall the condition that defines a Lorentz transformation, expressed in Eq. (6.361), namely

$$\begin{aligned} g_{\mu\nu} &= g_{\alpha\beta}\Lambda^{\alpha}{}_{\mu}\Lambda^{\beta}{}_{\nu} \\ &= \Lambda_{\beta\mu}\Lambda^{\beta}{}_{\nu}, \end{aligned} \tag{6.374}$$

where we used the fact that $g_{\alpha\beta}\Lambda^{\alpha}{}_{\mu} = \Lambda_{\beta\mu}$. When Λ is a (proper and orthochronous) Lorentz transformation infinitesimally close to the identity, one can write

$$\Lambda^{\alpha}{}_{\mu} = g^{\alpha}{}_{\mu} + \omega^{\alpha}{}_{\mu}, \tag{6.375}$$

where the components of ω are infinitesimally small, and one should note that $g^{\alpha}{}_{\mu} = \delta^{\alpha}{}_{\mu} = \delta^{\alpha\mu} = \delta_{\alpha\mu}$, so that Eq. (6.361) becomes

$$\begin{aligned} g_{\mu\nu} &= \left(g_{\beta\mu} + \omega_{\beta\mu}\right)\left(g^{\beta}{}_{\nu} + \omega^{\beta}{}_{\nu}\right) \\ &= g_{\mu\nu} + \omega_{\mu\nu} + \omega_{\nu\mu}, \end{aligned} \tag{6.376}$$

having used the contractions with the metric to lower the indices, and having dismissed subleading terms, from which we obtain

$$\omega_{\mu\nu} = -\omega_{\nu\mu}. \tag{6.377}$$

Let us now consider a boost along the x-axis; in the limit when η is infinitesimally small, Eq. (6.373) takes the form

$$\Lambda = \begin{pmatrix} 1 & -\eta & 0 & 0 \\ -\eta & 1 & 0 & 0 \\ 0 & 0 & 1 & 0 \\ 0 & 0 & 0 & 1 \end{pmatrix} = \mathbb{1} + \begin{pmatrix} 0 & -\eta & 0 & 0 \\ -\eta & 0 & 0 & 0 \\ 0 & 0 & 0 & 0 \\ 0 & 0 & 0 & 0 \end{pmatrix}, \tag{6.378}$$

having dismissed terms $O(\eta^2)$. More precisely, recalling Eq. (6.359), we note that Eq. (6.378) can be written in component form precisely as Eq. (6.375), i.e., $\Lambda^\alpha{}_\mu = g^\alpha{}_\mu + \omega^\alpha{}_\mu$, where $g^\alpha{}_\mu$ equals $\delta_{\alpha\mu}$, the matrix element of the identity matrix appearing on the right-hand side of Eq. (6.378), while $\omega^\alpha{}_\mu$ can be identified with the generic matrix element of the other matrix on the right-hand side of Eq. (6.378), namely

$$\omega^0{}_1 = \omega^1{}_0 = -\eta. \tag{6.379}$$

Lowering the upper indices of these components, we obtain

$$\omega_{01} = -\omega_{10} = -\eta, \tag{6.380}$$

in agreement with the expected antisymmetry of $\omega_{\mu\nu}$ (with *two* lower indices) according to Eq. (6.377). Factoring out the infinitesimal term η and an i factor, it is then convenient to define the generator of boosts along the x-axis, that we denote as K^1, as the matrix with $(K^1)^0{}_1 = (K^1)^1{}_0 = i$ as the only nonvanishing elements.

The generators of boosts along the y- and z-axes (that we denote as K^2 and K^3, respectively) can be worked out in a similar way, obtaining

$$K^1 = \begin{pmatrix} 0 & i & 0 & 0 \\ i & 0 & 0 & 0 \\ 0 & 0 & 0 & 0 \\ 0 & 0 & 0 & 0 \end{pmatrix},$$

$$K^2 = \begin{pmatrix} 0 & 0 & i & 0 \\ 0 & 0 & 0 & 0 \\ i & 0 & 0 & 0 \\ 0 & 0 & 0 & 0 \end{pmatrix},$$

$$K^3 = \begin{pmatrix} 0 & 0 & 0 & i \\ 0 & 0 & 0 & 0 \\ 0 & 0 & 0 & 0 \\ i & 0 & 0 & 0 \end{pmatrix}. \tag{6.381}$$

Note that the generators of boosts are anti-Hermitian.

In contrast to the generators of rotations, the generators of boosts do not form a closed subalgebra themselves; instead

$$[K^a, K^b] = -i\epsilon_{abc}J^c. \tag{6.382}$$

Moreover, the commutator of a rotation generator and a boost generator reads

$$[J^a, K^b] = i\epsilon_{abc}K^c. \tag{6.383}$$

- Time reversal: The Lorentz transformation described by the matrix

$$\Lambda = \begin{pmatrix} -1 & 0 & 0 & 0 \\ 0 & 1 & 0 & 0 \\ 0 & 0 & 1 & 0 \\ 0 & 0 & 0 & 1 \end{pmatrix} \tag{6.384}$$

is an improper, non-orthochronous Lorentz transformation that reverses the sign of time, leaving the spatial coordinates, as well as the $(\Delta s)^2$ invariant, unchanged.

Like the parity transformation, time reversal is a discrete transformation which squares to the identity and, hence, equals its inverse.

• Arbitrary products of all of the previous cases are still Lorentz transformations.

In passing, we note that the Lorentz group is non-compact.

While Eq. (6.382) and Eq. (6.383) imply that the generators of rotations and the generators of boosts do not form separate algebras, one can prove that, defining the following combinations of generators

$$N^a = \frac{1}{2}\left(J^a + iK^a\right), \qquad M^a = \frac{1}{2}\left(J^a - iK^a\right), \qquad \text{for } a \in \{1,2,3\}, \qquad (6.385)$$

one obtains two independent su(2) algebras

$$\left[N^a, N^b\right] = i\epsilon_{abc}N^c, \qquad (6.386)$$

$$\left[M^a, M^b\right] = i\epsilon_{abc}M^c, \qquad (6.387)$$

$$\left[N^a, M^b\right] = 0. \qquad (6.388)$$

Accordingly, the irreducible representations of the Lorentz group can be labeled by the labels of these two su(2) algebras, (p,q), where p and q are integer or half-integer nonnegative numbers related to the quadratic Casimir operators constructed from the N^a and the M^a generators as

$$N^a N^a = p(p+1)\mathbb{1}, \qquad (6.389)$$

$$M^a M^a = q(q+1)\mathbb{1}. \qquad (6.390)$$

Let us now consider the effect of Lorentz transformations on different types of *fields*, i.e., on functions of the spacetime coordinates $\phi(x^\mu)$. Let us now restrict ourselves to the subset of continuous, proper, orthochronous Lorentz transformations, i.e., rotations and boosts. As we discussed above, they depend on six parameters (three rotation angles and three components of the boost velocity). Let us consider a Lorentz transformation that is infinitesimally close to the identity

$$\Lambda^\beta{}_\alpha = \delta^\beta{}_\alpha + \epsilon^\beta{}_\alpha + O(\epsilon^2). \qquad (6.391)$$

Substituting Eq. (6.391) into Eq. (6.361)

$$\begin{aligned} g_{\mu\nu} &= g_{\alpha\beta}\left(\delta^\alpha{}_\mu + \epsilon^\alpha{}_\mu + O(\epsilon^2)\right)\cdot\left(\delta^\beta{}_\nu + \epsilon^\beta{}_\nu + O(\epsilon^2)\right) \\ &= g_{\alpha\beta}\left(\delta^\alpha{}_\mu\delta^\beta{}_\nu + \epsilon^\alpha{}_\mu\delta^\beta{}_\nu + \delta^\alpha{}_\mu\epsilon^\beta{}_\nu + O(\epsilon^2)\right) \\ &= g_{\mu\nu} + \epsilon_{\nu\mu} + \epsilon_{\mu\nu} + O(\epsilon^2), \end{aligned} \qquad (6.392)$$

from which it follows that $\epsilon_{\mu\nu} = -\epsilon_{\nu\mu}$, implying that the infinitesimal generator of Lorentz transformations can be thought of as an antisymmetric 4×4 tensor with six independent components.

Consider the differential operators defined as

$$L_{\mu\nu} = ix_\mu\frac{\partial}{\partial x^\nu} - ix_\nu\frac{\partial}{\partial x^\mu} = i\left(x_\mu\partial_\nu - x_\nu\partial_\mu\right); \qquad (6.393)$$

they satisfy the *Lorentz algebra*

$$[L^{\alpha\beta}, L^{\gamma\delta}] = i\left(g^{\beta\gamma}L^{\alpha\delta} + g^{\alpha\delta}L^{\beta\gamma}\right) - i\left(g^{\beta\delta}L^{\alpha\gamma} + g^{\alpha\gamma}L^{\beta\delta}\right) \qquad (6.394)$$

and allow one to write the variation of x^ν under an infinitesimal Lorentz transformation as

$$\delta x^\nu = \frac{i}{2}\epsilon^{\alpha\beta}L_{\alpha\beta}x^\nu. \tag{6.395}$$

For a Lorentz transformation that can be expressed as in Eq. (6.391), we can thus define the operator

$$U(\Lambda) = \exp\left(\frac{i}{2}\epsilon^{\alpha\beta}L_{\alpha\beta}\right), \tag{6.396}$$

which can be used to classify the different types of fields, based on their transformation properties under a Lorentz transformation $x \to x'$. These transformation properties must account for the nature of the field and for the fact that it depends on spacetime coordinates, which are not invariant under Lorentz transformations. For example:

- a scalar field $\phi(x)$ transforms as

$$\phi(x') = U(\Lambda)\phi(x)U^{-1}(\Lambda); \tag{6.397}$$

- a vector field $\phi^\alpha(x)$ transforms as

$$\phi^\beta(x') = \Lambda^\beta{}_\alpha\left(U(\Lambda)\phi^\alpha(x)U^{-1}(\Lambda)\right). \tag{6.398}$$

In fact, there could be many more types of fields, which could be constructed, for example, by considering the representations of the su(2) algebras generated by the N^α and $N^{\alpha\dagger}$, that we can denote as (p, q), and their products. For instance:

- $(0, 0)$ corresponds to a scalar field;
- $(1/2, 0)$ corresponds (conventionally) to a left-handed, spin-half field; conversely, $(0, 1/2)$ represents a right-handed, spin-half field;
- the product $(1/2, 0) \otimes (0, 1/2) = (1/2, 1/2)$ corresponds to a spin-one vector field;
- the $(1/2, 0) \otimes (1/2, 0)$ product, instead, can be decomposed as $(0, 0) \oplus (1, 0)$, where $(0, 0)$ is the scalar representation, while $(1, 0)$ is a two-index, antisymmetric, self-dual tensor; conversely, $(0, 1/2) \otimes (0, 1/2) = (0, 0) \oplus (0, 1)$, where $(0, 1)$ is a two-index, antisymmetric, anti-self-dual tensor. The field strength tensor of electromagnetism is a physical example of a field transforming in the $(1, 0) \oplus (0, 1)$ representation of the Lorentz group.

6.12.2 The Lorentz–Poincaré Group

As we discussed in Section 6.12.1, the Lorentz group consists of the *linear* coordinate transformations that are consistent with the principles of the special relativity theory. However, as the physical laws depend only on *differences* of physical coordinates, one can extend the set of coordinate transformations to include also spacetime translations, generalizing Eq. (6.359) to

$$x'^\beta = \Lambda^\beta{}_\alpha x^\alpha + a^\beta, \tag{6.399}$$

where a^β is a constant four-vector. The transformations defined in Eq. (6.399) can be concisely denoted as (Λ, a) and form the *Lorentz–Poincaré group*, which is a ten-dimensional group.

What is the group product of the Poincaré group? As we mentioned in Section 6.12.1, for Lorentz transformations, the group product can be identified with matrix multiplication of Λ matrices. For pure translations, on the other hand, the natural group "product" is the vector sum. The group product of two generic elements (Λ_1, a_1) and (Λ_2, a_2) of the Lorentz–Poincaré group can be evaluated by writing first (for arbitrary x) $x' = \Lambda_2 x + a_2$, then $x'' = \Lambda_1 x' + a_1$, from which one deduces $x'' = \Lambda_1 \Lambda_2 x + \Lambda_1 a_2 + a_1$. Hence the group product for the Lorentz–Poincaré group is given by

$$(\Lambda_1, a_1) \cdot (\Lambda_2, a_2) = (\Lambda_1 \Lambda_2, \Lambda_1 a_2 + a_1). \tag{6.400}$$

For this reason, the Lorentz–Poincaré group is said to be the *semidirect product* of the Lorentz group and the group of translations.

The generators of the Lorentz–Poincaré group include the six generators of the Lorentz group $(L^{\alpha\beta})$, and the four generators of translations P^{α}. They satisfy the commutation relations:

$$\left[L^{\alpha\beta}, L^{\gamma\delta} \right] = i \left(g^{\beta\gamma} L^{\alpha\delta} + g^{\alpha\delta} L^{\beta\gamma} \right) - i \left(g^{\beta\delta} L^{\alpha\gamma} + g^{\alpha\gamma} L^{\beta\delta} \right), \tag{6.401}$$

$$\left[L^{\alpha\beta}, P^{\gamma} \right] = i \left(g^{\alpha\gamma} P^{\beta} - g^{\beta\gamma} P^{\alpha} \right), \tag{6.402}$$

$$\left[P^{\alpha}, P^{\beta} \right] = 0. \tag{6.403}$$

The Lorentz–Poincaré group has two Casimir operators.

1. $P_{\alpha} P^{\alpha}$ is a Casimir operator, because it is a scalar (i.e., an invariant) under Lorentz transformations, and because, being built only from the P_{α}s, it also commutes with the translation generators, by virtue of Eq. (6.403). In the context of quantum field theory, in which the P^{α}s are identified as the components of the four-momentum of a particle, its eigenvalue is nothing but the squared mass of the particle:

$$P_{\alpha} P^{\alpha} = m^2 \mathbb{1}. \tag{6.404}$$

2. $W_{\alpha} W^{\alpha}$, i.e., the square of the *Pauli–Lubanski axial vector*, defined as

$$W^{\alpha} = \frac{1}{2} \epsilon^{\alpha\beta\gamma\delta} P_{\beta} L_{\gamma\delta}, \tag{6.405}$$

where $\epsilon^{\alpha\beta\gamma\delta}$ is the totally antisymmetric four-index tensor, with $\epsilon^{0123} = 1$, is also a Casimir operator. Physically, its eigenvalue is $-m^2$ times the square of the angular momentum of a particle:

$$W_{\alpha} W^{\alpha} = -m^2 j(j+1), \tag{6.406}$$

with j integer or half-integer, and nonnegative. For massive particles, the number of physical states is $2j+1$. For massless particles, instead, the Pauli–Lubanski axial vector is proportional to the momentum four-vector, $W_{\alpha} = h P_{\alpha}$, where $h = \pm j$ is the *helicity* of the particle, and there are two physical states.

Finally, we note that, like the Lorentz group, also the Lorentz–Poincaré group is non-compact.

Problems

6.1   Find all elements of the matrix $\exp(i\alpha A)$, where

$$A = \begin{pmatrix} 0 & 0 & 1 \\ 0 & 0 & 0 \\ 1 & 0 & 0 \end{pmatrix}. \tag{6.407}$$

6.2   Given two matrices A and B satisfying the commutation relation

$$[A, B] = B, \tag{6.408}$$

calculate

$$\exp(i\alpha A)B\exp(-i\alpha A). \tag{6.409}$$

6.3   The generators X_a (with $1 \le a \le N$) of a Lie algebra, defined as in Section 6.5, satisfy the commutation relations

$$[X_a, X_b] = if_{abc}X_c. \tag{6.410}$$

(i) Show that they satisfy the Jacobi identity

$$[X_a, [X_b, X_c]] + [X_b, [X_c, X_a]] + [X_c, [X_a, X_b]] = 0. \tag{6.411}$$

(ii) Show that the Jacobi identity implies the following identity for the structure constants f_{abc}:

$$f_{ade}f_{bcd} + f_{bde}f_{cad} + f_{cde}f_{abd} = 0. \tag{6.412}$$

(iii) In the adjoint representation the generators are represented by the $N \times N$ matrices T_a with elements

$$(T_a)_{bc} = -if_{abc}. \tag{6.413}$$

Verify that the T_a matrices satisfy the same commutation relations described by Eq. (6.410), i.e., verify that

$$[T_a, T_b] = if_{abc}T_c. \tag{6.414}$$

6.4   As we know, all matrices in the defining representation of the SU(2) group can be written in the exponential form using the Pauli spin matrices σ_i:

$$g = \exp(i\sigma_i x_i). \tag{6.415}$$

This is not the case for all Lie groups. We consider $SL(2, \mathbb{R})$ as an example. In this exercise, keep the imaginary units out of the exponent.

(i) The dimension of the $SL(2, \mathbb{R})$ group is three. Find three real and linearly independent generators X_i, i.e. find a basis for the Lie algebra $sl(2, \mathbb{R})$. Hint: Use $\det(\exp(A)) = \exp(\operatorname{tr}(A))$ and find a basis that is similar to Pauli matrices.

(ii) Calculate $\exp(a_i X_i)$ and show that there are elements of $SL(2, \mathbb{R}))$ that are not in the image of exp with any $\{a_i\}$. Hint: Try diagonal matrices.

6.5   Let D_1 and D_2 be two irreducible representations of a Lie algebra, of dimensions $\dim D_1$ and $\dim D_2$, and with Dynkin indices λ_{D_1} and λ_{D_2}, respectively.

(i) Prove that the Dynkin index of the direct-sum representation $D_1 \oplus D_2$ is given by

$$\lambda_{D_1 \oplus D_2} = \lambda_{D_1} + \lambda_{D_2}. \tag{6.416}$$

(ii) Prove that the Dynkin index of the tensor-product representation $D_1 \otimes D_2$ is given by

$$\lambda_{D_1 \otimes D_2} = (\dim D_1)\lambda_{D_2} + (\dim D_2)\lambda_{D_1}. \tag{6.417}$$

6.6 Given the generators X_a (with $a = 1, \ldots, N$) of a Lie algebra, prove that the second-order Casimir operator $C^{(2)} = \sum_{a=1}^{N} X_a X_a$ commutes with all the generators.

6.7 Let the matrix elements of the generators of the su(2) Lie algebra in the adjoint representation J_a (for $a \in \{1, 2, 3\}$) be $(J_a)_{bc} = -i\epsilon_{abc}$. Moreover, let $|J_a\rangle$ (again, for $a \in \{1, 2, 3\}$) denote the column vectors with three elements with entries $|J_a\rangle_b = \delta_{a,b}$. Given the vectors

$$|J^+\rangle = \frac{|J_1\rangle + i|J_2\rangle}{\sqrt{2}}, \qquad |J^-\rangle = \frac{|J_1\rangle - i|J_2\rangle}{\sqrt{2}}, \tag{6.418}$$

the matrix

$$U = \begin{pmatrix} \frac{1}{\sqrt{2}} & 0 & \frac{1}{\sqrt{2}} \\ \frac{i}{\sqrt{2}} & 0 & -\frac{i}{\sqrt{2}} \\ 0 & 1 & 0 \end{pmatrix}, \tag{6.419}$$

and the vectors

$$|j = 1, m = +1\rangle = U^\dagger |J^+\rangle,$$
$$|j = 1, m = 0\rangle = U^\dagger |J_3\rangle,$$
$$|j = 1, m = -1\rangle = U^\dagger |J^-\rangle, \tag{6.420}$$

show that the entries of the matrices

$$\tilde{J}_a = U^\dagger J_a U \tag{6.421}$$

are

$$\left(\tilde{J}_a\right)_{mm'} = \langle j = 1, m | \tilde{J}_a | j = 1, m'\rangle. \tag{6.422}$$

6.8 Starting from the relation $[H_i, E_\alpha] = \alpha_i E_\alpha$ and its adjoint, and using the Hermiticity of the Cartan generators, show that the E_αs are non-Hermitian and appear in pairs with opposite eigenvalues $E_{\pm\alpha}$, with $E_\alpha^\dagger = E_{-\alpha}$.

6.9 Show that $[E_\alpha, E_\beta]$ is proportional to $E_{\alpha+\beta}$. What if $\alpha + \beta$ is not a root?

6.10 Suppose that

$$[E_\alpha, E_\beta] = N E_{\alpha+\beta} \tag{6.423}$$

for some $N \neq 0$. Calculate

$$[E_\alpha, E_{-\alpha-\beta}]. \tag{6.424}$$

6.11   Compute the structure constants f_{147} and f_{458} of the algebra of generators of the SU(3) group.

6.12   Show that λ_2, λ_5 and λ_7 generate an su(2) subalgebra of the su(3) Lie algebra.

6.13   Suppose X_a are $N \times N$ matrices satisfying

$$[X_a, X_b] = if_{abc}X_c, \tag{6.425}$$

let $b_i^\dagger$, for $i = 1, \ldots, N$, be a set of creation operators, and, similarly, let b_i, again for $i = 1, \ldots, N$, be annihilation operators satisfying

$$[b_i, b_j] = [b_i^\dagger, b_j^\dagger] = 0, \quad [b_i, b_j^\dagger] = \delta_{ij}. \tag{6.426}$$

Show that the operators

$$\chi_a \equiv \sum_{i,j=1}^{N} (X_a)_{ij} b_i^\dagger b_j \tag{6.427}$$

satisfy

$$[\chi_a, \chi_b] = if_{abc}\chi_c. \tag{6.428}$$

6.14   su(2) **representations.** Decompose the following tensor products of spin representations into a direct sum of spin representations.
   (i) $1 \otimes \frac{1}{2}$
   (ii) $1 \otimes \frac{1}{2} \otimes 1$
   (iii) $\underbrace{\frac{1}{2} \otimes \frac{1}{2} \otimes \frac{1}{2} \otimes \ldots \otimes \frac{1}{2}}_{n \text{ factors}}$

6.15   su(3) **representations.** Decompose the tensor product $\mathbf{3} \otimes \mathbf{6}$ of su(3) representations
   (i) with tensor methods, by manipulating the tensor product $u^i v^{jk}$,
   (ii) with the help of Young tableaux.

6.16   Check that the fundamental roots μ^i satisfy

$$2\frac{\alpha^i \cdot \mu^j}{(\alpha^i)^2} = \delta^{ij}.$$

6.17   Construct the Young tableaux of the following su(4) representations and compute their dimensions.
   (i) $(1, 0, 0)$
   (ii) $(1, 1, 1)$
   (iii) $(2, 1, 2)$

6.18   Decompose the following tensor product of su(5) irreducible representations. Check that the dimensions multiply/add up and match correctly.

$$(0, 0, 1, 0) \otimes (1, 0, 0, 0) = ? \tag{6.429}$$

6.19 Show that the weight vectors defined in Eq. (6.301) satisfy

$$\omega^i \cdot \omega^j = \frac{1}{2}\left(\delta_{ij} - \frac{1}{N}\right) = \begin{cases} \frac{N-1}{2N} & \text{if } i = j \\ -\frac{1}{2N} & \text{if } i \neq j \end{cases}. \qquad (6.430)$$

6.20 Given the weight vectors ω^i of a generic su(N) Lie algebra, show that the simple roots are

$$\alpha^i \equiv \omega^i - \omega^{i+1}, \quad \text{for } i = 1, \ldots, N-1, \qquad (6.431)$$

and that their explicit form, in the representation defined in Section 6.9, is:

$$\alpha^1 = (1, 0, 0, 0, \ldots, 0, 0),$$

$$\alpha^2 = \left(-\frac{1}{2}, \frac{\sqrt{3}}{2}, 0, 0, \ldots, 0, 0\right),$$

$$\alpha^3 = \left(0, -\frac{1}{\sqrt{3}}, \sqrt{\frac{2}{3}}, 0, \ldots, 0, 0\right),$$

$$\vdots$$

$$\alpha^m = \left(0, 0, 0, 0, \ldots, -\sqrt{\frac{m-1}{2m}}, -\sqrt{\frac{m+1}{2m}}, \ldots, 0, 0\right),$$

$$\vdots$$

$$\alpha^{N-1} = \left(0, 0, 0, 0, \ldots, -\sqrt{\frac{N-2}{2(N-1)}}, -\sqrt{\frac{N}{2(N-1)}}\right). \qquad (6.432)$$

6.21 The relation between the spacetime coordinates in an inertial reference frame A, and those in another inertial reference frame B, moving at constant velocity v along the x_A^1 axis, can be expressed as

$$\begin{cases} t_B = \dfrac{t_A}{\sqrt{1-\frac{v^2}{c^2}}} - \dfrac{vx_A^1}{c\sqrt{c^2-v^2}} \\ x_B^1 = \dfrac{x_A^1}{\sqrt{1-\frac{v^2}{c^2}}} - \dfrac{vt_A}{\sqrt{1-\frac{v^2}{c^2}}} \\ x_B^2 = x_A^2 \\ x_B^3 = x_A^3 \end{cases}, \qquad (6.433)$$

where c denotes the speed of light in vacuum.

(i) After introducing suitable four-vectors with homogeneous components (i.e., with components that can be measured in the same physical units) to express the coordinates in the two frames, write the relations in Eqs. (6.433) in terms of a suitable matrix Λ relating the two four-vectors.

(ii) Work out the relation between the components of the velocities measured in the inertial frames A and B.

(iii) Given a pointlike particle moving at constant velocity V along the x_B axis, write the relation between the spacetime position four-vectors in the reference frame of the particle, to be denoted by D, and in the reference frame B.

(iv) Prove that the relation between the coordinates in the reference frame D and in the reference frame A can be expressed by a boost, characterized by a velocity consistent with the relativistic composition law of velocities derived above.

(v) Do the boosts along a given direction form a group?

Appendix A **Problem Solutions**

A.1 A Note on the Problem Solutions

The next sections present the solutions for a (proper) subset of the problems proposed in this book. The solutions are not necessarily unique.

Additional solutions are made available to instructors on the publisher's website.

A.2 Solutions to Problems of Chapter 2

2.1 We note that the elements of G, i.e., the 3×3 matrices

$$e = \begin{pmatrix} 1 & 0 & 0 \\ 0 & 1 & 0 \\ 0 & 0 & 1 \end{pmatrix}, \quad x_1 = \begin{pmatrix} 0 & 1 & 0 \\ 1 & 0 & 0 \\ 0 & 0 & 1 \end{pmatrix}, \quad x_2 = \begin{pmatrix} 0 & 0 & 1 \\ 0 & 1 & 0 \\ 1 & 0 & 0 \end{pmatrix},$$

$$x_3 = \begin{pmatrix} 1 & 0 & 0 \\ 0 & 0 & 1 \\ 0 & 1 & 0 \end{pmatrix}, \quad x_4 = \begin{pmatrix} 0 & 1 & 0 \\ 0 & 0 & 1 \\ 1 & 0 & 0 \end{pmatrix}, \quad x_5 = \begin{pmatrix} 0 & 0 & 1 \\ 1 & 0 & 0 \\ 0 & 1 & 0 \end{pmatrix},$$

have exactly one nonvanishing element, equal to 1, in each row and in each column. Therefore, it is easy to verify that multiplying them by the three vectors

$$v_1 = \begin{pmatrix} 1 \\ 0 \\ 0 \end{pmatrix}, \quad v_2 = \begin{pmatrix} 0 \\ 1 \\ 0 \end{pmatrix}, \quad v_3 = \begin{pmatrix} 0 \\ 0 \\ 1 \end{pmatrix}$$

yields one of these vectors again. This suggests that the elements of G may be interpreted as *permutations* of the three elements of the set $\{v_1, v_2, v_3\}$, i.e., that they are isomorphic to the group of permutations of three elements, the symmetric group S_3. This can be verified explicitly by computing the products of these matrices: Besides the products involving the e matrix as a factor, which are trivial, as e is the 3×3 identity matrix, and thus $ex_i = x_i e = x_i$ for every $i \in \{1, \ldots, 5\}$, one has, for example:

$$x_1 x_1 = \begin{pmatrix} 0 & 1 & 0 \\ 1 & 0 & 0 \\ 0 & 0 & 1 \end{pmatrix} \begin{pmatrix} 0 & 1 & 0 \\ 1 & 0 & 0 \\ 0 & 0 & 1 \end{pmatrix} = \begin{pmatrix} 1 & 0 & 0 \\ 0 & 1 & 0 \\ 0 & 0 & 1 \end{pmatrix} = e,$$

$$x_1 x_2 = \begin{pmatrix} 0 & 1 & 0 \\ 1 & 0 & 0 \\ 0 & 0 & 1 \end{pmatrix} \begin{pmatrix} 0 & 0 & 1 \\ 0 & 1 & 0 \\ 1 & 0 & 0 \end{pmatrix} = \begin{pmatrix} 0 & 1 & 0 \\ 0 & 0 & 1 \\ 1 & 0 & 0 \end{pmatrix} = x_4,$$

etc. From the matrix products one can build the Cayley table:

	e	x_1	x_2	x_3	x_4	x_5
e	e	x_1	x_2	x_3	x_4	x_5
x_1	x_1	e	x_4	x_5	x_2	x_3
x_2	x_2	x_5	e	x_4	x_3	x_1
x_3	x_3	x_4	x_5	e	x_1	x_2
x_4	x_4	x_3	x_1	x_2	x_5	e
x_5	x_5	x_2	x_3	x_1	e	x_4

$$\text{(A.1)}$$

We note that, by defining the bijection $f: G \to \{e, a, b, aba, ab, ba\}$ such that

$$
f: \begin{cases}
e \to e \\
x_1 \to a \\
x_2 \to b \\
x_3 \to aba \\
x_4 \to ab \\
x_5 \to ba
\end{cases},
\tag{A.2}
$$

the Cayley table (A.1) is mapped to the Cayley table of S_3 (2.12), proving that the group G is isomorphic to the symmetric group S_3.

Note that the isomorphism is not unique; for example, the bijection $g: G \to S_3$ defined as

$$
g: \begin{cases}
e \to e \\
x_1 \to b \\
x_2 \to a \\
x_3 \to aba \\
x_4 \to ba \\
x_5 \to ab
\end{cases},
\tag{A.3}
$$

which differs from f by the fact that it interchanges the rôles of a and b (note that $aba = bab$) is also an isomorphism between G and S_3.

2.3   The sphere of unit radius, centered at the origin of the three-dimensional real vector space, is the set of points (x, y, z) whose distance from the origin equals one:

$$S^2 = \{(x, y, z) \mid x^2 + y^2 + z^2 = 1\}. \tag{A.4}$$

Since orthogonal transformations in three dimensions preserve $x^2 + y^2 + z^2$, the group of transformations that leave S^2 invariant is O(3).

The sphere of unit radius, centered at the origin of three-dimensional space, with a point on its surface, to be denoted as $S^2_{(1)}$ (that we can call a *punctured sphere*) is, in general, not invariant under a generic transformation of O(3), because orthogonal transformations do not leave the particular point (sometimes called the *puncture*) invariant. However, the subset of orthogonal transformations that act as rotations only on the directions perpendicular to the vector that defines the position of the puncture on $S^2_{(1)}$ do leave $S^2_{(1)}$ invariant. These transformations form the O(2) subgroup of O(3).

2.5 O(2) is a non-Abelian group. To prove it, consider, for example, the O(2) elements represented by the matrices

$$a = \begin{pmatrix} 1 & 0 \\ 0 & -1 \end{pmatrix}, \tag{A.5}$$

which can be interpreted as a transformation of the reference frame in $\mathbb{R}^2$ leaving the first axis unchanged and flipping the orientation of the second axis, and

$$b = \frac{1}{\sqrt{2}} \begin{pmatrix} 1 & -1 \\ 1 & 1 \end{pmatrix}, \tag{A.6}$$

which describes a rotation of the reference frame in $\mathbb{R}^2$ by $\pi/4$. One can easily verify that a and b do not commute:

$$ab = \begin{pmatrix} 1 & 0 \\ 0 & -1 \end{pmatrix} \frac{1}{\sqrt{2}} \begin{pmatrix} 1 & -1 \\ 1 & 1 \end{pmatrix} = \frac{1}{\sqrt{2}} \begin{pmatrix} 1 & -1 \\ -1 & -1 \end{pmatrix}, \tag{A.7}$$

$$ba = \frac{1}{\sqrt{2}} \begin{pmatrix} 1 & -1 \\ 1 & 1 \end{pmatrix} \begin{pmatrix} 1 & 0 \\ 0 & -1 \end{pmatrix} = \frac{1}{\sqrt{2}} \begin{pmatrix} 1 & 1 \\ 1 & -1 \end{pmatrix}, \tag{A.8}$$

hence $ab \neq ba$, thus O(2) is a non-Abelian group.

2.6 The symmetric group of a set of N elements, S_N, is the set of permutations of N elements. Each permutation $P \in S_N$ can be uniquely identified by a $2 \times N$ matrix of the form

$$P = \begin{pmatrix} 1 & 2 & 3 & \cdots & N \\ p_1 & p_2 & p_3 & \cdots & p_N \end{pmatrix}, \tag{A.9}$$

in which both rows contain all natural numbers from 1 to N, but, while in the first row they are listed in strictly increasing order, each element of the second row, generically denoted as p_i, is the ith element in the permutation. To determine the order of S_N, we count the number of such matrices. As the first row is fixed, this boils down to counting the number of possible choices for the elements of the second row. We note that, when starting to fill the second row of P, the p_1 element can be any number from 1 to N, i.e., there are N different ways to choose p_1. Then, the p_2 element can be any number from 1 to N, except for the one that has already been chosen for p_1, because all values from 1 to N must appear in the N entries of the second row, i.e., there cannot be any repeated values. This means that, after having already chosen p_1, there remain $N-1$ possible ways to choose p_2. Next, p_3 can take any value between 1 and N, except for the two values, which are necessarily different from each other, that have already been chosen for p_1 and for p_2; this leaves $N-2$ possible ways to choose p_3. Continuing in this way, the number of possible choices to fill a generic p_i entry in the second row is $N + 1 - i$, and, in particular, there is only one way to fill the last entry, corresponding to the only integer number in $\{1, 2, \ldots, N\}$ that at that point has not been chosen yet. We deduce that the total number of possible ways to fill the P matrices, i.e., the number of permutations of N elements, which is the order of S_N, is

$$|S_N| = N \cdot (N-1) \cdot (N-2) \cdots 2 \cdot 1 = N! . \tag{A.10}$$

2.8   Let us denote the elements of $\mathbb{Z}_3$ as e (the identity element), a and b. Recalling the Cayley table (2.6):

$$
\begin{array}{c|ccc}
 & e & a & b \\
\hline
e & e & a & b \\
a & a & b & e \\
b & b & e & a
\end{array},
$$

we note that $a^2 = b$, and $b^2 = a$. To prove that $\mathbb{Z}_3$ does not have proper subgroups, we observe that the order of every subgroup must necessarily be larger than or equal to 1, as every subgroup, being a group, must at least include the identity element. Hence, the only subgroup of order 1 must be $\{e\}$, which is a trivial subgroup. Moreover, the order of every subgroup must be smaller or equal to 3, which is the order of the group. The only subgroup of order 3 is the group itself, and is a trivial subgroup. Thus, the only possible nontrivial subgroups of $\mathbb{Z}_3$ should be of order 2, and should include the identity element. There are only two such subsets: $\{e, a\}$ and $\{e, b\}$. However, neither of them is a group, because the they are not closed under the group product; for $\{e, a\}$, we have $a^2 = b$, which is not an element of the subset, whereas for $\{e, b\}$, we have $b^2 = a$, which is not an element of the subset. We conclude that $\mathbb{Z}_3$ does not have any proper subgroup.

2.9   The groups $G = (\mathrm{Mat}(n, \mathbb{R}), +)$ and $J = \left(\mathbb{R}^{n^2}, +\right)$ can be proved to be isomorphic as follows. The elements of G are $n \times n$ matrices with real entries, that can be written as

$$
A = \begin{pmatrix}
a_{1,1} & a_{1,2} & \cdots & a_{1,n} \\
a_{2,1} & a_{2,2} & \cdots & a_{2,n} \\
\cdots & \cdots & \cdots & \cdots \\
a_{n,1} & a_{n,2} & \cdots & a_{n,n}
\end{pmatrix}. \tag{A.11}
$$

The group product operation in G, i.e., matrix addition, acts by element-wise addition of matrices. Conversely, the elements of J can be represented as n^2-component real vectors of the form

$$
v = \begin{pmatrix}
v_1 \\
v_2 \\
\cdots \\
v_{n^2}
\end{pmatrix}. \tag{A.12}
$$

Also in this case, the group product operation in J acts by element-wise addition of vectors. The function

$$
f : \mathrm{Mat}(n, \mathbb{R}) \rightarrow \mathbb{R}^{n^2}, \qquad A \rightarrow f(A), \tag{A.13}
$$

where $f(A)$ is a vector whose entry of index $(i-1)n + j$ is defined to be $a_{i,j}$, is an injection and a surjection, and preserves the group product; hence it is a group isomorphism. From the existence of a group isomorphism between $(\mathrm{Mat}(n, \mathbb{R}), +)$ and $(\mathbb{R}^{n^2}, +)$, it follows that the two groups are isomorphic.

2.12   The problem asks to define three different functions $f_i : \mathbb{R} \rightarrow \mathbb{R}$, with $i \in \{1, 2, 3\}$ in such a way that f_1 is an injection but not a surjection, f_2 is a surjection but not an injection, and f_3 is a bijection. Clearly, there are infinitely many examples of functions with the required properties, including the examples listed below.

- As an example of a real function that is an injection but not a surjection, one can take

$$f_1 : \mathbb{R} \to \mathbb{R}, \qquad x \to \exp(x), \tag{A.14}$$

which is an injection because the exponential is a strictly increasing function of its argument, hence $x < y$ implies $f(x) < f(y)$, i.e., $f(x) \neq f(y)$, but it is not a surjection, because no negative number is the exponential of a real number.

- As an example of a real function that is surjection but not an injection, one can choose

$$f_2 : \mathbb{R} \to \mathbb{R}, \qquad x \to x^3 - x. \tag{A.15}$$

Being a continuous real-valued function that is both unbounded from below and unbounded from above, f_2 is a surjection. However, f_2 is not an injection because, for example, $f(0) = f(1)$.

- As an example of a real function that is bijection, one can take

$$f_3 : \mathbb{R} \to \mathbb{R}, \qquad x \to x^3. \tag{A.16}$$

Like f_2, also this is a continuous real-valued function that is unbounded from below and from above and takes all values in $\mathbb{R}$, i.e., f_3 is a surjection. In addition, f_3 is also strictly increasing for all points in its domain, except at $x = 0$ (where it is stationary), hence for every $x < y$ one has $f_3(x) < f_3(y)$, implying that f_3 is an injection. Being simultaneously an injection and a surjection, f_3 is a bijection.

2.13   The function $f : \mathbb{R} \to \mathbb{R}^3$ is defined as

$$f : x \to \begin{pmatrix} f_1(x) \\ f_2(x) \\ f_3(x) \end{pmatrix}, \tag{A.17}$$

where $f_1 : \mathbb{R} \to \mathbb{R}$ is an injection but not a surjection, $f_2 : \mathbb{R} \to \mathbb{R}$ is a surjection but not an injection, and $f_3 : \mathbb{R} \to \mathbb{R}$ is a bijection. We can make the following observations.

- f is an injection, because, for every $x \neq y \in \mathbb{R}$, one necessarily has $f(x) \neq f(y)$, because the first component of $f(x)$, which is $f_1(x)$, is different from the first component of $f(y)$, which is $f_1(y)$, and f_1 is an injection.
- f is not a surjection. To prove this, consider a given $x \in \mathbb{R}$, and the element of $\mathbb{R}^3$ defined as

$$\vec{v} = \begin{pmatrix} f_1(x) \\ f_2(x) + 1 \\ f_3(x) \end{pmatrix}. \tag{A.18}$$

There exists no element $y \in \mathbb{R}$ such that $f(y) = \vec{v}$, because, if it existed, it would be either equal to x or different from x. If y were equal to x, then $f(y)$

would be equal to $f(x)$, but then, in particular, the second components of $f(x)$ and $f(y)$ should be equal, $f_2(x) = f_2(x) + 1$, which is absurd. If y were different from x, then the first component of $f(y)$ should be equal to $f_1(y)$, but, from the definition of $\vec{v}$, it should also be equal to $f_1(x)$, in contradiction with the assumption that f_1 is an injection. Therefore we conclude that not every element of the co-domain ($\mathbb{R}^3$) of f is the image of some element of its domain ($\mathbb{R}$), namely f is not a surjection. In fact, f cannot be a surjection because it is a mapping from a lower-dimensional to a higher-dimensional real vector space.

- f is not a bijection because it is an injection but not a surjection.

2.14   We consider the group of permutations of a set $X = \{1, 2, 3, 4, 5, 6\}$, which has six elements, denoting the trivial permutation as E. Remembering that in each product the permutation on the right is the one that acts first, the products can be worked out as follows.

- The product $(235)(46) \cdot (14)(265)$ can be expanded as

$$(235)(46) \cdot (14)(265) = \begin{pmatrix} 1 & 2 & 3 & 4 & 5 & 6 \\ 1 & 3 & 5 & 6 & 2 & 4 \end{pmatrix} \cdot \begin{pmatrix} 1 & 2 & 3 & 4 & 5 & 6 \\ 4 & 6 & 3 & 1 & 2 & 5 \end{pmatrix}$$
$$= \begin{pmatrix} 1 & 2 & 3 & 4 & 5 & 6 \\ 6 & 4 & 5 & 1 & 3 & 2 \end{pmatrix}$$
$$= (1624)(35). \tag{A.19}$$

- The product $(1635)(24) \cdot (1536)(24)$ can be expanded as

$$(1635)(24) \cdot (1536)(24) = \begin{pmatrix} 1 & 2 & 3 & 4 & 5 & 6 \\ 6 & 4 & 5 & 2 & 1 & 3 \end{pmatrix} \cdot \begin{pmatrix} 1 & 2 & 3 & 4 & 5 & 6 \\ 5 & 4 & 6 & 2 & 3 & 1 \end{pmatrix}$$
$$= \begin{pmatrix} 1 & 2 & 3 & 4 & 5 & 6 \\ 1 & 2 & 3 & 4 & 5 & 6 \end{pmatrix}$$
$$= E. \tag{A.20}$$

- For the $(26)(35) \cdot (24536)$ product, we obtain

$$(26)(35) \cdot (24536) = \begin{pmatrix} 1 & 2 & 3 & 4 & 5 & 6 \\ 1 & 6 & 5 & 4 & 3 & 2 \end{pmatrix} \cdot \begin{pmatrix} 1 & 2 & 3 & 4 & 5 & 6 \\ 1 & 4 & 6 & 5 & 3 & 2 \end{pmatrix}$$
$$= \begin{pmatrix} 1 & 2 & 3 & 4 & 5 & 6 \\ 1 & 4 & 2 & 3 & 5 & 6 \end{pmatrix}$$
$$= (243). \tag{A.21}$$

- The $(165)(23) \cdot (13)(26)$ product can be expanded as

$$(165)(23) \cdot (13)(26) = \begin{pmatrix} 1 & 2 & 3 & 4 & 5 & 6 \\ 6 & 3 & 2 & 4 & 1 & 5 \end{pmatrix} \cdot \begin{pmatrix} 1 & 2 & 3 & 4 & 5 & 6 \\ 3 & 6 & 1 & 4 & 5 & 2 \end{pmatrix}$$
$$= \begin{pmatrix} 1 & 2 & 3 & 4 & 5 & 6 \\ 2 & 5 & 6 & 4 & 1 & 3 \end{pmatrix}$$
$$= (125)(36). \tag{A.22}$$

• Finally, for the $(1326) \cdot (154)$ product, one obtains

$$(1326) \cdot (154) = \begin{pmatrix} 1 & 2 & 3 & 4 & 5 & 6 \\ 3 & 6 & 2 & 4 & 5 & 1 \end{pmatrix} \cdot \begin{pmatrix} 1 & 2 & 3 & 4 & 5 & 6 \\ 5 & 2 & 3 & 1 & 4 & 6 \end{pmatrix}$$

$$= \begin{pmatrix} 1 & 2 & 3 & 4 & 5 & 6 \\ 5 & 6 & 2 & 3 & 4 & 1 \end{pmatrix}$$

$$= (154326). \tag{A.23}$$

2.16 Given the $(n+1)$-dimensional real vector space without the point at the origin, $A = \mathbb{R}^{n+1}\backslash\{0\}$, consider the equivalence relation $\sim$ among its elements that is defined as

$$a \sim b \Leftrightarrow \exists k \in \mathbb{R} \mid a = kb. \tag{A.24}$$

Note that, since both a and b are different from zero, k, too, is different from zero, otherwise, taking the modulus of both terms in the equality, one would have $0 \neq |a| = |kb| = 0$, which would be absurd, and, as a consequence, k^{-1} exists. The equivalence relation (A.24) can be interpreted by saying that a is a (nonzero) element of the vector space spanned by b. We note that $\sim$ is obviously a reflexive relation: $\forall a \in A$, one has $a = ka$ for $k = 1$. We can also say that $\sim$ is a symmetric relation: If $a \sim b$, then $a = kb$; using the fact that k^{-1} exists, one can multiply this equality by k^{-1}, finding $k^{-1}a = b$, which proves that also $b \sim a$. Finally, $\sim$ is a transitive relation: If $a \sim b$ and $b \sim c$, then there exist real numbers k and q such that $a = kb$ and $b = qc$; but then, substituting the latter equation into the former, one obtains $a = kqc$, where kq is a real number, that proves that $a \sim c$. Being simultaneously reflexive, symmetric, and transitive, $\sim$ is an equivalence relation among elements of A; as such, it partitions A into mutually disjoint equivalence classes. To choose a representative of each equivalence class, one can first consider the elements of A having norm equal to 1, $|a| = 1$; note that, by the definition of the n-dimensional unit sphere S^n, these elements of A are also elements of S^n. Each equivalence class of $\sim$ contains exactly two, distinct, such elements, which are opposite to each other. As representative of each equivalence class one can then choose, for example, the element of the pair of opposite elements with unit norm whose first nonvanishing component in a coordinate basis for $\mathbb{R}^{n+1}$ is positive (a and $-a$ are nonzero, hence they have at least one nonvanishing component, and the values of such component for a and for $-a$ must necessarily be opposite of each other, like all the other components). Thus the quotient space $X = A/\sim$ is the set of equivalence classes for which we have chosen as representative one element of each pair of opposite points of S^n, i.e., $X = A/\sim$ coincides with the n-dimensional unit sphere with antipodal points identified.

2.23 The inverse of each permutation can be obtained by reversing the order of the elements in each cycle (and then, possibly, rearranging each cycle by a cyclic permutation of its elements to make the cycle start from its smallest element, and ordering the cycles in increasing order of their first element). Moreover, the inverse of a product of permutations is given by the product of the inverses of the factors, in the *inverse order*. In this problem, we consider the permutations of the five-element set $X = \{1, 2, 3, 4, 5\}$.

- Given $X = (24)(35)$, the inverse permutation can be written as $X^{-1} = (42)$ (53), or, reordering the elements in each cycle by cycling permutations: $X^{-1} = (24)(35)$, which turns out to be equal to X. Hence $X^{-1} = X$ or, equivalently, $X^2 = E$, where E denotes the identity permutation.
- Given $Y = (14235)$, the inverse permutation is $Y^{-1} = (53241)$, or, reordering the elements by a cycling permutation: $Y^{-1} = (15324)$.
- Given $Z = (1)(2)(34)$, the inverse permutation can be written as $Z^{-1} = (1)$ $(2)(43)$, or, reordering the elements in each cycle by cycling permutations: $Z^{-1} = (1)(2)(34) = Z$.
- Given $T = (253)$, the inverse permutation is given by $T^{-1} = (352)$, or, reordering the elements by a cycling permutation: $T^{-1} = (235)$.
- Given the product of permutations $W \cdot R = (15)(24) \cdot (154)(23)$, the inverse permutation is $(W \cdot R)^{-1} = R^{-1} \cdot W^{-1} = (451)(32) \cdot (51)(42) = (145)(23) \cdot (15)(24)$. Note that, by explicitly working out the permutation product (remembering that the permutation on the right is the one that acts first), one finds $W \cdot R = (2345)$, and thus $(W \cdot R)^{-1} = (5432) = (2543)$, which is indeed equal to $(145)(23) \cdot (15)(24)$.

A.3   Solutions to Problems of Chapter 3

3.1   A linear transformation $f : V \to W$ between two (complex) vector spaces V and W is characterized by the fact that, for every $\vec{v}_1$ and $\vec{v}_2$ in V, and for every α_1 and α_2 in $\mathbb{C}$, one has

$$f(\alpha_1 \vec{v}_1 + \alpha_2 \vec{v}_2) = \alpha_1 f(\vec{v}_1) + \alpha_2 f(\vec{v}_2). \tag{A.25}$$

According to Eq. (3.2), the image of f is the subset of W defined as

$$\mathrm{Im}\, f = f(V) = \{\vec{w} \in W \mid \exists \vec{v} \in V : f(\vec{v}) = \vec{w}\}.$$

Let $\vec{w}_1$ and $\vec{w}_2$ be two arbitrary elements of $\mathrm{Im}\, f$. Then $\exists \vec{v}_1 \in V$ such that $f(\vec{v}_1) = \vec{w}_1$ and $\exists \vec{v}_2 \in V$ such that $f(\vec{v}_2) = \vec{w}_2$. Now consider two arbitrary complex numbers α_1 and α_2; since V is a vector space, the linear combination $\alpha_1 \vec{v}_1 + \alpha_2 \vec{v}_2$ is still an element of V. Its image through the f function is

$$\begin{aligned} f(\alpha_1 \vec{v}_1 + \alpha_2 \vec{v}_2) &= \alpha_1 f(\vec{v}_1) + \alpha_2 f(\vec{v}_2) \\ &= \alpha_1 \vec{w}_1 + \alpha_2 \vec{w}_2, \end{aligned} \tag{A.26}$$

having used the linearity of f in the first equality, and the fact that $f(\vec{v}_1) = \vec{w}_1$ and $f(\vec{v}_2) = \vec{w}_2$ at the second step. Thus $\alpha_1 \vec{w}_1 + \alpha_2 \vec{w}_2$, which is an arbitrary linear combination of elements of $\mathrm{Im}\, f$, is equal to the image of $\alpha_1 \vec{v}_1 + \alpha_2 \vec{v}_2$, which is an element of V. We conclude that any linear combination of elements of $\mathrm{Im}\, f$ is also an element of $\mathrm{Im}\, f$, i.e., $\mathrm{Im}\, f$ is a vector space.

Similarly, the kernel of f is defined as the subset of V that includes all vectors $\vec{v}$ such that $f(\vec{v}) = \vec{0}$. Given two arbitrary elements $\vec{w}_1$ and $\vec{w}_2$ of $\mathrm{Ker}\, f$ and two arbitrary complex numbers α_1 and α_2, one has

$$f(\alpha_1 \vec{v}_1 + \alpha_2 \vec{v}_2) = \alpha_1 f(\vec{v}_1) + \alpha_2 f(\vec{v}_2)$$
$$= \alpha_1 \vec{0} + \alpha_2 \vec{0}$$
$$= \vec{0}, \tag{A.27}$$

having used first the linearity of f, then the fact that $\vec{v}_1$ and $\vec{v}_2$ are elements of Ker f. Comparing the first and the last term in the chain of equalities (A.27), we conclude that an arbitrary linear combination of elements of Ker f is again an element of Ker f, namely that Ker f is a vector space.

3.4 Let $L(g)$ denote the matrix representing a generic element g of a finite group G in the left-regular representation. According to Eq. (3.70), the entries of $L(g)$, with the row and column indices labeled in terms of group elements i and j, are given by

$$L(g)_{i,j} = \delta_{i,gj}.$$

Thus we have

$$(L(g)L(g'))_{i,k} = \sum_{j \in G} L(g)_{i,j} L(g')_{j,k}$$
$$= \sum_{j \in G} \delta_{i,gj} \delta_{j,g'k}$$
$$= \sum_{j \in G} \delta_{g^{-1}i,j} \delta_{j,g'k}$$
$$= \delta_{g^{-1}i,g'k}$$
$$= \delta_{i,gg'k}$$
$$= (L(gg'))_{i,k}. \tag{A.28}$$

3.8 Let a_{ij}, with i and $j \in \{1, 2, \ldots, N\}$, denote the elements of the linear operator A acting on a vector space V of finite dimension N in a chosen orthonormal basis. Similarly, let b_{kl}, with k and $l \in \{1, 2, \ldots, M\}$, denote the elements of the linear operator B acting on the vector space W of finite dimension M in a chosen orthonormal basis. The tensor product $A \otimes B$ can then be represented as a matrix, acting on a vector space of dimension NM, whose entries are defined as

$$(A \otimes B)_{(i-1)M+k,(j-1)M+l} = a_{ij} b_{kl}. \tag{A.29}$$

Note that Eq. (A.29) can be rewritten as

$$(A \otimes B)_{(i-1)M+k,(j-1)M+l}$$
$$= \sum_{m=1}^{M} \sum_{n=1}^{N} a_{im} \delta_{m,j} \delta_{k,n} b_{nl}$$
$$= \sum_{m=1}^{M} \sum_{n=1}^{N} (a_{im} \delta_{m,j})(\delta_{k,n} b_{nl})$$
$$= \sum_{m=1}^{M} \sum_{n=1}^{N} (A \otimes \mathbb{1})_{(i-1)M+m,(m-1)M+j} (\mathbb{1} \otimes B)_{(k-1)M+n,(n-1)M+l}$$
$$= ((A \otimes \mathbb{1})(\mathbb{1} \otimes B))_{(i-1)M+k,(j-1)M+l}. \tag{A.30}$$

From Eq. (A.30) it follows that

$$\det (A \otimes B) = \det ((A \otimes \mathbb{1}) (\mathbb{1} \otimes B)) = \det (A \otimes \mathbb{1}) \cdot \det (\mathbb{1} \otimes B). \quad (A.31)$$

In turn, we note that $(\mathbb{1} \otimes B)$ has a block-diagonal structure, with N blocks equal to B:

$$(\mathbb{1} \otimes B) = \begin{pmatrix} B & 0 & \dots & 0 \\ 0 & B & \dots & 0 \\ \dots & \dots & \dots & \dots \\ 0 & 0 & \dots & B \end{pmatrix}, \quad (A.32)$$

which implies

$$\det (\mathbb{1} \otimes B) = (\det B)^N. \quad (A.33)$$

Similarly, the $(A \otimes \mathbb{1})$ matrix has the block structure

$$(A \otimes \mathbb{1}) = \begin{pmatrix} a_{11}\mathbb{1} & a_{12}\mathbb{1} & \dots & a_{1N}\mathbb{1} \\ a_{21}\mathbb{1} & a_{22}\mathbb{1} & \dots & a_{2N}\mathbb{1} \\ \dots & \dots & \dots & \dots \\ a_{N1}\mathbb{1} & a_{N2}\mathbb{1} & \dots & a_{NN}\mathbb{1} \end{pmatrix}, \quad (A.34)$$

which, by means of the following permutation of the row and column indices

$$p : \{1, 2, \dots, NM\} \to \{1, 2, \dots, NM\}, \quad (i-1)M+j \to (j-1)M+i, \quad (A.35)$$

can also be brought to a block-diagonal form with M equal blocks

$$\begin{pmatrix} A & 0 & \dots & 0 \\ 0 & A & \dots & 0 \\ \dots & \dots & \dots & \dots \\ 0 & 0 & \dots & A \end{pmatrix}, \quad (A.36)$$

which implies

$$\det (A \otimes \mathbb{1}) = (\det A)^M. \quad (A.37)$$

Combining Eq. (A.31) with Eq. (A.33) and with Eq. (A.37) we finally obtain

$$\det (A \otimes B) = (\det A)^M \cdot (\det B)^N. \quad (A.38)$$

A.4 Solutions to Problems of Chapter 4

4.1   The usual topology in $\mathbb{R}^n$ is defined as the set of Cartesian products of n open intervals and their unions: $\tau = \{ \,]a_1, b_1[\times \dots \times]a_n, b_n[\text{ and their unions}\}$. To prove that $(\mathbb{R}^n, \tau)$ is a Hausdorff space, consider two arbitrary, distinct elements $x = (x_1, x_2, \dots, x_n)$ and $y = (y_1, y_2, \dots, y_n)$ of $\mathbb{R}^n$. Since $x \neq y$, there exists at least one index i, between 1 and n, for which $x_i \neq y_i$. Hence, the quantity defined as

$$d = \max_{1 \le i \le n} |x_i - y_i| \quad (A.39)$$

is strictly positive and so is $r = d/5$. Consider the set defined as

$$N = \,]x_1 - 2r, x_1 + 2r[\,\times\,]x_2 - 2r, x_2 + 2r[\,\times \cdots \times\,]x_n - 2r, x_n + 2r[;$$

since

$$x \in \,]x_1 - r, x_1 + r[\,\times\,]x_2 - r, x_2 + r[\,\times \cdots \times\,]x_n - r, x_n + r[\,\subset N, \quad \text{(A.40)}$$

N is a neighborhood of x. Similarly,

$$M = \,]y_1 - 2r, y_1 + 2r[\,\times\,]y_2 - 2r, y_2 + 2r[\,\times \cdots \times\,]y_n - 2r, y_n + 2r[$$

has the open set $]y_1 - r, y_1 + r[\,\times\,]y_2 - r, y_2 + r[\times \cdots \times\,]y_n - r, y_n + r[$ containing y as a subset, and hence is a neighborhood of y. By construction, N and M are disjoint, $N \cap N' = \emptyset$, hence $\mathbb{R}^n$ with the usual topology is a Hausdorff space.

4.4 To prove that the homeomorphism relation between topological spaces (that we can denote as $\approx$) is an equivalence relation, we prove that it is reflexive, symmetric, and transitive. Let (A, τ), (B, σ), and (C, ρ) be topological spaces. By definition of homeomorphism, two topological spaces are homeomorphic when there exists a continuous and invertible function $f : A \to B$, such that its inverse f^{-1} is also continuous.

To prove that the homeomorphism between topological spaces is a reflexive relation, we note that the identity mapping id $: X \to X$ is a continuous and invertible function that is equal to its inverse; hence there exists a continuous and invertible function mapping (A, τ) to itself, such that its inverse is also continuous, thus $(A, \tau) \approx (A, \tau)$, namely $\approx$ is a reflexive relation.

To prove that the homeomorphism between topological spaces is a symmetric relation, we note that if $(A, \tau) \approx (B, \sigma)$, then there exists a continuous and invertible function $f : A \to B$, such that its inverse f^{-1} is also continuous. Then, defining $g = f^{-1}$, we have that $g : B \to A$ is a continuous and invertible function. Moreover, its inverse $g^{-1} = f$ is also continuous, thus $(A, \tau) \approx (B, \sigma)$ implies that also $(B, \sigma) \approx (A, \tau)$, i.e., $\approx$ is a symmetric relation.

Finally, to prove that the homeomorphism relation between topological spaces is transitive, we note that if $(A, \tau) \approx (B, \sigma)$ and $(B, \sigma) \approx (C, \rho)$, then there exist continuous and invertible functions $f : A \to B$ and $g : B \to C$, such that also their inverse functions f^{-1} and g^{-1} are continuous. Then, the function $h = g \circ f : A \to C$ is a continuous and invertible function, whose inverse is also continuous, therefore $(A, \tau) \approx (C, \rho)$. Thus the homeomorphism relation is a transitive relation.

Being simultaneously reflexive, symmetric, and transitive, the homeomorphism relation is an equivalence relation.

4.12 Given the $(1, 0)$-tensor

$$R = a \frac{\partial}{\partial x^1} + a^2 \frac{\partial}{\partial x^2} + a^3 \frac{\partial}{\partial x^3}$$

and the $(0, 1)$-tensor

$$S = b dx^1 + c dx^2 + d dx^3,$$

the components of the $(1, 1)$-tensor $T = R \otimes S = T^{\mu}{}_{\nu}\frac{\partial}{\partial x^{\mu}} \otimes dx^{\nu}$ can be obtained using Eq. (4.40), i.e., are given by $T^{\mu}{}_{\nu} = R^{\mu}S_{\nu}$. Explicitly, they read

$$
\begin{aligned}
T^1{}_1 &= ab, & T^1{}_2 &= ac, & T^1{}_3 &= ad, \\
T^2{}_1 &= a^2b, & T^2{}_2 &= a^2c, & T^2{}_3 &= a^2d, \\
T^3{}_1 &= a^3b, & T^3{}_2 &= a^3c, & T^3{}_3 &= a^3d.
\end{aligned}
\tag{A.41}
$$

4.14 Consider the smooth vector fields $X = X^{\mu}\partial_{\mu}$, $Y = Y^{\nu}\partial_{\nu}$ and $Z = Z^{\rho}\partial_{\rho}$ on the differentiable manifold M and the smooth function f defined in M.

To prove that $[X, fY] = (Xf)Y + f[X, Y]$, we can, for example, apply $[X, fY]$ to a generic smooth function g defined on M. This gives

$$
\begin{aligned}
[X, fY]g &= XfYg - fYXg \\
&= X^{\mu}\partial_{\mu}\left(fY^{\nu}\partial_{\nu}g\right) - fY^{\nu}\partial_{\nu}\left(X^{\mu}\partial_{\mu}g\right) \\
&= X^{\mu}\left(\partial_{\mu}f\right)Y^{\nu}\left(\partial_{\nu}g\right) + X^{\mu}f\left(\partial_{\mu}Y^{\nu}\right)\left(\partial_{\nu}g\right) + X^{\mu}fY^{\nu}\left(\partial_{\mu}\partial_{\nu}g\right) \\
&\quad - fY^{\mu}\left(\partial_{\mu}X^{\nu}\right)\left(\partial_{\nu}g\right) - fY^{\nu}X^{\mu}\left(\partial_{\mu}\partial_{\nu}g\right) \\
&= \left(X^{\mu}\partial_{\mu}f\right)\left(Y^{\nu}\partial_{\nu}g\right) + f\left(X^{\mu}\partial_{\mu}Y^{\nu} - Y^{\mu}\partial_{\mu}X^{\nu}\right)\partial_{\nu}g \\
&= \left(\left(X^{\mu}\partial_{\mu}f\right)Y^{\nu}\partial_{\nu} + f\left(X^{\mu}\left(\partial_{\mu}Y^{\nu}\right) - Y^{\mu}\left(\partial_{\mu}X^{\nu}\right)\right)\partial_{\nu}\right)g,
\end{aligned}
\tag{A.42}
$$

hence

$$
[X, fY] = (Xf)Y + f[X, Y].
\tag{A.43}
$$

To prove that the commutator of vector fields satisfies the Jacobi identity, we first recall that, in components,

$$
[X, Y] = \left(X^{\mu}(\partial_{\mu}Y^{\nu}) - Y^{\mu}(\partial_{\mu}X^{\nu})\right)\partial_{\nu}.
\tag{A.44}
$$

Let us denote $A^{\nu} = \left(X^{\mu}(\partial_{\mu}Y^{\nu}) - Y^{\mu}(\partial_{\mu}X^{\nu})\right)$, so that $[X, Y] = A^{\nu}\partial_{\nu}$. Then, the double commutator $[Z, [X, Y]]$ can be written as

$$
[Z, [X, Y]] = [Z, A] = \left(Z^{\rho}(\partial_{\rho}A^{\nu}) - A^{\rho}(\partial_{\rho}Z^{\nu})\right)\partial_{\nu},
\tag{A.45}
$$

where

$$
\partial_{\rho}A^{\nu} = (\partial_{\rho}X^{\mu})(\partial_{\mu}Y^{\nu}) + X^{\mu}(\partial_{\rho}\partial_{\mu}Y^{\nu}) - (\partial_{\rho}Y^{\mu})(\partial_{\mu}X^{\nu}) - Y^{\mu}(\partial_{\mu}\partial_{\rho}X^{\nu}),
\tag{A.46}
$$

from which one immediately finds

$$
\begin{aligned}
Z^{\rho}\partial_{\rho}A^{\nu} &= Z^{\rho}(\partial_{\rho}X^{\mu})(\partial_{\mu}Y^{\nu}) + Z^{\rho}X^{\mu}(\partial_{\rho}\partial_{\mu}Y^{\nu}) \\
&\quad - Z^{\rho}(\partial_{\rho}Y^{\mu})(\partial_{\mu}X^{\nu}) - Z^{\rho}Y^{\mu}(\partial_{\mu}\partial_{\rho}X^{\nu}).
\end{aligned}
\tag{A.47}
$$

On the other hand

$$
A^{\rho}(\partial_{\rho}Z^{\nu}) = X^{\mu}(\partial_{\mu}Y^{\rho})(\partial_{\rho}Z^{\nu}) - Y^{\mu}(\partial_{\mu}X^{\rho})(\partial_{\rho}Z^{\nu}).
\tag{A.48}
$$

Hence we get

$$
\begin{aligned}
[Z, [X, Y]] &= \left(Z^{\rho}(\partial_{\rho}X^{\mu})(\partial_{\mu}Y^{\nu}) + Z^{\rho}X^{\mu}(\partial_{\rho}\partial_{\mu}Y^{\nu})\right. \\
&\quad - Z^{\rho}(\partial_{\rho}Y^{\mu})(\partial_{\mu}X^{\nu}) - Z^{\rho}Y^{\mu}(\partial_{\mu}\partial_{\rho}X^{\nu}) \\
&\quad \left. - X^{\mu}(\partial_{\mu}Y^{\rho})(\partial_{\rho}Z^{\nu}) + Y^{\mu}(\partial_{\mu}X^{\rho})(\partial_{\rho}Z^{\nu})\right)\partial_{\nu}.
\end{aligned}
\tag{A.49}
$$

Applying the same reasoning to $[X, [Y, Z]]$ and to $[Y, [Z, X]]$, and denoting

$$[Z, [X, Y]] + [X, [Y, Z]] + [Y, [Z, X]] = S = S^\nu \partial_\nu, \qquad (A.50)$$

we obtain

$$
\begin{aligned}
S^\nu = {} & Z^\rho (\partial_\rho X^\mu)(\partial_\mu Y^\nu) + Y^\rho (\partial_\rho Z^\mu)(\partial_\mu X^\nu) + X^\rho (\partial_\rho Y^\mu)(\partial_\mu Z^\nu) \\
& + Z^\rho X^\mu (\partial_\rho \partial_\mu Y^\nu) + Y^\rho Z^\mu (\partial_\rho \partial_\mu X^\nu) + X^\rho Y^\mu (\partial_\rho \partial_\mu Z^\nu) \\
& - Z^\rho (\partial_\rho Y^\mu)(\partial_\mu X^\nu) - Y^\rho (\partial_\rho X^\mu)(\partial_\mu Z^\nu) - X^\rho (\partial_\rho Z^\mu)(\partial_\mu Y^\nu) \\
& - Z^\rho Y^\mu (\partial_\mu \partial_\rho X^\nu) - Y^\rho X^\mu (\partial_\mu \partial_\rho Z^\nu) - X^\rho Z^\mu (\partial_\mu \partial_\rho Y^\nu) \\
& - X^\mu (\partial_\mu Y^\rho)(\partial_\rho Z^\nu) - Z^\mu (\partial_\mu X^\rho)(\partial_\rho Y^\nu) - Y^\mu (\partial_\mu Z^\rho)(\partial_\rho X^\nu) \\
& + Y^\mu (\partial_\mu X^\rho)(\partial_\rho Z^\nu) + X^\mu (\partial_\mu Z^\rho)(\partial_\rho Y^\nu) + Z^\mu (\partial_\mu Y^\rho)(\partial_\rho X^\nu), \qquad (A.51)
\end{aligned}
$$

where the first term on each line on the right-hand side is one of those contributing to $[Z, [X, Y]]$, while the other two come from $[X, [Y, Z]]$ and $[Y, [Z, X]]$, and are obtained from the first by cyclic permutations of X, Y, and Z. By interchanging the dummy summation indices μ and ρ in the fourth line, in the fifth line and in the sixth line, Eq. (A.51) can be rewritten as

$$
\begin{aligned}
S^\nu = {} & Z^\rho (\partial_\rho X^\mu)(\partial_\mu Y^\nu) + Y^\rho (\partial_\rho Z^\mu)(\partial_\mu X^\nu) + X^\rho (\partial_\rho Y^\mu)(\partial_\mu Z^\nu) \\
& + Z^\rho X^\mu (\partial_\rho \partial_\mu Y^\nu) + Y^\rho Z^\mu (\partial_\rho \partial_\mu X^\nu) + X^\rho Y^\mu (\partial_\rho \partial_\mu Z^\nu) \\
& - Z^\rho (\partial_\rho Y^\mu)(\partial_\mu X^\nu) - Y^\rho (\partial_\rho X^\mu)(\partial_\mu Z^\nu) - X^\rho (\partial_\rho Z^\mu)(\partial_\mu Y^\nu) \\
& - Z^\mu Y^\rho (\partial_\rho \partial_\mu X^\nu) - Y^\mu X^\rho (\partial_\rho \partial_\mu Z^\nu) - X^\mu Z^\rho (\partial_\rho \partial_\mu Y^\nu) \\
& - X^\rho (\partial_\rho Y^\mu)(\partial_\mu Z^\nu) - Z^\rho (\partial_\rho X^\mu)(\partial_\mu Y^\nu) - Y^\rho (\partial_\rho Z^\mu)(\partial_\mu X^\nu) \\
& + Y^\rho (\partial_\rho X^\mu)(\partial_\mu Z^\nu) + X^\rho (\partial_\rho Z^\mu)(\partial_\mu Y^\nu) + Z^\rho (\partial_\rho Y^\mu)(\partial_\mu X^\nu), \qquad (A.52)
\end{aligned}
$$

which vanishes, because on the right-hand side the terms on the first line cancel against those on the fifth line, those on the second line cancel against those on the fourth line, and those on the third line cancel against those on the sixth line. We deduce that $S^\nu = 0$, thus $S^\nu \partial_\nu = 0$, namely

$$[X, [Y, Z]] + [Z, [X, Y]] + [Y, [Z, X]] = 0. \qquad (A.53)$$

4.22 Given the differential forms in $\mathbb{R}^3$

$$\alpha = xdx + ydy + zdz, \qquad \beta = zdx + xdy + ydz, \qquad \gamma = xydz,$$

the exterior derivative of α is

$$
\begin{aligned}
d\alpha &= \frac{\partial \alpha_\mu}{\partial x^\nu} dx^\nu \wedge dx^\mu \\
&= \frac{\partial x^\mu}{\partial x^\nu} dx^\nu \wedge dx^\mu \\
&= \delta^\mu_\nu dx^\nu \wedge dx^\mu \\
&= dx^\mu \wedge dx^\mu \\
&= 0, \qquad (A.54)
\end{aligned}
$$

by the antisymmetry of the $\wedge$ operation. Since $d\alpha$ is identically vanishing, α is closed. In fact, α is also exact, since it can be written as

$$\alpha = df, \qquad \text{with } f = \frac{1}{2}(x^2 + y^2 + z^2).$$

The exterior derivative of γ can be computed by writing $\gamma = \gamma_\mu dx^\mu$ (having identified $x^1 = x$, $x^2 = y$, and $x^3 = z$), where $\gamma_1 = \gamma_2 = 0$ and $\gamma_3 = x^1 x^2$. It follows that

$$
\begin{aligned}
d\gamma &= \frac{\partial \gamma_\mu}{\partial x^\nu} dx^\nu \wedge dx^\mu \\
&= \frac{\partial (xy)}{\partial x} dx \wedge dz + \frac{\partial (xy)}{\partial y} dy \wedge dz \\
&= y dx \wedge dz + x dy \wedge dz.
\end{aligned}
$$

Since $d\gamma$ is not identically vanishing, γ is not closed. Since it is not closed, it cannot be exact, either.

Next, we have

$$
\begin{aligned}
\alpha \wedge \beta &= (xdx + ydy + zdz) \wedge (zdx + xdy + ydz) \\
&= (x^2 - yz)dx \wedge dy + (xy - z^2)dx \wedge dz + (y^2 - xz)dy \wedge dz. \quad \text{(A.55)}
\end{aligned}
$$

Finally,

$$
\begin{aligned}
(\alpha + \gamma) \wedge (\alpha + \gamma) &= (xdx + ydy + (xy + z)dz) \wedge (xdx + ydy + (xy + z)dz) \\
&= xydx \wedge dy + (x^2 y + xz)dx \wedge dz + xydy \wedge dx \\
&\quad (xy^2 + yz)dy \wedge dz + (x^2 y + xz)dz \wedge dx + (xy^2 + yz)dz \wedge dy \\
&= (xy - xy)dx \wedge dy + (x^2 y + xz - x^2 y - xz)dx \wedge dz \\
&\quad + (xy^2 + yz - xy^2 - yz)dy \wedge dz \\
&= 0, \quad \text{(A.56)}
\end{aligned}
$$

having used the fact that $dx \wedge dy = -dy \wedge dx$, etc. In fact, the $\wedge$ product of every form of odd rank with itself vanishes identically.

A.5 Solutions to Problems of Chapter 5

5.1   We parameterize the two-dimensional torus T^2 in terms of the two real variables $x^1 \in [0, 2\pi)$ and $x^2 \in [0, 2\pi)$. Given the embedding of T^2 into the three-dimensional Euclidean space $\mathbb{R}^3$ with the metric $\delta = \text{diag}(1, 1, 1)$ defined as

$$
f : T^2 \mapsto \mathbb{R}^3,
$$

$$
(x^1, x^2) \mapsto \begin{pmatrix} A \cos x^1 + B \cos x^1 \cdot \cos x^2 \\ A \sin x^1 + B \sin x^1 \cdot \cos x^2 \\ B \sin x^2 \end{pmatrix}, \quad \text{(A.57)}
$$

with $0 < B < A$, the components of the induced metric $g = f^* \delta$ on T^2 can be obtained using Eq. (5.34), and in this case read

$$
g_{\mu\nu}(x) = \delta_{\alpha\beta} (f(x)) \frac{\partial f^\alpha}{\partial x^\mu} \frac{\partial f^\beta}{\partial x^\nu} = \frac{\partial f^\alpha}{\partial x^\mu} \frac{\partial f^\alpha}{\partial x^\nu}. \quad \text{(A.58)}
$$

Noting that

$$\frac{\partial f^1}{\partial x^1} = -A \sin x^1 - B \sin x^1 \cos x^2,$$

$$\frac{\partial f^2}{\partial x^1} = A \cos x^1 + B \cos x^1 \cos x^2,$$

$$\frac{\partial f^3}{\partial x^1} = 0,$$

$$\frac{\partial f^1}{\partial x^2} = -B \cos x^1 \sin x^2,$$

$$\frac{\partial f^2}{\partial x^2} = -B \sin x^1 \sin x^2,$$

$$\frac{\partial f^3}{\partial x^2} = B \cos x^2,$$

one obtains

$$
\begin{aligned}
g_{11}(x) &= \frac{\partial f^\alpha}{\partial x^1} \frac{\partial f^\alpha}{\partial x^1} \\
&= \sin^2 x^1 \left(A + B \cos x^2\right)^2 + \cos^2 x^1 \left(A + B \cos x^2\right)^2 \\
&= \left(A + B \cos x^2\right)^2, \tag{A.59}
\end{aligned}
$$

while

$$
\begin{aligned}
g_{22}(x) &= \frac{\partial f^\alpha}{\partial x^2} \frac{\partial f^\alpha}{\partial x^2} \\
&= B^2 \cos^2 x^1 \sin^2 x^2 + B^2 \sin^2 x^1 \sin^2 x^2 + B^2 \cos^2 x^2 \\
&= B^2 \left[\left(\cos^2 x^1 + \sin^2 x^1\right) \sin^2 x^2 + \cos^2 x^2\right] \\
&= B^2 \left[\sin^2 x^2 + \cos^2 x^2\right] \\
&= B^2, \tag{A.60}
\end{aligned}
$$

and finally

$$
\begin{aligned}
g_{12}(x) = g_{21}(x) &= \frac{\partial f^\alpha}{\partial x^1} \frac{\partial f^\alpha}{\partial x^2} \\
&= \sin x^1 \left(A + B \cos x^2\right) B \cos x^1 \sin x^2 \\
&\quad - \cos x^1 \left(A + B \cos x^2\right) B \sin x^1 \sin x^2 \\
&= B \sin x^1 \cos x^1 \sin x^2 \left(A + B \cos x^2 - A - B \cos x^2\right) \\
&= 0. \tag{A.61}
\end{aligned}
$$

Thus, the induced metric reads

$$g(x^1, x^2) = \left(A + B \cos x^2\right)^2 dx^1 \otimes dx^1 + B^2 dx^2 \otimes dx^2. \tag{A.62}$$

5.8   From the metric of the S^2 sphere of unit radius

$$g = d\theta \otimes d\theta + \sin^2 \theta \, d\phi \otimes d\phi, \tag{A.63}$$

we read off the components:

$$g_{\theta\theta} = 1, \tag{A.64}$$

$$g_{\theta\phi} = 0, \tag{A.65}$$

$$g_{\phi\theta} = 0, \tag{A.66}$$

$$g_{\phi\phi} = \sin^2 \theta. \tag{A.67}$$

The only nonvanishing partial derivative is $\partial_\theta g_{\phi\phi} = 2 \sin \theta \cos \theta = \sin(2\theta)$.

The equation defining a Killing vector field is $\mathcal{L}_X g_{\mu\nu} = 0$, namely

$$X^\alpha \partial_\alpha g_{\mu\nu} + \partial_\mu X^\alpha g_{\alpha\nu} + \partial_\nu X^\alpha g_{\mu\alpha} = 0. \tag{A.68}$$

We can verify that L_1, L_2, and L_3 satisfy Eq. (A.68) as follows.

For

$$L_1 = -\cos\phi\partial_\theta + \cot\theta\sin\phi\partial_\phi, \tag{A.69}$$

we have

$$X^\theta = -\cos\phi, \tag{A.70}$$

$$X^\phi = \cot\theta\sin\phi, \tag{A.71}$$

thus

$$\partial_\theta X^\theta = 0, \tag{A.72}$$

$$\partial_\phi X^\theta = \sin\phi, \tag{A.73}$$

$$\partial_\theta X^\phi = -\frac{1}{\sin^2\theta}\sin\phi, \tag{A.74}$$

$$\partial_\phi X^\phi = \cot\theta\cos\phi. \tag{A.75}$$

Therefore for $X = L_1$ the left-hand side of Eq. (A.68), for the different choices of μ and ν, can be rewritten as

$$\mu = \theta, \ \nu = \theta: \quad 0 + 0 \cdot 1 + 0 \cdot 1 = 0, \tag{A.76}$$

$$\mu = \theta, \ \nu = \phi: \quad -\frac{1}{\sin^2\theta}\sin\phi\sin^2\theta + \sin\phi = 0, \tag{A.77}$$

$$\mu = \phi, \ \nu = \theta: \quad \sin\phi - \frac{1}{\sin^2\theta}\sin\phi\sin^2\theta = 0, \tag{A.78}$$

$$\mu = \phi, \ \nu = \phi: \quad -\cos\phi \cdot 2\sin\theta\cos\theta + \sin^2\theta \cdot \frac{\cos\theta}{\sin\theta}\cos\phi$$
$$+ \sin^2\theta \cdot \frac{\cos\theta}{\sin\theta}\cos\phi = 0. \tag{A.79}$$

These relations show that L_1 is a Killing vector field.

Similarly, for $L_2 = \sin\phi\partial_\theta + \cot\theta\cos\phi\partial_\phi$, one has

$$X^\theta = \sin\phi, \tag{A.80}$$

$$X^\phi = \cot\theta\cos\phi, \tag{A.81}$$

therefore

$$\partial_\theta X^\theta = 0, \tag{A.82}$$

$$\partial_\phi X^\theta = \cos\phi, \tag{A.83}$$

$$\partial_\theta X^\phi = -\frac{1}{\sin^2\theta}\cos\phi, \tag{A.84}$$

$$\partial_\phi X^\phi = -\cot\theta\sin\phi. \tag{A.85}$$

Thus, in this case the left-hand side of Eq. (A.68), for the different possible values that μ and ν can take, can be rewritten as

$$\mu=\theta,\ \nu=\theta:\quad 0+0\cdot1+0\cdot1=0, \tag{A.86}$$

$$\mu=\theta,\ \nu=\phi:\quad -\frac{1}{\sin^2\theta}\cos\phi\sin^2\theta+\cos\phi=0, \tag{A.87}$$

$$\mu=\phi,\ \nu=\theta:\quad \cos\phi-\frac{1}{\sin^2\theta}\cos\phi\sin^2\theta=0, \tag{A.88}$$

$$\mu=\phi,\ \nu=\phi:\quad \sin\phi\cdot2\sin\theta\cos\theta-\sin^2\theta\cdot\frac{\cos\theta}{\sin\theta}\sin\phi$$
$$-\sin^2\theta\cdot\frac{\cos\theta}{\sin\theta}\sin\phi=0. \tag{A.89}$$

From these relations we deduce that L_2 is a Killing vector field also.

Finally, for $L_3 = \partial_\phi$ we have

$$X^\theta = 0, \tag{A.90}$$

$$X^\phi = 1, \tag{A.91}$$

from which one obtains

$$\mu=\theta,\ \nu=\theta:\quad 1\cdot\partial_\phi1+(\partial_\theta1)\cdot0+(\partial_\theta1)\cdot0=0, \tag{A.92}$$

$$\mu=\theta,\ \nu=\phi:\quad 1\cdot\partial_\phi0+(\partial_\theta1)\sin^2\theta+(\partial_\theta1)\cdot0=0, \tag{A.93}$$

$$\mu=\phi,\ \nu=\theta:\quad 1\cdot\partial_\phi0+(\partial_\phi1)\cdot0+(\partial_\theta1)\sin^2\theta=0, \tag{A.94}$$

$$\mu=\phi,\ \nu=\phi:\quad 1\cdot\partial_\phi\sin^2\theta+(\partial_\phi1)\cdot\sin^2\theta+(\partial_\phi1)\cdot\sin^2\theta=0, \tag{A.95}$$

thus L_3 is a Killing vector field also.

The commutation relations between these Killing vector fields can be worked out as follows. First, for the commutator of L_1 with L_2, we have

$$[L_1,L_2]=\left[-\cos\phi\frac{\partial}{\partial\theta}+\cot\theta\sin\phi\frac{\partial}{\partial\phi},\ \sin\phi\frac{\partial}{\partial\theta}+\cot\theta\cos\phi\frac{\partial}{\partial\phi}\right]$$
$$=-\sin\phi\cos\phi\frac{\partial^2}{\partial\theta^2}+\cos^2\phi\frac{1}{\sin^2\theta}\frac{\partial}{\partial\phi}-\cos^2\phi\cot\theta\frac{\partial^2}{\partial\theta\partial\phi}$$
$$+\cot\theta\sin\phi\cos\phi\frac{\partial}{\partial\theta}+\cot\theta\sin^2\phi\frac{\partial^2}{\partial\theta\partial\phi}-\cot^2\theta\sin^2\phi\frac{\partial}{\partial\phi}$$
$$+\cot^2\theta\sin\phi\cos\phi\frac{\partial^2}{\partial\phi^2}+\sin\phi\cos\phi\frac{\partial^2}{\partial\theta^2}+\sin^2\phi\frac{1}{\sin^2\theta}\frac{\partial}{\partial\phi}$$
$$-\sin^2\phi\cot\theta\frac{\partial^2}{\partial\theta\partial\phi}+\cot\theta\cos^2\phi\frac{\partial^2}{\partial\theta\partial\phi}-\cot^2\theta\sin\phi\cos\phi\frac{\partial^2}{\partial\phi^2}$$
$$-\cot\theta\cos\phi\sin\phi\frac{\partial}{\partial\theta}-\cot^2\theta\cos^2\phi\frac{\partial}{\partial\phi}. \tag{A.96}$$

Simplifying all terms that cancel against each other, we obtain

$$
\begin{aligned}
[L_1, L_2] &= \left(\cos^2\phi + \sin^2\phi\right)\frac{1}{\sin^2\theta}\frac{\partial}{\partial\phi} - \left(\cos^2\phi + \sin^2\phi\right)\cot^2\theta\frac{\partial}{\partial\phi} \\
&= \left(\frac{1}{\sin^2\theta} - \frac{\cos^2\theta}{\sin^2\theta}\right)\frac{\partial}{\partial\phi} \\
&= \frac{1 - \cos^2\theta}{\sin^2\theta}\frac{\partial}{\partial\phi} \\
&= \frac{\sin^2\theta}{\sin^2\theta}\frac{\partial}{\partial\phi} \\
&= \frac{\partial}{\partial\phi} \\
&= L_3.
\end{aligned}
\tag{A.97}
$$

Next, the commutator of L_1 with L_3 can be worked out as

$$
\begin{aligned}
[L_1, L_3] &= \left[-\cos\phi\frac{\partial}{\partial\theta} + \cot\theta\sin\phi\frac{\partial}{\partial\phi}, \frac{\partial}{\partial\phi}\right] \\
&= -\cos\phi\frac{\partial^2}{\partial\theta\partial\phi} + \cot\theta\sin\phi\frac{\partial^2}{\partial\phi^2} - \sin\phi\frac{\partial}{\partial\theta} \\
&\quad + \cos\phi\frac{\partial^2}{\partial\theta\partial\phi} - \cot\theta\cos\phi\frac{\partial}{\partial\phi} - \cot\theta\sin\phi\frac{\partial^2}{\partial\phi^2} \\
&= -\sin\phi\frac{\partial}{\partial\theta} - \cot\theta\cos\phi\frac{\partial}{\partial\phi} \\
&= -\left(\sin\phi\frac{\partial}{\partial\theta} + \cot\theta\cos\phi\frac{\partial}{\partial\phi}\right) \\
&= -L_2.
\end{aligned}
\tag{A.98}
$$

Finally, we can compute the commutator of L_2 with L_3 as follows:

$$
\begin{aligned}
[L_2, L_3] &= \left[\sin\phi\frac{\partial}{\partial\theta} + \cot\theta\cos\phi\frac{\partial}{\partial\phi}, \frac{\partial}{\partial\phi}\right] \\
&= \sin\phi\frac{\partial^2}{\partial\theta\partial\phi} + \cot\theta\sin\phi\frac{\partial^2}{\partial\phi^2} - \cos\phi\frac{\partial}{\partial\theta} \\
&\quad - \sin\phi\frac{\partial^2}{\partial\theta\partial\phi} + \cot\theta\sin\phi\frac{\partial}{\partial\phi} - \cot\theta\cos\phi\frac{\partial^2}{\partial\phi^2} \\
&= -\cos\phi\frac{\partial}{\partial\theta} + \cot\theta\sin\phi\frac{\partial}{\partial\phi} \\
&= L_1.
\end{aligned}
\tag{A.99}
$$

Equations (A.97), (A.98), and (A.99) can be written concisely as

$$
[L_a, L_b] = \epsilon_{abc}L_c,
\tag{A.100}
$$

i.e., L_1, L_2 and L_3 satisfy the so(3) algebra, thus the symmetry associated with these Killing vector fields is the SO(3) symmetry.

A.6 Solutions to Problems of Chapter 6

6.1 Given the matrix

$$A = \begin{pmatrix} 0 & 0 & 1 \\ 0 & 0 & 0 \\ 1 & 0 & 0 \end{pmatrix}, \tag{A.101}$$

the elements of

$$\exp(i\alpha A) = \sum_{n=0}^{\infty} \frac{i^n \alpha^n A^n}{n!} \tag{A.102}$$

can be obtained by noting that, for every $k \in \mathbb{N}$, one has

$$i^{4k} = 1,$$
$$i^{4k+1} = i,$$
$$i^{4k+2} = -1,$$
$$i^{4k+3} = -i,$$

and by observing that

$$A^2 = \begin{pmatrix} 0 & 0 & 1 \\ 0 & 0 & 0 \\ 1 & 0 & 0 \end{pmatrix}\begin{pmatrix} 0 & 0 & 1 \\ 0 & 0 & 0 \\ 1 & 0 & 0 \end{pmatrix} = \begin{pmatrix} 1 & 0 & 0 \\ 0 & 0 & 0 \\ 0 & 0 & 1 \end{pmatrix}, \tag{A.103}$$

from which follows, again for every $k \in \mathbb{N}$,

$$A^{2k+1} = \begin{pmatrix} 0 & 0 & 1 \\ 0 & 0 & 0 \\ 1 & 0 & 0 \end{pmatrix} \tag{A.104}$$

and

$$A^{2k} = \begin{pmatrix} 1 & 0 & 0 \\ 0 & 0 & 0 \\ 0 & 0 & 1 \end{pmatrix}. \tag{A.105}$$

Thus we obtain

$$\exp(i\alpha A) = \sum_{n=0}^{\infty} \frac{i^n \alpha^n A^n}{n!}$$

$$= \left(\sum_{j=0}^{\infty} \frac{(-1)^j \alpha^{2j}}{(2j)!}\right)\begin{pmatrix} 1 & 0 & 0 \\ 0 & 0 & 0 \\ 0 & 0 & 1 \end{pmatrix} + i\left(\sum_{j=0}^{\infty} \frac{(-1)^j \alpha^{2j+1}}{(2j+1)!}\right)\begin{pmatrix} 0 & 0 & 1 \\ 0 & 0 & 0 \\ 1 & 0 & 0 \end{pmatrix}$$

$$= \cos\alpha \begin{pmatrix} 1 & 0 & 0 \\ 0 & 0 & 0 \\ 0 & 0 & 1 \end{pmatrix} + i\sin\alpha \begin{pmatrix} 0 & 0 & 1 \\ 0 & 0 & 0 \\ 1 & 0 & 0 \end{pmatrix}$$

$$= \begin{pmatrix} \cos\alpha & 0 & i\sin\alpha \\ 0 & 0 & 0 \\ i\sin\alpha & 0 & \cos\alpha \end{pmatrix}. \tag{A.106}$$

6.2   Given two matrices A and B satisfying the commutation relation

$$[A, B] = B, \tag{A.107}$$

and another, arbitrary, matrix C of the same size, we first recall that

$$\begin{aligned}[AC, B] &= ACB - BAC \\ &= ACB - ABC + ABC - BAC \\ &= A[C, B] + [A, B]C. \end{aligned} \tag{A.108}$$

Using Eq. (A.108), we first compute $[A^n, B]$ for a generic natural number n. Let us start by observing the results that one obtains for the lowest values of n. For $n = 0$ we have

$$[A^0, B] = [\mathbb{1}, B] = \mathbb{1}B - B\mathbb{1} = B - B = 0; \tag{A.109}$$

for $n = 1$, instead, we have Eq. (A.107). For $n = 2$ we have

$$\begin{aligned}[A^2, B] &= A[A, B] + [A, B]A \\ &= AB + BA, \end{aligned} \tag{A.110}$$

while for $n = 3$ one obtains

$$\begin{aligned}[A^3, B] &= A[A^2, B] + [A, B]A^2 \\ &= A(AB + BA) + BA^2 \\ &= A^2B + ABA + BA^2, \end{aligned} \tag{A.111}$$

and for $n = 4$ we find

$$\begin{aligned}[A^4, B] &= A[A^3, B] + [A, B]A^3 \\ &= A(A^2B + ABA + BA^2) + BA^3 \\ &= A^3B + A^2BA + ABA^2 + BA^3. \end{aligned} \tag{A.112}$$

From Eq. (A.107), Eq. (A.109), Eq. (A.110), Eq. (A.111), and Eq. (A.112), one may guess

$$[A^n, B] = \sum_{k=0}^{n-1} A^{n-1-k} B A^k. \tag{A.113}$$

To prove that Eq. (A.113) holds in general, we proceed by induction on the natural number n. We first note that Eq. (A.113) is true for $n = 0$, when the sum on the right-hand side does not include any term, and Eq. (A.113) simply reduces to Eq. (A.109). Next, if Eq. (A.113) holds for a value n, then

$$\begin{aligned}[A^{n+1}, B] &= A[A^n, B] + [A, B]A^n \\ &= A\left(\sum_{k=0}^{n-1} A^{n-1-k} B A^k\right) + BA^n \end{aligned}$$

$$= \left(\sum_{k=0}^{n-1} A^{n-k} B A^k \right) + B A^n$$

$$= \sum_{k=0}^{n} A^{n-k} B A^k$$

$$= \sum_{k=0}^{(n+1)-1} A^{(n+1)-1-k} B A^k, \tag{A.114}$$

i.e., if Eq. (A.113) holds for a value n, then it also holds for the value $n + 1$. Thus Eq. (A.113) is true for every natural number n.

Now we consider the $\exp(i\alpha A) B \exp(-i\alpha A)$ product: We use Eq. (A.113) to rewrite $\exp(i\alpha A) B$ in terms of a product in which $\exp(i\alpha A)$ appears as the factor on the right, that will cancel against its inverse, $\exp(-i\alpha A)$. Since

$$\exp(i\alpha A) = \sum_{n=0}^{\infty} \frac{(i\alpha)^n A^n}{n!} \tag{A.115}$$

is defined in terms of a series of nonnegative powers of A, we first consider the steps involved in "moving B to the left" in a generic term of the form $A^n B$. As before, considering the lowest values of n (excluding the trivial one, $n = 0$) and using Eq. (A.113), we note that

$$
\begin{aligned}
AB &= BA + [A, B] \\
&= BA + B \\
&= B(A + \mathbb{1}), \\
A^2 B &= BA^2 + [A^2, B] \\
&= BA^2 + AB + BA \\
&= BA^2 + BA + B + BA \\
&= BA^2 + 2BA + B \\
&= B(A^2 + 2A + \mathbb{1}), \\
A^3 B &= BA^3 + [A^3, B] \\
&= BA^3 + A^2 B + ABA + BA^2 \\
&= BA^3 + BA^2 + 2BA + B + ABA + BA^2 \\
&= BA^3 + 2BA^2 + 2BA + B + (AB)A \\
&= BA^3 + 2BA^2 + 2BA + B + (BA + B)A \\
&= BA^3 + 2BA^2 + 2BA + B + BA^2 + BA \\
&= BA^3 + 3BA^2 + 3BA + B \\
&= B(A^2 + 3A^2 + 3A + \mathbb{1}), \\
A^4 B &= BA^4 + [A^4, B] \\
&= BA^4 + A^3 B + A^2 BA + ABA^2 + BA^3 \\
&= BA^4 + BA^3 + 3BA^2 + 3BA + B \\
&\quad + (BA^2 + 2BA + B)A + (BA + B)A^2 + BA^3
\end{aligned}
$$

$$\tag{A.116}$$
$$\tag{A.117}$$
$$\tag{A.118}$$

$$\begin{aligned}
&= BA^4 + BA^3 + 3BA^2 + 3BA + B + BA^3 + 2BA^2 \\
&\quad + BA + BA^3 + BA^2 + BA^3 \\
&= BA^4 + 4BA^3 + 6BA^2 + 4BA + B \\
&= B(A^4 + 4A^3 + 6A^2 + 4A + \mathbb{1}),
\end{aligned} \tag{A.119}$$

from which one can guess

$$A^n B = B(A + \mathbb{1})^n. \tag{A.120}$$

In order to prove that Eq. (A.120) holds in general, we proceed by induction on the natural number n. We first note that Eq. (A.120) holds for $n = 0$, when it reduces to the tautology $B = B$. Then, assuming that Eq. (A.120) holds for a certain value n, we have

$$\begin{aligned}
A^{n+1} B &= A(A^n B) \\
&= AB(A + \mathbb{1})^n \\
&= (BA + B)(A + \mathbb{1})^n \\
&= B(A + \mathbb{1})(A + \mathbb{1})^n \\
&= B(A + \mathbb{1})^{n+1}.
\end{aligned} \tag{A.121}$$

Thus, assuming that Eq. (A.120) holds for a natural number n, it also holds for $n + 1$. Having already shown that it also holds for $n = 0$, we conclude that Eq. (A.120) holds in general, for every $n \in \mathbb{N}$.

Then, using Eq. (A.120), we can rewrite the product $\exp(i\alpha A)B$ as

$$\begin{aligned}
\exp(i\alpha A)B &= \sum_{n=0}^{\infty} \frac{(i\alpha)^n}{n!} A^n B \\
&= \sum_{n=0}^{\infty} \frac{(i\alpha)^n}{n!} B(A + \mathbb{1})^n \\
&= B \sum_{n=0}^{\infty} \frac{(i\alpha)^n}{n!} (A + \mathbb{1})^n \\
&= B \exp(i\alpha(A + \mathbb{1})) \\
&= B \exp(i\alpha) \exp(i\alpha A).
\end{aligned} \tag{A.122}$$

Finally, using Eq. (A.122), we obtain

$$\begin{aligned}
\exp(i\alpha A)B \exp(-i\alpha A) &= B \exp(i\alpha) \exp(i\alpha A) \exp(-i\alpha A) \\
&= B \exp(i\alpha).
\end{aligned} \tag{A.123}$$

6.12 We recall that, according to Eq. (6.170), the Gell-Mann matrices λ_2, λ_5, and λ_7 are defined as

$$\lambda_2 = \begin{pmatrix} 0 & -i & 0 \\ i & 0 & 0 \\ 0 & 0 & 0 \end{pmatrix}, \quad \lambda_5 = \begin{pmatrix} 0 & 0 & -i \\ 0 & 0 & 0 \\ i & 0 & 0 \end{pmatrix}, \quad \lambda_7 = \begin{pmatrix} 0 & 0 & 0 \\ 0 & 0 & -i \\ 0 & i & 0 \end{pmatrix}. \tag{A.124}$$

It then follows that

$$[\lambda_2, \lambda_5] = \begin{pmatrix} 0 & -i & 0 \\ i & 0 & 0 \\ 0 & 0 & 0 \end{pmatrix} \begin{pmatrix} 0 & 0 & -i \\ 0 & 0 & 0 \\ i & 0 & 0 \end{pmatrix} - \begin{pmatrix} 0 & 0 & -i \\ 0 & 0 & 0 \\ i & 0 & 0 \end{pmatrix} \begin{pmatrix} 0 & -i & 0 \\ i & 0 & 0 \\ 0 & 0 & 0 \end{pmatrix}$$

$$= \begin{pmatrix} 0 & 0 & 0 \\ 0 & 0 & 1 \\ 0 & 0 & 0 \end{pmatrix} - \begin{pmatrix} 0 & 0 & 0 \\ 0 & 0 & 0 \\ 0 & 1 & 0 \end{pmatrix} = \begin{pmatrix} 0 & 0 & 0 \\ 0 & 0 & 1 \\ 0 & -1 & 0 \end{pmatrix}$$

$$= i \begin{pmatrix} 0 & 0 & 0 \\ 0 & 0 & -i \\ 0 & i & 0 \end{pmatrix} = i\lambda_7. \tag{A.125}$$

Similarly,

$$[\lambda_2, \lambda_7] = \begin{pmatrix} 0 & -i & 0 \\ i & 0 & 0 \\ 0 & 0 & 0 \end{pmatrix} \begin{pmatrix} 0 & 0 & 0 \\ 0 & 0 & -i \\ 0 & i & 0 \end{pmatrix} - \begin{pmatrix} 0 & 0 & 0 \\ 0 & 0 & -i \\ 0 & i & 0 \end{pmatrix} \begin{pmatrix} 0 & -i & 0 \\ i & 0 & 0 \\ 0 & 0 & 0 \end{pmatrix}$$

$$= \begin{pmatrix} 0 & 0 & -1 \\ 0 & 0 & 0 \\ 0 & 0 & 0 \end{pmatrix} - \begin{pmatrix} 0 & 0 & 0 \\ 0 & 0 & 0 \\ -1 & 0 & 0 \end{pmatrix} = \begin{pmatrix} 0 & 0 & -1 \\ 0 & 0 & 0 \\ 1 & 0 & 0 \end{pmatrix}$$

$$= -i \begin{pmatrix} 0 & 0 & -i \\ 0 & 0 & 0 \\ i & 0 & 0 \end{pmatrix} = -i\lambda_5. \tag{A.126}$$

Finally,

$$[\lambda_5, \lambda_7] = \begin{pmatrix} 0 & 0 & -i \\ 0 & 0 & 0 \\ i & 0 & 0 \end{pmatrix} \begin{pmatrix} 0 & 0 & 0 \\ 0 & 0 & -i \\ 0 & i & 0 \end{pmatrix} - \begin{pmatrix} 0 & 0 & 0 \\ 0 & 0 & -i \\ 0 & i & 0 \end{pmatrix} \begin{pmatrix} 0 & 0 & -i \\ 0 & 0 & 0 \\ i & 0 & 0 \end{pmatrix}$$

$$= \begin{pmatrix} 0 & 1 & 0 \\ 0 & 0 & 0 \\ 0 & 0 & 0 \end{pmatrix} - \begin{pmatrix} 0 & 0 & 0 \\ 1 & 0 & 0 \\ 0 & 0 & 0 \end{pmatrix} = \begin{pmatrix} 0 & 1 & 0 \\ -1 & 0 & 0 \\ 0 & 0 & 0 \end{pmatrix} =$$

$$= i \begin{pmatrix} 0 & -i & 0 \\ i & 0 & 0 \\ 0 & 0 & 0 \end{pmatrix} = i\lambda_2. \tag{A.127}$$

Defining

$$J_1 = \lambda_2, \quad J_2 = \lambda_5, \quad J_3 = \lambda_7, \tag{A.128}$$

we see that Eqs. (A.125), (A.126), and (A.127) can be rewritten as

$$[J_a, J_b] = i\epsilon_{abc}J_c, \tag{A.129}$$

which shows that λ_2, λ_5 and λ_7 satisfy the commutation rules of an su(2) Lie algebra.

6.14 Tensor products of irreducible representations of the su(2) Lie algebra can be decomposed using Eq. (6.251).

Thus, denoting the representation of spin j as $D^{(j)}$, for $D^{(1)} \otimes D^{(\frac{1}{2})}$ we have

$$D^{(1)} \otimes D^{(\frac{1}{2})} = \sum_{k=|1-\frac{1}{2}|}^{1+\frac{1}{2}} D^{(k)} = \sum_{k=\frac{1}{2}}^{\frac{3}{2}} D^{(k)} = D^{(\frac{1}{2})} \oplus D^{(\frac{3}{2})}. \tag{A.130}$$

Recalling that $\dim D^{(k)} = 2k + 1$, we note that, as expected, the dimensions of the representations in the tensor product match those in the direct sum: $3 \cdot 2 = 2 + 4$.

Next, for $D^{(1)} \otimes D^{(\frac{1}{2})} \otimes D^{(1)}$, we can use the decomposition of $D^{(1)} \otimes D^{(\frac{1}{2})}$ obtained in Eq. (A.130), as well as the associativity, distributivity, and commutativity (up to isomorphisms) of the tensor product, to write

$$\begin{aligned}
D^{(1)} \otimes D^{(\frac{1}{2})} \otimes D^{(1)} &= \left(D^{(1)} \otimes D^{(\frac{1}{2})} \right) \otimes D^{(1)} \\
&= \left(D^{(\frac{1}{2})} \oplus D^{(\frac{3}{2})} \right) \otimes D^{(1)} \\
&= \left(D^{(\frac{1}{2})} \otimes D^{(1)} \right) \oplus \left(D^{(\frac{3}{2})} \otimes D^{(1)} \right) \\
&= \left(D^{(1)} \otimes D^{(\frac{1}{2})} \right) \oplus \left(\sum_{k=|\frac{3}{2}-1|}^{\frac{3}{2}+1} D^{(k)} \right) \\
&= \left(D^{(\frac{1}{2})} \oplus D^{(\frac{3}{2})} \right) \oplus \left(\sum_{k=\frac{1}{2}}^{\frac{5}{2}} D^{(k)} \right) \\
&= \left(D^{(\frac{1}{2})} \oplus D^{(\frac{3}{2})} \right) \oplus \left(D^{(\frac{1}{2})} \oplus D^{(\frac{3}{2})} \oplus D^{(\frac{5}{2})} \right) \\
&= 2D^{(\frac{1}{2})} \oplus 2D^{(\frac{3}{2})} \oplus D^{(\frac{5}{2})}. \tag{A.131}
\end{aligned}$$

As before, the total dimension of the representations in the tensor product equals the sum of those in the direct sum (including their multiplicities): $3 \cdot 2 \cdot 3 = 2 \cdot 2 + 2 \cdot 4 + 6$.

Finally, let us compute the decomposition of a tensor product of the form $\left(\frac{1}{2} \otimes \frac{1}{2} \otimes \frac{1}{2} \otimes \cdots \otimes \frac{1}{2} \right)$ with n factors, which is sometimes denoted as $\left(\frac{1}{2} \right)^{\otimes n}$. Let us consider separately the case in which n is even ($n = 2k$, with $k \in \mathbb{N}$) and the case in which n is odd ($n = 2k + 1$, with $k \in \mathbb{N}$).

For even n, we use the fact that $D^{(\frac{1}{2})} \otimes D^{(\frac{1}{2})} = D^{(0)} \oplus D^{(1)}$, hence

$$\left(\frac{1}{2} \right)^{\otimes 2k} = \left(D^{(\frac{1}{2})} \otimes D^{(\frac{1}{2})} \right)^{\otimes k} = \left(D^{(0)} \oplus D^{(1)} \right)^{\otimes k}. \tag{A.132}$$

As the last term in Eq. (A.132) contains only irreducible representations of integer spin, so does its tensor-sum decomposition. Furthermore, the representation of lowest spin that will appear in the tensor-sum decomposition is $D^{(0)}$, which is obtained, for example, by taking the products of all $D^{(0)}$ factors in the expansion of the right-hand side of Eq. (A.132). Conversely, the representation of highest spin that can appear in the tensor-sum decomposition is $D^{(k)}$, which is obtained, along with lower-spin representations, when taking the product of all $D^{(1)}$ factors in the expansion of the right-hand side of Eq. (A.132). This is a consequence of the fact that the highest spin appearing in the tensor-sum decomposition of the product of representations with spins j_1 and j_2 is $j_1 + j_2$, hence the largest spin value that can appear in the decomposition of the right-hand side of Eq. (A.132) is k. We denote the multiplicities with which the different irreducible representations appear in the decomposition of $\left(\frac{1}{2}\right)^{\otimes 2k}$ as $c_{k,i}$:

$$\left(\frac{1}{2}\right)^{\otimes 2k} = \sum_{i=0}^{k} c_{k,i} D^{(i)}. \tag{A.133}$$

To compute the $c_{k,i}$ multiplicity coefficients, we first establish a recursion relation expressing the coefficients appearing in the decomposition of $\left(\frac{1}{2}\right)^{\otimes 2(k+1)}$ in terms of those appearing in the decomposition of $\left(\frac{1}{2}\right)^{\otimes 2k}$, then we solve the recursion.

To define the recursive relation between the coefficients $c_{k,i}$ and $c_{k+1,i}$, we write $\left(\frac{1}{2}\right)^{\otimes 2(k+1)}$ as the tensor product of $\left(\frac{1}{2}\right)^{\otimes 2k}$ with $\left(\frac{1}{2}\right)^{\otimes 2}$:

$$
\begin{aligned}
\left(\frac{1}{2}\right)^{\otimes 2(k+1)} &= \left(\frac{1}{2}\right)^{\otimes 2k} \otimes \left(\frac{1}{2}\right)^{\otimes 2} \\
&= \left(\sum_{i=0}^{k} c_{k,i} D^{(i)}\right) \otimes \left(D^{(0)} \oplus D^{(1)}\right) \\
&= \left(c_{k,0} D^{(0)} \oplus \sum_{i=1}^{k} c_{k,i} D^{(i)}\right) \otimes \left(D^{(0)} \oplus D^{(1)}\right) \\
&= c_{k,0} D^{(0)} \oplus \sum_{i=1}^{k} c_{k,i} \left(D^{(i)} \otimes D^{(0)}\right) \\
&\quad \oplus c_{k,0} D^{(1)} \oplus \sum_{i=1}^{k} c_{k,i} \left(D^{(i)} \otimes D^{(1)}\right) \\
&= c_{k,0} D^{(0)} \oplus \sum_{i=1}^{k} c_{k,i} D^{(i)} \oplus c_{k,0} D^{(1)} \\
&\quad \oplus \sum_{i=1}^{k} c_{k,i} \left(D^{(i-1)} \oplus D^{(i)} \oplus D^{(i+1)}\right). \tag{A.134}
\end{aligned}
$$

Collecting like terms together, we obtain

$$
\begin{aligned}
\left(\frac{1}{2}\right)^{\otimes 2(k+1)} &= \left(c_{k,0} + c_{k,1}\right) D^{(0)} \oplus \sum_{i=1}^{k-1} \left(c_{k,i-1} + 2c_{k,i} + c_{k,i+1}\right) D^{(i)} \\
&\quad \oplus \left(c_{k,k-1} + 2c_{k,k}\right) D^{(k)} \oplus c_{k,k} D^{(k+1)}, \tag{A.135}
\end{aligned}
$$

from which we read off

$$
\begin{cases}
c_{k+1,0} = c_{k,0} + c_{k,1} \\
c_{k+1,i} = c_{k,i-1} + 2c_{k,i} + c_{k,i+1} & \text{for } 0 < i < k \\
c_{k+1,k} = c_{k,k-1} + 2c_{k,k} \\
c_{k+1,k+1} = c_{k,k}
\end{cases} \tag{A.136}
$$

We now prove that the solution of the system of recursive relations (A.136) is

$$
\begin{cases}
c_{k,i} = P_{2k,k-i} - P_{2k,k-i-1} & \text{for } i < k, \\
c_{k,k} = 1
\end{cases} \tag{A.137}
$$

$$
\begin{array}{ccccccccc}
 & & & & 1 & & & & \\
 & & & 1 & & 1 & & & \\
 & & 1 & & 2 & & 1 & & \\
 & 1 & & 3 & & 3 & & 1 & \\
1 & & 4 & & 6 & & 4 & & 1 \\
\end{array}
$$

1   5   10   10   5   1

1   6   15   20   15   6   1

...   .

Fig. A.1   Pascal's triangle.

where

$$
P_{n,i} = \binom{n}{i} = \frac{n!}{i!\,(n-i)!}, \tag{A.138}
$$

where both i and n take values starting from zero, is the ith element of the nth row of Pascal's triangle, named after Blaise Pascal, which is shown in Fig. A.1. Note that in some countries this triangle is alternatively named after Niccolò Fontana Tartaglia, after Omar Khayyām, after Yang Hui, or after Michael Stifel.

We recall that the nth row, counting from 0 and from the top to the bottom, in Pascal's triangle contains $n + 1$ numbers, which are the coefficients of the monomials appearing in the expansion of the nth power of a binomial:

$$
(x+y)^n = \sum_{p=0}^{n} \frac{n!}{p!\,(n-p)!} x^p y^{n-p}, \tag{A.139}
$$

so that each coefficient in Pascal's triangle counts the number of possible ways to reach it, starting from $P_{0,0} = 1$, by a sequence of n diagonal moves downwards. For example, the second number in the fourth row, i.e., $P_{3,1} = 3$, is the number of ways to reach it, starting from $P_{0,0}$ and going downwards through a sequence of diagonal moves. In this case, the three possible ways to do so are:

1. $P_{0,0} \to P_{1,0} \to P_{2,0} \to P_{3,1}$,
2. $P_{0,0} \to P_{1,0} \to P_{2,1} \to P_{3,1}$,
3. $P_{0,0} \to P_{1,1} \to P_{2,1} \to P_{3,1}$.

Note that Pascal's triangle is symmetric with respect to the vertical axis through $P_{0,0}$ and has the property that each of its elements is the sum of the two immediately above it, $P_{n,p} = P_{n-1,p} + P_{n-1,p-1}$, which is a consequence of the identity

$$
\frac{n!}{p!\,(n-p)!} = \frac{(n-1)!}{p!\,(n-1-p)!} + \frac{(n-1)!}{(p-1)!\,(n-p)!}. \tag{A.140}
$$

Coming back to the problem of computing the decomposition of $\left(\frac{1}{2}\right)^{\otimes n}$ into a tensor sum of representations, we note that, in analogy with the property of the coefficients in Pascal's triangle, the multiplicity by which each irreducible representation appears in the direct sum is given by the number of possible ways it can be obtained by applying (or not applying) a lowering operator, out of n possible times. A crucial difference with respect to Pascal's triangle, however, is that representations can take only nonnegative spin values: Thus, since the maximum spin value appearing in the decomposition of $\left(\frac{1}{2}\right)^{\otimes n}$ is $j_{max} = \frac{n}{2}$, and the spin values appearing in the decomposition differ by integer values, if we would like to represent the multiplicities with which they appear, they could only be represented by $\lceil\frac{n}{2}\rceil$ coefficients, where $\lceil x\rceil$, the *ceiling function* of x, is the least integer greater than or equal to x, i.e., by as many coefficients as those that do not lie strictly to the right of the vertical symmetry axis in Pascal's triangle, whereas coefficients lying on the vertical symmetry axis are allowed. Moreover, the multiplicities must also, in general, differ from the coefficients in Pascal's triangle because the number of ways to reach the location of each coefficient cannot include any path going through at least one element on the right half of the triangle; in the example above, the second coefficient in the fourth row could only be reached through the first or through the second path, because the third path goes through the element $P_{1,1}$, which lies to the right of the symmetry axis. It turns out that this reduces the possible ways of obtaining the irreducible representation $D^{(i)}$ in the decomposition of $\left(\frac{1}{2}\right)^{\otimes 2k}$ to a number that is precisely given by Eq. (A.137).

To prove this claim, we proceed by induction on k, first by verifying that the claim is true for the smallest value(s) of k, and then by verifying that, assuming that it is true for k, then it is also true for $k+1$.

We start showing that Eq. (A.137) is indeed the solution of the recursion relations (A.136) for $k=0$ and for $k=1$. We verify this for both values, because, since $0 \le i \le k$, for $k=0$ only the second relation in Eq. (A.137) can be applied, whereas for $k=1$ the first relation can also be verified.

For $k=0$, we note that the zeroth power of $\left(\frac{1}{2}\right)$ is simply the trivial representation, $\left(\frac{1}{2}\right)^{\otimes 2k} = D^{(0)}$, hence $c_{0,0}=1$, in agreement with the claim $c_{k,k}=1$.

For $k=1$, using Eq. (6.251) we obtain $\left(\frac{1}{2}\right)^{\otimes 2} = D^{(0)} \otimes D^{(1)}$, i.e., $c_{1,0}=1$ and $c_{1,1}=1$. These values agree with the *Ansatz* (A.137), which predicts $c_{1,0} = P_{2,1} - P_{2,0} = \binom{2}{1} - \binom{2}{0} = 2-1 = 1$ and $c_{1,1}=1$.

We now prove that, if the claim is true for $\left(\frac{1}{2}\right)^{\otimes 2k}$, then it holds for $\left(\frac{1}{2}\right)^{\otimes 2(k+1)}$, too. We note that, if $c_{k,k}=1$, i.e., if the representation $D^{(k)}$ appears exactly once in the tensor-sum decomposition of $\left(\frac{1}{2}\right)^{\otimes 2k}$, then, using the fact that

$$\left(\frac{1}{2}\right)^{\otimes 2(k+1)} = \left(\frac{1}{2}\right)^{\otimes 2k} \otimes \left(\frac{1}{2}\right)^{\otimes 2}$$

$$= \left(\sum_{i=0}^{k} c_{k,i}D^{(i)}\right) \otimes \left(D^{(0)} \otimes D^{(1)}\right), \qquad (A.141)$$

we note that the representation $D^{(k+1)}$ can be obtained only once from the tensor product $D^{(k)} \otimes D^{(1)}$, each of which appears only once in the two factors on the last line of Eq. (A.141). Thus, assuming $c_{k,k} = 1$, we also have $c_{k+1,k+1} = 1$. Next, we prove that, assuming that the claim (A.137) holds for a certain k and for an arbitrary i, then it also holds for $c_{k+1,i}$ for arbitrary i different from $k + 1$, in addition to $i = k + 1$.

Let us start from the first equation in the system (A.136), which implies that $c_{k+1,0}$ must satisfy:

$$
\begin{aligned}
c_{k+1,0} &= c_{k,0} + c_{k,1} \\
&= P_{2k,k} - P_{2k,k-1} + P_{2k,k-1} - P_{2k,k-2} \\
&= P_{2k,k} - P_{2k,k-2} \\
&= \frac{(2k)!}{k!\,k!} - \frac{(2k)!}{(k-2)!\,(k+2)!} \\
&= \frac{(2k)!}{k!\,(k-2)!} \left[\frac{1}{k(k-1)} - \frac{1}{(k+1)(k+2)} \right] \\
&= \frac{(2k)!}{k!\,(k-2)!} \cdot \frac{k^2 + 3k + 2 - k^2 + k}{(k-1)k(k+1)(k+2)} \\
&= \frac{(2k)!}{(k+2)!\,k!} \cdot (4k+2) \\
&= \frac{(2k)!}{(k+2)!\,k!} \cdot 2(2k+1) \\
&= \frac{(2k)!}{(k+2)!\,k!} \cdot \frac{2(2k+1)(k+1)}{(k+1)} \\
&= \frac{(2k)!}{(k+2)!\,k!} \cdot \frac{(2k+1)(2k+2)}{(k+1)} \\
&= \frac{(2k+2)!}{(k+2)!\,(k+1)!}.
\end{aligned}
\tag{A.142}
$$

This value is consistent with the one defined according to Eq. (A.137) for $k + 1$, which is

$$
\begin{aligned}
P_{2k+2,k+1} - P_{2k+2,k} &= \frac{(2k+2)!}{(k+1)!\,(k+1)!} - \frac{(2k+2)!}{k!\,(k+2)!} \\
&= \frac{(2k+2)!}{k!\,(k+1)!} \left(\frac{1}{k+1} - \frac{1}{k+2} \right) \\
&= \frac{(2k+2)!}{k!\,(k+1)!} \cdot \frac{k+2-k-1}{(k+1)(k+2)} \\
&= \frac{(2k+2)!}{(k+1)!\,(k+2)!},
\end{aligned}
\tag{A.143}
$$

in agreement with Eq. (A.142).

Let us now consider the second equation of the system (A.136), which, in the decomposition of $\left(\frac{1}{2}\right)^{\otimes 2(k+1)}$, holds for $0 < i < k$:

$$
\begin{aligned}
c_{k+1,i} &= c_{k,i-1} + 2c_{k,i} + c_{k,i+1} \\
&= P_{2k,k-i+1} - P_{2k,k-i} + 2P_{2k,k-i} - 2P_{2k,k-i-1} + P_{2k,k-i-1} - P_{2k,2k,k-i-2} \\
&= P_{2k,k-i+1} + P_{2k,k-i} - P_{2k,k-i-1} - P_{2k,k-i-2}.
\end{aligned}
\tag{A.144}
$$

Defining

$$A = P_{2k,k-i+1} + P_{2k,k-i}, \tag{A.145}$$

$$B = P_{2k,k-i-1} + P_{2k,k-i-2}, \tag{A.146}$$

Eq. (A.144) can be rewritten as

$$c_{k+1,i} = A - B. \tag{A.147}$$

From Eq. (A.145), we have

$$
\begin{aligned}
A &= P_{2k,k-i+1} + P_{2k,k-i} \\
&= \frac{(2k)!}{(k-i+1)!\,(k+i-1)!} + \frac{(2k)!}{(k-i)!\,(k+i)!} \\
&= \frac{(2k)!}{(k-i)!\,(k+i-1)!}\left(\frac{1}{k-i+1} + \frac{1}{k+i}\right) \\
&= \frac{(2k)!}{(k-i)!\,(k+i-1)!} \cdot \frac{2k+1}{(k+i)(k-i+1)} \\
&= \frac{(2k+1)!}{(k-i+1)!\,(k+i)!}.
\end{aligned} \tag{A.148}
$$

From its definition (A.146), we note that B can be obtained from A, by substituting i with $i+2$, thus:

$$B = \frac{(2k+1)!}{(k-i-1)!\,(k+i+2)!}. \tag{A.149}$$

Substituting Eq. (A.148) and Eq. (A.149) into Eq. (A.147), we get

$$
\begin{aligned}
c_{k+1,i} &= A - B \\
&= \frac{(2k+1)!}{(k-i+1)!\,(k+i)!} - \frac{(2k+1)!}{(k-i-1)!\,(k+i+2)!} \\
&= \frac{(2k+1)!}{(k-i-1)!\,(k+i)!}\left[\frac{1}{(k-i)(k-i+1)} - \frac{1}{(k+i+1)(k+i+2)}\right] \\
&= \frac{(2k+1)!}{(k-i-1)!\,(k+i)!} \cdot \frac{2+4ki+2k+4i}{(k-i)(k-i+1)(k+i+1)(k+i+2)} \\
&= \frac{(2k+1)!}{(k-i+1)!\,(k+i+2)!} \cdot (2k+2)(2i+1) \\
&= \frac{(2k+2)!}{(k-i+1)!\,(k+i+2)!} \cdot (2i+1).
\end{aligned} \tag{A.150}
$$

We can verify that this equals the value defined according to Eq. (A.137) for $k+1$, namely

$$
\begin{aligned}
P_{2k+2,k+1-i} - P_{2k+2,k-i} &= \frac{(2k+2)!}{(k+1-i)!\,(k+1+i)!} - \frac{(2k+2)!}{(k-i)!\,(k+i+2)!} \\
&= \frac{(2k+2)!}{(k-i)!\,(k+1+i)!}\left(\frac{1}{k+1-i} - \frac{1}{k+i+2}\right)
\end{aligned}
$$

$$= \frac{(2k+2)!}{(k-i)!\,(k+1+i)!} \cdot \frac{2i+1}{(k+1-i)(k+i+2)}$$

$$= \frac{(2k+2)!}{(k+1-i)!\,(k+2+i)!} \cdot (2i+1), \qquad (A.151)$$

in agreement with Eq. (A.150).

The third equation in the system (A.136) requires $c_{k+1,k}$ to be

$$
\begin{aligned}
c_{k+1,k} &= c_{k,k-1} + 2c_{k,k} \\
&= P_{2k,1} - P_{2k,0} + 2 \\
&= \frac{(2k)!}{1!\,(2k-1)!} - \frac{(2k)!}{0!\,(2k)!} + 2 \\
&= \frac{(2k)(2k-1)!}{(2k-1)!} - 1 + 2 \\
&= 2k+1.
\end{aligned}
\qquad (A.152)
$$

According to Eq. (A.137) for $k+1$, this should equal

$$
\begin{aligned}
P_{2k+2,1} - P_{2k+2,0} &= \frac{(2k+2)!}{1!\,(2k+2-1)!} - \frac{(2k+2)!}{0!\,(2k+2)!} \\
&= \frac{(2k+2)(2k+1)!}{(2k+1)!} - 1 \\
&= 2k+2-1 \\
&= 2k+1,
\end{aligned}
\qquad (A.153)
$$

which is indeed consistent with Eq. (A.152). This concludes our proof that the decomposition of $\left(\frac{1}{2}\right)^{\otimes n}$ for even values of $n = 2k$ is given by

$$
\begin{aligned}
\left(\frac{1}{2}\right)^{\otimes 2k} &= \sum_{i=0}^{k} c_{k,i} D^{(i)} \\
&= \sum_{i=0}^{k-1} \left(\frac{(2k)!}{(k-i)!\,(k+i)!} - \frac{(2k)!}{(k-i-1)!\,(k+i+1)!} \right) D^{(i)} \oplus D^{(k)} \\
&= \sum_{i=0}^{k-1} \frac{(2k)!}{(k-i-1)!\,(k+i)!} \left(\frac{1}{k-i} - \frac{1}{k+i+1} \right) D^{(i)} \oplus D^{(k)} \\
&= \sum_{i=0}^{k-1} \frac{(2k)!}{(k-i-1)!\,(k+i)!} \cdot \frac{2i+1}{(k-i)(k+i+1)} D^{(i)} \oplus D^{(k)} \\
&= \sum_{i=0}^{k-1} \frac{(2i+1)\cdot(2k)!}{(k-i)!\,(k+i+1)!} D^{(i)} \oplus D^{(k)}.
\end{aligned}
\qquad (A.154)
$$

Finally, let us consider the case of odd $n = 2k+1$, with $k \in \mathbb{N}$. Then

$$
\begin{aligned}
\left(\frac{1}{2}\right)^{\otimes 2k+1} &= \left(\frac{1}{2}\right)^{\otimes 2k} \otimes \left(\frac{1}{2}\right) \\
&= \left(\sum_{i=0}^{k} c_{k,i} \left(D^{(i)} \right) \otimes D^{\left(\frac{1}{2}\right)} \right)
\end{aligned}
$$

$$= c_{k,0}D^{\left(\frac{1}{2}\right)} \oplus \sum_{i=1}^{k}\left(c_{k,i}D^{\left(i-\frac{1}{2}\right)} \oplus c_{k,i}D^{\left(i+\frac{1}{2}\right)}\right)$$

$$= c_{k,0}D^{\left(\frac{1}{2}\right)} \oplus \sum_{j=1}^{k}c_{k,j}D^{\left(j-\frac{1}{2}\right)} \oplus \sum_{i=1}^{k}c_{k,i}D^{\left(i+\frac{1}{2}\right)}$$

$$= c_{k,0}D^{\left(\frac{1}{2}\right)} \oplus \sum_{i=0}^{k-1}c_{k,i+1}D^{\left(i+\frac{1}{2}\right)} \oplus \sum_{i=1}^{k}c_{k,i}D^{\left(i+\frac{1}{2}\right)}. \tag{A.155}$$

Thus the decomposition of $\left(\frac{1}{2}\right)^{\otimes 2k+1}$ into a tensor sum of irreducible representations will only contain half-integer-spin representations, from $D^{\left(\frac{1}{2}\right)}$ to $D^{\left(k+\frac{1}{2}\right)}$. Denoting their multiplicities as $f_{k,i+\frac{1}{2}}$, with $0 \leq i \leq k$:

$$\left(\frac{1}{2}\right)^{\otimes 2k+1} = \sum_{i=0}^{k}f_{k,i+\frac{1}{2}}D^{\left(i+\frac{1}{2}\right)}, \tag{A.156}$$

from Eq. (A.155) we read off

$$\begin{cases} f_{k,\frac{1}{2}} = c_{k,0} + c_{k,1} \\ f_{k,i+\frac{1}{2}} = c_{k,i} + c_{k,i+1} & \text{for } 0 < i < k. \\ f_{k,k+\frac{1}{2}} = c_{k,k} = 1 \end{cases} \tag{A.157}$$

Noting that the first two equations can be combined together, we obtain:

$$\begin{cases} f_{k,i+\frac{1}{2}} = c_{k,i} + c_{k,i+1} & \text{for } 0 \leq i < k \\ f_{k,k+\frac{1}{2}} = c_{k,k} = 1 \end{cases} \tag{A.158}$$

Thus, for $i < k - 1$ we obtain

$$\begin{aligned} f_{k,i+\frac{1}{2}} &= c_{k,i} + c_{k,i+1} \\ &= P_{2k,k-i} - P_{2k,k-i-1} + P_{2k,k-i-1} - P_{2k,k-i-2} \\ &= P_{2k,k-i} - P_{2k,k-i-2} \\ &= \frac{(2k)!}{(k-i)!\,(k+i)!} - \frac{(2k)!}{(k-i-2)!\,(k+i+2)!} \\ &= \frac{(2k)!}{(k-i-2)!\,(k+i)!}\left[\frac{1}{(k-i-1)(k-i)} - \frac{1}{(k+i+1)(k+i+2)}\right] \\ &= \frac{(2k)!}{(k-i)!\,(k+i+2)!} \cdot 2(2k+1)(i+1) \\ &= \frac{(2k+1)!}{(k-i)!\,(k+i+2)!} \cdot 2(i+1). \end{aligned} \tag{A.159}$$

In principle, the $i = k - 1$ case has to be treated separately, because for $i = k - 1$ the second term appearing on the right-hand side of the first line in Eq. (A.159) reduces to $c_{k,k}$, which is not given by a difference of combinatorial terms. For $i = k - 1$, we obtain

$$
\begin{aligned}
f_{k,k-\frac{1}{2}} &= c_{k,k-1} + c_{k,k} \\
&= P_{2k,k-k+1} - P_{2k,k-k+1-1} + 1 \\
&= P_{2k,1} - P_{2k,0} + 1 \\
&= P_{2k,1} \\
&= \frac{(2k)!}{1!\,(2k-1)!} \\
&= 2k.
\end{aligned}
\tag{A.160}
$$

However we note that, in fact, Eq. (A.160) is consistent with Eq. (A.159) evaluated for $i = k - 1$:

$$
\begin{aligned}
\left. \frac{(2k+1)!}{(k-i)!\,(k+i+2)!} \cdot 2(i+1) \right|_{i=k-1} &= \frac{(2k+1)!}{(k-k+1)!\,(k+k-1+2)!} \cdot 2(k-1+1) \\
&= \frac{(2k+1)!}{1!\,(2k+1)!} \cdot 2k \\
&= 2k,
\end{aligned}
\tag{A.161}
$$

so that Eq. (A.159) actually holds for every $0 \le i < k$. Thus we finally obtain

$$
\left(\frac{1}{2} \right)^{\otimes 2k+1} = \sum_{i=0}^{k-1} \frac{2(i+1) \cdot (2k+1)!}{(k-i)!\,(k+i+2)!} D^{(i+\frac{1}{2})} \oplus D^{(k+\frac{1}{2})}.
\tag{A.162}
$$

6.18 The $(0,0,1,0)$ irreducible representation of the su(5) Lie algebra can be associated with the following Young tableau

$$
(0,0,1,0) = \quad\text{}
\tag{A.163}
$$

and its dimension, which can be computed using the factors-over-hooks rule, is given by $\dim(0,0,1,0) = F/H = (5 \cdot 4 \cdot 3)/(3 \cdot 2 \cdot 1) = 60/6 = 10$. In fact, since this representation is the complex conjugate of $(0,1,0,0)$, one could denote it as $\overline{\mathbf{10}}$.

In turn, the $(1,0,0,0)$ irreducible representation can be associated with the Young tableau

$$
(1,0,0,0) = \square
\tag{A.164}
$$

and is the defining fundamental representation, of dimension 5.

The tensor product $(0,0,1,0) \otimes (1,0,0,0)$ can be decomposed into a direct sum of irreducible representations by using the rules discussed in Section 6.10:

$$
\begin{aligned}
(0,0,1,0) \otimes (1,0,0,0) &= \quad\text{} \\
&= \quad\text{} \\
&= (1,0,1,0) \oplus (0,0,0,1).
\end{aligned}
\tag{A.165}
$$

Using the factors-over-hooks rule, we find that $\dim(1, 0, 1, 0) = 45$, while $\dim(0, 0, 0, 1) = 5$. Noting that $(0, 0, 0, 1)$ is actually the complex conjugate representation of the defining fundamental representation, we can write

$$\overline{10} \otimes 5 = 45 \oplus \overline{5}. \tag{A.166}$$

We observe that, as expected, the dimensions on the two sides match correctly, both being 50.

6.21 We note that, by multiplying the time coordinate by the speed of light in vacuum (which is a relativistic invariant), the spacetime coordinates can be arranged into four-vectors with homogeneous components, with dimension length, e.g.,

$$x_A^\mu = \begin{pmatrix} ct_A \\ x_A^1 \\ x_A^2 \\ x_A^3 \end{pmatrix}. \tag{A.167}$$

Then, the relation between the spacetime coordinates in the inertial reference frames A and B can be rewritten as

$$\begin{pmatrix} ct_B \\ x_B^1 \\ x_B^2 \\ x_B^3 \end{pmatrix} = \begin{pmatrix} \gamma & -\beta\gamma & 0 & 0 \\ -\beta\gamma & \gamma & 0 & 0 \\ 0 & 0 & 1 & 0 \\ 0 & 0 & 0 & 1 \end{pmatrix} \begin{pmatrix} ct_A \\ x_A^1 \\ x_A^2 \\ x_A^3 \end{pmatrix}, \tag{A.168}$$

with $\beta = v/c$ and $\gamma = 1/\sqrt{1 - \beta^2}$. For later convenience, we note that the inverse boost, expressing the spacetime coordinates of the reference frame A in terms of those of B, can be associated with the inverse of the matrix appearing in Eq. (A.168), which is simply obtained by changing the sign of β, in agreement with the fact that an observer in the reference frame B sees the reference frame A moving along the x_B^1-axis in the *negative* direction with a velocity of modulus v.

The relation between the components of the velocities measured in the reference frames A and B (to be denoted as u_A^i and u_B^i) can be obtained by noting that

$$u_A^i = \frac{dx_A^i}{dt_A}, \qquad u_B^i = \frac{dx_B^i}{dt_B}. \tag{A.169}$$

Computing the differentials from Eq. (A.168), one obtains

$$dt_B = \gamma \, dt_A - \frac{1}{c}\beta\gamma \, dx_A^1, \tag{A.170}$$

$$dx_B^1 = -c\beta\gamma \, dt_A + \gamma \, dx_A^1, \tag{A.171}$$

$$dx_B^2 = dx_A^2, \tag{A.172}$$

$$dx_B^3 = dx_A^3, \tag{A.173}$$

thus we get

$$u_B^1 = \frac{dx_B^1}{dt_B} = \frac{-c\beta\gamma dt_A + \gamma dx_A^1}{\gamma dt_A - \frac{\beta}{c}\gamma dx_A^1} = \frac{\gamma\left(\frac{dx_A^1}{dt_A} - c\beta\right)}{\gamma\left(1 - \frac{\beta}{c}\frac{dx_A^1}{dt_A}\right)} = \frac{u_A^1 - v}{1 - \frac{vu_A^1}{c^2}}, \tag{A.174}$$

$$u_B^2 = \frac{dx_B^2}{dt_B} = \frac{dx_A^2}{\gamma dt_A - \frac{\beta}{c}\gamma dx_A^1} = \frac{\frac{dx_A^2}{dt_A}}{\gamma\left(1 - \frac{\beta}{c}\frac{dx_A^1}{dt_A}\right)} = \frac{u_A^2}{\gamma\left(1 - \frac{vu_A^1}{c^2}\right)}, \tag{A.175}$$

$$u_B^3 = \frac{dx_B^3}{dt_B} = \frac{dx_A^3}{\gamma dt_A - \frac{\beta}{c}\gamma dx_A^1} = \frac{\frac{dx_A^3}{dt_A}}{\gamma\left(1 - \frac{\beta}{c}\frac{dx_A^1}{dt_A}\right)} = \frac{u_A^3}{\gamma\left(1 - \frac{vu_A^1}{c^2}\right)}. \tag{A.176}$$

The relation between the spacetime coordinates in the reference frame D of a pointlike particle moving at constant velocity V along the x_B^1-axis, and those of the reference frame B is of the same form as Eq. (A.168),

$$\begin{pmatrix} ct_D \\ x_D^1 \\ x_D^2 \\ x_D^3 \end{pmatrix} = \begin{pmatrix} \gamma_V & -\beta_V\gamma_V & 0 & 0 \\ -\beta_V\gamma_V & \gamma_V & 0 & 0 \\ 0 & 0 & 1 & 0 \\ 0 & 0 & 0 & 1 \end{pmatrix} \begin{pmatrix} ct_B \\ x_B^1 \\ x_B^2 \\ x_B^3 \end{pmatrix}, \tag{A.177}$$

but this time with $\beta_V = V/c$ and $\gamma_V = 1/\sqrt{1 - \beta_V^2}$.

The relation between the coordinates in the reference frame D and those in the reference frame A can be expressed by combining Eq. (A.177) with Eq. (A.168):

$$\begin{pmatrix} ct_D \\ x_D^1 \\ x_D^2 \\ x_D^3 \end{pmatrix} = \begin{pmatrix} \gamma_V & -\beta_V\gamma_V & 0 & 0 \\ -\beta_V\gamma_V & \gamma_V & 0 & 0 \\ 0 & 0 & 1 & 0 \\ 0 & 0 & 0 & 1 \end{pmatrix} \begin{pmatrix} \gamma & -\beta\gamma & 0 & 0 \\ -\beta\gamma & \gamma & 0 & 0 \\ 0 & 0 & 1 & 0 \\ 0 & 0 & 0 & 1 \end{pmatrix} \begin{pmatrix} ct_A \\ x_A^1 \\ x_A^2 \\ x_A^3 \end{pmatrix}$$

$$= \begin{pmatrix} \gamma_V\gamma(1 + \beta_V\beta) & -\gamma_V\gamma(\beta_V + \beta) & 0 & 0 \\ -\gamma_V\gamma(\beta_V + \beta) & \gamma_V\gamma(1 + \beta_V\beta) & 0 & 0 \\ 0 & 0 & 1 & 0 \\ 0 & 0 & 0 & 1 \end{pmatrix} \begin{pmatrix} ct_A \\ x_A^1 \\ x_A^2 \\ x_A^3 \end{pmatrix}$$

$$= \begin{pmatrix} \gamma_{AD} & -\beta_{AD}\gamma_{AD} & 0 & 0 \\ -\beta_{AD}\gamma_{AD} & \gamma_{AD} & 0 & 0 \\ 0 & 0 & 1 & 0 \\ 0 & 0 & 0 & 1 \end{pmatrix} \begin{pmatrix} ct_A \\ x_A^1 \\ x_A^2 \\ x_A^3 \end{pmatrix}, \tag{A.178}$$

where the matrix appearing in the last line has again the same structure as those in Eq. (A.168) and in Eq. (A.177), with

$$\beta_{AD} = \frac{\beta_V + \beta}{1 + \beta_V\beta}, \qquad \gamma_{AD} = \gamma_V\gamma(1 + \beta_V\beta), \tag{A.179}$$

which satisfy the condition

$$\frac{1}{\sqrt{1 - \beta_{AD}^2}} = \frac{1}{\sqrt{1 - \frac{(\beta_V + \beta)^2}{(1 + \beta_V \beta)^2}}} = \frac{1 + \beta_V \beta}{\sqrt{(1 + \beta_V \beta)^2 - (\beta_V + \beta)^2}}$$

$$= \frac{1 + \beta_V \beta}{\sqrt{1 + 2\beta_V \beta + \beta_V^2 \beta^2 - \beta_V^2 - \beta^2 - 2\beta_V \beta}}$$

$$= \frac{1 + \beta_V \beta}{\sqrt{(1 - \beta_V^2)(1 - \beta^2)}}$$

$$= \gamma_V \gamma (1 + \beta_V \beta) = \gamma_{AD}, \qquad (A.180)$$

meaning that the relation between the coordinate frames of A and D is written as a boost with velocity

$$v_{AD} = c\beta_{AD} = c\frac{\beta_V + \beta}{1 + \beta_V \beta} = \frac{V + v}{1 + \frac{Vv}{c^2}} \qquad (A.181)$$

along the x_A^1-axis, in agreement with the velocity composition law (A.174), noting that, for the reference frame B, the reference frame A is in motion at speed $-v$ along the x_B^1-axis.

As the discussion above can be generalized to any spatial axis, by a rotation of the spatial axes, we conclude that the boosts along any fixed axis form a group: The identity, corresponding to zero relative velocity between the frames is an element of the group, the inverse of each boost exists and is obtained by inverting the sign of the non-diagonal elements, and the composition of boosts along a fixed axis is still a boost, with parameters defined according to Eq. (A.179). This group is a continuous, one-parameter, non-compact Abelian group.

Note, however, that in general the composition of boosts along different axes is not a boost. This is a consequence of the fact that, according to Eq. (6.382), the boost generators are not a subalgebra of the Lorentz group.

Notes

References

[1] S. M. Carroll, *Spacetime and Geometry* (Cambridge: Cambridge University Press, 2019).

[2] H. Georgi, *Lie Algebras in Particle Physics: from Isospin to Unified Theories* Frontiers in Physics (New York: Avalon Publishing, 1999).

[3] H. Georgi and S. L. Glashow, Unity of all elementary particle forces. *Phys. Rev. Lett.*, **32** (1974), 438–441.

[4] M. Hamermesh, *Group Theory and its Application to Physical Problems*. Addison Wesley Series in Physics (New York: Dover Publications, 1989).

[5] F. Iachello, *Lie Algebras and Applications*. Lecture Notes in Physics (Berlin, Heidelberg: Springer, 2007).

[6] H. F. Jones, *Groups, Representations and Physics* (Boca Raton, FL: CRC Press, 2020).

[7] J. M. Lee, *Manifolds and Differential Geometry*. Graduate Studies in Mathematics (Providence, RI: American Mathematical Society, 2009).

[8] J. M. Lee, *Introduction to Smooth Manifolds*. Graduate Texts in Mathematics (Berlin, Heidelberg: Springer, 2003).

[9] C. W. Misner, K. S. Thorne, and J. A Wheeler, *Gravitation* (Princeton: Princeton University Press, 2017).

[10] M. Nakahara, *Geometry, Topology and Physics*, 2nd ed., Graduate Student Series in Physics (Oxfordshire: Taylor & Francis, 2003).

[11] C. Nash, and S. Sen, *Topology and Geometry for Physicists*. Dover Books on Mathematics (New York: Dover Publications, 2011).

[12] M. Srednicki, *Quantum Field Theory* (Cambridge: Cambridge University Press, 2007).

[13] M. Tinkham, *Group Theory and Quantum Mechanics*. Dover Books on Chemistry and Earth Sciences (New York: Dover Publications, 2003).

[14] R. M. Wald, *General Relativity* (Chicago, IL: University of Chicago Press, 2010).

Index